A TEXTBOOK OF
ENGINEERING CHEMISTRY

A TEXTBOOK OF
ENGINEERING CHEMISTRY

Dr. Savita Garg

Ph.D., B.Ed., M.Sc. (Inorganic and Physical)
Former Head Department of Chemistry
Gopalan College of Engineering and Management (Bangalore)

Associate Professor
Department of Chemistry
East Point College of Engineering & Technology
Bangalore

I.K. International Pvt. Ltd.

NEW DELHI

Published by

I.K. International Pvt. Ltd.
4435-36/7, Ansari Road, Daryaganj
New Delhi–110 002 (India)
E-mail: info@ikinternational.com
Website: www.ikbooks.com

ISBN: 978-93-90620-19-7

Published by Krishan Makhijani for I.K. International Pvt. Ltd., 4435-36/7, Ansari Road, Daryaganj, New Delhi–110 002 and Printed by Rekha Printers Pvt. Ltd., Okhla Industrial Area, Phase II, New Delhi–110 020.

Dedicated to My Loving Parents

MRS. RUKMANI & MR. ARUN KUMAR GUPTA

For their inspiration, support and encouragement
Throughout my academic excellence
and for
What I am today

Preface

Preparing a textbook of engineering chemistry which is based on the latest syllabus of several universities and which also gives the student sufficient information to understand the principles and practice of the subject has been a fascinating experience. Engineering students will hopefully find the book easy to read and understand.

This book has been divided into 10 chapters, covering all the disciplines of engineering chemistry such as inorganic, organic, organometallic, synthetic, physical, applied, industrial, spectroscopic and titrimetric chemistry. Each chapter consists of methodical introduction, historical background, in depth discussion, practical applications, and significance. A large number of numerical problems drawn from various university examinations are given at the end of theoretical part of each chapter.

The book aims at-

- Developing the techniques of interpretation so as to enhance the practicability of experiments.
- Developing analytical capabilities of students so that they can characterize, transform, and use materials in engineering and apply information gained in solving related engineering problems.
- Developing the habit of scientific reasoning in the students so that they can work with open and inquiring mind.
- Imparting intensive and extensive information of the subject so that readers can understand the role of chemistry in the field of engineering.

I hope this book will prove a useful tool for students of undergraduate engineering classes and also to teachers in preparing for their lectures. Any suggestion for further improvement of the book is most welcome.

Dr. Savita Garg
Savitagarg81@gmail.com

Acknowledgements

"At time our own light goes out and is rekindled by a spark from another person. Each of us has cause to think with deep gratitude of those who have lighted the flame within us."

– Albert Schweitzer

With the grace of **Almighty God**, I would like to express my gratitude to many people who saw me through this book; to all those who provided support, talked things over read, wrote offered comments allowed me to quote their remarks and assisted in the editing, proofreading and design.

I would like to thank **Dr. T.K. Sateesh** (Principal), **Dr. G. Maruti** (Dean of Basic Science) and **Dr. Ashok Kumar Shukla** (HoD and Professor Department of Chemistry), East Point College of Engineering and Technology, Bangalore, who supported, encouraged and guided me time to time for enabling me to publish this book.

Above all I would also like to express my sincere gratitude to **Dr. Anjali Goel**, Reader Department of Chemistry, Kanya Gurukul Mahavidyalaya, Jwalapur, Haridwar whose constant inspiration and encouragement had been a guiding force to write this book.

My family has extended more patience than I deserve. That patience is greatly appreciated and these acknowledgements are just a token to their faith in me. My words are not enough to express my feelings for my younger sister **Ms. Sulekha Gupta** and my sister-in-law **Smt. Anika Gupta** for her sincere help, immaculate affection, constant encouragement and unfettering support. She with all comity and alacrity helped me in this endeavour.

I know, I have no words, no language to express my love and regards for my husband **Mr. Piyush Garg** and my younger brother **Mr. Sanjay Kumar Gupta** for his support and cheerful enthusiastic efforts. I will be failing in my duties, if I do not acknowledge the help and support received.

I am not only the one who thinks their parents are beyond comparison in the whole world. **My parents** have been the strongest influence in my life always being there a guidance star, when I lost direction. Thanks to my parents for the wonderful gift of a fresh and painted canvas of my life.

Last and not least I beg forgiveness of all those who have been with me over the course of the years and whose names I have failed to mention. I humbly acknowledge gratitude to them without which it would not have been possible to perform the work.

Dr. Savita Garg
Savitagarg81@gmail.com

Contents

1

Electrochemistry and Energy Storage System

A. ELECTROCHEMISTRY

1.1 INTRODUCTION

Electrochemistry is the study of relationship between electrical energy and chemical energy and vice versa. It has important applications in everyday life stretching from the battery that powers your portable radio to the electron refining that produces the copper pipes carrying your drinking water. These electrochemical processes utilize oxidation and reduction reactions. Oxidation involves the loss of one or more electrons from a chemical species while reduction is the gain of one or more electrons by a chemical species. When oxidation and a reduction are paired together in a redox reaction, electrons can flow from the oxidized species, to the reducing species, and vice versa. That electron flow can either be spontaneously produced by the reaction and converted into electricity, as in a galvanic cell, or it can be imposed by an outside source to make a non-spontaneous reaction proceed, as in an electrolytic cell.

1.2 ELECTROCHEMICAL CELL

An electrochemical cell is a device in which chemical energy is converted into electrical energy or electrical energy into chemical energy. Thus, there are two types of electrochemical cells.

1. The one which converts chemical energy into electrical energy $\longrightarrow$ Galvanic cell
2. The one which converts electrical energy into chemical energy $\longrightarrow$ Electrolytic cell

The device in which electrical energy is produced from a chemical reaction is called electrochemical cell or galvanic cell or voltaic cell.

(or)

A galvanic cell is a device in which the free energy of a physical or chemical process is converted into electrical energy.

1.3 COMPARISON BETWEEN ELECTROLYTIC CELL AND ELECTROCHEMICAL/GALVANIC CELL

Electrolytic cell	Electrochemical cell/Galvanic cell
1. Electrical energy is converted into chemical energy.	1. Chemical energy is converted into electrical energy.
2. The cell reaction is non-spontaneous.	2. The cell reaction is spontaneous.
3. ${}_{-1}e^0$ are supplied to the cell from an external battery.	3. Electrons are drawn from it.
4. ${}_{-1}e^0$ move from cathode to anode through the electrolyte.	4. Electrons move from anode to cathode through external current.
5. The anode carries a positive charge, while cathode carries negative charge.	5. The anode carries a negative charge while cathode carries a positive charge.
6. The extent of the chemical reaction taking place in it is governed by the Faraday's law of electrolysis.	6. The E.M.F. of the cell depends on the concentrate(s) of the electrolyte as well as the chemical nature of the electrodes.
7. The quantity of electricity passed during electrolysis is measured by coulomb meter. **Example:** Electrolysis, electroplating, electroless plating, etc.	7. The E.M.F. generated in the cell is measured by potentiometer **Example:** Daniel cell, fuel cell, etc.

1.4 SALT BRIDGE

A salt bridge is used for making electrical contact between electrolyte solutions. It is usually an inverted U-tube filled with concentrated solution of KCl or NH_4NO_3. The inclusion of a salt bridge is commonly indicated by a vertical double bar (||).

Essential Requirements of an electrolyte in a salt bridge

1. The mobility of anion and cation of the electrolyte should be almost the same.
2. The ions of the electrolyte are not involved in the electrochemical change.
3. The ions do not react chemically with the species of the cell. The jelly keeps the electrolyte in semisolid phase and thus prevents mixing.

Functions

1. Salt bridge completes the electrical circuit, which connects the two solutions of the half cells.
2. Salt bridge maintains electrical neutrality of two half cell solutions.
3. It allows the flow of current.

1.5 CELL NOTATIONS AND CONVENTIONS

An electrochemical cell or galvanic cell is represented by the following conventions.

Convention 1: The half cell in which oxidation (removal of an electron) occurs is placed on the left and the half cell in which reduction (gain of electrons) occurs is placed on the right.

Convention 2: The metal and the ionic phases are separated by a vertical line (|). The inclusion of a salt bridge is commonly indicated by two vertical lines (||), which separates the two half cells.

Convention 3: $Zn \mid \overrightarrow{Zn^{2+} (1M) \parallel Cu^{2+}} (1M) \mid Cu$

The arrow indicates the direction of the flow of electron. The cell potential is given a positive sign if the cell reaction is spontaneous. This means that if the electron travels from left to right in the external circuit, the cell potential has a positive value. So, the reaction proceeds spontaneously from left to right.

Suppose the Daniell cell is written in the order

$$Cu(Cu^{2+} \overleftarrow{(1M) \parallel Zn^{2+}} (1M) \mid Zn$$

Its cell potential is written as (–1.10 V) then the cell reaction is non-spontaneous. This means that if the electron travels from right to left in the external circuit, the cell reaction has a negative value. So the reaction proceeds non-spontaneously. This convention is summarized as follows:

Direction of flow of electrons	Nature of cell reaction	E_{cell}
Left to right Right of left	Spontaneous Non-spontaneous	(+) ve (> 1) (–) ve (< 1)

Some common cells may be represented as:

(i) $Cd \mid Cd^{2+} (aq) \parallel Cu^{2+}(aq) \mid Cu$

Oxidation Reduction

The electrode reactions occurring in the cell are:

At the anode $Cd \longrightarrow Cd^{2+} + 2e^-$

At the cathode $Cu^{2+} + 2e^- \longrightarrow Cu(s)$

Overall cell reaction $Cd + Cu^{2+} \longrightarrow Cd^{2+} + Cu$

(ii) $Ni|Ni^{2+}(aq)||Ag^+(aq)|Ag$

Oxidation Reduction

The electrode reactions occurring in the cell are:

At the anode $Ni \longrightarrow Ni^{2+} + 2e^-$

At the cathode $2Ag^+ + 2e^- \longrightarrow 2Ag(s)$

Overall cell reaction $Ni + 2Ag^+ \longrightarrow Ni^{2+} + 2Ag$

(iii) $Cu(s) \mid Cu^{2+} (aq) \parallel Ag^+ (aq) \mid Ag$

Oxidation Reduction

The electrode reactions occurring in the cell are:

At the anode $Cu(s) \longrightarrow Cu^{2+} + 2e^-$

At the cathode $2Ag^+ + 2e^- \longrightarrow 2Ag(s)$

Overall cell reaction $Cu + 2Ag^+ \longrightarrow Cu^{2+} + 2Ag$

(iv) $Zn(s) \mid Zn^{2+}(aq) \| Cu^{+2}(aq) \mid Cu(s)$

Oxidation Reduction

The electrode reactions occurring in the cell are:

At the anode $Zn(s) \longrightarrow Zn^{2+} + 2e^-$

At the cathode $Cu^{2+} + 2e^- \longrightarrow Cu(s)$

Overall cell reaction $Zn + Cu^{2+} \longrightarrow Zn^{2+} + Cu$

(v) $Zn(s) \mid Zn^{2+}$ (1M $ZnSO_4$) $\|$ H^+ (1M HCl) $\mid$ H_2(1 atm), Pt

Oxidation Reduction

The electrode reactions occurring in the cell are:

At the anode $Zn(s) \longrightarrow Zn^{2+} + 2e^-$

At the cathode $2H^+ + 2e^- \longrightarrow H_2$

Overall cell reaction $Zn + 2H^+ \longrightarrow Zn^{2+} + H_2$

1.6 SINGLE ELECTRODE POTENTIAL

For example, to understand the phenomena of electrode potential consider a half cell of copper metal.

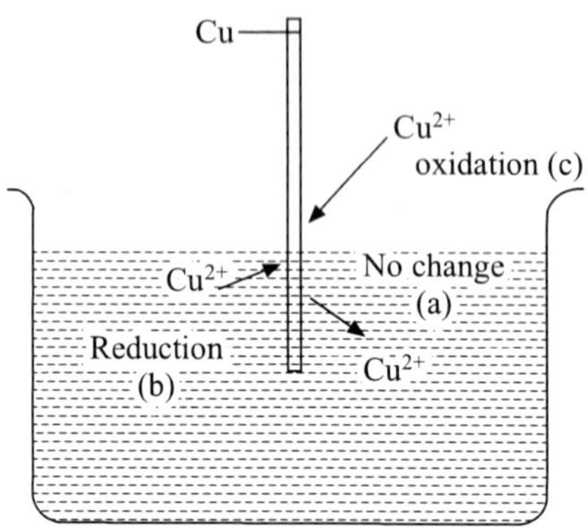

Fig. 1.1 *Electrode potential*

Half Cell of Copper Metal

When a copper electrode is immersed in a solution of its ions, then either of the three possibilities can takes place.

1. The metal ions Cu^{2+} may collide with the electrode and do not undergo any change.
2. Cu^{2+} ions may collide with the electrode to gain electrons and are reduced into metal (Cu) atoms

$$Cu + 2e^- \longrightarrow Cu \quad \text{(reduction)}$$

3. Cu atoms on the electrode may lose electrons to the electrode and become Cu^{2+} ions and go into the solution

$$Cu \longrightarrow Cu^{2+} + 2e^- \quad \text{(oxidation)}$$

Conclusion

If the metal has relatively high tendency to get oxidized its atoms will lose electrons readily and form Cu^{2+} ions which go into the solution. The electrons lost on the electrode would be accumulated on the metal electrode and the electrode acquires a slight negative (–ve) charge with respect to the solution. Slightly, some Cu^{2+} ions from the solution will take up electrons and become Cu atoms. After some time an equilibrium will be reached

$$Cu \rightleftharpoons Cu^{2+} + 2e^-$$

When such an equilibrium is attained, it results in separation of charges. Similarly, if the metal ions have relatively greater tendency to get reduced they will take electrons from the electrode as a result net positive (+ve) charge will be developed on the electrode with respect to solution. This will also result into separation of charges. Due to separation of charges between the electrode and solution, an electrical potential is set up between the metal electrode and its solution. The electrical potential difference set up between the metal and its solution is known as electrode potential. **The difference between the electrode potentials of the two electrodes constituting an electrochemical cell is known as emf** or **the potential difference between the two electrodes of a galvanic cell which causes the flow of current from one electrode (higher potential) to the other (lower poential).**

$$E_{cell} = E_{cathode} - E_{anode}$$

In terms of reduction potential

$$E_{cell} = E_{right} - E_{left}$$

The electrode potential depends on the following factors:

- The nature of metal and it ions
- Concentration of the ions in the solution
- Temperature

WORKED PROBLEMS

Problem 1.1

Calculate the standard reduction electrode of Ni^{+2}/Ni electrode when the cell potential for the cell is:

$$Ni/Ni^{+2}\ (1M)\ ||\ Cu^{+2}\ |\ Cu \text{ is } 0.59\ V \text{ and } E^\circ_{cu^{+2}/cu} = 0.34\ V.$$

Solution:

$$E^\circ_{cell} = E^\circ_{cu^{+2}/cu} - E^\circ_{Ni^{+2}/Ni}$$

$$E^\circ_{Ni^{+2}/Ni} = 0.344 - 0.59 = \mathbf{-0.246\ V}$$

Problem 1.2

Derive the expression to calculate the emf of the following half cell.

$$H^+ \mid H_2, Pt$$

Solution: $E_{cell} = E^\circ_{cell} - 0.0591/2 \log [H_2]/[H^+]^2$

$E_{cell} = E^\circ_{cell} - 0.0591/2 \log 1[H^+]^2$ since $[H_2] = 1$

Problem 1.3

A cell uses $Ag^+|Ag$ and $Zn|Zn^{2+}$ electrodes. Given that $E_{Ag^+|Ag} = 0.85$ V and $E_{zn^{2+}/Zn} = -0.70$ V

(a) Write the net cell reaction.

(b) Which electrode of the cell is negative?

(c) Calculate the cell voltage.

Solution: Cell representation

$$Zn(s)\ |Zn^{2+}(aq)||\ Ag^+(aq)|Ag(s)$$

Half-cell reaction: $Zn \longrightarrow Zn^{2+} + 2e$ anode

$[Ag + e \longrightarrow Ag] \times 2$ cathode

Net-cell reaction: $Zn + 2Ag^+ \rightleftharpoons Zn^{2+} + 2Ag$

(a) $Zn \mid Zn^{2+}$ electrode of the cell is negative since electrons are lost at this electrode.

(b)
$$
\begin{aligned}
E_{cell} &= E_{cathode} - E_{anode} \\
&= E_{Ag^+/Ag} - E_{zn^{2+}/Zn} \\
&= 0.85 - (-0.70) \\
&= \mathbf{1.55\ V}
\end{aligned}
$$

Problem 1.4

Represent a cadmium-copper cell. Give the electrode and net cell reactions. If the standard reduction potentials of Cd and Cu are –0.40 and 0.34 V respectively. Calculate the standard emf of the cell.

Solution: $E^\circ_{Cu^{2+}/Cu} > E^\circ_{Cd^{2+}/Cd}$

Cd acts as the anode and Cu behaves as the cathode.

(i) Cell representations $Cd|Cd^{2+}||Cu^{2+}|Cu$

(ii) Electrode reaction

At anode: $Cd \longrightarrow Cd^{2+} + 2e$ $E^\circ_{Cd^{2+}/Cd} = 0.40$ V

At cathode: $Cu^{2+} + 2e \longrightarrow Cu$ $E^\circ_{Cu^{2+}/Cu} = 0.34$ V

(iii) The net cell reaction:

$$Cd + Cu^{2+} \longrightarrow Cd^{2+} + Cu$$

(iv) Calculation of standard emf of the cell.

$$E^\circ_{cell} = E^\circ_{cathode} - E^\circ_{anode}$$

$$= E^\circ_{Cu^{2+}/Cu} - E^\circ_{Cd^{2+}/Cd}$$

$$= \mathbf{0.34 - (-0.40) = 0.74\ V}$$

1.6.1 Origin of Single Electrode Potential

When a metal is in contact with a solution of its own ions two types of reactions are possible. The metal shows the tendency to go into the solution as metal ion by losing electrons, i.e., oxidation

$$M \rightarrow M^{n+} + n\bar{e}$$

At the same time, metal ions in the solution show the tendency to get deposited as metal atoms, i.e., reduction.

$$M^{n+} + n\bar{e} \rightarrow M$$

Due to these reactions the metal acquires negative or positive charge and attracts oppositely charged ions from the solution and form a dynamic equilibrium.

$$M^{n+} + n\bar{e} \rightleftharpoons M$$

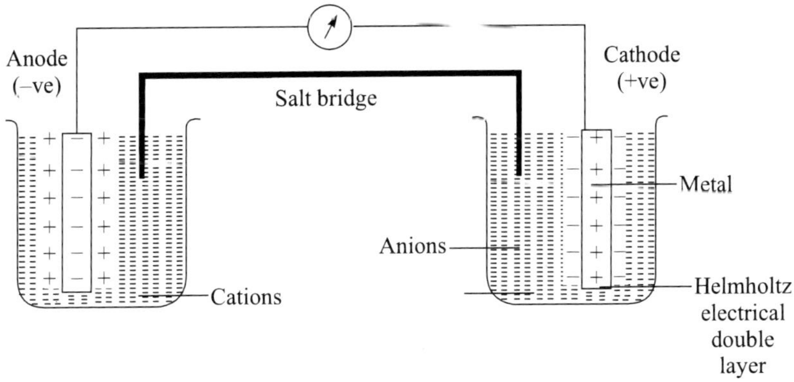

Fig. 1.2

Once the dynamic equilibrium is established, positive or negative ions remain quite close to the metal and form Helmholtz electrical double layer. This process creates potential difference between the metal and solution and is known as single electrode potential. At equilibrium, the potential difference b/w metal and solution remains constant until the charge is not allowed to flow.

1.7 MEASUREMENT OF SINGLE ELECTRODE POTENTIAL

It is not possible to measure the absolute value of a single electrode because neither the oxidation nor the reduction reaction can occur by itself. We need two electrodes to measure the potential difference between two points. In order to have an operative value for electrode potential. Electrode potential values are assigned to different electrodes with respect to the standard hydrogen electrode (SHE). The electrode potential of the standard hydrogen electrode is arbitrarily fixed at zero and the numerical values of the electrode potentials of various electrodes are relative values with respect to the standard hydrogen electrode. Therefore, single electrode potentials of electrodes are referred to as potentials of the hydrogen scale.

Examples:

1. Measurement of Electrode Potential of Cu^{2+}/Cu

Suppose we want to measure the electrode potential of the half cell consisting of Zn immersed in 1.0 M $ZnSO_4$ solution at 298 K. This half cell is suitably coupled with SHE through potassium chloride – agar agar salt bridge to have a working or spontaneous cell as shown in Fig. 1.3.

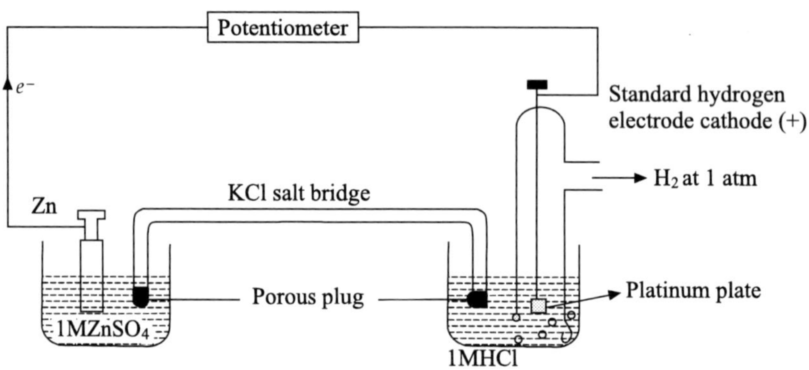

Fig. 1.3 *EMF measurement of Zn – H_2 galvanic cell*

The electron flow from zinc electrode to hydrogen electrode and therefore, the zinc electrode acts as anode and SHE acts as a cathode.

The emf of the resulting cell is

$$Zn \mid Zn^{2+}(1M\ ZnSO_4) \parallel H^+ (1MHCl) \mid H_2 (1 \text{ atm}), Pt$$

Cell reaction:

$$Zn(s) \longrightarrow Zn^{2+} + 2e^- \text{ (Oxidation)}$$

$$2H^+ + 2e^- \longrightarrow H_2 \quad \text{(Reduction)}$$

$$Zn + 2H \longrightarrow Zn^{2+} + H_2 \text{ (Redox)}$$

$$E^\circ_{cell} = E^\circ_{right} - E^\circ_{left}$$

$$E^\circ_{cell} = E^\circ_{H^+/H_2} - E^\circ_{Zn^{2+}/Zn}$$

$$0.76\ V = 0.0\ V - E^{\circ}_{Zn^{2+}/Zn}$$

$$E^{\circ}_{Zn^{2+}/Zn} = -\ 0.76\ V$$

Thus, by combining with standard hydrogen electrode reduction potential of Zn^{2+}/ Zn is found to be – 0.76 V.

2. Measurement of Electrode Potential of Cu^{2+}/Cu

To measure electrode potential of copper – copper ions, a cell consisting of copper electrode immersed in 1M $CuSO_4$ solution and SHE is set up as shown in Fig. 1.4. The cell may be represented as

Pt, H_2 (1 atm) H^+ (1 MHCl) || Cu^{2+}(1M $CuSO_4$)| Cu

$$H_2 \longrightarrow 2H^+ + 2e^- \text{ (Oxidation)}$$

$$Cu^{2+} + 2e^- \longrightarrow Cu \qquad \text{(Reduction)}$$

$$Cu^{2+} + H_2 \longrightarrow Cu + 2H^+ \text{ (Redox)}$$

$$E_{cell} = E^{\circ}_{right} - E^{\circ}_{left}$$

$$E^{\circ}_{cell} = E^{\circ}_{Cu^{2+}/Cu} - E^{\circ}_{H^+/H_2}$$

$$0.34\ V = E^{\circ}_{Cu^{2+}/Cu} - 0.00\ V$$

$$E^{\circ}_{Cu^{2+}} = +\ 0.34\ V$$

Thus, the standard electrode potential of copper is + 0.34 V.

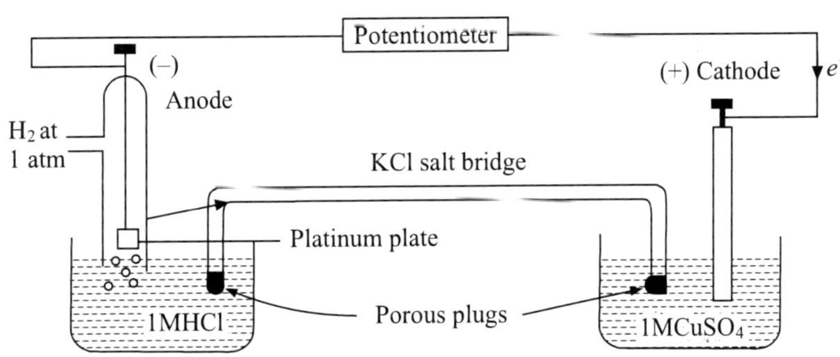

Fig. 1.4 *EMF measurement of Cu – H_2 galvanic cell*

1.7.1 Significance of Sign of Single Electrode Potential

The positive value of a single electrode potential indicates that the given electrode is more positive than the standard hydrogen electrode. In other words when coupled with the standard hydrogen electrode the given electrode acts as a cathode and standard hydrogen electrode acts as an anode of the working cell so obtained. The negative value of a single electrode potential of an electrode shows that the given electrode is more negative than the standard hydrogen electrode. The given electrode, therefore when combined with standard hydrogen electrode acts as anode. Whereas standard hydrogen electrode acts as cathode for the resulting cell to be working as spontaneous. For example, when copper electrode is combined with standard hydrogen electrode, copper electrode acts as cathode and undergoes reduction and hydrogen electrode acts as anode.

$$H_2(g) \longrightarrow 2H^+ + 2e \quad \text{(Oxidation)}$$
$$Cu^{2+} + 2e^- \longrightarrow Cu \quad \text{(Reduction)}$$

Hence, electrode potential of copper is assigned a positive sign. Its standard electrode potential is 0.34 V.

When zinc is coupled with standard hydrogen electrode, zinc electrode ats as the anode and hydrogen electrode acts as the cathode.

$$Zn \longrightarrow Zn^{2+} + 2e^-$$
$$2H + 2e^- \longrightarrow H_2$$

Hence, the electrode potential of zinc is negative.

1.8 STANDARD ELECTRODE POTENTIAL (E°)

Single electrode potential of an electrode depends on the purity of the metal, activities of various species involved in electrode reaction and the temperature. Hence, to have comparable values, standard electrode potentials are used. So, **standard electrode potential is the electrode potential of an electrode when all the species involved in the given half cell (or electrode reaction) are in their standard state with reference to the standard hydrogen electrode.** It is denoted by $E°$ and expressed in volts. **The emf of a cell when the reactants and products of the cell reaction are at unit concentration or unit activity at 298 K and at 1 atmospheric pressure.**

The standard states or standard conditions may be taken as:

1. 1.0 M concentration of each ion in solution.
2. 1 atm for each gas.
3. A temperature of 298 K.

1.8.1 Significance of Sign of Standard Electrode Potential (E°)

If $E°$ for an electrode is positive, it means that the electrode under standard conditions is more positive than SHE so that reduction occurs at this electrode (cathode) and oxidation occurs at SHE (anode) when combined to have spontaneous reaction or working cell.

If $E°$ for an electrode is negative, it shows that this electrode under standard conditions is more negative than SHE. Hence, for a spontaneous reaction to occur so that the resulting cell is a source of electrical energy, the given electrode will act as anode when coupled with

SHE (cathode), i.e., oxidation occurs at the given electrode (with $E°$ negative) whereas the reduction occurs at SHE in the cell.

1.9 EMF OF THE CELL AND FREE ENERGY CHANGE

In a galvanic cell, oxidation reaction occurs at the anode and reduction occurs at the cathode. Both the electrodes are at different potential values. The difference in potential of the electrodes is known as the electromotive force (EMF). It is also referred to as the vigor with which the cell works and is the measure of chemical reactions occurring in the cell. In a spontaneous reaction, change in free energy (ΔG) decreases. Larger the decrease in free energy greater would be the EMF. Mathematically, it can be expressed as

$$-\Delta G = nEF$$

where negative sign indicates the decrease in the free energy

$n \rightarrow$ Number of electrons involved in the redox reaction

$E \rightarrow$ EMF of the cell

$F \rightarrow$ Faraday's constant

1.10 ENTROPY OF THE REACTION

The standard cell potential is related to standard free energy or cell function through equation

$$\Delta G = -nFG \qquad \ldots(1)$$

The thermodynamic relation

$$\left(\frac{\partial(\Delta G)}{\partial T}\right)_P = -\Delta S$$

By substituting the value of $-\Delta G$ from equation

$$\left(\frac{\partial(-nFE)}{\partial T}\right)_P = -\Delta S$$

$$-nF\left(\frac{\partial E}{\partial T}\right)_P = -\Delta S$$

$$\boxed{\Delta S = nF\left(\frac{\partial E}{\partial T}\right)_P}$$

The temperature coefficient of the cell potential gives the energy of the cell reaction.

1.11 ENTHALPY OF THE REACTION

The enthalpy of the reaction is related at ΔG and ΔS are

$$\Delta H = \Delta G + T\Delta S$$

Substituting the values of ΔG and ΔS

$$\Delta H = -nFE + nFT\left(\frac{\partial E}{\partial T}\right)_P$$

$$\Delta H = -nF\left[E - T\left(\frac{\partial E}{\partial T}\right)_P\right]$$

$$\boxed{E_{\text{cell}} = \frac{-\Delta H}{nF} + T\left(\frac{\partial E}{\partial T}\right)}$$

Whether the electrical work done is greater or less than the heat of the reaction depends upon the temperature coefficient of the cell that is E_{cell} varies with T.

If $\frac{\partial E}{\partial T}$ is negative then $\Delta G > \Delta H$ and heat is absorbed from the surroundings during the working of the cell.

If $\frac{\partial E}{\partial T}$ is positive then $\Delta H > \Delta G$ and heat is given out to the surroundings of the cell during working.

Thus ΔG, ΔH and ΔS can be calculated for the cell reaction from measurements of EMF and temperature coefficient of EMF.

1.12 EMF OF THE CELL AND EQUILIBRIUM CONSTANT OF A CELL REACTION

Consider the general electrochemical cell reaction

$$aA + bB \rightleftharpoons cC + dD$$

The decrease in free energy $-\Delta G$ for the process is given by the thermodynamic equation.

$$-\Delta G = -\Delta G^\circ - RT \ln Q \qquad \ldots(1)$$

where $-\Delta G^\circ$ is decrease in free energy of system when the substances are in their standard state of unit activity Q = Reaction quotient.

$$Q = \frac{[C]^c\,[D]^d}{[A]^a\,[B]^b}$$

The electrical energy produced during the reaction by the cell is nEF volt

$$-\Delta G = -nFE$$

Put the value in Eq. (1)

$$nFE = -\Delta G^\circ - RT \ln Q \quad \ldots(2)$$

we know that

$$-\Delta G^\circ = RT \ln K$$

where $K \rightarrow$ equilibrium constant for the cell reaction

Substitute the value in Eq. (2)

$$nFE = RT \ln K - RT \ln Q$$

$$nFE = RT \ln K - RT \ln \frac{[C]^c\,[D]^d}{[A]^a\,[B]^b}$$

changing ln to $\log_{10}$ we get

$$E_{\text{cell}} = \frac{2.303\,RT}{nF} \log K - \frac{2.303\,RT}{nF} \log \frac{[C]^c\,[D]^d}{[A]^a\,[B]^b} \quad \ldots(3)$$

At 25°C, the value of $\frac{2.303\,RT}{nF}$ is

$$= \frac{2.303 \times 8.313\ \text{JK}^{-1}\text{mol}^{-1} \times 298\ \text{K}}{n \times 96500\ \text{C mol}^{-1}}$$

$$= \frac{0.0591}{n}\ \text{volt}$$

Putting value in Eq. (3)

$$E_{\text{cell}} = \frac{0.0591}{n} \log \text{K}$$

$$= \frac{0.0591}{n} \log \frac{[C]^c\,[D]^d}{[A]^a\,[B]^b}$$

The equilibrium constant of the cell can be calculated from this equations if we know the EMF of the cell and the concentration of reactions and products of the cell reaction.

1.13 NERNST EQUATION

Walther Nernst in 1889 derived the mathematical relationship to calculate the potential of half cell, *E* from the standard electrode potential $E°$ and the temperature of the cell. It is also a thermodynamic equation which relates the change in free energy and cell potential with concentration.

Consider a reduction electrode reaction:

$$M^{n+} + ne^- \longrightarrow M$$

The thermodynamic relationship between the change in decrease in free energy ($-\Delta G$) and equilibrium constant is given by

$$-\Delta G = -\Delta G° - RT \ln K \quad ...(1)$$

where $-\Delta G \rightarrow$ Change in decrease in free energy

$\Delta G \rightarrow$ Standard change in decrease in free energy accompanying the same process

$K \rightarrow$ Ratio of concentration of product and reactants at the equilibrium

Substituting the value of $K = [M]/[M^{n+}]$,

We have, $$-\Delta G = -\Delta G° - RT \ln [M]/[M^{n+}] \quad ...(2)$$

The potential developed at the electrode is at the expense of the decrease in free energy of the electrode process, therefore,

$$-\Delta G = nFE \text{ and } -\Delta G° = nFE°$$

where, $E \rightarrow$ Electrode potential

$F \rightarrow$ Faraday's constant

$N \rightarrow$ Number of electrons

Hence, $$nFE = nFE° - RT \ln [M]/[M^{n+}] \quad ...(3)$$

Since $[M] = 1$, ln = 2.303 log hence Eq. (3), becomes

$$nFE = nFE° - 2.303\ RT \log 1/[M^{n+}] \quad ...(4)$$

Dividing Eq. (4) by nF on both sides, we get

$$E = E° - 2.303\ RT/nF \quad \log 1/[M^{n+}] \quad ...(5)$$

R is the gas constant = 8.314 kJ/Mol/K

T is the temperature in Kelvin

F is the Faraday constant = 96500 C/mol

Equation (5) becomes:

$$E = E° - 0.0591/n \log 1/M^{n+}] \text{ at } 298 \text{ K} \quad ...(6)$$

This equation is known as Nernst equation for single electrode potential.

WORKED PROBLEMS

Problem 1.5

Calculate the standard electrode potential of Cu^{2+}/Cu if its electrode potential at 25 °C is 0.296 V and $[Cu^{2+}]$ is 0.015 M.

Solution: $Cu^{2+} + 2e^- \longrightarrow Cu$

Use Equation

$$E = E^o - 0.059/n \log [Cu]/[Cu^{2+}]$$

$$E = 0.296 \text{ V}, n = 2, [Cu^{2+}] = 0.015 \text{ M}$$

$$0.296 \text{ V} = E^o_{Cu^{2+}/Cu} + 0.059/2 \log 0.015 \text{ [as (Cu) = 1]}$$

$$E^o_{Cu^{2+}/Cu} = 0.296 \text{ V} + 0.059/2 \log 0.015$$

$$= 0.296 \text{ V} + 0.0540$$

$$= \mathbf{0.35\ V}$$

Problem 1.6

Calculate the emf of a cadmium – copper cell in which cadmium is in contact with 0.002 M Cd SO_4 and copper is in contact with 0.02 M $CuSO_4$ solution. The standard emf of the cell is 0.74 V at 298 K.

Solution: The cell reaction is,

$$Cd + Cu^{2+} \longleftrightarrow Cd^{2+} + Cu$$

$$[Cu^{2+}] = 0.02 \text{ M}$$

$$[Cd^{2+}] = 0.002 \text{ M}$$

$$E^o_{cell} = 0.74 \text{ V} \quad n = 2$$

At 298 K, emf of the cell,

$$E = E^o_{cell} - 0.0591/n \log [Cd^{2+}] [Cu]/[Cd] [Cu^{2+}]$$

$$= E^o_{cell} - 0.0591/n \log [Cd^{2+}]/Cu^{2+}] \text{ (as [Cu] = [Cd] = 1)}$$

$$= 0.74 - 0.0591/2 \log 0.002/0.02$$

$$= 0.74 - 0.02955 \log 0.1$$

$$= 0.74 + 0.02955$$

$$= \mathbf{0.7695\ V}$$

Problem 1.7

A zinc rod is dipped in 0.1 M solution of $ZnSO_4$. The salt is 95% dissociated at this dilution at 298K. Calculate the electrode potential, given that

$$E^o_{zn^{+2}/zn} = -0.76 \text{ V}.$$

Solution: The electrode reaction is

$$Zn^{+2} + 2e^- \rightleftharpoons Zn(s)$$

According to Nernst equation, at 298K

$$E\,(_{Zn^{+2}/Zn}) = E^o{}_{Zn^{+2}/Zn} - 0.059/2 \log [Zn^{2+}(aq)]$$

$$E^o_{Zn^{+2}/Zn} = -\,0.76 \text{ V} \qquad [Zn] = 1\ [Zn^{2+}(aq)] = 0.1 \times 95/100 = 0.095 \text{ M}$$

$$E^o_{Zn^{+2}/Zn} = -\,0.76 - 0.059/2 \log 1/0.095$$

$$= -\,0.76 - 0.03$$

$$= \mathbf{-\,0.076\ V}$$

Problem 1.8

From the following electrode potential at 25 °C

$$Cu^{+2} + 2e^- \longrightarrow Cu \qquad E^o = +\,0.34V$$

$$Cu^{+2} + 2e^- \longrightarrow Cu^+ \qquad E^o = +\,0.15V$$

Calculate equilibrium constant of the reaction.

Solution:

$$E^o{}_{cell} = E_{cathode} - E_{anode}$$

$$= 0.34 - 0.15$$

$$= +\,0.19$$

$$E^o_{cell} = 0.0591/n \log K_C$$

$$0.19 = 0.0591/2 \log K_C$$

$$\log K_C = 2 \times 0.19/0.0591$$

$$\log K_C = 6.429$$

$$K_C = \text{antilog}\ (6.429)$$

$$K_C = 2.68 \times 10^6$$

Problem 1.9

What is the maximum amount of work that can be obtained under standard conditions from the discharge of Ag_2O cell if 1.00 g of Ag_2O reacts. The half cell reaction for Ag_2O is

$$Ag_2O + H_2O + 2e^- \longrightarrow 2Ag + 2OH^-$$

E^o for the cell reaction is 1.50 V and the molecular mass of Ag_2O is 232 g mol^{-1}, $F = 96500$ C mol^{-1}.

Solution:

$$\Delta G^o = nFE^o$$

n = mol electrons = 2/232

$F = 96500$ C mol^{-1},

$E^o = 1.50$ V $= 1.5$ JC^{-1} = **12481 J**

Problem 1.10

The E^o_{cell} of an aluminium – air battery is 2.73 volts and it involves a 12-electron process. Calculate the free energy change (ΔG^o) of the battery in kJ.

Solution:

$$\Delta G^o = -nFE^o = -12 \times 96500 \times 2.73 \text{ J}$$
$$= -3161340 \text{ J} = \mathbf{-3161.34 \text{ kJ}}$$
$$= 0.74 + 0.02955$$
$$= \mathbf{0.7695 \text{ V}}$$

1.14 TYPES OF ELECTRODES

In electrochemical work, the cells encountered involve various types of electrodes depending on the purpose for which they are used. Some of the important types may be grouped into the following.

1. Metal – Metal ion Electrode (Metal – insoluble salt electrode)

These electrodes consist of a metal in contact with a sparingly soluble salt of the same metal dipped in a solution containing anion of the salt.

Example: Calomel electrode ($Hg/Hg_2Cl_2/Cl^-$)
Silver – Silver chloride electrode ($Ag/AgCl/Cl^-$)
Lead – Lead sulphate electrode ($P/PbSO_4/SO_4^{2-}$)

2. Gas Electrode

It consists of a gas bubbling about an inert metal wire or foil, immersed in a solution containing ions to which the gas is reversible. The metal provides electrical contact and facilitates the establishment of equilibrium between the gas and its ions.

Example: Hydrogen electrode ($Pt/H_2/H^+$)
Chlorine electrode ($Pt/Cl_2/Cl^-$)

3. Amalgam Electrode

It is similar to metal – metal ion electrode except for the fact that in amalgam electrode, metal amalgam is in contact with a solution containing its own metal ions.

Example: Lead amalgam electrode ($Pb - Hg/Pb^{2+}$)

4. Oxidation – Reduction Electrode

An oxidation-reduction electrode is the one in which the electrode potential arises from the presence of oxidized and reduced forms of the same substance in solution. The potential arises from the tendency of one form to change into the other more stable form. The potential developed is picked up by an inert electrode like platinum.

Example: $Pt/Fe^{2+}, Fe^{3+}$

$Pt/Ce^{3+}, Ce^{4+}$

$Pt/Sn^{2+}, Sn^{4+}$

5. Ion Selective Electrode (Membrane Electrode)

In an ion selective electrode, a membrane is in contact with a solution, with which it can exchange ions.

Example: Glass electrode (exchanges H^+ ions with the solution), various metal ion selective electrodes, which exchange metal ions with solution.

1.15 REFERENCE ELECTRODES

Reference electrodes are the electrodes whose potentials are known and are used for the determination of potential of other electrodes. The purpose of the reference electrode is to complete the measuring circuit and provide a stable and reproducible potential against which the indicator electrode is compared and measured. Reference electrodes are of two types.

1.15.1 Primary Reference Electrodes

1.15.1.1 Standard Hydrogen Electrode, SHE

A common reference electrode used to measure the potential of another electrode is the standard hydrogen electrode (SHE) whose potential is arbitrarily taken as zero at all temperatures, i.e., $E° = 0$.

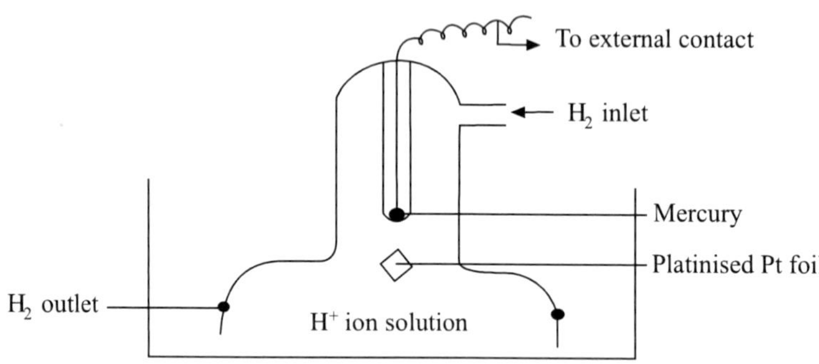

Fig. 1.5 *Normal hydrogen electrode*

The hydrogen electrode consists of a glass tube with a platinised platinum foil fused to it at the bottom. Mercury is placed at the bottom of the copper wire dipped in it provides external electrical contact. The glass tube is surrounded by a bell-shaped glass jacket fused at the top to the glass tube. There is a hydrogen gas inlet at the top. The gas outlets are provided at the bottom. The platinum foil is dipped in a solution containing H^+ ions and hydrogen gas is bubbled about the electrode through the gas inlet. Hydrogen gas gets adsorbed on the platinum foil and establishes an equilibrium with H^+ ions. The electrode is represented as follows:

$$Pt/H_2/H^+$$

The electrode reaction is $2H^+ + 2e^- \rightleftharpoons H_2(g)$

If the concentration of H^+ ions in 1M and hydrogen gas is bubbled at 1 atmospheric pressure at 298 K, then it is standard hydrogen electrode. Its electrode potential has been arbitrarily chosen as zero.

Uses

A standard hydrogen electrode is used as a primary reference electrode to determine the electrode potential of other electrodes. The test electrode is coupled with a standard hydrogen electrode and the emf of the so formed cell is determined experimentally by the potentiometer method. Since the electrode potential of the standard hydrogen electrode is zero, the electrode potential of the test electrode can be calculated.

Limitations

1. It cannot be used in the presence of oxidizing agents.
2. The hydrogen electrode gets poisoned even in the presence of traces of impurities.
3. Construction and working is difficult on account of difficulties involved in maintaining the concentration at unity and in keeping the pressure of the gas uniform at one atmosphere.

1.15.2 Secondary Reference Electrode

To overcome these difficulties, secondary reference electrodes of the types Calomel electrode, silver-silver chloride electrodes have been successfully designed. They have the following advantages over primary reference electrodes.

1. They have a constant and stable potential values at a particular temperature.
2. They are calibrated with respect to SHE (Standard Hydrogen Electrode).
3. They have the least potential gradient with temperature.
4. The potential values are reliable and measurements can be made with great accuracy.

1.15.2.1 Calomel Electrode

Calomel electrode is a metal-metal salt ion electrode. It consists of mercury, mercurous chloride and a solution of potassium chloride. Mercury is placed at the bottom of a glass tube having a side tube on each side. Mercury is covered by a paste of mercurous chloride (calomel) with mercury and potassium chloride solution. A solution of potassium chloride is introduced above the paste through the side tube. A platinum wire is dipped into mercury to provide external electrical contact. The side tube is used for making electrical contact with a salt bridge. The concentration of potassium chloride solution used is either decinormal, normal or saturated. Correspondingly, the electrode is known as decinormal, normal or saturated calomel electrode, respectively.

Calomel electrodes can be represented as:

$$Hg \mid Hg_2Cl_2(s) \mid Cl^-$$

The calomel electrode can act as anode or cathode depending on the nature of the other electrode of the cell.

When it acts as **anode**, the electrode reaction is

$$2Hg \longrightarrow Hg_2^{2+} + 2e^-$$
$$Hg_2^{2+} + 2Cl^- \longrightarrow Hg_2Cl_2$$
$$2Hg + 2Cl^- \longrightarrow Hg_2Cl_2 + 2e^-$$

When it acts as cathode, the electrode reaction is

$$Hg^{2+} + 2e^- \longrightarrow 2Hg$$
$$Hg_2Cl_2 \longrightarrow Hg_2^{2+} + Cl^-$$
$$Hg_2Cl_2 + 2e^- \longrightarrow 2Hg + 2Cl^-$$

The net reversible electrode reaction is

$$Hg_2Cl_2 + 2e^- \rightleftharpoons 2Hg(l) + 2Cl^-$$

Electrode potential, $$E = E° - \frac{2.303\,RT}{F} \log [Cl^-]^2$$

$$E = E° - \frac{2.303\,RT}{F} \log [Cl^-]$$

$$E = E° - 0.0591 \log [Cl^-] \text{ at } 298K$$

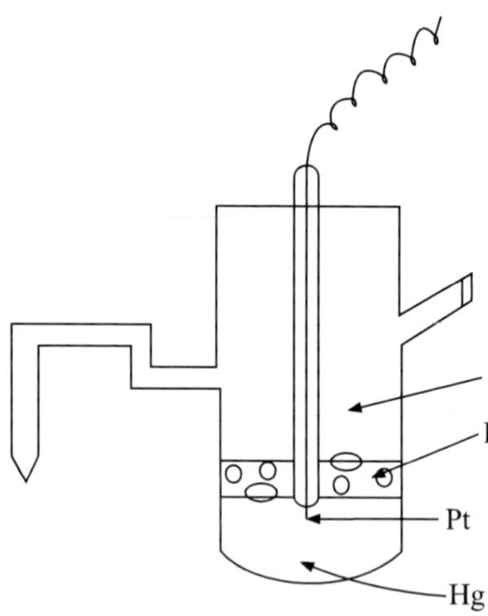

Fig. 1.6 *Saturated Calomel electrode*

The electrode potential is decided by the concentration of chloride ions and the electrode is reversible with chloride ions. At 298 K, the electrode potentials are as follows:

KCl concentration	0.1 N	1.0 N	Saturated
Electrode potential (in volts)	0.3335	0.2810	0.2422

When the KCl solution is 0.1 N, 1.0 N and saturated, it is called decinormal, normal and saturated calomel electrode respectively.

Advantages

1. It is simple to construct.
2. The cell potential is reproducible.
3. The cell potential does not vary with temperature.
4. The cell potential is constant over a long period.

Uses

It is used as a secondary reference electrode in the measurement of single electrode potential. It is the most commonly used reference electrode in all potentiometer determinations.

Demerits of Calomel Electrode

Calomel electrodes should not be used above 50 °C because mercurous chloride breaks down, yielding unstable readings. It is less prone to contamination because mercury/mercurous chloride interface is protected inside a tube and is not in direct contact with the electrolyte.

Reproducibility of a calomel electrode is much better than a silver/silver chloride. Recently, calomel has fallen into disfavour because of its toxicity.

Determinations of Single Electrode Potential

The test electrode [Ex: Zn/Zn^{2+}) is coupled with a saturated calomel electrode.

$$Zn(s) \mid Zn^{2+} \parallel Cl^- \mid Hg_2Cl_2(s) \mid Hg(l)$$

The emf of the so formed cell is determined by the potentiometric method. Then,

$$\begin{aligned} E_{cell} &= E_{cathode} - E_{anode} \\ &= E_{calomel} - E_{Zn} \\ &= 0.2422 - E_{Zn} \\ E_{Zn} &= 0.2422 - E_{cell} \end{aligned}$$

Determination of pH Value of a Solution

The electrode can be coupled with hydrogen electrode containing solution of unknown pH.

$$Pt, H_2\ (1\ atm) \parallel H^+ = ? \parallel Hg_2Cl_2(s) \mid Hg$$

The emf of the cell, $E_{cell} = E_{right} - E_{left}$

$$= 0.2422 \text{ V} + 0.0592 \text{ V pH}$$

$$\text{pH} = \frac{E_{cell} - 0.2422 \text{ V}}{0.0592 \text{ V}}$$

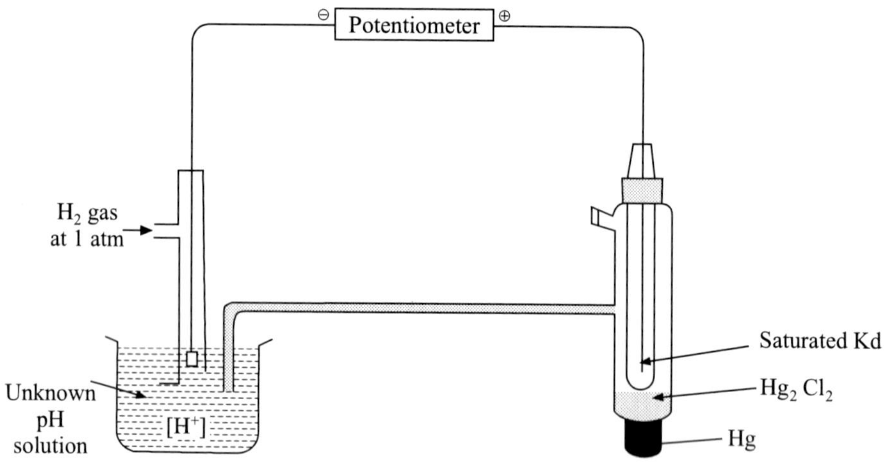

Fig. 1.7 *Determination of pH value of a solution using calomel electrode*

1.16 ION SELECTIVE ELECTRODES

Ion selective electrode has the ability to respond only to certain specific ion in a mixture and develop a potential proportional to the concentration of the ions. In other words, the potential developed by an ion – selective electrode depends only on the concentration of species/ions of interest. The electrode generally consists of a membrane which is capable of exchanging the specific ions with the solution with which it is in contact. Therefore, these electrodes are also referred to as membrane electrodes.

Various types of ion selective electrodes are listed below.

A. Crystalline membrane electrodes:

1. Single crystal, for example: LaF_3 for F^-
2. Polycrystalline or mixed crystal, for example, Ag_2S for S^{2-} and Ag^+

B. Non-crystalline membrane electrodes:

1. Liquid, for example, liquid ion exchangers for Ca^{2+}, K^+, etc.
2. Immobilized liquid in a rigid polymer, for example, polyvinyl chloride matrix for Ca^{2+}, NO_3^-, etc.
3. Glass, for example, silicate glasses for Na^+ and H^+

1.16.1 Determination of Concentration of Metal Ions

An ion selective electrode generally consists of ion selective membrane in contact with an analyte solution on one side and an internal reference solution on the other side. An internal reference electrode is constituted in contact with the reference solution. The electrode can be represented as follows:

Solution to be analysed $[M^{n+}] = C_1$	Membrane	Internal standard solution $[M^{n+}] = C_2$	Internal reference electrode

The mechanism of potential development in ion selective electrodes is different from that in other types of electrodes. The observed potential of an ion selective electrode is a kind of junction potential that develops across a membrane that separates the analyte solution from reference solution. The potential developed across the membrane is a function of concentration of the analyte and reference solutions.

$$E_j = \frac{2.303\,RT}{nF} \log \frac{C_1}{C_2}$$

where C_1 and C_2 are the concentrations of the analyte and reference solution, respectively. If the concentration of the inner standard solution is kept constant, then

$$E_j = k + \frac{2.303\,RT}{nF} \log C_1$$

where k is a constant, $k = -\,2.303\,RT/nF \log C_2$

The overall potential of the electrode also includes that of the internal reference electrode, and thus,

$$\begin{aligned} E_m &= E_j + E_{\text{ref}} \\ &= K + 2.303\,RT/nF \log C_1 + E_{\text{ref}} \\ &= E^\circ_{\text{M}} + 2.303\,RT/nF \log C_1 \end{aligned}$$

where $E^\circ_{\text{M}} = K + E_{\text{ref}}$

At 298 K, $\quad E_m = E^\circ_{\text{M}} + 0.0591/n \log C_1$

The electrode is combined with an external reference electrode, and emf of the so formed cell is determined by the potentiometric method. Knowing the electrode potential of the reference electrode, the electrode potential of the ion selective electrode is calculated, which, in turn, gives the ion concentrations. The potentiometers are generally automated to read the ion content directly.

Advantage

It is used to determine the concentrations of a number of cations and anions such as H^+, Li^+,Na^+, K^+, Pb^{2+}, Cu^{2+}, Mg^{2+}, CN^-, NO_3^-, F^-, etc.

Disadvantage

Simple potentiometers cannot be used with these electrodes, because of high resistance of the membranes. This necessitates special potentiometers which are expensive.

1.17 GLASS ELECTRODE

Glass electrode is a type of ion-selective electrode, and is most widely used for pH determination.

Construction

The glass electrode consists of a glass bulb made up of a special type of glass with high electrical conductance. The glass bulb is filled with a solution of constant pH (0.1 MHCl), and is inserted with a Ag – AgCl electrode, which is the internal reference electrode and also serves for the external electrical contact. The electrode is dipped in a solution containing H^+ ions.

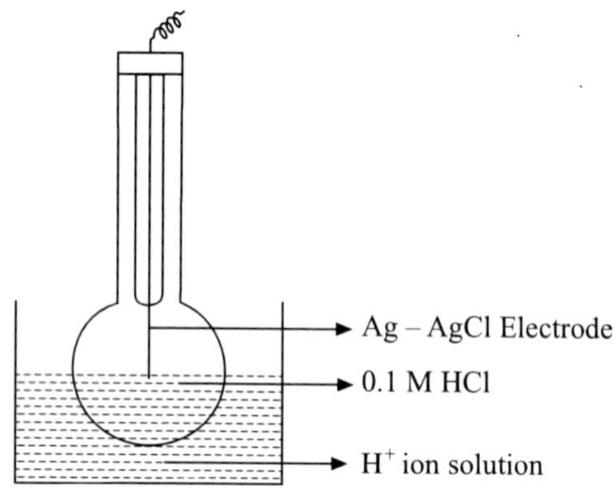

Fig. 1.8 *Glass electrode*

Electrode Potential of Glass Electrode

The glass electrode system can be represented as follows

$$Ag \mid AgCl(s) \mid 0.1\ M\ HCl \mid glass$$

When dipped in a solution it can be represented as

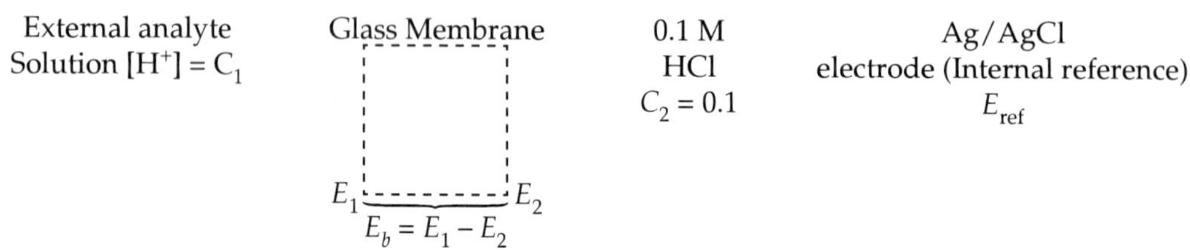

The glass membrane exchanges ions with the solution

$$H^+_{(Solution)} + MGl_{(Membrane)} \rightleftharpoons M^+_{(Solution)} + HGl_{(Membrane)}$$

The exchange of ions by the inner and outer membranes gives rise to a boundary potential. This boundary potential consists of two potentials, E_1 and E_2, which are associated with the outer and inner membrane respectively.

Therefore, $$E_b = E_1 - E_2$$

From thermodynamic consideration it has been shown that E_1 and E_2 are related to the hydrogen ion activities at each face, by Nernst like equation,

$$E_1 = j_1 - 0.0591/n \log a'_1/a_1$$
$$E_2 = j_2 - 0.0591/n \log a'_2/a_2$$

where j_1 and j_2 are constants and a_1 and a_2 are activities of H^+ in the outer and inner solution respectively. a'_1 and a'_2 are the activities of H^+ ions at the external and internal surfaces of the glass membrane.

Since the glass at the inner and outer membranes is identical, $j_1 = j_2$ and also $a'_1 = a'_2$.

Then
$$E_b = E_1 - E_2$$
$$= 0.0591 \log a_1/a_2 \ (n = 1 \text{ for } H^+ \text{ ion})$$

Thus, the boundary potential E_b depends only on the hydrogen ion activities of the solutions. Since the concentration of the inner solution is constant, a_2 is constant. The above equation simplifies to

$$E_b = L' + 0.0591 \log a_1 = L' + 0.0591 \log C_1$$
$$= L' - 0.0591 \text{ pH}$$

where $L' = -0.591 \log a_2 = -0.0591 \log C_2$

The potential of a glass electrode E_G has three components:

1. The boundary potential
2. The potential of the internal Ag/AgCl reference electrode
3. A small symmetry potential

$$E_G = E_b + E_{ref} + E_{asy} = L' + 0.0591 \log C_1 + E_{ref} + E_{asy}$$
$$= E^\circ_G + 0.0591 \log C_1$$
$$= E^\circ_G - 0.0591 \text{ pH}$$

where E°_G is a combination of three constant terms,

$$E^\circ_G = L' + E_{ref} + E_{asy}$$

Asymmetric potential arises due to the difference in responses of the inner and outer surfaces of the glass bulb to changes in H^+ ion activity. This may originate as a result of differing conditions of strain in the two glass surfaces (due to difference in curvature).This asymmetric potential varies with time and therefore, glass electrode must be standardized frequently by placing a solution of known H^+ activity (pH).

Applications

Glass electrode is a pH sensitive electrode, most widely used for pH determination.

Determination of pH

To determine the pH of a given solution, the glass electrode is dipped in the solution whose pH is to be determined. It is combined with a saturated calomel electrode.

$$\text{Hg} \mid Hg_2Cl_2 \mid Cl^- \parallel \text{Solution (pH = ?)} \mid \text{glass} \mid 0.1\text{N HCl} \mid \text{AgCl} \mid \text{Ag}$$

HCl in the bulb furnishes a constant H^+ ion concentration. Thus, it is a silver - silver chloride electrode, reversible with respect of chloride ions.

Glass electrode is used as the "internal reference electrode". For determining the pH of solutions, specially coloured solutions containing oxidizing or reducing agents. Usually, calomel electrode is used as the second electrode. In order to determine the pH of a solution, the glass electrode is placed in the solution under test and this half cell is coupled with saturated calomel electrode. The emf of the cell is measured.

The emf of the complete cell is given by

$$E_{cell} = E_{right} - E_{left} \text{ (at 25 °C)} = E_G - E_{cal}$$
$$= [E^\circ_G - 0.0591 \text{ pH}] - E_{cal}$$

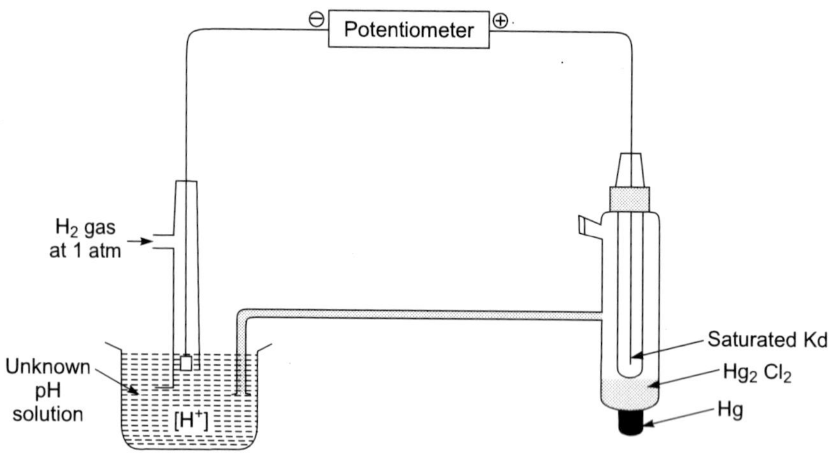

Fig. 1.9 *Determination of pH value of a solution using calomel electrode*

Advantages

- Can be used in oxidizing and reducing environments.
- Accurate results are obtained between pH range 1 – 9. However, by using special glass electrodes, pH 1 – 14 can be measured.
- Electrode does not get poisoned.
- Are very good indicator electrodes in potentiometer.

Disadvantages

Because of the high resistance of glass, simple potentiometers cannot be used. It requires sensitive electronic potentiometers for emf measurement.

1.18 CONCENTRATION CELL

Concentration cells are electrochemical cells in which emf arises from a transfer of material from one half cell to another half cell due to difference in concentration of the species. It is of two types.

1. Electrolyte concentration cell
2. Electrode concentration cell

1.18.1 Electrolyte Concentration Cell

It is a galvanic cell, which consist of two identical electrodes immersed in same electrolytic solution but of different concentration.

Metal immersed in dilute solution will have lower potential and is taken as anode, whereas, metal immersed in concentrated solution will have higher potential and it is taken as cathode.

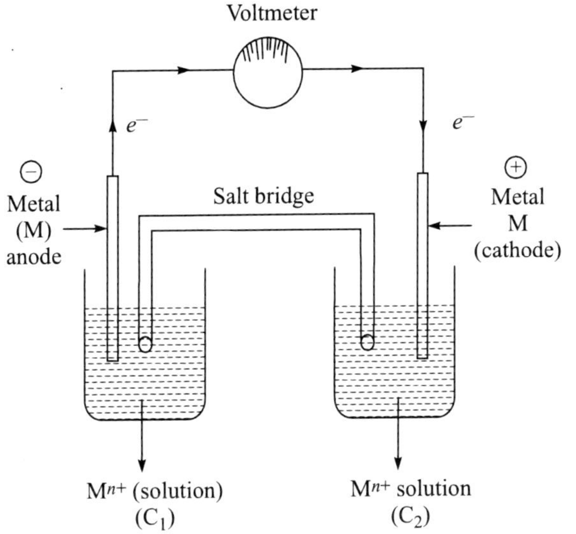

Fig. 1.10 *Electrolyte concentration cell*

Potential of Concentration Cell

It is represented as

$$\mathrm{M}|\mathrm{M}^{n+}_{(\mathrm{C}_1)}|\mathrm{M}^{n+}_{(\mathrm{C}_2)}|\mathrm{M} \quad \text{where} \quad \mathrm{C}_1 > \mathrm{C}_2$$

Metal immersed in dilute solution acts as anode, where oxidation occurs

$$\mathrm{M} \rightarrow \mathrm{M}^{n+}_{(\mathrm{C}_1)} + ne^-$$

Metal immersed in concentration solution acts as cathode where reduction occurs

$$\mathrm{M}^{n+}_{(\mathrm{C}_2)} + ne^- \rightarrow \mathrm{M}$$

Net cell reaction is mere change in concentration of the electrolyte concentration of metal ions at anode increases whereas it decrease at cathode.

$$\mathrm{M}^{n+}_{(\mathrm{C}_2)} \rightarrow \mathrm{M}^{n+}_{(\mathrm{C}_1)}$$

As both half cells are made of same electrode, E°_{cell} for concentration cell is zero.

Hence, Nernst equation for electrolyte concentration cell is

$$E_{cell} = E^\circ_{cell} + \frac{2.303}{nF}\log\frac{[M^{n+} \text{ at cathode}]}{[M^{n+} \text{ at anode}]}$$

For concentration cell

$$E^\circ_{cell} = E^\circ_{cathode} - E^\circ_{anode} = 0$$

$$E_{cell} = \frac{2.303\,RT}{nF}\log\frac{[M^{n+} \text{ at cathode}]}{[M^{n+} \text{ at anode}]}$$

$$= \frac{2.303\,RT}{nF}\log\frac{C_2}{C_1}$$

$$E_{cell} = \frac{0.0591}{n}\log\frac{C_2}{C_1} \text{ at 298 K}$$

E_{cell} is +ve and reaction is spontaneous only when $C_2 > C_1$.

1.18.2 Electrode Concentration Cell

It is a galvanic cell in which electrode and electrolyte present in both half cells are same but only the concentration of electrode is different.

According to Nernst equation

$$E = E^\circ + \frac{2.303\,RT}{nF}\log\frac{[M^{n+}]}{[M]}$$

Electrode containing higher concentration of metal will have lower potential and it is taken as anode whereas electrode containing lower concentration of metal will have higher potential and it is taken as cathode. Electrodes with different concentration of metal are generally obtained by dissolving different amount of metal in liquid mercury. This type of electorde is commonly called as amalgam electrode.

Potential of Concentration Cell

$$M_{(C_1)}(Hg)|M^{n+}|M_{(C_2)}(Hg) \quad \text{where} \quad C_1 > C_2$$

Electrode containing higher concentration of metal acts as anode, where oxidation occurs.

$$M_{(C_1)} \rightarrow M^{n+} + ne^-$$

Electrode containing lower concentration of metal acts as cathode, where reduction occurs

$$M^{n+} + ne^- \rightarrow M_{(C_2)}(Hg)$$

Net cell reaction is mere change in concentration of metal concentration of metal at anode decreases whereas at cathode it increases.

$$M_{(C_1)}Hg \rightarrow M_{(C_2)}Hg$$

As both half cells are made of same electrode. E°_{cell} for a concentration cell is zero. Hence, according to Nernst equation for electrode concentration cell is

$$E_{cell} = (E^\circ_{cathode} - E^\circ_{anode}) + \frac{2.303\,RT}{nF}\log$$

$$\left[\frac{[\text{Metal ion at cathode}]}{[\text{Matal at cathode}]} - \frac{[\text{Metal ion at anode}]}{[\text{Metal at anode}]}\right]$$

For an electrode concentration cell

$$E^\circ_{cathode} = E^\circ_{anode}$$

and $$[\text{Metal ion at cathode}] = [\text{Metal ion at anode}]$$

$$E_{cell} = \frac{2.303\,RT}{nF}\log\left[\frac{\frac{[\text{Metal ion at cathode}]}{[\text{Metal at cathode}]}}{\frac{[\text{Metal ion at anode}]}{[\text{Metal at anode}]}}\right]$$

$$= \frac{2.303\,RT}{nF}\log\left[\frac{[\text{Metal at anode}]}{[\text{Metal at cathode}]}\right]$$

$$E_{cell} = \frac{0.0591}{n}\log\frac{C_1}{C_2} \text{ at } 298\,\text{K}$$

E_{cell} is +ve and reaction is spontaneous only when $C_1 > C_2$.

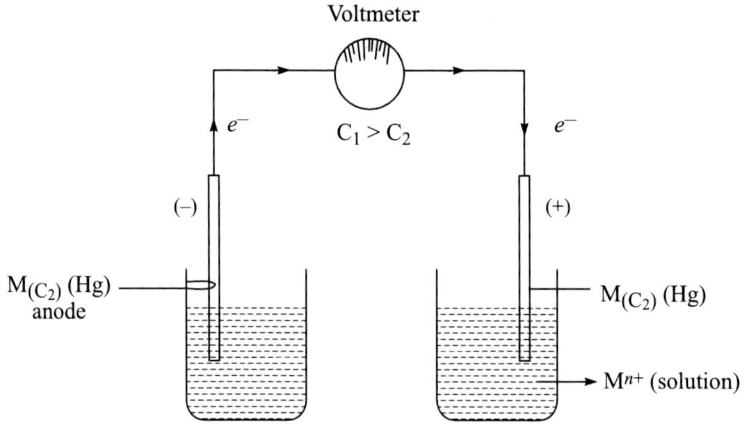

Fig. 1.11 *Electrode concentration cell*

WORKED PROBLEMS

Problem 1.11

A concentration cell is constructed by immersing two iron electrodes in 0.01 M and 0.1 M $FeSO_4$ solution. Represent the cell and calculate EMF of the cell at 298 K.

The concentration cell is represented as

$$Fe_{(S)} \,|\, FeSO_4(0.01\text{ M}) \,||\, FeSO_4(0.1\text{ M} \,|\, Fe^+$$

EMF of the cell

We have
$$E_{cell} = \frac{0.0591}{n}\log\frac{[C_2]}{[C_1]}$$

$$= \frac{0.0591}{2}\log\frac{0.1}{0.01}$$

$$= \frac{0.0591}{2}\log 10$$

$$E_{cell} = 0.02955\text{ V}$$

Problem 1.12

Calculate the EMF of the cell

$$Al \,|\, AlCl_3(0.015\text{ M}) \,||\, AlCl_3(0.15\text{ M}) \,|\, Al \text{ at } 298\text{ K}$$

$$E_{cell} = \frac{0.0591}{n}\log\frac{C_2}{C_1}$$

$$= \frac{0.0591}{3} \log \frac{0.015}{0.015}$$

$$= 0.0197 \times \log 10$$

$$E_{cell} = 0.0197 \text{ V}$$

Problem 1.13

Calculate the emf of the following cell at 25 °C.

$$\underset{2 \text{ atm}}{\text{Pt } H_2}|HCl|\underset{10 \text{ atm}}{H_2 \text{ Pt}}$$

Solution:

$$E_{cell} = 0.0591/2 \log P_1/P_2$$

$$= 0.0591/2 \log 2/10$$

$$= -0.0206 \text{ volt}$$

Problem 1.14

Two copper rods are placed in copper sulphate solution of concentration 0.1 M are 0.01 M separately in the form of a cell. Write the scheme of the cell and calculate its emf at 298 K.

Solution: The cell is

$$Cu(s) \mid (Cu^{2+} (0.01 \text{ M}) \parallel Cu^{2+} (0.1 \text{ M})|Cu(s)$$

$$E_{cell} = 0.059/n \log C_2/C_1$$

$$E_{cell} = 0.059 \tfrac{1}{2} \log 0.1/0.01$$

$$= 0.0296 \text{ V}$$

Example 1.15

Calculate the EMF of the given concentration cell at 298 K.

$$Ag(s) \mid Ag\,NO_3 \,(0.018 \text{ M}) \parallel AgNO_3 \,(1.2 \text{ M}) \mid Ag(s)$$

Solution:

$$E_{cell} = 0.0591/n \log C_2/C_1$$

$$= 0.0591 \log 66.66$$

$$= 0.1078 \text{ V}$$

B. ENERGY STORAGE SYSTEM

1.19 INTRODUCTION

Battery is a compact device consisting of two or more galvanic cells, which convert the chemical energy into electrical energy by means of an electrochemical reaction called oxidation reduction (redox reaction). Thus, a battery acts as portable source of electrical energy.

Demand for portable power sources is ever increasing to meet diverse range of new applications such as watches, hearing aids, flashlights, radiosets, electric clocks, photoflash devices, emergency lighting, distress signalling, tape recorders, alarm systems, hand tools, toys, calculators, cordless appliances, telephone systems, electric bells, space vehicles, motor ignition and a host of other applications.

1.20 BATTERY

"Battery is a device that constitutes two or more number of galvanic cells connected in series or parallel or both which convert chemical energy into electrical energy through electrochemical reaction".

or

The arrangement of two or more cells normally coupled in series to produce higher potential is called battery.

Example: Lead – Acid battery, Nickel – Cadmium battery, etc.

1.21 BASIC COMPONENTS OF A BATTERY

A battery cell consists of four major components.

1.21.1 Anode

An anode is a negative electrode, where the oxidation reaction occurs which liberates electron into the external circuit during electrochemical reaction

$$M \longrightarrow M^{n+} + ne$$

The following materials can be used for the negative electrode in batteries.

1. **Metals:** Li, Na, Cd, Pb, etc.
2. **Metal alloys** (for example, Lithium alloys where the alloying elements can be Al, Sn, Si, etc. and metal hydrides $LaNi_5hx$, Mg_2NiHx, $ZrMn_2Hx$, etc.
3. **Carbons:** Graphite carbons, non-graphite carbons.
4. Redox polymers with negative redox potentials.

1.21.2 Cathode

A cathode is a positive electrode, where the reduction reaction occurs, which accepts electron into the external circuit during electrochemical reaction

$$M^{n+} + ne \longrightarrow M$$

The cathode must be an efficient oxidizing agent, high stability when it is in contact with the electrolyte with useful working voltage. Following are the commonly used cathode materials in batteries.

1. **Metal oxides:** MnO_2, PbO, etc., metal sulphides (such as TiS_2, NbS_3, MoS_2)
2. Insertion or host materials like metal oxides
   Example: $LiMo_2$ where M = Co, Ni, Cr, Fe, V and Mn

3. Metal oxyhydroxides (NiOOH, PbOOH, MnOOH)
4. Redox polymers with positive redox potential.

1.21.3 Electrolyte

The electrolyte is an ionic conductor which provides the medium for transfer of ions from one electrode to another electrode and also prevent short circuit. The electrolyte must have good ionic conductivity, non-reactivity, with electrode materials small property change with respect to temperature variation, safety in handling, etc.

In majority of conventional batteries, the electrolytes are aqueous solutions such as $ZnCl_2$, KOH, etc. But some advanced batteries are fabricated with non-aqueous solutions, ion conducting solids or polymers as electrolytes.

1.21.4 Separator

It is used to prevent electrical contact between the anode and cathode.

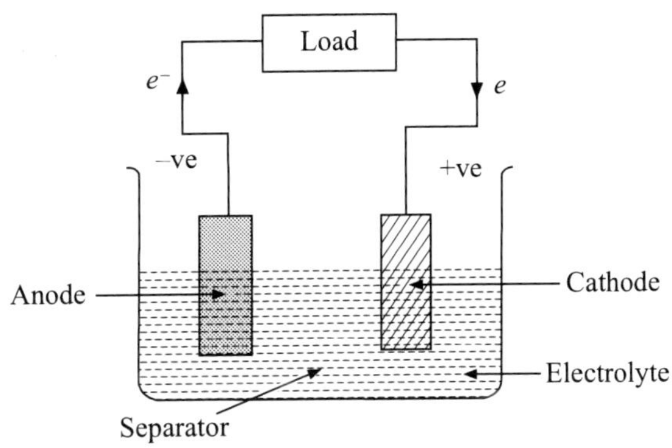

Fig. 1.12

Example: Cellulose, cellophane, etc.

In addition, cathode current collector, anode current collector, terminals seal and container make a battery complete.

1.22 CLASSIFICATION OF BATTERIES

Electrochemical cells or batteries are identified as primary (non-rechargeable) or secondary (rechargeable) depending on their capability of being electrically recharged. The batteries are classified as:

1. Primary batteries (non-rechargeable)
2. Secondary batteries (chargeable)
3. Reserve batteries (chargeable)

1.22.1 Primary Batteries

These are not rechargeable batteries, because the cell reactions are irreversible, i.e., when most of the reactant parts have been converted to products, no more electricity is produced and the battery becomes dead) and are to be discarded after the use. Such batteries are called primary batteries.

Example: Zn–MnO_2, Li–MnO_2, etc.

1.22.2 Secondary Batteries

These are rechargeable batteries, because the cell reaction are reversible. They are also called storage batteries, as they are the storage devices for electrical energy. The discharged cell can be recharged by passing current through it in the direction opposite to that of discharge current.

Example: Lead – Acid Battery, Mi–Cd battery, etc.

1.22.3 Reserve Batteries

Reserve batteries have long shell life and extreme storage. In this type of battery, one of the components is stored separately and is incorporated into the battery when required; usually the electrolyte is the component that is isolated.

Example: Magnesium batteries activated by water (Mg –AgCl and Mg –CuCl)

It is used to deliver high power for relatively short periods of time in applications such as missiles, torpedoes and other weapon systems.

1.23 CHARACTERISTICS OF BATTERIES

A single battery may not exhibit all good characteristics, but a battery can be designed and constructed with specific characteristics for a particular application.

1.23.1 Capacity

The capacity of a battery is expressed as the total amount of electricity or charge that may be obtained from the battery. It is given in ampere hours (Ah). It depends on the size of the battery. It is determined by the Faraday's relation.

$$C = WnF/M$$

where, $W \rightarrow$ Weight of the active material

$M \rightarrow$ Molecular weight of the material

$n \rightarrow$ No. of $_{-1}e^{\circ}$ transferred per mole of reaction

$F \rightarrow$ Faraday's constant

$C \rightarrow$ Capacity in Ah

Higher the capacity, higher is the efficiency of the battery.

It is measured by plotting time against the voltage at a fixed current discharge. More is the length of the flat portion of the curve, better is the capacity of the battery.

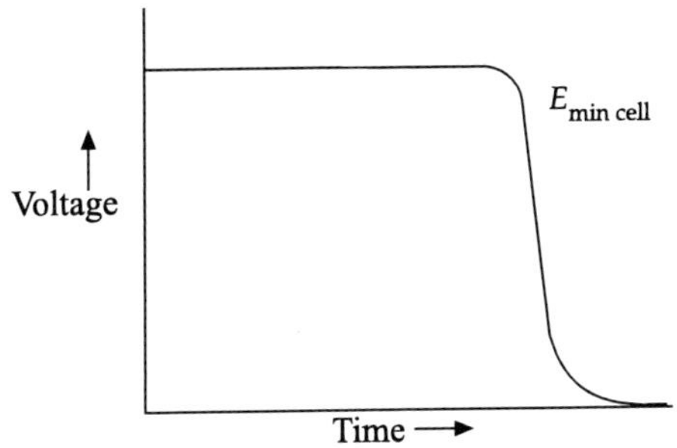

Fig. 1.13 *Voltage against time for fixed current discharge.*

1.23.2 Current

The rate of discharge of battery is called current of battery. Higher the potential difference in active materials faster the reaction and higher will be the rate of discharge of battery.

It is determined by the Ohm's law

$$I = V/R$$

where, $I \rightarrow$ Current
$V \rightarrow$ A difference in potential between anode and cathode
$R \rightarrow$ Resistance offered by the cell.

At higher resistance more potential difference is required to force the current through the cell.

1.23.3 Power Density

It measures how quickly the battery can deliver energy. It is measured in W/kg

$$\text{Power density} = \frac{I_E}{W}$$

where, $I \rightarrow$ Current drawn from the battery
$E \rightarrow$ Voltage
$W \rightarrow$ Weight of the battery

If a battery is operated at high power level, then energy available will be less. To obtain maximum amount of energy, battery should be operated at a low power level.

Higher the power density, higher is the efficiency of the battery.

1.23.4 Energy Density

Energy density is a measure of how much energy a battery can store in a given size or mass. So, a battery with a higher energy density can power a load for longer than one with a low energy density and the same physical size or mass. It is measured in W/kg.

So, $$\text{Energy density} = I\frac{Et}{W}$$

where, $I \rightarrow$ Current drawn from the battery at fixed voltage
E $\rightarrow$ Time duration t
W $\rightarrow$ Weight of the battery

Therefore, better is the electricity storage density of the battery, higher is the efficiency of the battery.

1.23.5 Cycle Life

It is the number of charge and discharge cycles that are possible before the failure of the battery. It is applicable to rechargeable batteries.

The discharge/charge cycle depends on chemical composition, morphological changes and distribution of active materials in the cell.

1.23.6 Energy Efficiency

It is a measure of energy output. It is applicable to rechargeable battery. It is given by,

$$\%\ \text{energy efficiency} = \frac{\text{Energy released on discharge}}{\text{Energy required for recharge}} \times 100$$

Higher is the energy efficiency, better is the battery.

1.23.7 Shelf Life

The storage capacity of a battery without self discharge.

A good battery should possess a long shelf life.

1.23.8 EMF or Voltage

It depends on the total number of cells of a battery constitute and potential difference between electrode of each cell or in other words, it depends on the change in free energy of the reaction.

$$\Delta G = nF\,E_{cell}$$

Larger the change in free energy, higher will be the emf.

For example:

$$M_1 + M_2^{n+} \rightleftharpoons M_1^{n+} + M_2$$

According to Nernst equation

$$E_{cell} = E_{cell} \frac{2.303\,RT}{nF} \log \frac{[M_1^{n+}]}{[M_2^{2+}]}$$

$$E_{cell} = E_{cell} + \frac{0.0591}{n} \frac{[M_2^{n+}]}{[M_1^{n+}]}$$

More the number of the cells, higher is emf and more efficient is the battery, similarly, higher the potential difference between electrodes of each cell, higher is the emf of the cell and more efficient is the battery.

1.23.9 Electricity Storage Density

It is a measure of the charge per unit mass stored in the battery.

The mass of the battery includes masses of the electrolyte, current collectors, terminals and other subsidiary elements. Lighter subsidiary elements lead to high storage density.

For example: 7g of lithium anode gives 96500 C whereas, for the same charge, 65 g of zinc is required.

1.24 WORKING OF A BATTERY

It involves discharging (delivering power) and recharging.

1.24.1 Discharging

A battery is packed with active materials at anode and cathode. During discharge, the anode undergoes oxidation to form cations and release electrons.

$$\text{At anode } M_1 \rightarrow M_1^{n+} + ne$$

Hence, cationic concentration increases in the anodic compartment. The electron liberated at the anode flows to cathode through external circuit and thus constitutes the flow of current.

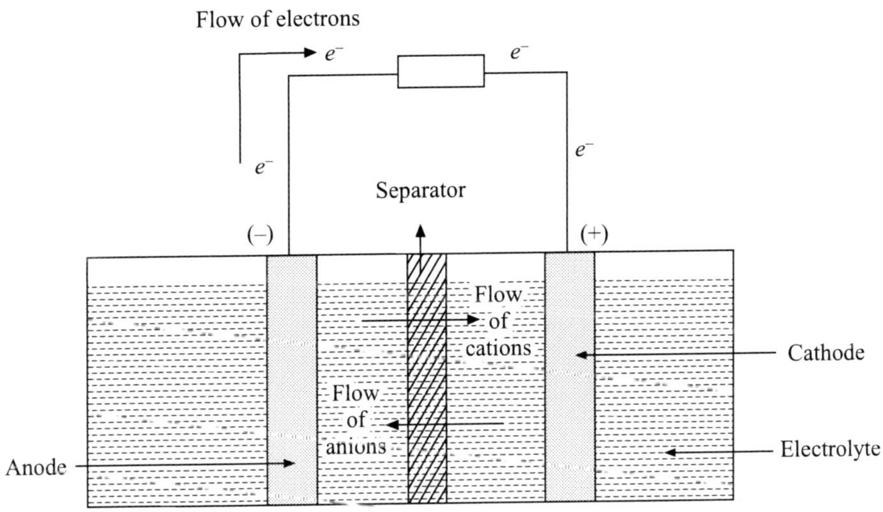

Fig. 1.14 *Discharging of a battery*

At cathode, electrons are accepted for the reduction of the metal ions (cations) to metal atoms (which get deposited on cathode) and hence anionic concentration becomes more in the cathodic compartment.

At cathode: $$M_1^{n+} + ne^- \rightarrow M_2$$

Thus, during discharge concentration of active species at anode gradually decreases hence the emf of the battery decreases.

1.24.2 Charging

In this process, the current flow is reversed with the aid of a dc power supply. The negative terminal of the power supply is connected to the anode of the battery and positive terminal to the cathode of the battery. Under this condition metal ions are reduced to metal atoms (which get deposited on anode) at the anode and anionic concentration increases in anodic compartment.

At anode: $M_1^{n+} + ne^- \rightarrow M_1$

At cathode, oxidation occurs by releasing electrons and hence, cationic concentration increases in cathodic compartment.

At anode: $M_2 \rightarrow M_2^{n+} + ne^-$

Thus, during recharge, the positive electrode acts as anode and the negative electrode acts as cathode. It means electrode reactions are reversed. Hence, original emf of the battery will be restored.

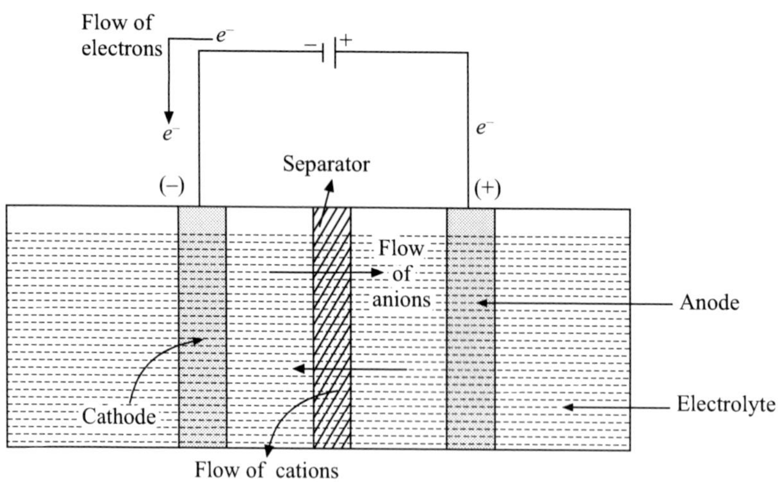

Fig. 1.15 *Charging of a battery*

1.25 MODERN BATTERIES

Many commercial batteries are not able to meet the performance requirements of many applications. Thus, a continual need exists for both conventional battery technology with improved performance and advance battery technologies with high energy level, long life, low cost, little or no maintenance and safety.

1.25.1 Nickel-metal Hydride Batteries

Nickel – metal hydride batteries are alkaline rechargeable batteries and were commercialized in 1990. It has better cycle life and high power density which makes it highly useful in automobiles.

Construction:

The anode: Porous nickel grid pasted with hydrides of metals like VH_2, ZrH_2 and TiH_2 with a hydrogen storage metal alloy such as $TiNi_2$ or $LaNi_5$.

The anode material has:

1. A good hydrogen storage capacity capable of adsorbing and desorbing hydrogen as the battery is discharged and charged.
2. High resistance to chemical oxidation and corrosion.
3. High electrochemical reactivity.

The cathode: Porous nickel grid pasted with $Mi(OH)_2$.

The electrolyte: Aqueous KOH solution used as an electrolyte.

Electrode materials are separated by a synthetic non-woven material which serves as separator as well as medium for absorbing the electrolyte.

The cell is represented as:

$$MH_2\ M \mid KOH \mid NiO\ (OH)\ Ni\ (OH)_2$$

Reaction:

At the anode:

$$MH_2 + 2OH^- \underset{\text{Charge}}{\overset{\text{Discharge}}{\rightleftharpoons}} 2H_2O + M + 2e^-$$

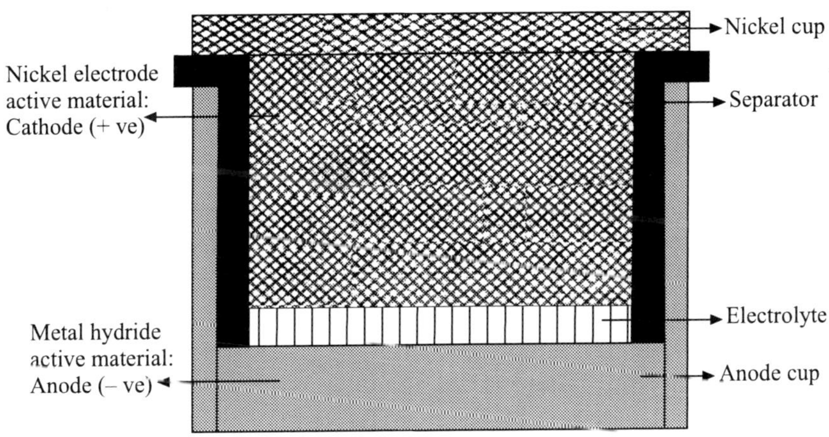

Fig. 1.16 *Nickel-Metal hydride cell*

At the cathode:

$$2\ NiO(OH) + 2H_2O + 2e^- \underset{\text{Charge}}{\overset{\text{Discharge}}{\rightleftharpoons}} 2\ Ni(OH)_2 + 2OH$$

The overall cell reaction is:

$$MH_2 + 2OH^- \longrightarrow M + 2H_2O + 2e^-$$

$$(NiO\ (OH) + H_2O + e- \longrightarrow Mi\ (OH)_2 + OH^-) \times 2$$

$$MH_2 + 2NiO\ (OH) \longrightarrow M + 2Ni\ (OH)_2$$

From the above reaction, we can say that in Mi-MH battery hydrogen moves from +ve to –ve electrode during charge and in reverse order during the discharge with the electrolyte taking no part in the reaction.

The emf of the battery is 1.25 – 1.35 V.

Advantages

1. High capacity
2. Sealed construction
3. Long cycle life
4. Rapid recharge capability
5. Cadmium free-minimal environmental problems, long shelf life

Disadvantage

Performance is not as good as of nickel – cadmium batteries.

Applications

1. In electric vehicles, computers, cellular phone, spacecraft
2. Portable electronic equipment.

1.26 LITHIUM BATTERIES

Lithium has the following characteristics:

1. It is light
2. It has good electrical conductivity
3. It has high electrochemical equivalence
4. It has high standard electrode potential.

So, it has all the properties required to become the most attractive anodic material for the design of a battery with high energy density.

A large number of lithium batteries are available which have lithium as anode, but they differ in choice of cathode and electrolyte. These may also be classified as primary or secondary.

Primary, non-rechargeable batteries which use lithium ion and commonly called Li-ion batteries.

1.26.1 Li–ion Battery

In this type of battery, lithium ions are used instead of lithium ion through electrolyte takes place from one electrode to another electrode.

In general, the battery consists of soluble lithium anode as ion into carbon, and the cathode material is made up of lithium liberating compounds.

Example: Lithium cobalt oxide ($LiCoO_2$) battery.

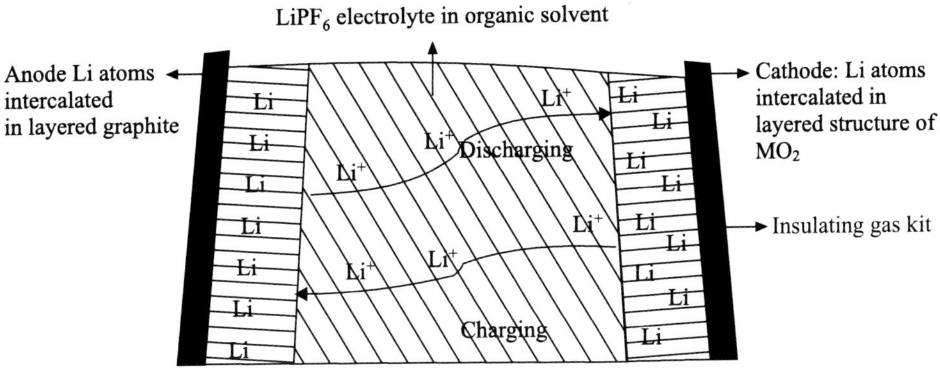

Fig. 1.17 *Lithium-ion rechargeable battery*

Construction

1. **The anode:** Anode is made up of carbon electrode with thin copper foil as current collector.
2. **The cathode:** It is made up of lithium metal oxide compound (Li–MO_2) where M is commonly Co or Mn.
3. **Electrolyte:** A lithium salt such as $LiPF_6$ dissolved in binary organic solvent mixture such as ethylene carbonate – dimethyl carbonate. (Fig. 1.17).

Cell Representation

$$Li \mid Li^+, C \mid LiPF_6 \text{ in ethylene carbonate} \mid Li\text{–}MO_2$$

Working

During discharging, lithium atoms are oxidized, liberating electrons and lithium ions which migrate through electrolyte to cathode.

$$Li - C_6 \longrightarrow Li^+ + 6C + e^-$$

At cathode, lithium ions are reduced to lithium atoms and are inserted into the layered structure of metal oxide.

$$Li^+ + e^- + MO_2 \longrightarrow Li\text{–}MO_2$$

During charging, lithium atoms of metal oxide are oxidized, liberating electrons and lithium ions. Electrons flow through the external circuit and lithium ions flow through the electrolyte towards graphite carbon electrode

$$Li - MO_2 \longrightarrow Li^+ + e^- + MO_2$$

At graphite electrode, lithium ions are reduced to lithium atoms and are inserted into the layered structure of graphite.

$$Li^+ + 6C + e^- \longrightarrow Li - C_6$$

The emf of the battery is 4.0 V.

Advantages

- They have high energy density than other rechargeable batteries.
- They are less weight.

- They produce high voltage out about 4 V as compared with other batteries.
- They have improved safety, i.e., more resistance to overcharge.

Disadvantages

- They are not available in standard cell types.
- They are expensive.

Applications

1. Electrical vehicles
2. Cardiac pacemakers
3. Aerospace applications
4. Power tools
5. Laptops, cell phone.

1.27 RECYCLING OF LITHIUM-ION BATTERY

As lithium-ion battery technology matures, it is widely used in various industries, resulting in a rapid increase in production and sales. However, since lithium-ion batteries generally only have a service life of about 5 to 8 years. It will inevitably bring about an explosive growth of waste lithium-ion batteries. The technical and safety requirements for the recycling of lithium batteries are becoming increasingly demanding and the correct handling and resolution of the cost and safety risks in the recycling process is an important issue in the development of lithium batteries today

As we know, lithium batteries contain about 15% cobalt, 25% iron, 0.1% lithium, 14% copper and 4.7% aluminium and these metals have a high recycling value. Since high value-added metals such as lithium, cobalt, nickel, etc., are mainly found in cathode materials in used lithium-ion batteries, the current research on recycling of lithium-ion batteries mainly focuses on cathode materials, so recycling process of used lithium-ion batteries is the separation, purification and reuse of high value-added metal elements in cathode materials.

Scrap lithium-ion battery contains a large number of cobalt, lithium, copper, manganese and other metal elements as well as lithium hexafluorophosphate, polyvinylidene fluoride and other harmful toxic substances, resource recovery and harmless treatment is of great significance. Lithium-ion battery recycling can be divided into three parts.

1. Pretreatment
2. Secondary treatment
3. Deep treatment

1. **Pretreatment:** The lithium-ion battery still has some residual charge inside when it is recycled and toxic HF will be formed when it encounters moisture, therefore, the lithium battery must be pretreated to fully deplete its charge before recycling. Currently there are two main pretreatment methods.
    1. Immersion method
    2. Resistance method

Immersion mainly releases the charge by sodium chloride solution and in order to prevent the generation of toxic gases, the battery is placed in dilute alkaline water.

$$HF + NaOH \rightarrow NaF + H_2O$$

It is then crushed at low temperature and physically sorted according to the different material densities of the battery casing and separator, and the proportion of positive and negative electrodes.

2. **Secondary treatment:** The secondary treatment is to separate the anode and cathode active materials from the substrate, generally using heat treatment, electrolysis and organic solvent dissolution method, and the heat treatment method is a relatively simple and convenient operation. The cell is placed at a certain temperature and the material is separated by the volatilization and decomposition of PVDF. At 370°C to 400°C, PVDF decomposes and at 600°C to 700°C, the aluminium foil melts due to the combustion reaction of the conductive additive.

   The secondary treatment can also be performed by organic solvent dissolution and alkaline dissolution. Organic solvents can dissolve polyvinylidene fluoride (PVDF) very effectively and partially recover the electrolyte at the same time generally N-methyl-2-pyrrolidone (NMP), Dimethylformamide (DMF), Dimethylacetamide (DMAC), Dimethylsulfoxide (DMSO) and other organic solutions can be used for treatment at a temperature higher than 70°C. PVDF dissolution reaches the maximum degree, the solubility of 211 g/L, 177 g/L, 218 g/L, 241 g/L, respectively. After a comparative study, DMSO has the highest solubility for waste lithium-ion batteries and has the excellent performance of environmental protection, low cost and non-toxic, etc. When the temperature is below 65°C, the aluminium foil obtained after 90 min. treatment can be directly used for recycling.

3. **Deep treatment:** Deep processing is a key step in the recycling of lithium-ion batteries which mainly includes leaching and separation.

Leaching process: There are two main types:

1. Microbial leaching
2. Inorganic acid leaching

Microbial leaching: It is more commonly used because of its high efficiency, low cost and low difficulty, at present as the waste lithium-ion battery is mainly recycled by autotrophic bacteria including ferrous oxide microspirochetes, sulphur oxide-thiobacillus and ferrous oxide thiobacilllus, etc., through the physiological reaction of the bacteria to promote their own growth and regeneration process, the metabolic acid can be used to carry out the leaching process of waste lithium-ion power battery

Inorganic acid leaching: Inorganic acid leaching method is to leach the battery cathode material by inorganic acids such as hydrochloric acid, sulphuric acid, nitric acid and organic acids such as oxalic acid, citric acid and grapes acid. Oxalic acid, for example, is used to recycle the cobalt and lithium elements in the battery and leaching efficiency of the cobalt and lithium elements can be as high as 97%-98% after one and a half hours of stirring with $H_2C_2O_4$ at 95°C and a solid-to-liquid ratio of 16 g/L. In the inorganic acid treatment, the cobalt and lithium elements can be leached out of the cathode material by using inorganic acid. In the treatment of inorganic acid, the leaching efficiency of lithium battery was analyzed by using HCl acid, H_2SO_4 acid, HNO_3 acid and the results showed that the leaching efficiency of Al, Ni, CO and Li reached 82.1%, 92.1%, 85.3% and 82.3%, respectively, after 20 hours of stirring with 4 mol/L HCl at a solid-liquid ratio of 210 g/L at 90°C.

In general, the leaching efficiency of inorganic acid is relatively high but it produces a lot of toxic and harmful gases such as sulphur dioxide, nitrogen dioxide, etc., and the residual liquid is difficult to treat, often resulting in secondary pollution. The pH value of the organic acid effluent is lower and easier to handle but the relative cost is higher.

Separation: The final separation and purification is mainly the separation and purification of Li, Ni^{2+}, Mn^{2+}, Co^{2+}, etc., in the leachate, the main methods are solvent extraction, chemical precipitation and electrochemical methods.

Chemical precipitation is mainly the precipitation of metal ions by specific precipitants to form the corresponding metal compound products. Common precipitating agents include potassium permanganate, ammonium oxalate, sodium carbonate, etc., which precipitate the corresponding metal ions at different pH value.

For example, the combined use of NaOH, Na_2CO_3, $KMnO_4$, etc., will give precipitation products such as MnO_2, $LiCO_3$, $Ni(OH)_2$, $Co(OH)_2$, etc., and the recoveries of Li, Ni^{2+}, Mn^{2+}, CO^{2+} are 96.97%, 97.45%, 96.99%, and 98.25% respectively.

Review Questions

1. Define single electrode potential. Derive Nernst equation for single electrode potential.
2. What are ion selective electrodes? Describe the construction of a glass electrode and derive an expression relating glass electrode potential and pH.
3. What are concentration cells? Derive an expression for the emf of a concentration cell.
4. Explain a method for the determination of single electrode potential.
5. An electrochemical cell consists of magnesium electrode in 0.042 m $Mg(NO_3)_2$ solution and silver electrode in 0.35 m $AgNO_3$ solution. The standard electrode potential of Mg and Ag are –2.363 V and 0.80 V respectively. Represent the cell, write the cell reaction and calculate the emf of the cell.
6. What are reference electrodes? Describe the construction and working of calomel electrode. Write its advantages.
7. The emf of the cell

   Cu | $CuSO_4$(0.1 M) || $CuSO_4$(x M) | Cu is 0.0295 V at 25 °C. Calculate the value of x.
8. The emf of the cell

   Ag | $AgNO_3(C_1)$ || $AgNO_3(C_2 = 0.2)$ | Ag is 0.8 V at 25 °C. Calculate the value of C_1.

9. An electrochemical cell consists of metallic zinc immersed in 0.1 $MZn(NO_3)_2$ solution and metallic copper immersed in 0.2 M $CuSO_4$ solution. Calculate the emf of the cell at 25 °C and change in free energy of the cell reaction. $E°$ cell = 1.1 V.
10. An electrochemical cell consists of an iron electrode dipped in 0.1 M $FeSO_4$ and silver electrode dipped in 0.05 M $AgNO_3$ solution. Write the cell representation, cell reaction and calculate the emf of the cell at 298 K. Given that the standard electrode potential of Fe and Ag electrodes are –0.44 V and 0.80 V respectively.
11. Calculate the voltage of a cell which consists of a rod of iron immersed in a 1.0 M solution of $FeSO_4$ and a rod of manganese immersed in a 0.1 M solution of MnS_4 at 25 °C. Write the cell reaction. Give $E^{\circ}_{Fe^{2+}/Fe} = -0.44$ V and $E^{\circ}_{mn^{2+}/mn} = -1.18$ V.
12. Define standard electrode potential.
13. An electrochemical cell is formed from nickel and lead electrodes having 0.01 m $NiSO_4$ and 0.5 m $PbSO_4$ electrolytes. The standard electrode potentials of Ni and Pb electrodes are – 0.24 V and – 0.13 V respectively. Write the cell scheme, cell reaction and calculate emf of the cell at 298 K.
14. The emf of the cell

    Cu | $CdSO_4$ (0.0093 M)|| Cd (x M) SO_4 | Cd is 0.086 V at 25 °C. Find the value of x.
15. Derive an expression for single electrode potential.
16. A concentration cell was constructed by immersing two silver electrodes in 0.01 M and 10 M $AgNO_3$ solutions. Write the cell reactions and calculate the emf of the concentration cell.
17. A concentration cell was constructed by immersing two silver electrodes in 0.05 M and 0.1 M $AgNO_3$ solution. Write cell representation, cell reactions and calculate the emf of the concentration cell.
18. An electrochemical cell consists of iron electrode dipped in 0.1 M $FeSO_4$ and silver electrode dipped in 0.05 M $AgNO_3$. Write the cell representation, cell reaction and calculate emf of the cell at 298 K. Given the standard reduction potentials of iron and silver electrodes are – 0.44 and + 0.80 V.
19. Calculate the voltage of the cell Mg/Mg^{2+}(aq soln.)|| Cd^{2+}(aq soln.)/Cd at 25 °C. When $[Cd^{2+}] = 7.0 \times 10^{-11}$ M, $[Mg^{2+}] = 1.0$ M and $E^{\circ}_{Cell} = 1.97$ V.
20. A cell is constructed by coupling zinc electrode dipped in 0.5 M $ZnSO_4$ and nickel electrode dipped in 0.05 M $NiSO_4$. Write the cell representation, cell reaction. Calculate emf of the cell, given that standard reduction potential of Zn and Ni as –0.76 and –0.25 volt, respectively.
21. Define single electrode potential and standard electrode potential and explain the origin of electrode potential.
22. Describe the method of determining the pH of a solution.

23. Derive Nernst equation for single electrode potential. Explain the determination of single electrode potential using standard hydrogen electrode.
24. Write brief notes on: (1) calomel electrode (2) glass electrode.
25. Describe the working principle of a battery.
26. Describe the following with respect to a battery:
    (a) voltage (b) current
    (c) capacity (d) energy density efficiency
    (e) life cycle and shelf life
27. How do you classify batteries? What are the factors that affect the emf of a battery?
28. Describe the working of a Ni-MH battery. Give the electrode reactions during its discharge.
29. Discuss any three important characteristics of a battery.
30. What are reserve batteries? Give example.
31. Mention the anode, cathode and electrolyte in zinc-air battery. Give the anodic and cathodic reactions taking place during discharges.
32. Discuss on capacity and shelf life a battery. Explain the construction and working of LI-MnO_2 cell.
33. Give the half cell and net cell reactions in the following batteries.
    (a) Zinc–air cell (b) Nickel-metal hydride battery
    (c) Lithium-manganese dioxide cell.
34. Why secondary batteries are considered storage batteries? Explain the construction and working of Nickel-metal hydride battery? Give the reactions involved in discharge and recharge mode.
35. Describe the construction and working of zinc-air battery.
36. What are rechargeable and reserve batteries?
37. Explain the following battery characteristics:
    (a) Voltage (b) Cycle life (c) Energy efficiency
38. Give the classification of fuel cells on the basis of temperature and electrolyte.
39. Write short notes on the following:
    (a) Nickel-cadmium battery.
    (b) Zinc-air battery.
    (c) Nickel-metal hydride battery.
    (d) Lithium battery.
    (e) Rechargeable lithium-ion battery.
40. Give the characteristics of zinc-air battery.

EXERCISE

1. Calculate the potential of Cu^{2+}/Cu electrode, when the free energy change for the reaction, $Cu^{2+} + 2e^{-}$ Cu is (–) 62.7 kJ mol^{-1}. **[*Ans.* + 0.34 V]**
2. Calculate $E°$ Zn^{2+}/Zn, if the potential of the cell Zn(s) | Zn^{2+}(1M)|| Ni^{2+}(1M) | Ni(s) is 0.51 V, that of Ni (s)| Ni^{2+} (1M) || SCE is 0.49 V and ESCE = 0.24 V. **[*Ans.* – 0.76 V]**
3. Calculate the potential of the cell composed of Zn(s) | Zn^{2+} and Cu^{2+}/Cu if Zn^{2+} and Cu^{2+} concentrations are 1M and 1×10^{-9} M respectively at 25 °C. Given $E°$ Cu^{2+}/Cu = 0.34 V and $E°$ Zn^{2+}/Zn = 0.76 V. **[*Ans.* 0.83 V]**
4. Find out the solubility of AgCl if the emf of the following cell is 0.1182 volt.

   Ag | satd. AgCl soln. (C_2) || 0.001 N $AgNO_3$ (C_1) | Ag. **[*Ans.* 10^{-5} mole /litre]**
5. Calculate the emf of the following concentration cell at 25 °C.

   Ag|0.001 N $AgNO_3$|| 0.1 N $AgNO_3$| Ag

   Assuming the activities of Ag^{+} in the two solutions to be equal to the concentrations. **[*Ans.* 0.1182 volt]**
6. Calculate the emf of the cell, when the concentration of $ZnSO_4$ and $CuSO_4$ 0.001 m and 0.1 M respectively. The standard potential of the cell is 1.10 volts. **[*Ans.* 1.129 V]**
7. Write the cell reaction and calculate the emf of the cell at Fe| Fe^{2+} (0.01 M) || Ag^{+} (0.1 M) | Ag 298 K. If standard electrode potential of Fe and Ag electrodes are – 0.44 V and 0.8 V respectively. **[*Ans.* 1.24 V]**
8. Calculate the standard electrode potential of Cu^{2+}/Cu its electrode potential at 259 C is 0.297 V when [Cu^{2+}] = 0.015 M. **[*Ans.* 0.3508 V]**
9. Calculate the emf of the galvanic cell by using the standard reduction potentials given below.

   $Zn^{2+} + 2e \longrightarrow Zn(s)$ $E°$ = – 0.76 volt

   $Cd^{2+} + 2e \longrightarrow Cd$ (s) $E°$ = – 0.44 volt **[*Ans.* 0.32 V]**
10. Find the standard potential of the cell Cu(s)| Cu^{2+}(aq)|| Cl^{-} | AgCl(s) | Ag(s) and predict the direction of electron flow when the two electrodes are connected. The standard reduction potentials of Cu/Cu^{2+} = –0.34 V and Ag^{+}/Ag = 0.222 V respectively. **[*Ans.* 0.115 V]**
11. Find the $\Delta G°$ for the cell reaction in a zinc-aluminium galvanic cell. The standard reduction potentials of Zn and Ag are –0.76 V and + 1.66 V, respectively. **[*Ans.* – 521 kJ]**
12. Calculate the emf for the cell Sn(s)| Sn^{2+}(0.15 M)|| Ag^{+} (0.30 M) | Ag(s) at 25°2 C. The standard reduction potential of Sn and Ag are - 0.14 and + 0.80 V respectively. **[*Ans.* 0.95 V]**

13. Consider the given cell, Al(s)| $A|^{3+}$(aq)|| Mg^{2+}(aq) | Mg(s). Write the electrode and net cell reactions and calculate the standard electrode potential for the cell. The standard reduction potentials of Al and Mg are (–) 1.66 V and (–) 2.38 V respectively.

[*Ans.* 0.70 V]

14. The emf of the following cell is found to be 0.20 V at 298 K, Cd(s) | Cd^{2+} (aq)(?)|| Ni^{2+} (aq)(2.0 M) | Ni(s). What is the molar concentration on Cd^{2+} ions in solution.

[*Ans.* 0.0409 M]

15. Calculate the emf of the cell in which reaction is

$$Zn(s) + 2Ag^{+}(aq) \longrightarrow Mg^{2+}(aq) + 2Ag(s)$$

when [Mg^{2+} (aq)] = 0.130 M and [Ag^{+} (aq)] = 1×10^{-4} M. **[*Ans.* 2.96 V]**

16. Calculate the equilibrium constant for the reaction

$Zn(s) + Cd^{2+}(aq)\ Zn^{2+}(aq) + Cd(s)$. Given: $E^{\circ}_{cell} = 0.36$ V. **[*Ans.* 1.52 × 10^{12}]**

17. Calculate the standard free energy change and maximum work obtainable for the reaction occurring in a Daniel cell. $Zn + Cu^{2+} \quad Zn^{2+} + Cu$

Given: $E^{\circ}_{Zn^{2+}/Zn} = -0.76$ V, E° Cu^{2+}/Cu = 0.34 V and F = 96,500 C mol^{-1}. How is it related to the equilibrium constant for the reaction?

[*Ans.* ΔG° = –212.3 kJ, K = 1.6 × 1037]

18. Calculate the equilibrium constant for the following reaction

$$Cu(s) + 2Ag^{+} \longrightarrow Cu^{2+}(aq) + 2Ag(s)$$

Given: $E^{\circ}_{cell} = 0.46$ V. **[*Ans.* 3.69 × 1015]**

19. Consider the cell, Mg(s) | Mg^{2+} (aq)|| Ag^{+} (aq) |Ag(s). The standard reduction potential of Mg and Ag are –2.38 and 0.80 V, respectively. Calculate the equilibrium constant at 298 K and also the maximum work that can be obtained using the cell.

[*Ans.* K = 189 × 10^{107}]

20. Consider an electrochemical cell,

$$Al(s) \mid Al^{3+}\ (0.01\ M) \| Fe^{2+}(0.20\ M) \mid Fe(s)$$

The standard reduction potentials of Al and Mg are –1.66 and –0.44 V, respectively.

Calculate the $\frac{W_{max}}{1}$ that can be obtained by the cell.

[*Ans.* E°_{cell} = 1.22 V; $\frac{W_{max}}{1}$ = 705.8 kJ]

21. The E°_{cell} of an aluminium–air battery is 2.73 volts and it involves a 12 electron process. Calculate the free energy change (ΔG°) of the battery in kJ. **[*Ans.* –3161.34 kJ]**

22. Zinc rod is dipped in 0.1 M solution of $ZnSO_4$. The salt is 95% dissociated at this dilution at 298 K. Calculate the electrode potential given that Zn^{2+}/Zn = 0.76 V.

[*Ans.* – 0.79 V]

23. Two half cells of hydrogen–oxygen fuel cell under basic conditions can be depicted as $OH^- / O_2(g)$ / Pt and $OH^- / H_2(g)$ / Pt and their standard electrode potentials at 25 °C are 0.4009 and – 0.829 V respectively. Write the half cell reactions and the complete cell reaction. And calculate the emf of the cell. **[*Ans.* 1.2288 V]**

24. From the following electrode potential at 25°C.

$$Cu^{2+} + 2e^- \longrightarrow Cu \quad E° = +\,0.34\text{ V}$$

$$Cu^{2+} + e^- \longrightarrow Cu^+ \quad E° = +\,0.15\text{ V}$$

Calculate equilibrium constant of the reaction. **[*Ans.* 2.68 × 10^6]**

25. Calculate the emf of the following cell at 25°C.

$$Zn(s)\,|\,Zn^{2+}\,(0.1\text{ M})\,||\,Cu^{2+}\,(1.751\text{ M})\,|Cu(s)$$

Given that $E°_{Zn^{2+}/Zn} = -\,0.76$ V, $E°\ Cu^{2+}/Cu = 0.34$ V. **[*Ans.* 1.1368 V]**

2

Corrosion and Metal Finishing

A. CORROSION

2.1 INTRODUCTION

The word corrosion derives from the Latin word **corroders** meaning to gnaw to a piece that refers to the destruction and consequent loss of a metal due to its reaction with the surrounding gaseous or liquid environment. When metals are exposed to the environment, the metal surface starts decaying by conversion into more stable metal compounds more or less rapidly. It is exactly the reverse of extraction of metals and also known as *weeping of metals*. So, corrosion may be defined as *"the process of slow conversion of metals into their undesirable compounds by reaction with moisture and other gases present in environment"*. The environment could be of any type such as atmosphere, water, seawater, acids, alkalis, steam, gases, soils, liquid metals, etc., and the interaction of metal with the environment could take place at any temperature. Thus, according to modern definition *"corrosion is the deterioration of a substance because of its reaction with its environment"*.

Examples:

1. Rusting of iron – a reddish brown scale formation on iron and steel objects. It is due to the formation of hydrated ferric oxide.
2. Green scales formed on copper vessels. It is due to the formation of basic cupric carbonate.

$$[CuCO_3 + Cu\,(OH)_2]$$

2.2 CAUSES OF CORROSION

Causes of corrosion: In nature, metals occur in two different forms:

1. Native state
2. Combined state

1. Native State: Native state means free or uncombined state. These metals exist in the crust of the earth and are non-reactive in nature. They are noble metals which have very good corrosion resistance.

Examples: Av Pt, Ag, etc.

2. Combined State: Except noble metals, all other metals are highly reactive in nature which undergo reaction with their environment to form stable compounds called ores and minerals. This is the combined state of metals.

Examples: Fe_2O_3, ZnO, PbS, $CaCO_3$, etc.

3. Metallic Corrosion: Metals are extracted from their metallic compounds (ores). During the extraction, ores are reduced to their metallic states by applying energy in the form of various processes. In pure metallic state the metals are unstable as they are considered in excited state (higher energy state).

Therefore, as soon as metals are extracted from their ores, the reverse process begins and form metallic compounds which are thermodynamically stable (lower energy state). Hence, when metals are used in various forms they are exposed to environment, the exposed metal surface begins to decay (conversion to more stable compound). This is the basic reason for metallic corrosion.

$$\text{Metal} \underset{\text{Metallurgy reduction}}{\overset{\text{Corrosion - Oxidation}}{\rightleftarrows}} \text{Metallic compound + Energy}$$

Although corroded metal is thermodynamically more stable than the pure metal but due to corrosion useful properties of a metal like malleability, ductility, hardness, lustre and electrical conductivity are lost.

2.3 EFFECTS OF CORROSION

1. Loss of useful properties of metal and thus loss of efficiency.
2. Decrease in production rate, efficiency is less and replacement of corroded equipment or machinery is time consuming.
3. Increase in maintenance and production rate.
4. Contamination of product.

2.4 IMPORTANCE OF CORROSION

To reduce the *economic impact* of corrosion engineers with the support of corrosion scientists, aim to reduce material losses, as well as the accompanying economic losses, that result from the corrosion of piping, tanks, metal components of machines, ships, bridges, marine structures, and so on. For example, pressure vessels, boilers, metallic containers for toxic chemicals, turbine blades and rotors, bridges, airplane components, and automotive steering mechanisms. **Safety** is a critical consideration in the design of equipment for nuclear power plants and for disposal of nuclear wastes. Loss of metal by corrosion is a waste not only of the metal, but also of the energy, the water, and the human effort that was used to produce and fabricate the metal structures in the first place. In addition, rebuilding corroded equipment requires further investment of all these resources – metal, energy, water, and human. **Economic losses** are divided into (1) direct losses and (2) indirect losses.

2.4.1 Direct Losses

Direct losses include the costs of replacing corroded structures and machinery or their components, such as condenser tubes, pipelines, and metal roofing, including necessary labour. Other examples are (a) repainting structures where prevention of rusting is the prime objective; and (b) the capital costs plus maintenance of cathodic protection systems for underground pipelines. Sizable direct losses are illustrated by the necessity to replace several million domestic hot-water tanks each year because of failure by corrosion and the need for replacement of millions of corroded automobile mufflers. Direct losses include the extra cost of using corrosion-resistant metals and alloys instead of carbon steel where the latter has adequate mechanical properties but not sufficient corrosion resistance. There are also the costs of galvanizing or nickel plating of steel, of adding corrosion inhibitors to water, and of dehumidifying storage rooms for metal equipment.

2.4.2 Indirect Losses

Indirect losses are more difficult to assess, but a brief survey of typical losses of this kind compels the conclusion that they add huge amount of money to the direct losses already outlined. Examples of indirect losses are as follows:

1. **Shutdown:** The replacement of a corroded tube in an oil refinery may cost a few hundred dollars, but shutdown of the unit while repairs are underway may cost a large amount production loss.
   Similarly, replacement of corroded boiler or condenser tubes in a large power plant also requires a large amount per day for power purchased from interconnected electric systems to supply customers while the boiler is down.
2. **Loss of Product:** Losses of oil, gas, or water occur through a corroded pipe system until repairs are made. Antifreeze may be lost through a corroded auto-radiator; or gas leaking from a corroded pipe may enter the basement of a building, causing an explosion.
3. **Loss of Efficiency:** It may occur because of diminished heat transfer through accumulated corrosion products, or because of clogging of pipes with rust necessitating increased pumping capacity. It has been estimated that, in the United States, increased pumping capacity, made necessary by partial clogging of water mains with rust, costs many millions of dollars per year.
4. **Contamination of Product:** A small amount of copper picked up by slight corrosion of copper piping or of brass equipment that is otherwise durable may damage an entire batch of soap. Copper salts accelerate rancidity of soaps and shorten the time that they can be stored before use. Traces of metals may similarly alter the colour of dyes. Lead equipment, otherwise durable, is not permitted in the preparation of foods and beverages because of the toxic properties imparted by small quantities of lead salts. The poisonous effects of small amounts of lead have been known for a long time. In a letter to Benjamin Vaughn dated July 31, 1786, Benjamin Franklin warned against possible ill effects of drinking rain water collected from lead roofs or consuming alcoholic beverages exposed to lead. The symptoms were called in his time "dry bellyache" and were accompanied by paralysis of the limbs. The disease originated because New England rum distillers used lead coil condensers. On recognizing the cause, the Massachusetts Legislature passed an Act outlawing use of lead for this purpose.

5. **Overdesign:** It is common in the design of reaction vessels, boilers, condenser tubes, oil-well sucker rods, pipelines transporting oil and gas at high pressure, water tanks, and marine structures. Equipment is often designed many times heavier than normal operating pressures or applied stresses would require in order to ensure reasonable life. With adequate knowledge of corrosion, more reliable equipment can be made, and design can be simplified in terms of materials and labour. For example, oil-well sucker rods are normally overdesigned to increase service life before failure occurs due to corrosion fatigue. If the corrosion factor were eliminated, losses would be cut at least in half. There would be further savings because less power would be required to operate a lightweight rod, and the expense of recovering a lightweight rod after breakage would also be lower.

2.5 TYPES OF CORROSION

The process of corrosion of metals depends on the environment to which the metals are exposed. The corrosion of metals can be either dry corrosion or wet corrosion.

1. Dry corrosion or direct chemical corrosion.
2. Wet corrosion or immersed corrosion or electrochemical corrosion.

1. Dry Corrosion or Direct Chemical Corrosion: Dry corrosion occurs through direct attack on metals by dry gases such as oxygen, hydrogen, sulphur dioxide, hydrogen sulphide, nitrogen or through chemical reaction which is of three types.

(i) Oxidation corrosion
(ii) Corrosion by other gases
(iii) Liquid metal corrosion

Oxidation Corrosion: This type of corrosion is brought by the direct attack of oxygen on the metal surface at low or high temperature usually in the absence of moisture. For example, alkali metals and alkaline earth metals are highly prone for oxidation even at low temperatures, while at high temperature; almost all metals except noble metals (Ag, Au, and Pt) and are oxidized.

Mechanism: First oxidation occurs at the surface of the metal and the formed metal oxide scale forms a barrier which restricts further oxidation of the metal surface. For oxidation to continue either the atmospheric oxygen must diffuse through the scale to the metal surface must come outwards through the scale. Outward diffusion is more rapid than inward diffusion of oxygen due to small size of metal ions. In oxidation corrosion, the oxide layer formed at the surface decides the further course of corrosion.

The oxide layer formed at the metal surface may be of (i) stable, (ii) unstable, (iii) volatile, or (iv) porous nature. If the film is of *stable nature,* it adheres lightly to the metal surface and acts as a protective coating, thereby shielding the metal. In case of metals such as Al, Sn, Pb, Cu, etc., the formed oxide layer is stable, abhorrent and impervious in nature, prevents further corrosion.

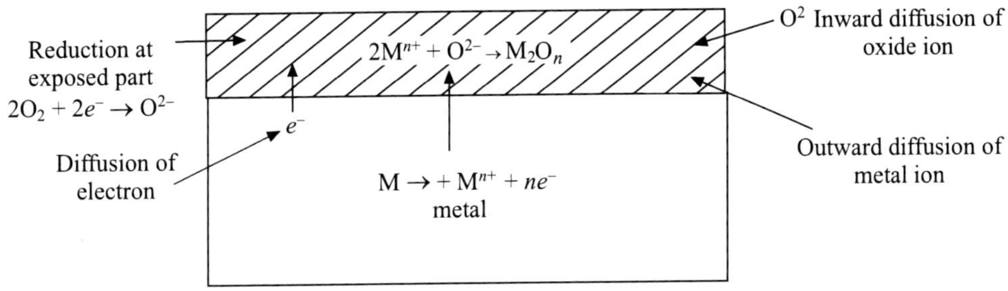

Fig. 2.1 *Oxidation mechanism of metals*

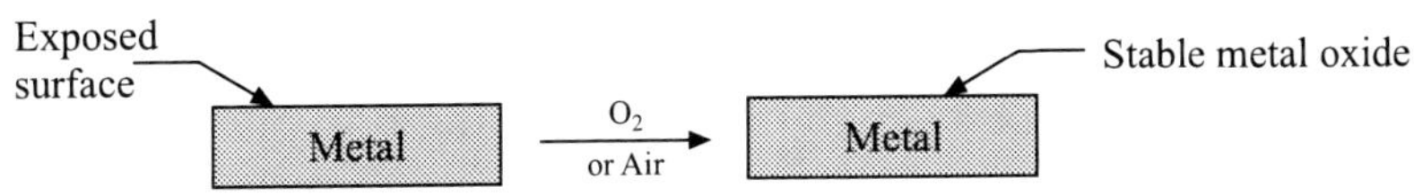

Fig. 2.2 *Formation of stable metal oxide*

An unstable oxide film at metal surface decomposes back into the metal and oxygen.

$$\text{Metal oxide} \rightleftharpoons \text{Metal} + \text{Oxygen}$$

Due to this reason, noble metals (Ag, Au, Pt) do not undergo oxidation corrosion.

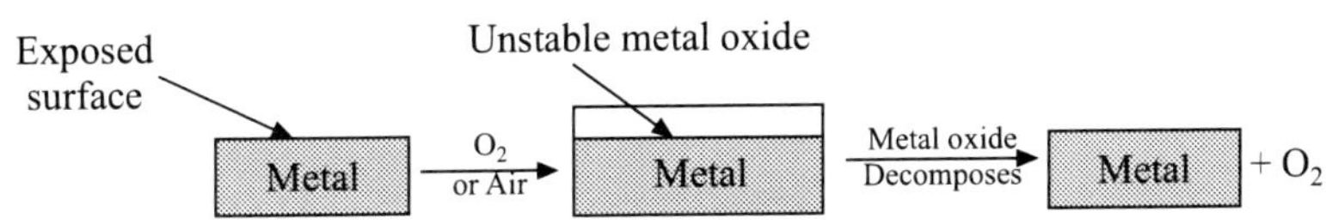

Fig. 2.3 *Formation of unstable metal oxide*

Volatile oxide film at the metal surface exposes it for further attack of oxygen resulting in continuous corrosion.

Example: Oxide of molybdenum (MoO_3) is highly volatile in nature.

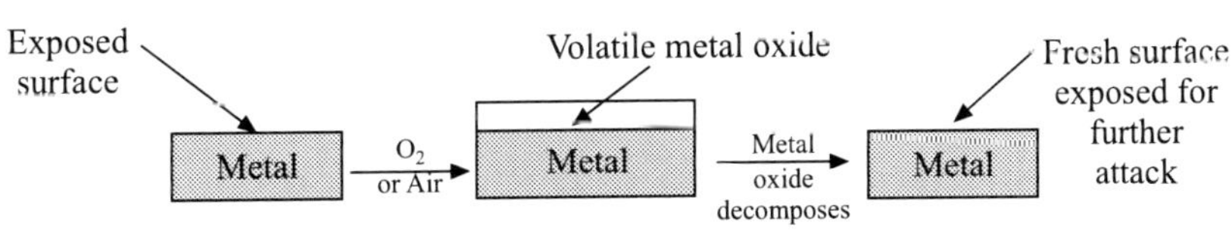

Fig. 2.4 *Formation of volatile metal oxide*

If the layer is porous in nature the atmospheric oxygen has easy access to the underlying metal surface though the pores or cracks in the layer resulting in continuous corrosion of metal surface till the entire metal block is converted into its oxide.

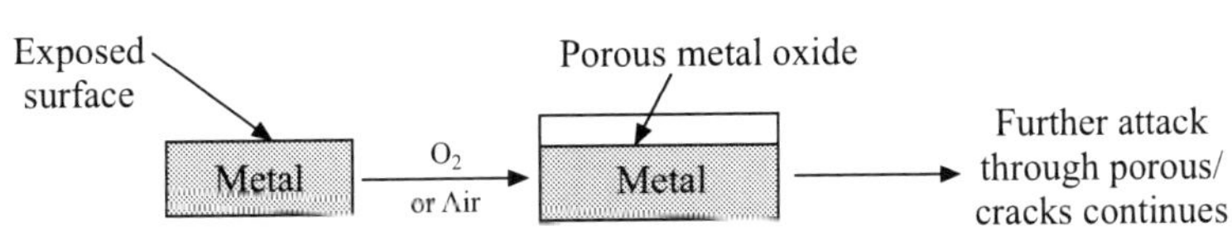

Fig. 2.5 *Formation of porous metal oxide*

Protective and non-protective nature of the oxide layer at the metal surface is decided by Pilling-Bedworth rule. According to this rule "an oxide is said to be protective or non-porous if the volume of the oxide is at least as great as the volume of the metal from which it is formed". On the other hand "if the volume of the oxide is less than the volume of metal, the oxide layer is porous and hence non-protective, because it cannot prevent the access of oxygen to the fresh metal surface below".

Thus, alkali and alkaline earth metals (like Li, K, Na, Mg) form oxides of volume less than the volume of metals. Consequently, the oxide layer faces stress and strains, thereby developing cracks and pores in its structure. Porous oxide scale permits free access of oxygen to the underlying metal surface for fresh action and thus corrosion continues non-stop. Metals like aluminium forms oxide, whose volume is greater than the volume of the metal. Consequently, an extremely tightly adhering non-porous layer is formed. Due to the absence of any pores or cracks in the oxide film, the rate of oxidation rapidly decreases to zero.

Corrosion by other gases: In the absence of moisture of atmosphere, dry gases such as SO_2, CO_2, Cl_2, H_2S, F_2, etc., also cause corrosive effect on the metals. The degree of corrosion depends on the formation of the thin film which may be protective and non-protective on the surface of the metal. For example:

1. Dry Cl_2 attacks silver metal and forms AgCl as a *thin protective* and *non-porous layer* on the metal. As a result of this protective layer on the metal surface, the intensity of corrosion decreases.
2. However, it is observed, that in the case of tin, the corrosion attack increases and the entire metal gets destroyed in due course. This is due to the fact that the thin layer ($SnCl_4$) formed on the tin surface is non-protective and porous.

Liquid metal corrosion: This type of corrosion has been found to occur in devices used for nuclear power. It occurs due to the chemical action of flowing liquid metal at high temperature on solid metal or alloy. This behaviour may be due to the:

(a) dissolution of a solid metal by a liquid metal; and
(b) penetration of the internal liquid metal into the solid metal.

2. Wet Corrosion or Immersed Corrosion or Electrochemical Corrosion: Most of the corrosion cases are electrochemical in nature, taking place by an electrochemical attack on the metal. It happens in the presence of moisture/conducting medium called wet corrosion. This theory involves oxidation and reduction reactions. This type of corrosion can be observed when:

(a) a metal is in contact with an acid solution, and
(b) dissimilar metals are dipped partially in a corrosive environment.

2.6 ELECTROCHEMICAL THEORY OF CORROSION

According to this theory, corrosion of metals takes place due to the formation of anodic and cathodic regions on the same metal surface or when two different metals are in contact with each other in the presence of a conducting medium. At the anodic region oxidation reaction takes place and the metal gets converted into its ions, liberating electrons. Consequently, metal undergoes corrosion at the anodic region.

$$\text{At anode (oxidation reaction): } M \rightarrow M^{n+} + ne'$$

At the cathodic region, reduction reaction takes place. Since the metal cannot be reduced further, metal atoms at the cathodic region are unaffected by the cathodic reaction. So the cathodic part, dissolved constituents in the conducting medium accept the electrons to form some ions (like OH^-, O^{2-}). The metallic ions (at anodic part) and non-metallic ions (formed at cathodic part) diffuse towards each other through conducting medium and form a corrosion product somewhere between the anode and cathode. The electrons set free at the anode flow through the metal and is finally consumed in the cathodic reaction. Corrosion of metal continues as long as both anodic and cathodic reactions take place simultaneously. These cannot be anodic reactions without cathodic reactions and vice versa.

The reaction at cathode (reduction reaction) depends on the nature of the environment. If the medium is acidic:

(a) In the presence of dissolved oxygen $2H^+ + \frac{1}{2} O_2 + 2e^- \rightarrow H_2O$
(b) In the absence of dissolved oxygen $2H^+ + 2e^- \rightarrow H_2$

If the medium is alkaline / neutral:

(a) In the presence of dissolved oxygen $H_2O + \frac{1}{2} O_2 + 2e^- \rightarrow 2OH^-$
(b) In the absence of dissolved oxygen $2H_2O + 2e^- \rightarrow 2OH^- + H_2$

Example: *Rusting of iron:* It is generally caused by moisture, carbon dioxide and oxygen present in the air. It has been observed that rusting of iron takes place only when iron is in contact with moist air. Chemically rust is hydrated form of ferric oxide, i.e., $Fe_2O_3 \times H_2O$. The main features are:

1. The covered part of the metal surface acts as anode. Where iron changes to Fe^{2+} ions by giving electrons.
   At anode: $Fe(s) \longrightarrow Fe^{2+}(aq) + 2e^-$ (oxidation)
   The anodic area gets corroded. The liberated electrons move from the anodic area to the cathodic area. where it is available for the reduction reactions with the absorption of oxygen and forms OH^- ions $\frac{1}{2} O_2 + H_2O + 2e^- \rightarrow 2OH^-$ (reduction)
2. The Fe^{2+} ions (at anode) and OH^- ions (at cathode) diffuse and when they meet, ferrous hydroxide is precipitated.

$$Fe^{2+} + 2OH^- \longrightarrow Fe(OH)_2$$

   If the environment to which the metal is exposed is rich in oxygen, $Fe(OH)_2$ formed gradually gets oxidized to $Fe(OH)_3$. The net corrosion reaction due to rusting of iron is thus,

$$4Fe(OH)_2 + O_2 + 2H_2O \longrightarrow 4Fe(OH)_3 \longrightarrow 2Fe_2O_3.xH_2O$$

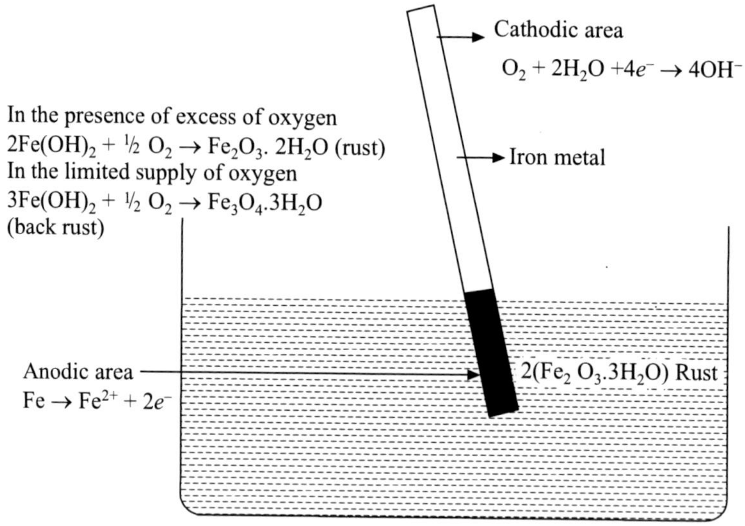

Fig. 2.6 *Rusting of iron*

2.7 TYPES OF WET CORROSION

In general, a metal gets destroyed due to corrosion. Depending on the conditions, metals undergo corrosion by different mechanisms, giving rise to different types of corrosion.

The following types of wet corrosion have been observed:

1. Differential metal corrosion or galvanic corrosion
2. Differential aeration or concentration cell corrosion
   Example: Pitting corrosion, water line corrosion
3. Stress corrosion. **Example:** Caustic embrittlement
4. Grain boundary corrosion.

2.7.1 Galvanic Corrosion or Differential Metal Corrosion

When two dissimilar metals are in contact with each other in a corrosive conducting medium, the metal with lower electrode potential or more active metal (higher in electrochemical series) acts as anode and the metal with higher electrode potential (lower in electrochemical series) acts as cathode. The potential difference between two metals is the cause or driving force for corrosion. If the potential difference between the electrodes is high, greater is the extent of corrosion. The anodic metal undergoes corrosion and the cathodic metal is generally unattacked, for example, Zn and Cu, Fe and Cu, Zn and Ag.

In Zn – Cu galvanic cell zinc metal has low reduction potential ($E° = -0.76$ V) acts as anode, where oxidation (corrosion) occurs and copper behaves as cathode, where reduction takes place.

Galvanic corrosion can be minimized by

(i) avoiding galvanic couple, and
(ii) providing insulating material between the two metals.

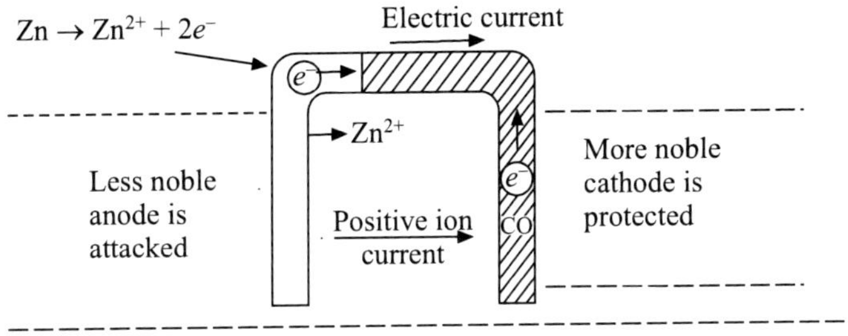

Fig. 2.7 *Galvanic corrosion or differential metal corrosion*

2.7.2 Concentration Cell Corrosion or Differential Aeration Corrosion

This type of corrosion occurs only when the metal surface is exposed to differential air or oxygen concentration. The part of metal exposed to higher concentration of oxygen acts as cathode and the part of the metal exposed to lower oxygen concentration acts as anode. Due to difference in concentration of oxygen over the surface of metal galvanic cell is formed. It acts as a driving force for corrosion.

Example: Let us consider a metal which is dipped in an electrolyte. The dipped area of the metal behaves as anode (due to lack of oxygen) while the remaining portion or area behaves as cathode (well oxygenated part).

At anode, oxidation occurs:

$$Zn \longrightarrow Zn^{2+} + 2e^- \text{ (Oxidation)}$$

At cathode, the oxygen takes up electrons of form OH^- ions

$$\frac{1}{2}O_2 + H_2O + 2e^- \longrightarrow 2OH^- \text{ (Reduction)}$$

$$Zn^{2+} + 2OH^- \longrightarrow Zn(OH)_2$$

The electrons liberated at the anode migrate towards the cathode. The metal cation and anions formed at different parts of the metal combine to form corrosion product which sink to the bottom.

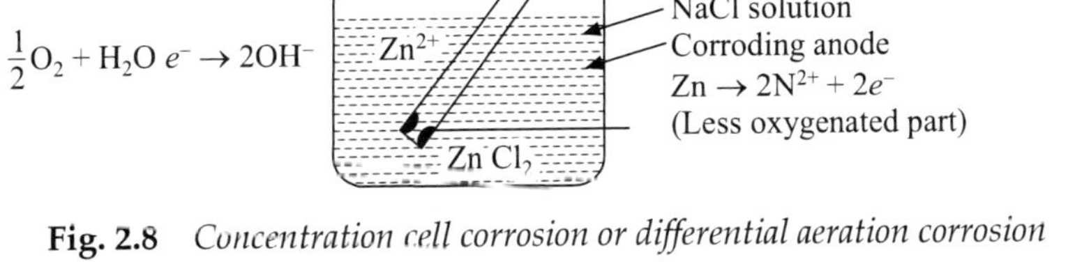

Fig. 2.8 *Concentration cell corrosion or differential aeration corrosion*

2.7.2.1 *Pitting Corrosion*

It is also a type of differential aeration corrosion. It occurs due to unwanted matter such as dust, sand scale, etc. This gives rise to the formation of small anodic and large cathodic areas. The metal below the deposit exposed to lower oxygen concentration, acts as anode whereas the remaining metal surface acts as cathode. Corrosion starts below the dust resulting in a pit. The pit thus formed becomes a smaller anodic area and the entire metal becomes the large cathodic area. Hence, corrosion is faster.

Example: Rusting of iron.

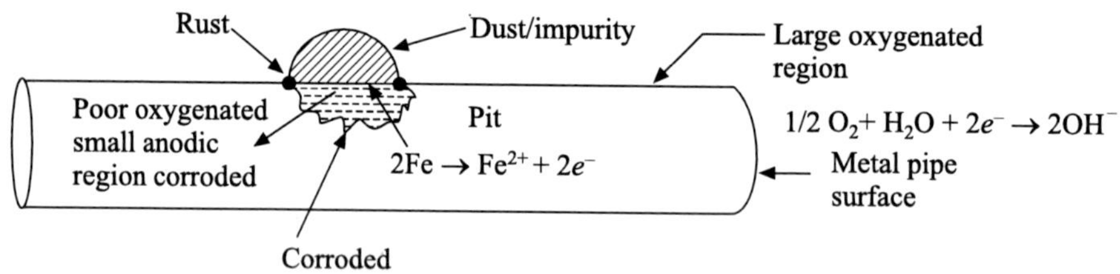

Fig. 2.9 *Pitting corrosion*

2.7.2.2 *Water Line Corrosion*

This type of corrosion results from differential aeration. When water is stored in a steel tank, it is usually found that the maximum amount of corrosion takes place along a line just below the level of water meniscus. This is due to the difference in the concentration of air or oxygen, i.e., the part of the tank above the water line acts as cathode (more aerated) and is completely unaffected by corrosion. Whereas the part of the tank below the water line acts as anode (less aerated). However, if the water is relatively free from acidity, little corrosion takes place.

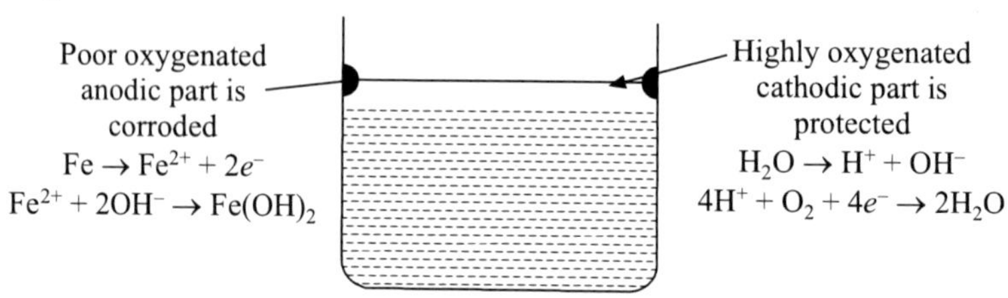

Fig. 2.10 *Water line corrosion occurs just underneath the meniscus and the water label*

2.7.3 Stress Corrosion

It is a highly localized attack in the metal. This corrosion occurs only in the presence of specific corrosive environment. It is commonly observed in fabricated articles of some alloys (zinc brasses, nickel brasses). It is due to the presence of stresses caused by heavy working like riveting, cold working, welding, bending, pressing and quenching. In a corrosive environment, the stressed portion of the metal acts as anode and unstressed portion of the metal behaves as cathode. The potential difference is thus developed. This results in corrosion of the strained portion of the metal, for example, caustic embrittlement.

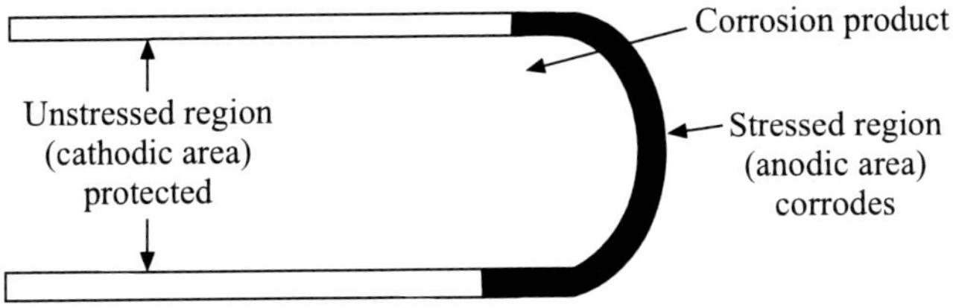

Fig. 2.11 *Stress corrosion in a bent metal strip*

Table 2.1 Corrosive environment for different metals

Metal / Alloy	Corrosive environment
Milid steel	NaOH solution
Low carbon steel	Solution of nitrates — $Ca(NO_3)_2$, $NaNO_3$
Stainless steel	Solution of metal chlorides — $MgCl_2$,$ZnCl_2$,liCl, NaCl
	Solution of NaOH + H_2O_2, brackish and seawater
Nickel	Strong solution of NaOH, KOH
Brass	Ammonia vapours and ammoniacal solution

2.7.3.1 *Embrittlement*

It is a form of stress corrosion which usually takes place in boilers operating at high temperature and pressure. Caustic embrittlement occurs at stressed part of boilers such as cracks, rivets, bends, joints, etc. The boiler fed water usually contains some residual sodium carbonate (used for softening process). At high pressure and temperature it undergoes hydrolysis to form sodium hydroxide.

$$Na_2CO_3 + H_2O \longrightarrow 2NaOH + CO_2$$

This alkali water flows through the minute cracks, crevices between the rivets and joints by capillary action. Inside the cracks, water gets evaporated leaving behind NaOH. This concentrated alkali dissolves iron as sodium ferroate in crevices, cracks and the metal under stress. The sodium ferrorate (Na_2FeO_2) decomposes into magnetite (Fe_3O_4) and alkali is regenerated in the process as per the following reaction.

$$NaOH + Fe \longrightarrow Na_2FeO_2, + H_2$$

$$3Na_2FeO_2 + 4H_2O \longrightarrow 6NaOH + Fe_3O_4$$

or $$6Na_2FeO_2 + 4H_2O + O_2 \longrightarrow 12NaOH + 2Fe_3O_4$$

Sodium hydroxide is regenerated in the process and its concentration keeps on increasing thus, maintaining the required environment. Thus, corrosion process develops cracks and makes the metal brittle by the deposition of the product. The corrosion cell can be represented as follows.

Fe under stress(anode)/Conc. NaOH/dilute NaOH/iron stress free (cathode)

Prevention: Caustic embrittlement can be prevented by the addition of compounds like sodium sulphite, tannin, lignin, phosphates, etc., which block the cracks thereby preventing infiltration of alkali.

2.7.4 Grain Boundary Corrosion

When stainless steel (18% Cr, 8% Ni) containing more than 0.1% carbon is heated to high temperature and cooled slowly or held at 650 °C for a short interval of time, a rapid reaction occurs between carbon and chromium forming CrC. The product precipitates at the grain boundaries and results in the formation of galvanic cells as shown in Fig. 2.18.

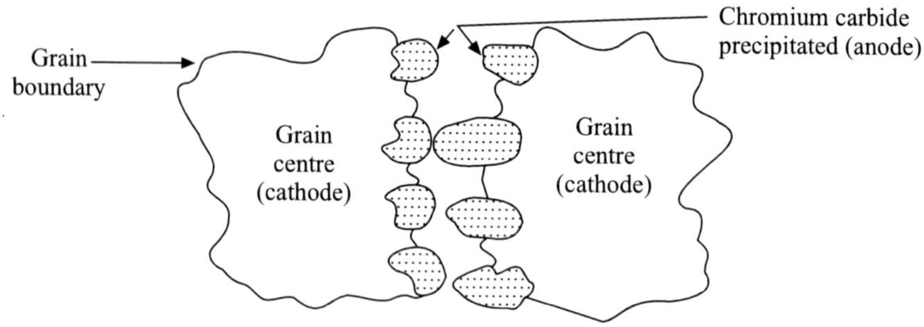

Fig. 2.12 *Grain boundary corrosion*

Precipitation of chromium carbide leads to a decrease in the concentration of chromium at the grain boundary. The boundary has a greater potential and acts as the anode relative to grain centres which are richer in chromium. As a result corrosion takes place at the grain boundary. Such a phenomenon is referred to as the grain boundary corrosion. The grain boundary corrosion is not observed in pure metals. In a steel structure, it may be avoided by using 18 – 8 steel containing less than 0.03% carbon.

2.8 COMPARISON BETWEEN CHEMICAL CORROSION AND ELECTROCHEMICAL CORROSION

The comparison between chemical corrosion and electrochemical corrosion is given in the tabular form as follows:

Chemical Corrosion	Electrochemical Corrosion
1. It occurs in dry condition	It occurs in wet condition (in the presence of electrolyte or moisture)
2. It involves direct chemical attack of the metal by environment	It involves the setting up of a large number of galvanic cells
3. It is explained by absorption mechanism	It is explained by mechanism of electrochemical reactions
4. It occurs on both homogeneous and heterogeneous surfaces	It occurs only on heterogeneous surface only
5. Corrosion is uniform	Corrosion is not uniform
6. It is a slow process	It is a fast process
7. Corrosion products accumulate at the same place	Corrosion occurs at the anode, but products accumulate near the cathode where corrosion occurs

2.9 GALVANIC SERIES

In the electrochemical series (reduction electrode potential arranged in an increasing order) a metal high in the series is more anodic and undergoes corrosion faster than the metal below it.

For example, Li corrodes faster than Mg, Zn corrodes faster than Fe.

However, some exceptions to this generalization are known. For example, Ti (above Ag in the electrochemical series) is less reactive than Ag. In Zn–Al couple, (Zn below Al in the electrochemical series) is corroded, while Al acts cathodic and is protected. These observations are exactly opposite to that predicted by the emf series due to the fact that metals like Ti and Al develop strongly adhering oxide layers on their surfaces thereby making their effective electrode potential more positive (or less negative). In order to overcome these limitations of electrochemical series in corrosion studies, a series called galvanic series was introduced. The corrosion studies on various metals and alloys were actually performed in various environments. Then the metals and alloys were arranged in the order of their corrosion tendencies or corrosion resistance. An arrangement of metals and alloys in the order of their corrosion resistance in the given environment is referred to as galvanic series.

2.10 COMPARISON BETWEEN ELECTROCHEMICAL SERIES AND GALVANIC SERIES

Electrochemical Series

1. E° values are measured by dipping pure metals in their salt solution without any oxide layer on them.
2. The position of a given metal in electrochemical series is fixed.
3. No information regarding position of alloys.
4. It predicts relative displacement tendency of metal.
5. Metals and non-metals are included in this series.
6. Electrode potentials are measured by using standard hydrogen electrode as reference electrode.

Galvanic Series

1. Developed by studying corrosion of metal or alloy in the environment without removal of oxide layer.
2. The position may shift.
3. Alloys are included in series and their corrosion can be studied.
4. It predicts relative corrosion tendency.
5. The galvanic series comprises both metals and alloys.
6. Electrode potentials are measured by using calomel electrode as reference electrode.

Table 2.2 Galvanic series

<table>
<tr>
<td>Mg
Mg alloys
Zn
Al
Al alloys
Low carbon steel (mild steel)
Cast iron
High Ni cast iron
Pb- Sn alloy (solder)
Pb
Sn
Brass
Silver solder
Cu
Ni
Cr stainless steel
18-8 stainless steel
18-8 Mo stainless steel
Ag (silver)
Ti (titanium)
Graphite
Au (gold)
Pt (platinum)</td>
<td>
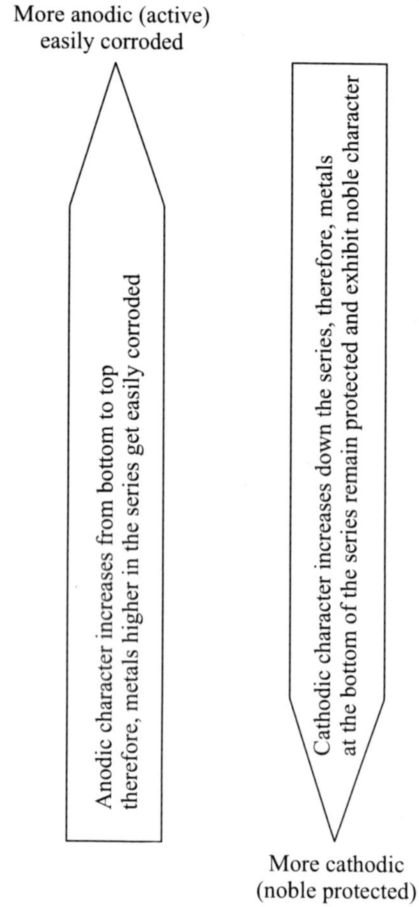

</td>
</tr>
</table>

- Galvanic series is the relative hierarchy of metals arranged in order from magnesium (least noble) at the anodic corroded end through platinum (most noble) at the cathodic, protected end.
- The greater difference in position between two metals or alloys, the greater the chances that corrosion will occur.
- Position of iron is higher in galvanic series than Ag and Cu, so silver and copper do not undergo much corrosion like iron in most atmosphere.

2.11 FACTORS INFLUENCING THE CORROSION RATE

Rate and extent of corrosion depends on the following factors:

1. Nature of metal (primary factor)
2. Nature of environment (secondary factor)

FACTORS

Primary Factors (Related to metal)	**Secondary Factors** (Related to environment)
1. Nature of metal	1. Temperature
2. Nature of corrosion product	2. pH of the medium
3. Anodic and cathodic area	3. Conducting medium
4. Hydrogen overvoltage	
5. Polarization at anodic and cathodic regions	

2.11.1 Primary Factors

Nature of the Metal: The tendency of the metal to undergo corrosion mainly depends on the nature of the metal. In general, the metal with lower electrode potential have more reactive and more susceptible for corrosion and metal with high electrode potential are less reactive and less susceptible for corrosion. For example, metals like K, Na, Mg, Zn, etc., have low electrode potential undergo corrosion easily, whereas noble metals hence, like Ag, Au, Pt have higher electrode potential, their corrosion rate is negligible. There are few exceptions for this general trend as some metals show the property of passivity like Al, Cr, Ti, Ta, etc. The rate of corrosion depends on the difference in the position of the metal in the galvanic series. Greater the difference, faster is the corrosion at anode.

Surface State of the Metal or Nature of the Corrosion Product (Passivity): The nature of the corrosion product largely decides the rate of further corrosion process. The corrosion product is usually the oxide of the metal, if the oxide layer, which forms on the surface, is stoichiometric, highly insoluble and non-porous in nature with low ionic and electronic conductivity, then that type of product layer effectively prevents further corrosion. For example, Al, Cr, Ti develop such a layer on their surface and become passive to further corrosion and some metals like Ta, Zr and Mo not only form such a protective layer but are capable of self-repairing oxide films when it is damaged. Hence, these are extremely passive metals if the oxide layer formed on the metal surface is non-stoichiometric, soluble, unstable and porous in nature and have an appreciable conductivity, they cannot control corrosion on the metal surface. For example, oxide layer formed on metals like Zn, Fe, Mg, etc.

Anodic and Cathodic Area: The rate of the corrosion is greatly influenced by the relative sizes of cathodic and anodic areas. If the metal has smaller anodic area and larger cathodic area exposed to corrosive atmosphere, more intense and faster is the corrosion at the anodic area. This is explained as follows. At anode, oxidation takes place and electrons are liberated. At the cathode these electrons are consumed. This process makes the anodic reaction to take place at its maximum rate, thus increasing the corrosion rate. If the cathode is smaller the consumption of electrons will be slower and the corrosion reaction as a whole will be slower.

Example: A broken coating of tin on iron surface enhances corrosion rate of iron and steel. Tin is cathodic to iron and the small exposed part of iron acts as anodic region and rest of the tin coated surface acts as cathodic region. Because of the small ratio of anodic to cathodic areas, the corrosion rate is high. For the same region iron bolts and nuts are not used in copper boilers. In general,

$$\text{Rate of corrosion of anodic region, } \alpha \ \frac{\text{Cathodic area}}{\text{Anodic area}}$$

Hydrogen Overvoltage: A metal with low hydrogen overvoltage on its surface is more susceptible to corrosion. When the cathodic reaction is hydrogen evolution type, liberation of H_2 gas is easier so the cathodic reaction is very fast, that makes anodic reaction faster hence overall corrosion process is very fast. If the H_2 overvoltage is high, cathodic reaction is slow hence corrosion reaction is also slow with high electrode potential are less reactive and less susceptible for fast corrosion. For example, when Zn metal in contact with $1NH_2SO_4$, it undergoes corrosion by the evolution of hydrogen gas. The rate of reaction is very slow because its overvoltage is high (~0.7 V). If a few drops of Cu solution are added the rate of corrosion increases since Cu, gets deposited on Zn forming minute cathodes, where the hydrogen overvoltage value is only 0.33 V.

Polarization at anodic and cathodic regions: In general polarization causes the potential of the cathode to become more anodic and that of anode to become more cathodic. Because of the anodic polarization, the anode reaction is slower and overall corrosion rate decreases, because the tendency of anodic metal surface to undergo oxidation decreases. Cathodic polarization retards the cathodic reaction. This can be due to chemical polarization of the cathode, i.e., hindering the combination of cathode reactant with the electrons. The retarded movement of the cathodic reactant to the cathode surface or retarded removal of the cathodic reaction product from the surface of the cathode also makes the cathodic reaction slower.

2.11.2 Secondary Factors (Related to Environment)

Temperature: The rate of a chemical reaction, in general, increases with rise in temperature. Corrosion process (redox reaction) is one such chemical reaction. Therefore, the rate of corrosion increases as the temperature increases. Increase in temperature increases the conductance of the corrosion medium, which also contributes to the increase in corrosion rate. If the medium is alkaline/neutral oxygen absorption takes place at the cathode, since solubility of the dissolved gases decreases with increase in temperature. The rate of corrosion also decreases. In the case of corrosion-resistant passive metals, the rise in temperature decreases the passive range and thereby increases the corrosion rate with increasing temperature, for example, caustic embrittlement in high pressure boiler.

pH of the Medium: Acidic media are generally more corrosive than alkaline/neutral media. The pH of the solution decides the type of cathodic reaction. In general, higher the acidic nature of the environment (low pH), higher is the corrosion rate. The corrosion rate of iron in oxygen-free water is slow, until the pH is below 5. The corresponding corrosion rate in presence of oxygen is much higher. However, some metals which are amphoteric in nature like Al, Zn, etc., undergo fast corrosion in a highly alkaline solution as complex ion.

Conducting Medium: As the conductivity of the corrosion medium increases, the corrosion rate also increases. Higher the conductivity of the medium, faster the ions can migrate between the anodic and cathodic regions of the corrosion cell, in turn, faster will be the exchange of electrons at the electrode surfaces. This facilitates higher corrosion rate. Therefore, corrosion problem is more in seawater than in fresh water.

2.12 PREVENTION OF CORROSION OR CORROSION CONTROL

To prevent corrosion of metals during manufacturing, the primary aim should be to obtain metal of uniform composition, corrosion reactions are the surface phenomenon of metal. Hence, the protection of surface of metal should be the main object. There are three main methods of protection of metal from corrosion.

1. Proper selection and designing of the material
2. Protective coatings.
   (i) Metal coatings — anodic, cathodic
   (ii) Inorganic coatings — anodizing, phosphating
   (iii) Organic coatings — paints and enamel
3. Cathodic protection
4. Anodic protection

2.12.1 Design and Selection of Materials

Corrosion can be controlled by selecting suitable materials which resist corrosion environment and designing them suitably during fabrication. Some important tips are as follows:

1. Use of dissimilar metals must be avoided to prevent galvanic corrosion. If their use is unavoidable the position of metal choosen should be of minimum difference in the potential in the galvanic series (Fig. 2.13b).
2. In order to minimize corrosion, the anode material must be as large as possible compared to cathode material.
3. The anodic material should not be painted or coated.
4. An insulating material should be preferred at the contact points of two dissimilar metals. However, avoid the use of insulator such as asbestos or wood because they absorb and hold liquids which act as electrolytic conductors.
5. When metals are joined care should be taken not to leave gaps between them, where some liquids or air can be trapped resulting in differential direction corrosion.
6. Stress corrosion can be controlled by annealing the equipment.
7. Proper design of the equipment will minimize the corrosion. It can be achieved by supporting legs to the equipment instead on large blocks to provide free circulation of air and prevention of the formation of damp area. (Fig. 2.13 a).
8. Design the storage containers in such a manner that they can be completely drained and cleaned (Fig. 2.13 c).

9. The equipment should be designed in such a way that it can be easily kept free from dust to reduce pitting corrosion.
10. Arrange profiles in such a manner that water is completely drained as far as possible to avoid crevice corrosion. For example, design electrical boxes (exposed to atmosphere) in such a manner that water does not collect at the top or in crevices.

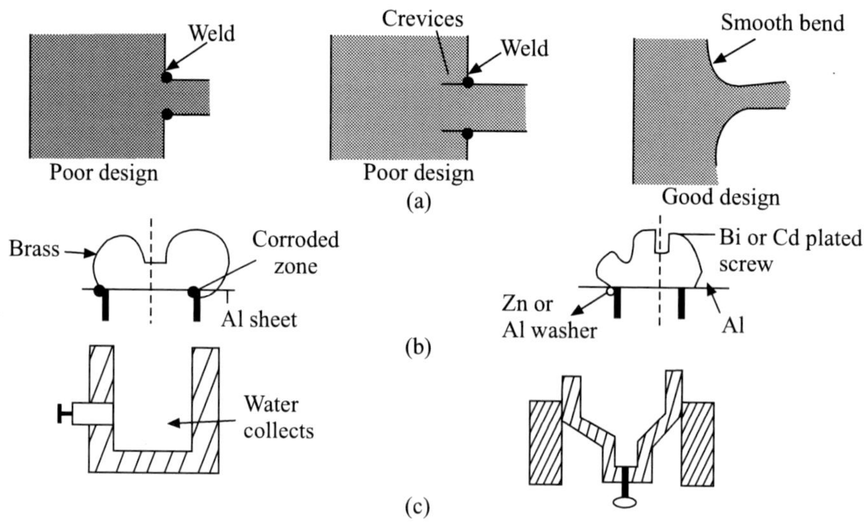

Fig. 2.13 *Control of corrosion by design and material selection: (a) sharp corners and recesses to be avoided, (b) use of insulating washers, (c) storage containers*

2.12.2 Protective Coatings

The protective coatings are a simple way to reduce corrosion, by limiting the exposure of the metal to a corrosive environment. For example, paint is a common protective coating, but tar, pitch, bitumen and plastics are also used. The principal types of coatings applied on the metal surface are:

(a) Metal coating
(b) Inorganic coating
(c) Organic coating

(a) Metal Coating: Deposition of a protective metal over the surface of the base metal by hot dipping process is called metallic coating. This method is used for producing a coating of low melting metals such as Zn, Al, Sn, etc., on iron steel metals which have relative high melting points. It is a valuable and well tried method for improving the corrosion life of the underlying metal. Metal coatings are applied by elector deposition, flame spraying, cladding, hot dipping, diffusion coating and vapour deposition. It is of two types.

(*i*) *Anodic Coating:* Anodic coatings are produced by coating a base metal with more active metals which are anodic to the base metal. For example, iron is coated with more active metals such as zinc, magnesium and aluminium. Galvanization is a familiar example for anodic coating and is extensively used to protect iron and steel objects.

Galvanization: The process of coating of zinc over the surface of iron to prevent corrosion is called galvanization. It is carried out by the hot dipping method.

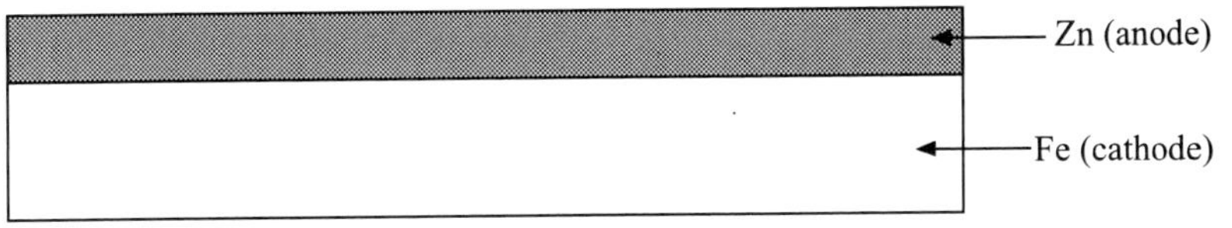

The galvanization process involves the following steps.

- The metal surface is washed with organic solvents to remove organic matter on the surface (oil, grease, fatty substances).
- Then the metal sheet is washed with dilute sulphuric acid at 60 – 90°C in order to remove the existing oxide scale.
- The metal sheets are passed through distilled water to remove traces of sulphuric acid and dried.
- The dried metal sheet is passed in molten zinc bath at 425 – 430°C. The surface of the bath is covered with flux NH_4Cl to prevent oxidation of zinc.
- The zinc coated metal sheet is passed through a pair of hot rollers to remove excess of zinc and produces a thin film of uniform thickness.
- Finally, the metal sheet is annealed.

Uses: Galvanized materials are used extensively in roof sheets, fencing wire, barbed wire, buckets, bolts, nuts, nails, screws, pipes, etc.

Note: Galvanized articles are, however, not used for preparing and storing foodstuffs, since zinc dissolves in dilute acids producing toxic zinc compounds.

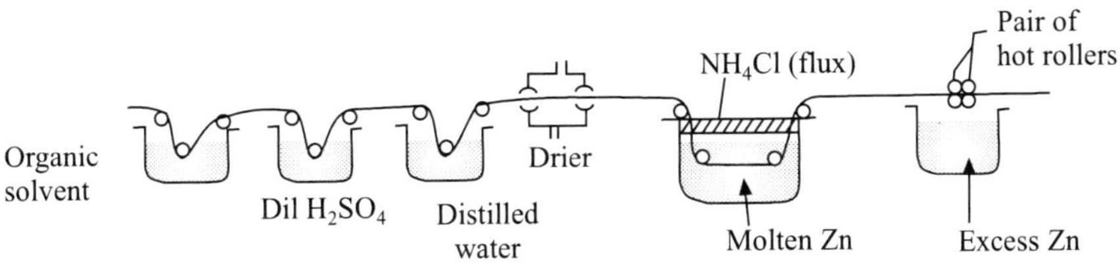

Fig. 2.14 *Galvanization of metal sheet*

(*ii*) *Cathodic Coating:* Cathodic coatings are produced by coating a base metal with a more noble metal which is cathodic to the base metal. The metal such as copper, nickel, tin, silver are commonly used as cathodic coatings on iron and its alloy like steel. Tinning is the most commonly used cathodic coating on iron and its alloy like steel.

Tinning: It is a process of coating a base metal iron and its alloy like steel with tin.

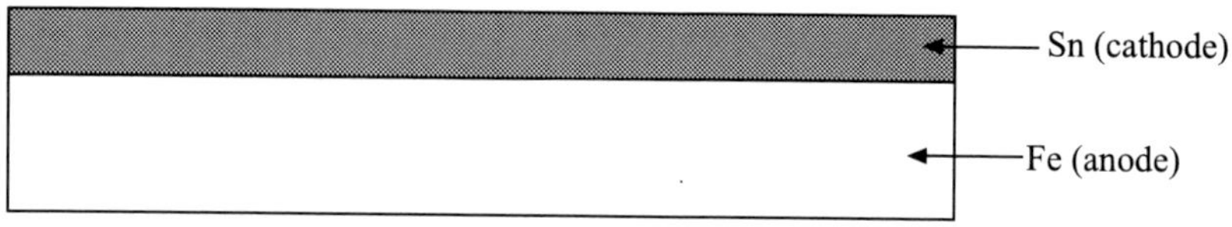

Tinning of iron or steel process involves the following steps:

- The metal surface is washed with organic solvents to remove any organic matter on the surface.
- Then the metal sheet is washed with dilute sulphuric acid at 60 – 90 °C in order to remove any existing oxide scale.
- The metal sheet is passed through distilled water to remove traces of sulphuric acid and dried.
- The clean and dry sheet is passed through molten zinc chloride flux to improve adhesive capacity of the metal.
- Now the metal sheet is passed through molten tin bath.
- Finally, the metal sheet is passed through palm oil containing series of rollers. The palm oil prevents oxidation of tin, the rollers remove excess of tin and produces a thin uniform tin coated metal sheet.

Applications: It is widely used in:

1. Manufacturing of food storage cans
2. Making of tin coated copper sheets for cooking utensils
3. Making of tin coated steel cans for kerosene.

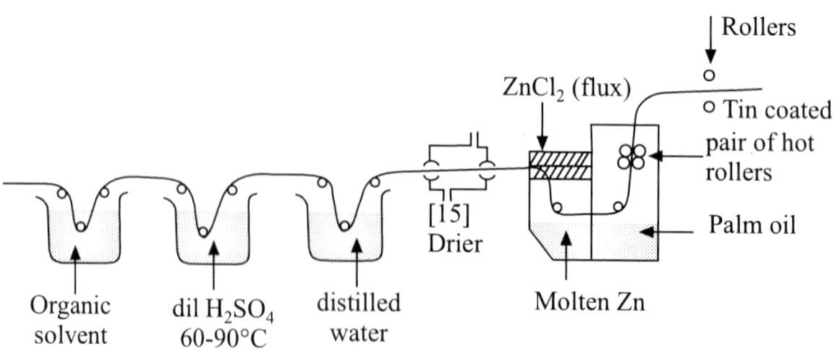

Fig. 2.15 *Tinning of metal sheet*

Comparison between Galvanizing and Tinning Processes

Galvanizing Process	Tinning Process
1. Coating of Zn or Fe or steel by pot dipping process	1. Coating of Sn or Fe or steel by hot dipping process
2. Coating metal is anodic to iron steel called anodic coating	2. Coating metal is cathodic to iron/steel called cathodic coating
3. The molten metal bath is covered with a flux of NH_4Cl which prevents the oxidation of the coated metal	3. The molten metal is covered by a flux bath which is further covered by a flux of zinc chloride
4. Galvanised sheet cannot be used for preparing/storing foodstuffs since Zn dissolves in acidic medium and forms toxic compounds	4. It is used for preparing/storing food stuffs
5. The article is dipped in molten bath of Zn. The excess of coated metal is removed by passing through pair of hot rollers and cooled gradually	5. The clean and dry sheet is passed through flux layer, molten tin, finally removed out through palm oil, which prevents the oxidation of the coated tin

(b) Inorganic Coating: Inorganic coatings are generally chemical conversion coatings carried out on the base metal. A surface layer of the metal is converted into a compound, by chemical or electrochemical reactions, which forms a barrier between the underlying surface and the corrosion environment. There are two types of inorganic coatings such as:

1. Anodized coating
2. Phosphate coating

1. *Anodized coating:* These coatings are generally produced on non-ferrous metals like Al, Zn, Mg and their alloys by anodic oxidation (electrochemical) process. The base metal is made the anode.

2. *Anodizing of Al:* It is carried out by electrolysis in a suitable acid electrolyte bath (H_2SO_4, chromic acid, oxalic acid, etc., at 35° to 40° and moderate current densities. It is an electrochemical process during which aluminium is the anode. When current is passed the oxygen liberated at the anode combines with it to form an oxide film. The outer part of the film is porous and considerably softer than the layer directly adjoining the metal surface. The anodized surface is made impermeable by treatment with hot water, this converts aluminium oxide into its monohydrate (Al_2O_3, H_2O) which occupies more volume thereby the pores are sealed. If chromic acid is used, the oxide layer of 2–8 μm thickness is obtained. For higher thickness of the oxide film, 10% sulphuric acid is used as electrolyte at 25 °C, which gives a colourless 22 μm thick film.

Uses:

Anodized articles are used as soap boxes, tiffin carriers, window frames, etc.

Anodizing is also used to prevent galling of threaded components.

These films are also used for a number of cosmetic effects.

Phosphate Coating: Phosphating is generally given to the metal surface before finishing. It gives better corrosion resistance and also improves the paint adhesion capacity. It is applied frequently on iron, steel and zinc and to a small extent, aluminum. The coating of mixed iron and zinc phosphate on steel sheet fabrication such as refrigerators, washing machines and car bodies provide a better paint adhesion in addition to corrosion resistance. Phosphate bath solution contains a mixture of phosphoric acid, a metal phosphate and accelerator. Accelerators are used to speed up the reaction and reduce the crystal size. The pH of the bath is maintained at about 1.8 – 3.2 and electrolysis is carried out at 30 – 35 °C. It occurs in the following stage:

(a) Dissolution of the metal as metal ions.

(b) Metal ions combine with phosphate ions to form a metal phosphate.

(c) Deposition of the metal phosphate on the surface of metal.

Uses: The most important application of phosphate coating is as an underlayer for paint finishes. In the case of car bodies, electrophoretic priming usually follows the phosphating stage. Phosphating is widely applied for the protection of metal parts which cannot be painted or readily plated, such as nuts, bolts, moving parts of machinery, etc.

(c) Organic Coating: Organic coatings are inert organic barriers applied on a metal surface and corrosion environment. The term organic coating is used when the thickness of the protective layer is less than 0.4 mm. Few important groups of organic coatings are paints and enamels, varnishes, etc.

1. **Paints:** Paint is a mechanical mixture consisting of pigments and fillers or extenders suspended in a vehicle. Vehicle is a film forming drying oil. Other liquids called thinners are added to the oil. Thinners are volatile organic solvents such as acetone, xylol, etc. Paint is usually applied to the surface of the metal by spraying or brushings.

   Properties of a good paint:

   - It must have high covering power and adhesion capacity.
   - It must form a tough, uniform coherent film.
   - It must resist corrosion.
   - Film must not get cracked during drying.
   - Colour should be stable, glossy and washable.

2. **Enamels:** An enamel is a pigmented varnish. It gives lustrous, hard and glossy finish to the film. The important constituents of enamels are pigments, vehicle, driers and thinner. These are used for painting metallic surfaces like bicycles, electrical devices, etc.

2.12.3 Cathodic Protection

Cathodic protection is a method of protecting the base metal by making it to behave like a cathode. Cathodic protection can be achieved by the following methods.

(i) Sacrificial anodic protection method
(ii) Impressed current cathodic protection method

(i) **Sacrificial Anode Protection Method:** In this method, the metallic structure to be protected (base metal) is connected by a wire to a more anodic metal so that all corrosion is concentrated at this more anodic metal. The more anodic metal itself gets corroded slowly. While the parent structure (cathodic) is protected. The more active metal so employed is called sacrificial anode. The corroded sacrificial anode is replaced by a fresh one when consumed completely. Metals commonly employed as sacrificial anode are Mg, Zn, Al and their alloys which possess low reduction potential and occupy higher end in electrochemical series.

Examples of sacrificial anode methods are:

- A magnesium block connected to a buried oil storage tank.
- Mg bars are fixed to the sides of ocean going ships to act as sacrificial anodes.
- Mg blocks are connected to buried pipelines.

Advantages

- The method is simple.
- Low installation cost.
- Minimum maintenance cost.
- Does not require power supply.

Disadvantages

- Involves recurring expenditure for replacement of consumed anodes.
- Involves:
  (a) Mg block connected to a buried oil storage tank.
  (b) Mg bars fixed to the sides of ocean going ship.
  (c) Mg blocks connected to buried pipelines.

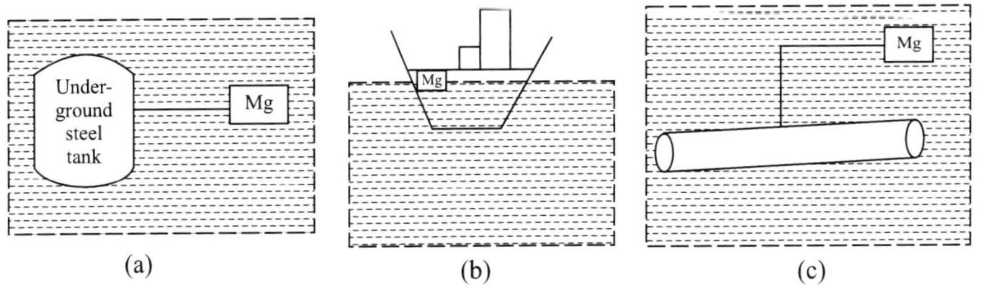

Fig. 2.16 *Cathodic protection by sacrificial anode method*

Impressed Current Cathodic Protection: In this method the metal to be protected is made cathode by connecting it to the negative terminal of a DC source.

There is electron flow from the anode to the cathode during corrosion. This is called corrosion current, but in this method the applied voltage sends current in opposite direction changing the anode to cathode. Hence, there is no destruction or loss of metal. Anode being inert remains unaffected. Usually graphite or platinum is used as inert anode along with back fill composed of coke, breeze or gypsum which provides proper electrical contact with the surrounding soil.

Advantages

- One installation can protect large area of metal.
- Low maintenance cost.

Limitations

- Rather expensive, since it needs high current for safe protection structure.
- If the impressed current is not uniform on the entire surface of the protected structure localized corrosion takes place on the protected metal.
- If hydrogen liberation is the cathodic reaction on the metal surface, then the protected metal may suffer hydrogen embrittlement.

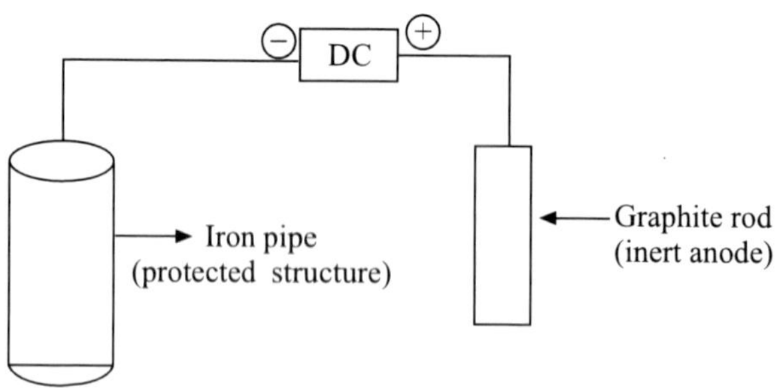

Fig. 2.17 *Cathodic protection by impressed current method*

Comparison between sacrificial anodic protection method and impressed current cathodic protection.

Sacrificial Anodic Protection	Impressed Current Cathodic Protection
1. The metallic structure to be protected is connected to a more anodic metal using a metallic wire 2. The more active metal gets corroded while the parent store is protected from correction 3. The more active metal so employed is called sacrificial anode 4. The sacrificial anodes to be replaced by fresh ones as and when it is required 5. Commonly used sacrificial anodes are Mg, Zn, Al, etc. 6. This method is generally used for the protection of buried pipelines, shiphulls, water tanks, etc.	1. The metallic structure to be protected is mode cathode by impressing 2. The current is applied in the opposite direction to nullify the corrosion current. 3. The impressed current is obtained from a source like battery 4. An insoluble anode (Ex: graphite high silicon content iron, etc.) is buried in the soil and connected to the structure to be protected 5. The anode is usually placed in a back fill to provide a better electrical contact with the surroundings 6. This method is suitable for large structure and for long-term operations

2.13 ANODIC PROTECTION

In contrast to cathodic protection, anodic protection is relatively new. The prevention of corrosion by the impressed anodic current is called anodic protection. The principle of anodic protection is the growth of oxide film occurring as a result of application of an anodic current on the metal specimen in a suitable oxidizing environment. The method, suggested by Edeleanu in 1954, is based on the formation of a protective film on metals by externally applied anodic current. Generally, when a metal is made an anode and current is applied, the dissolution rate of the metal tends to increase. But in the case of passive metals with active – passive transitions such as nickel, iron, chromium, titanium and their alloys, application of suitable current makes them passivated and decreases the rate of dissolution of metals. The active – passive transition of these metals with the application of potentials can be explained from the potential- current curves as shown in Fig. 2.18.

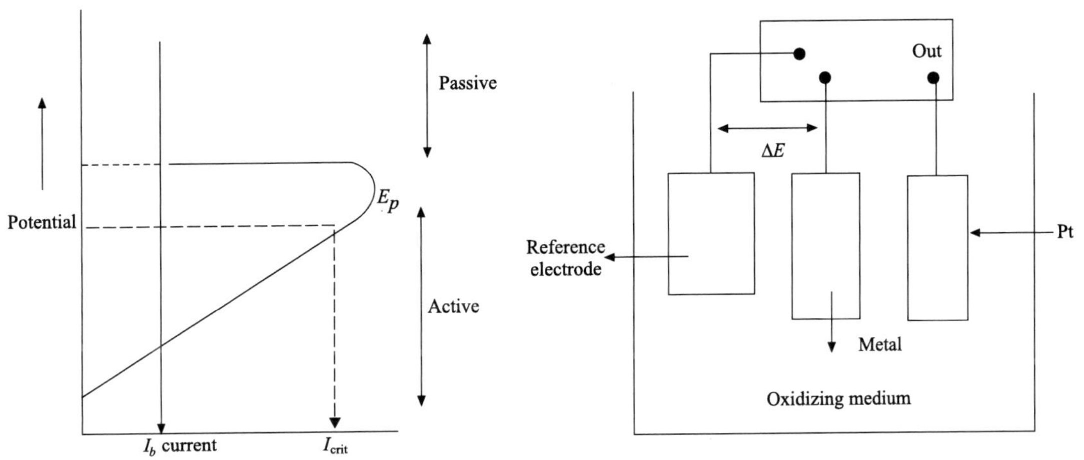

(a) Potential-current curve for passivation

(b) Anodic protection (potentiostatic)

Fig. 2.18

At a predetermined range of applied potential the changes in the observed potential and the corresponding changes in the current are studied. During the initial stages (*AB*) the current increases indicating the dissolution (corrosion) of the metal. When the current reaches a critical point, passivation, that is, the development of oxide layer sets in. The potential at i_{cir} is called the passivation potential (E_p). Above E_p (along *BC*) the current flow decreases to a very small value called the passivating current i_p. It is the minimum protective current density to maintain passivation. At this stage (at *C*), an increase in potential will not corrode the metal since the latter is in a highly passive state. This small current (l_p) is sufficient to maintain the passivity of the metal and bring down the corrosion rate.

A schematic arrangement for anodic protection is shown in Fig. 2.18(b). It consists of a generator capable of producing high current density, an auxiliary platinum electrode, a reference calomel electrode, and a potentiostat to maintain constant potential. The metal specimen to be protected kept in a suitable oxidizing atmosphere acts as the working electrode (anode). The potential is first increased to E_p for the initial corrosion to occur. The potentials are then slightly increased to E_f (Flade potential) when the current decreases to a very small value (I_p) indicating passivation of the metal specimen. Now the potential is kept constant at E_p and the current is maintained at I_p.

The advantage of the method is that it requires a small current. However, the corrosion rate cannot be reduced to zero as in cathodic protection. A drawback of the method is that it cannot be applied to metals which do not passivate. The principle of anodic protection is utilized in the transportation of concentrated acids. The stainless steel container is passivated by maintaining at its passivating potential.

Anodic Protection	Cathodic Protection
1. Small amount of the metal migrates into the coating film limiting its usefulness in corrosion protection 2. Anodic coatings are used in less severe environment and are generally economical and produce a nice surface finish 3. This technique is used in environment 4. Coating is anodic to the base metal 5. Rate of corrosion is less 6. Galvanization is the example 7. Large anodic and small cathodic area	1. Cathodic system produces a coating free of metal migration and is much more robust. These systems are useful in heavy corrosive environment 2. Cathodic protection uses a metal rod called sacrificial cathode to corrod in place of the protected metal 3. This technique is usually used in water 4. Cathodic coating on base metal 5. Rate of corrosion more 6. Tinning is the example 7. Small anodic and large cathodic area

2.14 CORROSION PENETRATION RATE (CPR)

The corrosion penetration rate is the speed at which any metal or alloy deteriorates in a specific corrosive environment through chemical or electrochemical reactions. It is also defined as the amount of weight loss per year in the thickness of metal or alloy due to corrosion. The corrosion penetration rate also referred as corrosion rate.

Corrosion penetration rate depends on following factors;

1. Nature of metal.
2. Nature of corrosive environment/medium.
3. Nature corrosion product.
4. Speed at which corrosion spreads in the inner portion of metal.

The Corrosion penetration rate is calculated using the following formula;

$$\text{CPR} = \frac{k \times W}{D(\rho) \times A \times T}$$

where,

W is weight loss after exposure time.

T is exposure time in corrosive medium.

D is the density of metal or alloy.
A is surface area of exposed specimen.
K is constant.

The CPR is conveniently expressed in terms of either mils (0.001 inch) per year (mpy) or millimeters per year (mmpy). $K = 534$ to give CPR in mpy when surface area of test specimen is in inch^2 and $K = 87.6$ for CPR in mmpy when surface area of test specimen is in cm^2

where, 1 mil = 0.001 inch

	CPR in mpy	CPR mmpy
k	534	87.6
W (wt loss)	mg	Mg
(*D*) ρ	g/cm^3	g/cm^3
A	inch^2	cm^2
T	hr	Hr

WORKED PROBLEMS

Problem 2.1

A sheet of carbon steel – one metre wide by three metres long – has lost 40 g to corrosion over the past six months. Convert that mass loss to the penetration rate of the steel in mm units and mpy units. What would be the corrosion rate? (Carbon steel density = 7.8 g/cm^3) To calculate CPR in mmpy

Solution: To calculate CPR in mmpy

	Given	CPR in mmpy
k		87.6
W (wt loss)	40 g	40 × 1000 mg
(*D*) ρ	7.8 g/cm^3	7.8 g/cm^3
A	1 m × 3 m = 3 m^2	3 × 100 × 100 cm^2
T	6 months	6 × 30 × 24 hr

$$\text{CPR} = \frac{k \times W}{D(\rho) \times A \times T}$$

$$= \frac{87.6 \times 40 \times 1000}{7.8 \times 3 \times 100 \times 100 \times 6 \times 30 \times 24}$$

$$\textbf{CPR} = \mathbf{3.466 \times 10^{-3}}\ \textbf{mmpy}$$

To calculate CPR in mpy

	Given	CPR in mpy
k		534
W (wt loss)	40 g	40×1000 mg
$(D)\ \rho$	$7.8\ g/cm^3$	$7.8\ g/cm^3$
A	$1\ m \times 3\ m = 3\ m^2$	$3 \times 1550\ inch^2$
T	6 months	$6 \times 30 \times 24$ hr

$$1 \text{ sq mt} = 1550 \text{ sq inch}$$

$$\text{CPR} = \frac{k \times W}{D(\rho) \times A \times T}$$

$$= \frac{534 \times 40 \times 1000}{7.8 \times 3 \times 100 \times 100 \times 6 \times 30 \times 24}$$

$$\textbf{CPR} = \mathbf{0.0503\ mpy}$$

Problem 2.2

Calculate the CPR in both mpy and mmpy for a thick steel sheet of area 100 inch^2 which experiences a weight loss of 485 g after one year. (Density of steel = 7.9 g/cm^3).

Solution: To calculate CPR in mmpy

	Given	CPR in mmpy
k		87.6
W (wt loss)	485 g	485×1000 mg
$(D)\ \rho$	$7.9\ g/cm^3$	$7.9\ g/cm^3$
A	$100\ inch^2$	$100 \times 6.45\ cm^2$
T	1 year	365×24 hr

$$1\ \text{inch}^2 = 6.45\ \text{cm}^2$$

$$1\ \text{cm}^2 = 0.155\ \text{inch}^2$$

$$\text{CPR} = \frac{k \times W}{D(\rho) \times A \times T}$$

$$= \frac{87.6 \times 485 \times 1000}{7.9 \times 100 \times 6.45 \times 365 \times 24}$$

$$\textbf{CPR} = \mathbf{0.9518\ mmpy}$$

To calculate CPR in mpy

	Given	CPR in mpy
k		534
$(D)\ \rho$	7.9 g/cm^3	7.9 g/cm^3
A	100 inch^2	100 inch^2
T	1 year	365×24 hr

$$\text{CPR} = \frac{k \times W}{D(\rho) \times A \times T}$$

$$= \frac{534 \times 485 \times 1000}{7.9 \times 100 \times 365 \times 24}$$

CPR = 37.424 mpy

Problem 2.3

A thickness of a brass sheet of area 400 inch^2 is exposed to air near the ocean. After two years it experienced a weight loss of 375 g due to corrosion. If the density of Brass is 8.73 g/cm^3. Calculate the CPR in mm/year mpy.

Solution: To calculate CPR in mmpy

	Given	CPR in mmpy
k		87.6
W (wt loss)	375 g	375×1000 mg
$(D)\ \rho$	8.73 g/cm^3	8.73 g/cm^3
A	400 inch^2	400×6.45 cm^2
T	2 years	$2 \times 365 \times 24$ hr

$$1 \text{ inch}^2 = 6.45 \text{ cm}^2$$

$$1 \text{ cm}^2 = 0.155 \text{ inch}^2$$

$$\text{CPR} = \frac{k \times W}{D(\rho) \times A \times T}$$

$$= \frac{87.6 \times 375 \times 1000}{8.73 \times 400 \times 6.45 \times 2 \times 365 \times 24}$$

CPR = 0.083 mmpy

To calculate CPR in mpy

	Given	CPR in mpy
k		534
W (wt loss)	375 g	375×1000 mg
$(D)\ \rho$	8.73 g/cm^3	8.73 g/cm^3
A	400 inch^2	400 inch^2
T	2 years	$2 \times 365 \times 24$ hr

$$\text{CPR} = \frac{k \times W}{D(\rho) \times A \times T}$$

$$= \frac{534 \times 375 \times 1000}{8.73 \times 400 \times 365 \times 24 \times 2}$$

$$\textbf{CPR = 3.27 mpy}$$

Problem 2.4

A metal iron plate was found in a vessel containing acidic media, it was estimated that the original area was 20 inch^2 that approximately 1.2 kg had corroded. Assuming a corrosion penetration rate of 400 mpy for this iron, calculate time in years, density of iron 7.87 g/cm^3.

Solution: To calculate time in years

$$\text{CPR} = \frac{k \times W}{D(\rho) \times A \times T}$$

$$T = \frac{k \times W}{D(\rho) \times A \times \text{CPR}}$$

	Given	CPR in mpy
k		534
W (wt loss)	1.2 kg	$1.2 \times 1000 \times 1000$ mg
$(D)\ \rho$	7.87 g/cm^3	7.87 g/cm^3
A	20 inch^2	20 inch^2
T	?	
CPR mpy	400	400

$$T = \frac{534 \times 1.2 \times 1000 \times 1000}{7.87 \times 20 \times 400}$$

$$T = 10177.89 \text{ hr}$$

Converted into years

$$T = \frac{10177.89}{365 \times 24} = 1.16 \text{ years}$$

Problem 2.5

A piece of corroded aluminium plate was found in a submerged ocean vessel, it was estimated that the original area was 64 cm^2 that approximately 350 g had corroded away during the submersion. Assuming a corrosion penetration rate of 2.4 mmy for this alloy in seawater, estimate the time in years, density of aluminium 2.7 g/cm^3.

Solution: To calculate time in years

$$\text{CPR} = \frac{k \times W}{D(\rho) \times A \times T}$$

$$T = \frac{k \times W}{D(\rho) \times A \times \text{CPR}}$$

	Given	CPR in mmpy
k		87.6
W (wt loss)	350 g	350 × 1000 mg
$(D)\ \rho$	2.7 g/cm^3	2.7 g/cm^3
A	64 cm^2	64 cm^2
T	?	
CPR mmy	2.4	2.4

$$T = \frac{87.6 \times 350 \times 1000}{2.7 \times 64 \times 2.4}$$

$$T = 73929.39 \text{ hr}$$

Converted into years

$$T = \frac{73929.39}{365 \times 24} = 8.44 \text{ years}$$

Problem 2.6

The alloy bronze exposed to different corrosive environments of area 50 cm^2, for a period of 6 months. The bronze alloy's loss of weight in acidic medium was 200 g, in neutral 120 g and in basic medium 48 g. Calculate the CPR in mpy. (Density of bronze = 8.73 g/cm^3)

Solution: To calculate CPR in mpy

In acidic medium

	Given	CPR in mpy
k		534
W (wt loss)	200 g	200×1000 mg
$(D)\ \rho$	8.73 g/cm^3	8.73 g/cm^3
A	50 cm^2	50×0.155 inch^2
T	6 months	$6 \times 30 \times 24$ hr

$$1 \text{ cm}^2 = 0.155 \text{ inch}^2$$

$$\text{CPR} = \frac{k \times W}{D(\rho) \times A \times T}$$

$$= \frac{534 \times 200 \times 1000}{8.73 \times 50 \times 0.155 \times 6 \times 30 \times 24}$$

$$\textbf{CPR = 365.40 mpy}$$

In neutral medium

	Given	CPR in mpy
k		534
W (wt loss)	120 g	120×1000 mg
$(D)\ \rho$	8.73 g/cm^3	8.73 g/cm^3
A	50 cm^2	50×0.155 inch^2
T	6 months	$6 \times 30 \times 24$ hr

$$\text{CPR} = \frac{534 \times 120 \times 1000}{8.73 \times 50 \times 0.155 \times 6 \times 30 \times 24}$$

$$\textbf{CPR = 219.24 mpy}$$

In basic medium

	Given	CPR in mpy
k		534
W (wt loss)	48 g	48×1000 mg
$(D)\ \rho$	8.73 g/cm^3	8.73 g/cm^3
A	50 cm^2	50×0.155 inch^2
T	6 months	$6 \times 30 \times 24$ hr

$$\text{CPR} = \frac{534 \times 48 \times 1000}{8.73 \times 50 \times 0.155 \times 6 \times 30 \times 24}$$

CPR = 87.70 mpy

Problem 2.7

A zinc rod is dipped in a NaCl solution of area 20 inch^2 each at different temperatures. The metal zinc undergone corroded of 50 g at 338 K, 30 g at 323 K, 10 g at 300 K over a period of one month. Determine the CPR in terms of mmpy. (Density of zinc = 7.12 g/cm^3).

Solution: To calculate CPR in mmpy

At 338K

	Given	**CPR in mmpy**
k		87.6
W (wt loss)	50 g	50 × 1000 mg
(D) ρ	7.12 g/cm^3	7.12 g/cm^3
A	20 inch^2	20 × 6.45 cm^2
T	1 month	30 × 24 hr

$$1 \text{ inch}^2 = 6.45 \text{ cm}^2$$

$$\text{CPR} = \frac{k \times W}{D(\rho) \times A \times T}$$

$$= \frac{87.6 \times 50 \times 1000}{7.12 \times 20 \times 6.45 \times 30 \times 24}$$

CPR = 6.60 mmpy

At 323K

	Given	**CPR in mmpy**
k		87.6
W (wt loss)	30 g	30 × 1000 mg
(D) ρ	7.12 g/cm^3	7.12 g/cm^3
A	20 inch^2	20 × 6.45 cm^2
T	1 month	30 × 24 hr

$$1 \text{ inch}^2 = 6.45 \text{ cm}^2$$

$$\text{CPR} = \frac{87.6 \times 30 \times 1000}{7.12 \times 20 \times 6.45 \times 30 \times 24}$$

CPR = 1.32 mmpy

At 300 K

	Given	CPR in mmpy
k		87.6
W (wt loss)	10 g	10×1000 mg
$(D)\ \rho$	$7.12\ g/cm^3$	$7.12\ g/cm^3$
A	$20\ inch^2$	$20 \times 6.45\ cm^2$
T	1 month	30×24 hr

$$1\ inch^2 = 6.45\ cm^2$$

$$CPR = \frac{87.6 \times 10 \times 1000}{7.12 \times 20 \times 6.45 \times 30 \times 24}$$

$$\mathbf{CPR = 3.97\ mmpy}$$

Problem 2.8

A copper plate of length of 2 cm and width of 5 cm were kept in a salt solution of pH – 3, loss of metal was 1 g, at pH – 8 of the solution was 0.8 g and at pH – 12 of the solution was 0.2 g in one year. Estimate the CPR of copper in terms of mmpy. (Density of the copper is $8.96\ g/cm^3$)

Solution: To calculate CPR in mmpy

At pH – 3 of the salt solution

	Given	CPR in mmpy
k		87.6
W (wt loss)	1 g	1×1000 mg
$(D)\ \rho$	$8.96\ g/cm^3$	$8.96\ g/cm^3$
A	$2\ cm \times 5\ cm$	$10\ cm^2$
T	1 year	365×24 hr

$$CPR = \frac{k \times W}{D(\rho) \times A \times T}$$

$$= \frac{87.6 \times 1 \times 1000}{8.96 \times 10 \times 365 \times 24}$$

$$\mathbf{CPR = 0.112\ mmpy}$$

At pH – 8 of the salt solution

	Given	CPR in mmpy
k		87.6
W (wt loss)	0.8 g	0.8×1000 mg
$(D)\ \rho$	8.96 g/cm^3	8.96 g/cm^3
A	2 cm $\times$ 5 cm	10 cm^2
T	1 year	365×24 hr

$$\text{CPR} = \frac{87.6 \times 0.8 \times 1000}{8.96 \times 10 \times 365 \times 24}$$

$$\textbf{CPR = 0.09 mmpy}$$

At pH – 12 of the salt solution

	Given	CPR in mmpy
k		87.6
W (wt loss)	0.2 g	0.2×1000 mg
$(D)\ \rho$	8.96 g/cm^3	8.96 g/cm^3
A	2 cm $\times$ 5 cm	10 cm^2
T	1 year	365×24 hr

$$\text{CPR} = \frac{87.6 \times 0.2 \times 1000}{8.96 \times 10 \times 365 \times 24}$$

$$\textbf{CPR = 0.022 mmpy}$$

Problem 2.9

Estimate the corrosion period (in years) of a metal tin bar exposed to environment of summer, the surface area of 20 inch^2, 0.8 mpy of CPR with a loss of 730 mg tin. The density of tin is 7.31 g/cc.

Solution: To calculate time in years

$$\text{CPR} = \frac{k \times W}{D(\rho) \times A \times T}$$

$$T = \frac{k \times W}{D(\rho) \times A \times \text{CPR}}$$

	Given	CPR in mmpy
k		534
W (wt loss)	730 mg	730 mg
$(D)\ \rho$	7.31 g/cm^3	7.31 g/cm^3
A	20 inch^2	20 inch^2
T	?	
CPR mpy	0.8	0.8

$$T = \frac{534 \times 730}{7.31 \times 20 \times 0.8}$$

$$T = 3335.93 \text{ hr}$$

Converted into years

$$\boldsymbol{T = \frac{3332.93}{365 \times 24} = 0.38 \text{ years}}$$

Problem 2.10

The thickness of an alloy sheet of area 100 inch^2 is exposed to air near the ocean. After 1 year it was found to experience a weight loss of 35 g due to corrosion. If the density of alloy is 8.4 g/cm^3. Calculate the CPR in mmy and mpy.

Solution: To calculate CPR in mmpy

	Given	CPR in mmpy
k		87.6
W (wt loss)	35 g	35 × 1000 mg
$(D)\ \rho$	8.4 g/cm^3	8.4 g/cm^3
A	100 inch^2	100 × 6.45 cm^2
T	1 year	365 × 24 hr

$$1 \text{ inch}^2 = 6.45 \text{ cm}^2$$

$$1 \text{ cm}^2 = 0.155 \text{ inch}^2$$

$$\text{CPR} = \frac{k \times W}{D(\rho) \times A \times T}$$

$$= \frac{87.6 \times 35 \times 1000}{8.4 \times 100 \times 6.45 \times 365 \times 24}$$

$$\textbf{CPR = 0.064 mmpy}$$

To calculate CPR in mpy

	Given	CPR in mpy
k		534
W (wt loss)	35 g	35 × 1000 mg
$(D)\ \rho$	8.4 g/cm^3	8.4 g/cm^3
A	100 inch^2	100 inch^2
T	1 year	365 × 24 hr

$$\text{CPR} = \frac{k \times W}{D(\rho) \times A \times T}$$

$$= \frac{534 \times 35 \times 1000}{8.4 \times 100 \times 365 \times 24}$$

CPR = 2.54 mpy

B. METAL FINISHING

2.15 INTRODUCTION

Surface modification of a metal is called metal finishing. It is the name given to all those processes employed to modify the surface properties of conducting or non-conducting materials by deposition of a layer of metal over them.

Many elements known to us are metals. A large number of them are in common use. Starting from gold and silver, which are used in ornaments, we use aluminium, copper and their alloys and stainless steel in household articles and steel in the manufacture of machines and machinery parts. But none of these metals possesses all properties such as (1) resistance to corrosion, (2) resistance to wear and tear, and (3) resistance to impact.

In order to satisfy the conditions of their applications, the surfaces of metals are suitably modified and any such surface modification is called metal finishing. For example, copper coating used in electronic and electrical components must have the highest possible electrical conductivity whereas, for copper coatings on the cylinders used in printing, surface hardness is more important.

2.16 DEFINITION

Metal finishing is defined as a surface phenomenon involving modification of surface properties of metals by the deposition of a layer of another metal or a polymer or by the formation of oxide film of the metal.

2.17 TECHNOLOGICAL IMPORTANCE

The technological importance of metal finishing is in imparting certain additional properties to the materials in addition to their intrinsic properties so that the utility of the material is increased. Following are some important technological applications of metal finishing:

1. To increase corrosion resistance.
2. To impart thermal and electrical conductivity.
3. To improve wear resistance.
4. Imparting hardness.
5. Imparting thermal resistance.
6. Electroforming of articles.
7. Electrochemical machining, electro-poilshing and electrochemical etching.
8. To increase the durability of the metal.
9. Manufacturing electrical and electronic components such as printed circuit board, capacitors, etc.
10. To produce reflectivity and appearance.
11. To increase the solderability.

Techniques of metal finishing:

1. Electroplating of metals and alloys.
2. Electroless plating of metals or alloys.
3. Electrophoretic plating.
4. Metal cladding.
5. Metal spraying.

Keeping in view of the syllabus the details of only first two techniques are discussed in this chapter.

2.18 ELECTROPLATING

The term electroplating refers to the deposition of a metal to the surface of another metal, alloy or any conductor, in general, by the process of electrolysis.

"Electroplating is a process in which a base metal is coated with a thin layer of a noble metal by electrolytic deposition".

The common coating metals and alloys are Zn, Cu, Ni, Cr, Ag, Pt, Av, etc.

Alloys: Cu – Zn, Cu – Sn, Pb – Sn, Sn – Ni, Ni – Co, Ni – Cr and Ni – Fe.

Process: The electroplating device is similar to that of an electrolytic cell, in which the base metal to be coated is made the cathode and coating metal the anode. If the anode is inert or insoluble, sufficient quantity of electrolytic solution of the soluble salt of the coating metal is added from outside in order to maintain constant metal ion concentration. The process of electroplating is made of four important parts.

Cathode is the negative electrode. During electrolysis, negative electrons are created and positive ions are discharged. It is the object, that is going to be plated.

Anode is the positive electrode. During electrolysis, negative ions are discharged and positive ions are created. It is of the same material as the plating metal.

Electrolyte is the conducting medium where the flow of current is with the movement of matter.

Direct current is the electricity that passes from anode to the cathode.

When a direct current is passed through an electrolyte solution, the following chemical changes take place in the cell.

Cations (metal ions) of the electrolytic solution move towards the cathode and get deposited over it. The process of deposition of metal on the cathodic material is called electroplating. Thus, concentration of cations decreases in the bulk of the solution

$$M^{n+} + ne^- \rightarrow M \quad \text{(at cathode)}$$

Anions move towards the anode, where they get oxidized. It can be inert.

$$M \rightarrow M^{n+} + ne^- \quad \text{(at anode)}$$

Thus, a thin layer of coating metal is obtained on the article made as the cathode shown in Fig. 2.19.

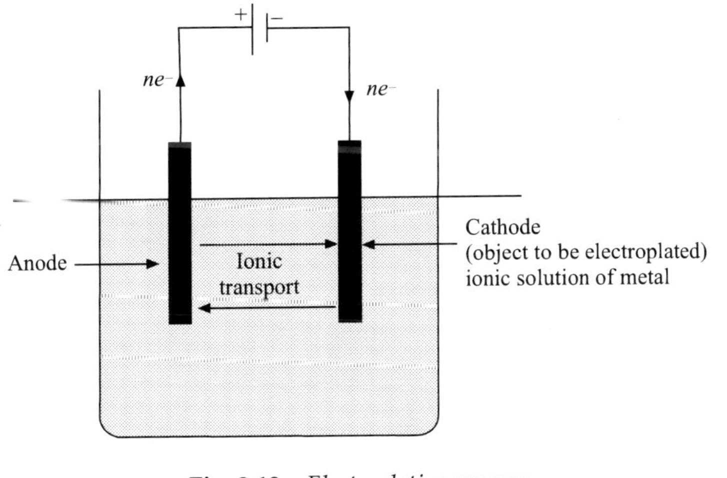

Fig. 2.19 *Electroplating process*

In order to get smooth, uniform, bright and lustrous deposits, favourable conditions are:

1. Low temperature
2. High current density
3. Low metal ion concentration.

Since the process of electroplating essentially involves electrochemical changes that occur in an electrolytic cell. The three important principles of electroplating are:

1. Decomposition potential
2. Overvoltage
3. Polarization

2.18.1 Decomposition Potential

In an electrolytic cell, the following processes occur during electroplating:

1. Cations (metal ions) are attracted towards cathode.
2. Anions are attracted towards anode.
3. Metal ions are reduced to metal atoms at the surface of cathode.
4. Anions are oxidized at the surface of anode.

Each of these steps has its own potential. If the externally applied potential is less than the potential of these processes, then electroplating of metal does not occur. So it can be defined as "The minimum external potential required to bring the electrolysis of an electrolyte is known as decomposition potential (E_D)".

Process: It consists of two platinum electrodes dipped in an electrolyte whose decomposition potential is to be measured and is connected to a source of variable voltage. The voltage is gradually increased and the current passing through the cell is measured using an ammeter because at low voltage ions do not move through the solution and there is slight flow of current through the cell. On applying sufficient voltage by moving the contact edge on the wire, ions freely move towards opposite electrodes. Now circuit is completed and electrode reactions occur smoothly. Hence, above this potential, current flowing through the cell increases proportionately to the applied voltage and electroplating takes place continuously. This potential above which electroplating occurs continuously is called decomposition potential.

A plot of current versus applied voltage is shown in Fig. 2.20. There is a slight increase in the current at low voltages. Beyond the decomposition voltage, electrolysis begins and thus there is an abrupt increase in the current. Intersection of the two straight lines gives decomposition potential (E_D).

The decomposition potential values enables:

1. calculation of emf required for continuous electrolysis; and
2. separation of metal ion mixture by electrolysis.

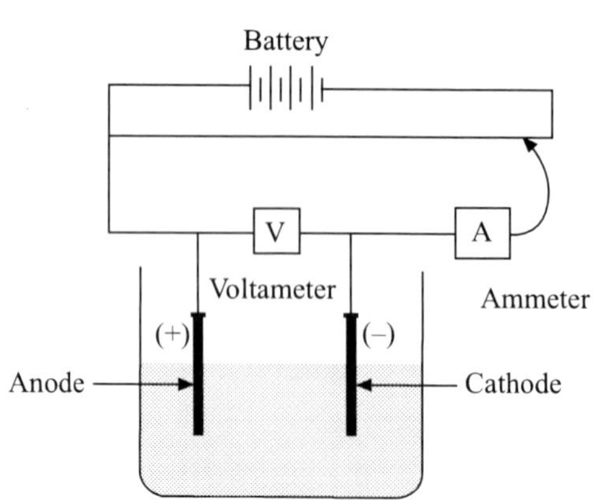

Fig. 2.20 *Measurement of decomposition potential*

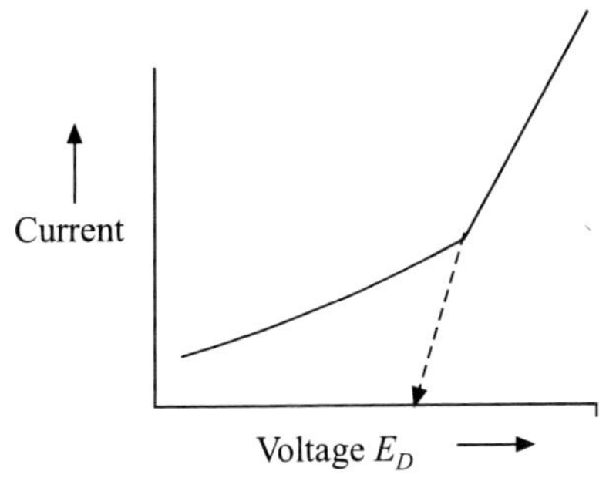

Fig. 2.21 *Plot of current flowing between electrodes and applied voltage*

Significance of decomposition potential:

1. It is possible to predict the order of the discharge of ions from a solution containing several ions.
   Example: From an electrolyte containing Cu^{2+}, Zn^{2+} and Cd^{2+} the order of deposition is copper, aluminium and then zinc. This is due to the fact that the $E°$ of Cu^{2+}, Cd^{2+} and Zn^{2+} are +0.34 V, –0.44 V, – 0.76 respectively.
2. Decomposition potential refers to the minimum voltage required for the electrolysis of an electrolyte.

2.18.2 Overvoltage or Overpotential

It plays a role in many technically important electrode processes.

For continuous electrolysis, the applied voltage should be at least slightly more than the decomposition potential (E_D). However, in a few cases, the applied voltage has to exceed the theoretical value for continuous electrolysis. This is known as overvoltage, which is represented by (η).

η = Experimental decomposition potential – Theoretical decomposition potential

The decomposition potential for such system is given by

$$E_D = E_{\text{cathode}} - E_{\text{anode}} + \eta$$

So, it may be defined as "The excess voltage that has to be applied above the theoretical decomposition potential for continuous electrolysis".

For example, the reversible electrode potential of oxygen gas with smooth platinum electrode under standard conditions is 1.23 V. But actual gas evolution takes place at a potential of 1.68 V. The excess 0.45 V is the overvoltage of oxygen on smooth platinum surface.

$$\begin{aligned}\text{Overvoltage of oxygen} &= \text{Actual discharge potential} \\ &\quad - \text{Theoretical electrode potential} \\ &= 1.68 - 1.23 \\ &= 0.45 \text{ V}\end{aligned}$$

Overvoltage varies from electrode to electrode and also gas to gas. The overvoltage of hydrogen on metals is important in electrolysis.

Table 2.3 Hydrogen overvoltage (current density = 10 A/m^2)

Electrode	η(v)
Platinum (Black)	0.01
Platinum (Smooth)	0.09
Nickel	0.33
Iron	0.40
Silver	0.44
Chromium	0.50
Lead	0.67
Mercury	4.04

Overvoltage is used in the electrolysis for deposition of metals. Higher electrode potential cations easily discharge at the electrode surface.

Factors affecting overvoltage:

Overvoltage or overpotential for a given electrolyte depends on several factors such as:

1. Nature and physical state of the electrodes.
2. Current density applied.
3. Nature of electrolyte.
4. Temperature.
5. Nature of the substance deposited.

Significance of Hydrogen Overvoltage

$$M^{n+}\ ne^- \longrightarrow M\ [\eta - 0]$$

Thus, metal deposition at the cathode takes place nearly at the same reversible M^{n+}/M electrode potential. In an aqueous solution there is a competition between H^+ ions and M^{n+} ions to get reduced. If the potential required for metal deposition is less than that for hydrogen, the metal gets deposited and in the reverse case, hydrogen gas is set free. Thus, according $E°$ values metal with a higher reduction potential than hydrogen, will be preferentially deposited before H_2 is set free. In the case of metals with lesser $E°$ than hydrogen we expect the liberation of H_2 in preference to the metals. But in many such cases, the metal gets deposited rather than H_2 due to the hydrogen overvoltage.

2.18.3 Polarization

It is an electrode phenomenon. During electrolysis, metal ions present at the cathode get reduced to metal atoms. Therefore, concentration of metal ions at the cathode decreases. For maintaining the same concentration, metal ions must be supplied from the bulk of the solution of the electrode. If metal ions are not supplied sufficiently, then the concentration of metal ions at the electrode surface decreases. This show decrease in the potential of the electrode.

The electrode potential is given by the Nernst's equation

$$E = e° + \frac{2.303RT}{nF} \log [M^{n+}]$$

where, $E^\circ \rightarrow$ Standard electrode potential
$[M^{n+}] \rightarrow$ Metal ion concentration.

This decrease in electrode potential due to inadequate supply of ions from the bulk of the solution to the surface of electrode is called polarization.

It occurs at anode also. Potential required to overcome polarization of anode and cathode is called **polarization potential**.

Factors affecting polarization

It depends on several factors such as:

1. Size, shape and the composition of the electrode
2. Conductivity of the solution
3. Temperature
4. Rate of stirring of the electrolyte solution
5. Products formed at the electrode
6. Large electrode surface, low concentrations and use of highly conducting solutions largely decrease the polarization effects.

2.19 FACTORS INFLUENCING THE NATURE OF THE ELECTRODEPOSIT

The nature of electrodeposit is affected by several factors which are called plating variables. These parameters need to be optimized and monitored throughout the electrode position process.

2.19.1 Current Density

It is defined as the current per unit area. It is expressed in milliamperes per square centimetre (mA/cm^2) or amperes per square metre (A/m^2) of the electrode surface.

At low current density: Diffusion of ions on the surface is fast as compared to the electron transfer. Hence, the metal is deposited slowly and takes longer for the deposition.

At high current density: Diffusion of ions on the surface slows down and the number of nuclei formed increases, the added atoms may not move towards favourable sites. This results in a less ordered macroscopic deposit.

At high current density (below limiting value): Mass transport of ions predominates resulting in a bad deposit which is loosely held to the surface.

At current densities above limiting value: Hydrogen evolution takes place resulting in depletion of H^+ ion concentration in the vicinity of the cathode. As a result metal hydroxide gets precipitated. When these get included in the electrodeposit, the deposit appears dark, burnt and spongy.

2.19.2 Temperature

Rate of electrodeposition increases with increase in temperature, because at high temperature solubility and dissociation of metal salt increases which, in turn, leads to higher conductivity of the solution. A high temperature increases the mobility of the metal ions and decreases the viscosity of the solution. But the disadvantages of maintaining the higher temperatures is evolution of H_2 gas, corrosion of equipment also increase and organic additives may be decomposed. At *lower temperature,* rate of electrodeposition is slow. Hence, plating is generally carried out at moderate temperature of 30–60°C.

2.19.3 Plating Bath Solution

Generally, it contains ionic solution of the metal to be coated on the material and other compounds like electrolytes, complexing agents and organic additives. So, it is important to maintain the concentration of the various species in the bath to obtain a good deposit.

1. **Metal ions:** The concentration of metal ion (either as a simple or complex ion) is normally 1–3 mol/dm^3. Higher concentration increases the mass transfer leading to poor deposit. Therefore, low metal ion concentration should be maintained.
2. **Complexing agent:** Metal ion is converted into a complex ion by the addition of suitable complexing agents like CN^-, OH^-, etc. This is done to obtain uniform and adherent deposits on the surface of the substrates

$$Cu^{2+} + 4CN^- \longrightarrow [Cu\,(CN)_4]^{2-}$$

Complexing agents are added for the following purposes:

(i) To improve the throwing power of the plating bath. **Example:** cyanides, hydroxide and sulphamates are complexing agents. Copper, silver, gold, zinc and cadmum are plated as smooth deposits in the presence of cyanide ions.

(ii) To increase the solubility of sparingly soluble salts. **Example:** $PbCl_2$ is less soluble in water. But addition of Cl^- ions converts its ions to more soluble complex ions, $[Pb\,Cl_4]^{2-}$.

(iii) To prevent direct reaction between the cathode metal and the plating ions. **Example:** When copper is coated on the surface of iron taken as cathode, copper ions react with iron instead of getting coated on it

$$Cu^{2+} + Fe \longrightarrow Cu + Fe^{2+}$$

To prevent this direct reaction, Cu^{2+} ion is converted into $[Cu(CN)_4)^{2-}$ complex ion.

(iv) To maintain low metal ion concentration in solution. (Higher concentration of metal ions leads to burnt deposit).

(v) To prevent the passivation of anodes so that they dissolve easily and thus increases the current efficiency. **Example:** Chloride ions in low concentration are used in many baths.

(vi) To make the potential of the plating ion more negative so that plating takes place at a lower potential.

3. **Additional Electrolytes:** These are the compounds which increase the conductivity of plating bath without taking part in electrode reaction. They do not participate in the electrode reactions but increase the conductivity of the plating and control the change in pH.

4. **Organic Additives:** To modify the structure and property of deposit a wide range of organic compounds are added in a small quantity to the plating bath. These are classified into several groups based on their mode of action.
   (a) **Brighteners:** Brighteners are the chemical substances that help in the formation of bright, lustrous and smooth and light reflecting deposit. The reflection takes place only when grain size of the deposit is lower than the wavelength of the incident light. Brighteners are used to produce microscopically fine deposit parallel to substrate surface.
   Example: Aromatic sulphones or sulphonates and compounds containing $C=N, N=C=S$ or $>C=O$ groups.
   (b) **Levellers:** Plating takes place rapidly on certain regions such as on dislocations (cracks, recesses, etc.) where the diffusion of ions is faster producing uneven thickness. This leads to uneven deposits. When levellers are used they get adsorbed at such regions and prevent electron transfer.
   Example: Sodium allyl sulphonate is used in Ni electrode position.
   (c) **Structure Modifiers (Stress Relievers):** Electroplating process involves internal stress. If the stress level is high, cracking takes place. Structural modifiers help in modifying the structure of the deposit and make it stress-free.
   Example: Saccharin.
   (d) **Wetting Agents:** During electroplating hydrogen gas is evolved at the cathode. These agents are added to release the gas bubbles from the surface. They also improve the levelling and uniformity of the deposit and reduce its brittleness. In the absence of wetting agents, the gas bubbles get entrapped within the plated metal and try to escape after the plating. This makes the substrate metal itself brittle when subsequently heat treated. This is called hydrogen embrittlement.
   Example: Sodium lauryl sulphate.

2.19.4 pH of Plating Bath

For a good electrodeposit, the pH of the bath must be properly maintained by using buffer.

At low pH, evolution of hydrogen gas is more which leads to burnt deposit.

$$2H^+ + 2e^- \longrightarrow H_2$$

At higher pH, OH^- ions form insoluble hydroxides

$$M^{2+} + 2OH^- \longrightarrow M(OH)_2$$

Therefore, in most of the platings, optimum pH value of 4–8 is used.

Example: Borate buffer is used in Ni plating

2.19.5 Throwing Power

The ability of a plating bath to produce a uniform and even deposit on the entire surface of the object is called throwing power. The property is particularly important when the article is of an irregular shape.

If the distribution of the deposit is uniform throughout the surface of the articles to be coated, the throwing power is said to be good. Higher the throwing power, more uniform will be the electrodeposit.

The throwing power of an electroplating bath is decided by the following factors:

(a) **Conductivity of the electrolyte:** In an electrolyte of high conductivity, the current distribution on the cathode surface is uniform and hence the metal is also deposited uniformly. Thus, the solutions with higher conductivity will have higher throwing power.

(b) **Presence of addition agents:** Throwing power is usually enhanced by the presence of additives like complexing agents and levellers.

(c) **Competing electrode reaction:** Occurrence of competing reactions or side reactions at the cathode like hydrogen evolution increases the throwing power.

Throwing power of plating bath can be determined by using the Haring-Blum cell.

Haring-Blum Process

It is an electrolyte cell containing plating bath solution whose throwing power is to be determined. Two cathodes are placed at two different distances from the anode. Cathode 1 is placed at a distance of a_1 from anode and cathode 2 is placed at a distance of a_2 from the anode in such a way that $a_1 > a_2$. All three electrodes are completely immersed in plating bath whose throwing power is to be determined.

The cathodes are weighed initially then electroplating is carried out for a certain period of time. If the weight of metal deposited over cathodes 1 and 2 is w_1 and w_2 respectively; w_1 is for a_1 and w_2 is for a_2. Then the percentage throwing power of plating bath can be calculated from the equation

$$\% \text{ throwing power of bath solution} = \frac{100 \times (x - y)}{(x + y - 2)}$$

where $x = a_1/a_2$ and $y = w_2/w_1$

If the weight of the metal deposited on both the cathodes is same (i.e., $w_1 = w_2$) then percentage throwing power of plating bath is 100%. Quality of such plating bath is said to be good. When percentage throwing power of plating bath is less than 100% then quality of such plating is said to be poor.

Throwing power of plating bath can be increased by:

1. Reducing side reaction like evolution of hydrogen gas
2. Adding complexing and levelling agents to bath
3. Increasing the conductivity of the plating bath
4. Maintaining optimum distance between electrodes.

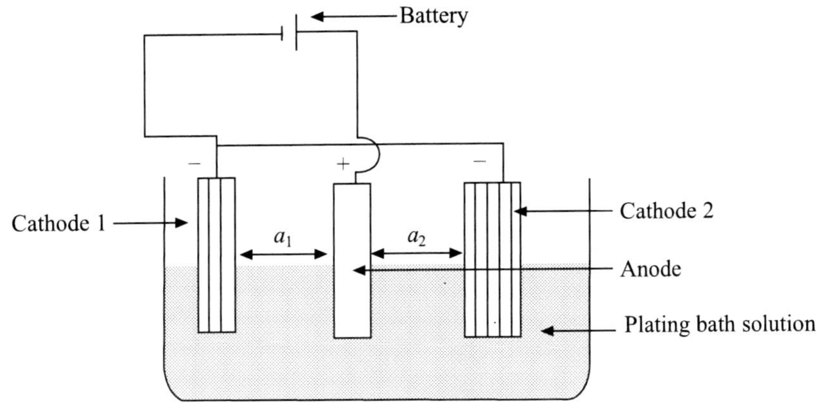

Fig. 2.22 *The Haring-Blum cell*

2.20 ELECTROPLATING OF CHROMIUM

Chromium is one of the widely used metals for electroplating. It is carried out as (i) decorative coating and (ii) hard plating. The main aspects of the two processes are the same except that for hard chromium plating higher current density is applied. The plating bath composition and conditions are given below:

Plating Bath Condition	For Decorative Plating	For Hard Chrome Plating
Plating bath	250 g of chromic acid (CrO_3 and H_2O) 2.5 g of conc. H_2SO_4 per/dm^3 (100 : 1 proportion) + 1 g of trivalent chromium which produce in situ by adding 4 g of oxalic acid	250 g of chromic acid (CrO_3 and H_2O) + 2.5 g of conc. H_2SO_4 per/dm^3 + 1 g of trivalent chromium per litre of the bath
Current density	145–430 A/ft^2	290-580 A/ft^2
Current efficiency	10–15%	17–21%
Operating temperature	45–55°C	45–55°C
Anode	Pb-Sb alloy or Pb-Sb alloy coated with thin layer of PbO_2	Pb-Sn alloy or Pb-Sb alloy coated with thin layer of PbO_2
Cathode	Substrate (object to be plated)	Substrate (object to be plated)

In chromium plating sulphate ion provided by the sulphuric acid is believed to act as a catalyst. Cr is present in the hexavalent state Cr^{6+} as CrO_3 in the bath solution. This is converted into trivalent Cr^{3+} by a complex anodic reaction in the presence of sulphate ions and then coated as Cr on the cathode surface. The amount of Cr^{3+} should be restricted in order to get proper deposit. Chromium anodes are not used in Cr plating because chromium anodes are not used in Cr plating because chromium metal passivates in strong acid medium and this increases deposits of chromium. Therefore, insoluble anodes are used to maintain the Cr^{3+} concentrations as they oxidize Cr^{3+} to Cr^{6+}.

$$\underset{\text{Chronic trioxide}}{CrO_3} + H_2O \longrightarrow \underset{\text{Chromic acid}}{H_2CrO_4}$$

$$2H_2CrO_4 \longrightarrow H_2Cr_2O_7 + H_2O$$

$$Cr_2O_7^{2-} + 14H^+ + 6e^- \xrightarrow[\text{(catalyst)}]{SO_4^{2-}} 2Cr^{3+} + 7H_2O$$

$$2Cr^{3+} + 6e^- \longrightarrow 2Cr$$

$$2Cr^{3+} + 3O_2 \xrightarrow[\text{(catalyst)}]{PbO_4^{2-}} 2CrO_3 + 6e^-$$

Applications

1. Chromium coatings are widely used in automobile, household fittings and surgical instruments.
2. Black chromium coatings are used on optical instruments, machine tools, electronic parts, solar energy collectors and non-glare finishes on automobiles.
3. Hard chromium coatings are used in printing industry, cutting tools, piston rings, hydraulic rams, cylinder linings and bearings, etc.

2.21 ELECTROPLATING OF GOLD

It was initially applied for decorative purpose on jewellery articles. But later it found applications for various purposes. Depending on the applications, plating is carried out in different pH ranges. It is divided into three types based on pH range.

1. **Acid cyanide bath:** pH of this bath is in the range of 3.8 – 4.3 which is maintained using citric acid. This bath is used for gold plating on printed circuit boards and connectors.

2. **Neutral cyanide bath:** pH of this bath is in the range of 6.0 – 8.0 which is maintained using monopotassium phosphate and potassium citrates. This bath is used for gold plating on semiconductors.

3. **Alkaline cyanide bath:** pH of this bath is in the range of 10 – 13 which is maintained using dipotassium phosphate and potassium cyanide. This bath is also used for gold plating on semiconductors.

Keeping in view of the syllabus the detail of only acid cyanide bath are discussed in this topic.

Acid cyanide Bath (pH 3.8 – 4.3) for Electroplating of Gold.

Plating Bath: Potassium gold cyanide and citric acid

Current density: 100 – 400 ASF (ampere per square feet)

Operating temperature: 40 – 60°C.

Anode: Platinised titanium, gold, niobium, graphite

Cathode: Object to be coated with gold

$$[Au(CN)_4]^- + 2e^- \rightarrow [Au(CN)_2]^- + 2CN^-$$

$$[Au\ (CN)_2]^- + e^- \rightarrow Au + 2CN^-$$

Applications:

1. It is used for decorative purposes like jewellery, watch cases, pen points, etc.
2. It is used in electrical industry like PCB, contacts/connectors.
3. It is used in electronic industry like transistors, integrated circuit parts, etc.

2.22 ELECTROLESS PLATING

It is also called autocatalytic plating. It is the deposition of a metal without the use of electrical energy. So, it is defined as: "Electroless plating is a technique in which metal is coated over a substrate (conductor or non-conductor) by controlled chemical reaction of metal ions by a strong reducing agent without using electrical energy".

The reducing agent converts the metal ions (M^{n+}) to the metal (M) which gets plated over a catalytic surface.

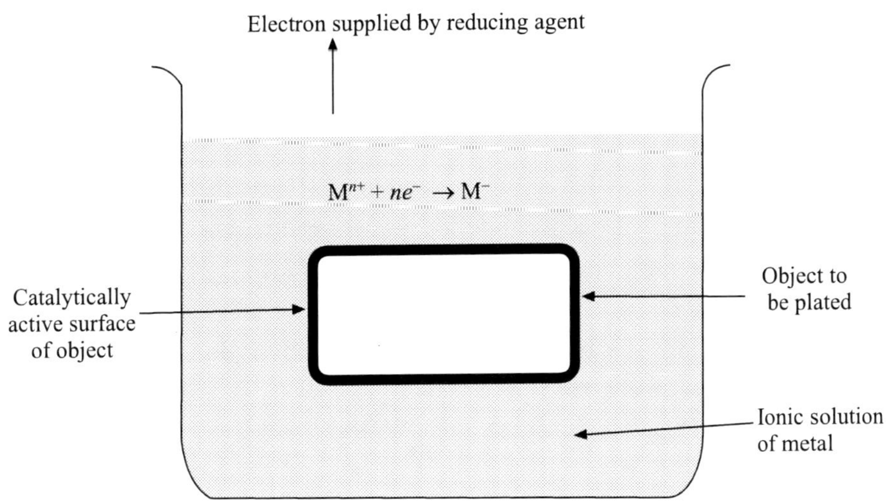

Fig. 2.23 *Electroless plating*

$$M^{n+} + \text{Reducing agent catalytic} \xrightarrow[\text{surface}]{\text{Catalytic}} \underset{\text{oxidized product}}{M^+}$$

It is based on the catalytic reduction of metal ions on the surface of the substrate be plated. The driving force is autocatalytic redox reaction on pretreated catalytic redox reaction on pretreated catalytic surface. It commonly generates more waste than other plating techniques, but it varies significantly in efficiency. Even irregular shapes can also be plated uniformly. Electroless plating is also used to deposit a conductive surface on a non-conductive object to improve its electroplating. **Example:** It is widely used for machine frames, base plates, fixtures, etc., machine parts where metal-to-metal wear applications are required which cannot be achieved by using conventional oil and greases.

Pretreatment of Article before Electroless Plating

Before electroless plating the surface of the substrate should be catalytically activated as follows:

1. Ni, Co, steel, Fe, Rh, Pd and Al can be plated directly without any activation. In these cases electroless plating is autocatalytic.
2. In the case of non-catalytic metals such as Cu, Brass, Ag, etc., need activation. This can be done by rendering them cathodic for a short period by passing direct current or by dipping in $Pd\ Cl_{2-}$ HCl solution.
   Example: Surface of stainless steel is activated by dipping in 1 : 1 H_2SO_4. Mg and its alloys are activated by coating thin layer of Cu and Zn.

In the case of non-conducting materials like glass, plastics, ceramics, printed circuit board (PCB), pretreatment is done as given below before subjecting to electroless plating:

1. Etching and conditioning of the surface by dipping in dilute sluphuric acid.
2. Activation of the surface by dipping in $SnCl_2$ containing HCl at 25 °C followed by $PdCl_2$. $SnCl_2$. $SnCl_2$ reduces palladium ions to deposit small amount of palladium metal on the surface of the material which catalyzes the redox reaction.

Composition of Electroless Plating Baths

An electroless plating bath consists of the following components:

1. Soluble electroactive metal salts usually chloride or sulphate to provide metal ions for deposition.
2. Reducing agent (**Example:** hypophosphites, formaldehyde, sodium borohydride, hydrazine) for reduction of metal ions to metal atoms.
3. Complexing agent (**Example:** citrate, tartrate, succinate) to complex metal ions to prevent bulk deposition.
4. Stabilizers (**Example:** lead, calcium, thallium ions, thiourea) to give more stability to the solution.
5. Accelerator to increase the rate of plating.

6. Buffer solution to control pH.
7. Brightness to get lustrous surface.

Advantages of Electroless Plating

1. Electrical power is not required.
2. It is applicable to conductors, semiconductors and non-conductors like plastics.
3. This method needs no levellers.
4. Electroless deposit is less porous thereby possesses better chemical, mechanical and magnetic properties.
5. Electroless coatings are pore-free, harder with better corrosion resistance.
6. Electroless plating baths have better throwing power and deposit a more uniform metal coating over an article irrespective of its shape or size.

2.23 COMPARISON BETWEEN ELECTROPLATING OR ELECTROLESS PLATING

Property	Electroplating	Electroless plating
Driving force	Electric current	Autocatalytic redox reaction
Anode	Separate anode	Catalytic surface of the substrate acts as an anode
Cathode	Object to be coated after removing surface impurities	Object to be coated after making its surface catalytically active
Anodic reaction	$M \rightarrow M^{n+} + ne^-$	Reducing agent
Cathodic reaction	$M^{n+} + ne^- \rightarrow m$	Oxidised product $M^{n+} + ne^- \rightarrow m$
Throwing power	Low	High irrespective of its shape or size
Nature of deposit	Pure metal or a definite alloy	Metal with reducing agent and oxidizing agent products
Reducing agent	Electrons	Chemical reagents
Applicability	Only to conductors	Conductors and non-conductors

2.24 ELECTROLESS PLATING OF COPPER

Copper from its solution gets deposited spontaneously in the presence of suitable reducing agents on metals such as gold, silver, palladium, iron, cobalt and nickel. Insulators like plastics and glass.

There are two kinds of copper plating in the printed circuit boards.

1. Electroless copper plating
2. Acid copper plating

1. Electroless copper plating: In electroless plating, the mechanism is similar but the source of electrons is a chemical reducing agent. Hence, non-conductive surfaces may be metallized by employing electroless baths since baths contain their own source of electrons.

2. Acid copper electroplating: Acid copper electroplating plates copper only on a conductive surface via electrochemistry method whereby electrons are obtained from a DC power source to reduce metal ions to the metallic state.

Pretreatment and Surface Activation

The plastic board is degreased and etched in acid. It is activated by dipping in $SnCl_2/HCl$ at 25°C followed by dipping in Pd Cl_2.

Plating Bath Solution

Metal ion solution: Copper sulphate (12 g/L)

Reducing agent: For maldehyde (8 g/L)

Complexing agent and exaltant: EDTA (20 g/L)

Buffer to maintain pH: Sodium hydroxide (15 g/L) and Rochelle salt (14 g/L)

Temperature: 25°C

pH: 11

Reactions:

At cathode: $Cu^{2+} + 2e^- \longrightarrow Cu$

At Anode: $2HCHO + 4OH^- \longrightarrow 2HCOO^- + 2\,H_2O + H_2 + 2e^-$

Net reaction:

$$Cu^{2+} + 2HCHO + 4OH^- \xrightarrow[\text{of article}]{\text{Catalytic surface}} Cu + 2HCOO^- + 2H_2O + H_2$$

Plating process: Main application of electroless plating of Cu is in the manufacturing of double sided and multilayer printed circuit boards (PCB) where plating through holes is required. Various steps involved in it are shown in Fig. 2.30.

Plating is carried out at room temperature by dipping the activated plastic board in the bath solution maintained at about 25°C. Plating is carried out at a rate of 1 – 5 μm/n till 5 – 100 μm thickness is obtained.

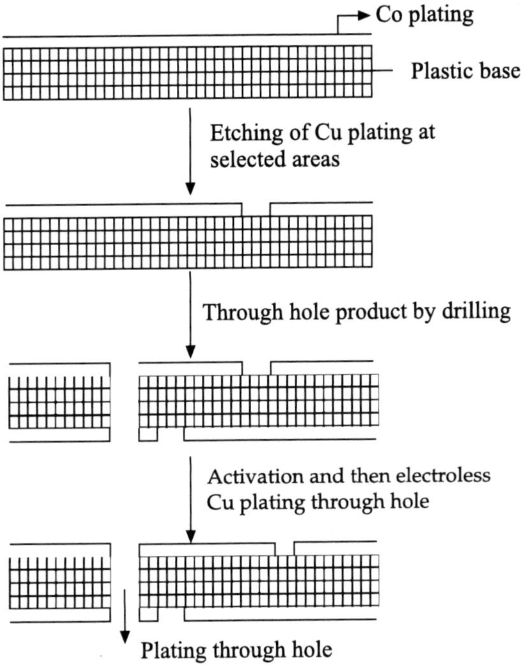

Fig. 2.24 *Schematic representation of electroless copper plating through hole in the manufacture of double sided PCB*

Selected areas are protected by an electroplated image and the rest of the Cu is etched to produce the circuit pattern. Double-sided copper clad laminates with tracks are preferred because of feasibility of packing of more components in a smaller space. The connection between both sides of track is made by drilling hole followed by plating through holes by electroless plating.

Applications

Typical application of electroless copperplating is printed circuit boards (PCB) particularly double-sided and multilayered boards in which plating through holes is required. Such holes, however, cannot be copper electroplated.

Review Questions

1. What is metallic corrosion? Explain the electrochemical theory of corrosion.
2. Discuss the effect of the following factors on corrosion rate.
   (a) Nature oxide film
   (b) Anodic to cathodic area
   (c) Polarization

3. Explain the following corrosion control methods.
4. Explain the corrosion control technique by cathodic protection.
5. Explain galvanization process.
6. Explain the type of corrosion occurring in the following cases.
   (a) Presence of small dust particles on iron surface for a long time.
   (b) Copper nut is contact with iron bolt.
7. What is anodizing ? Explain anodizing of aluminium.
8. What are corrosion inhibitors ? Explain how corrosion is controlled by inhibitors.
9. Explain the following types of corrosion:
   (a) Galvanic corrosion
   (b) Stress corrosion
10. What are metallic coatings ? Explain the galvanizing process.
11. Discuss the sacrificial anode and impressed current methods of corrosion control.
12. Explain the following types of corrosion:
    (a) Differential metal
    (b) Waterline
    (c) Stress
13. Describe the following processes:
    (a) Galvanizing
    (b) Tinning
14. Discuss (a) Stress corrosion and (b) Water line corrosion.
15. Explain the effect of the following factors on the rate of corrosion.
    (a) Nature of corrosion product
    (b) Hydrogen overvoltage
16. Explain what type of corrosion occurs when
    (a) Steel pipe is connected to copper plumbing
    (b) Presence of small dust particles on iron surface for a long time
17. Define metallic corrosion? Discuss the following factors which affect the rate of corrosion.
    (a) Relative areas of anode and cathode
    (b) pH of the medium
18. What is cathodic protection? Explain cathodic protection by impressed current technique.
19. Define corrosion inhibition ? Explain anodic and cathodic inhibition, anodic inhibitors with suitable examples.

20. What is scarified anode protection and how it is carried out?
21. Write short notes on the following:
    (a) Differential aeration corrosion
    (b) Stress corrosion
    (c) Dry corrosion
22. What is differential corrosion? Explain with an example.
23. Write a note on pitting corrosion? Give an example.
24. Distinguish between anodic and cathodic coatings.
25. Write in detail about galvanic corrosion and differential aeration corrosion.
26. What is cathodic protection? Explain the sacrificial anode and impressed current technique prevention of corrosion.
27. How do you control corrosion by proper designing and selection of materials?
28. Define corrosion and explain the various factors influencing corrosion of metal.
29. How do you control corrosion by proper designing and selection of materials?
30. Distinguish between dry corrosion and wet corrosion.
31. What is meant by metal finishing? Give a brief account of electroplating.
32. Define the term metal finishing? What are its technological advantages?
33. What is metal finishing? Enumerate five objectives of metal finishing.
34. What is decomposition potential? How is it determined; what is its significance?
35. Define polarization. Mention the factors that affect polarization.
36. What is meant by decomposition potential?
37. Explain the term overvoltage? What are the factors affecting overvoltage?
38. Explain how the following factors affect the nature of electrodeposit.
    (a) pH (b) Organic additives
    (c) Throwing power (d) Current density
39. What are the objectives of electroplating?
40. Explain why chromium anodes are not used during chromium plating.
41. What is meant by electroless plating? Briefly explain the electroless plating of copper on PCB.
42. Give a brief account of the electroless plating of copper.
43. (a) What is decomposition potential? How is it determined? What is its significance in electroplating?
    (b) Explain the principle involved in electroless plating. What are the advantages? Explain the electroless plating of copper?
    (c) Explain the role of the following factors on the nature of the electrodeposit
        (i) Current density (ii) Throwing power (iii) pH

44. (a) Explain four factors influencing the nature of an electrodeposit
    (b) Define polarization, decomposition potential and overvoltage. Mention their applications with reference to electrode composition.
    (c) Explain the electroless plating of nickel
45. (a) What is electroplating? Give the technological importance of metal finishing.
    (b) Explain the following factors influencing the nature of deposit
        (i) Complexing agents (ii) Brightners
        (iii) Levellers (iv) Wetting agents.
    (c) Discuss the electroless plating copper on PCB.
    (d) Write a note on overvoltage.
46. (a) Define polarization, decomposition potential and overvoltage. Mention their significance with reference to electrode position.
    (b) How do the following affect the nature of electroplating
        (i) Current density (ii) Temperature
        (iii) pH (iv) Organic additives
    (c) What is electroless plating? Mention any two advantages.
47. (a) What is electroless plating? Write two advantages of electroless plating of nickel.
    (b) Write briefly the process of electroplating of chromium.
    (c) Write notes on:
        (i) organic additives used in plating
        (ii) Throwing power of plating bath.
48. What are the advantages of electroless plating over electroplating?
49. Explain the differences between electroplating and electroless plating.
50. Discuss electroless plating of copper and its application to printed circuit boards.
51. Write an account on any three organic additives that can be used to improve electrodeposition.
52. Discuss electroplating of gold and mention its applications.
53. What is the significance of throwing power in electroplating.
54. What is electroless plating? Explain the principles involved. Mention its advantages.

55. Explain the following terms and their significance in electrolysis:
    (a) Polarization (b) Decomposition potential
56. What is decomposition potential? How is it determined experimentally?
57. Explain the process of electroplating of chromium for engineering applications. Indicate the reasons for not employing chromium as anode.
58. Explain the process of electroplating of chromium for decorative chromium.
59. Mention the differences between electroplating and electroless plating.
60. Discuss the electroless plating of copper on PCB.

3

Chemical Fuel and Solar Energy

A. CHEMICAL FUEL

3.1 INTRODUCTION

A chemical fuel is a combustible substance containing carbon as the main constituent which on proper burning gives large amount of heat, that can be used economically for domestic and industrial purposes. Wood, charcoal, coal, kerosene, petrol, diesel, etc., are some of the fuels. The principal element invariably present in chemical fuels are carbon and hydrogen.

During the process of combustion of a fuel (like coal), the atoms of carbon, hydrogen, etc., combine with oxygen with simultaneous liberation of heat at a rapid rate. This energy is liberated due to rearrangement of valence electron in these atoms, resulting in the formation of new compounds (like CO_2, H_2O, etc.). These new compounds have less energy in them and therefore, the energy released during the combustion process is the difference in the energy of the reactants (C, H and O, etc., of fuel) and that of the products formed.

$$\underset{\text{More heat energy content}}{\text{Fuel}} + O_2 \longrightarrow \underset{\text{Less heat energy content}}{\text{Products}} + \text{Heat}$$

The primary or main sources of fuels are coal and petroleum oils. These are stored fuels available in the earth's crust and are generally called fossil fuels.

3.2 CLASSIFICATION

Fuels may broadly be classified in two ways.

1. On the basis of their origin. Fuels are classified as primary and secondary fuels.
   Primary Fuels occur naturally such as coal, crude petroleum and natural gas.
   Secondary Fuels are derived from primary fuels. **Example:** Coke, gasoline, coal gas, etc.
2. On the basis of their state of aggregation or physical state like:

   (i) Solid fuel
   (ii) Liquid fuel
   (iii) Gaseous fuel

The classification of fuels is summarized as follows:

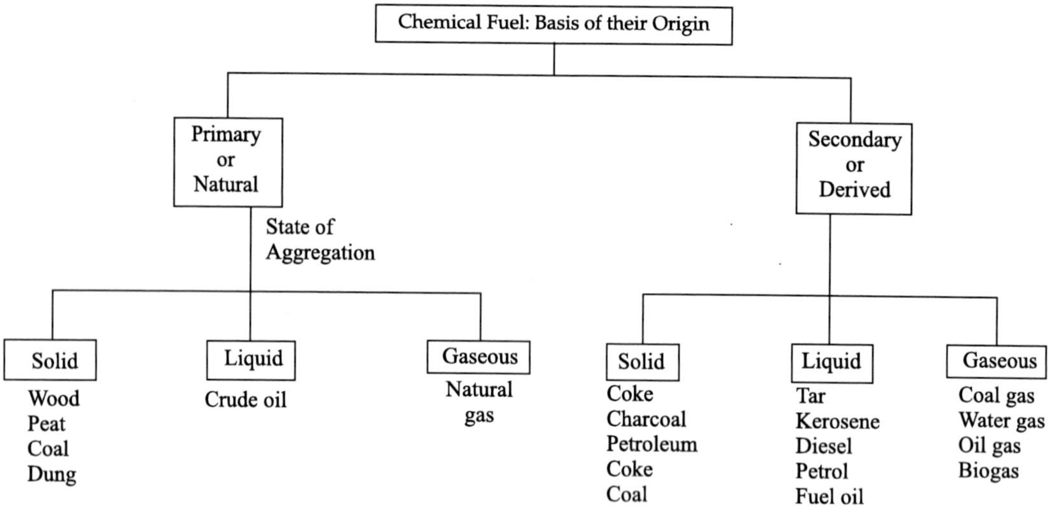

3.3 CHARACTERISTICS OF A GOOD FUEL

1. High Calorific Value: A good fuel should have high calorific value, i.e., it should produce large amount of heat on burning. Higher the calorific value, more efficient is the fuel.

2. Moderate Ignition Temperature: Ignition temperature is the minimum temperature to which a fuel is preheated so that it starts burning smoothly. If ignition temperature is low, can cause fire hazards during their handling, applications, storage and transportation. High temperature causes difficulty in kindling. So, a good fuel should have moderate ignition temperature.

3. Low Ash Content: A good fuel must therefore, be free from ash content. Formation of ash during combustion is due to the presence of inorganic matter in the fuel. High ash content, apart from lowering the calorific value, poses problems in its disposal.

4. Low Moisture Content: The presence of high percentage of moisture in the fuel decreases its efficiency. It also increases the ignition temperature and fuel cost. So, a good fuel should have low moisture content.

5. No Harmful Combustion Product: A good fuel should not produce harmful combustion products such as CO, SO_2, NO, H_2S smoke and clinkers during combustion. Formation of such by-products can cause harmful effects on health. A good fuel must therefore burn with clean flame without producing undesired by-products.

6. Moderate Velocity of Combustion: If the rate of combustion is low, then the required high temperature is not possible, because a part of the heat liberated gets radiated, instead of raising the temperature. On the other hand, too high combustion rates are also not required.

7. Low Cost: A good fuel should be cheap and readily available.

8. Easy to Transport: The fuel must be easy to handle, store and transport. **Example:** Solid and liquid fuels can easily be transported from one place to another. On the other hand, transportation of gaseous fuels is costly and can even cause fire hazards.

9. Combustion should be easily Controllable: Combustion of fuel should be easy to start or stop, when required.

10. Storage cost in bulk should be low.

11. Should burn in air with efficiency, without much smoke.

12. In case of solid fuel, the size should be uniform, so that combustion is regular.

3.4 CALORIFIC VALUE

The calorific value of a fuel is measured in terms of heating efficiency or the performance during its ignition in the presence of oxygen. The performance of a fuel is expressed in terms of its calorific value. So, it may be defined as "The amount of heat released when a unit quantity (mass or volume) of a fuel is burnt completely in air or oxygen".

The unit of calorific value of a solid or liquid is cal g^{-1}, or kcal kg^{-1} or Jkg^{-1} or kJ kg^{-1}. When describing a gaseous fuel, it is customary to refer to the temperature and pressure of the gas along with the heat quantity per unit volume, since volume of a gas varies with temperature and pressure. Hence, the unit of calorific value of a gaseous fuel is J cm^{-3} or kJ m^{-3} at a given temperature and pressure.

Generally, calorific value of fuels is expressed in two ways.

1. Gross (higher) calorific value, represented as GCV or HCV
2. Net (lower) calorific value, represented as NCV or LCV

3.4.1 Gross/Higher Calorific Value (GCV)

Gross calorific value is defined as "*The amount of heat released when a unit quantity (mass or volume) of a fuel is burnt completely in air and the products of combustion are cooled to room temperature*".

Explanation: Most of the fuels used are hydrocarbon fuels. During combustion carbon gets oxidized to CO_2 and hydrogen is converted into steam. On cooling the combustion products steam gets condensed to water and liberates its latent heat. The measured gross calorific value includes the latent heat of steam. Therefore, it is always higher than the net calorific value.

3.4.2 Net/Lower Calorific Value (NCV)

Net calorific value is defined as "*The amount of heat released when a unit quantity (mass or volume) of a fuel is burnt completely in air and the products of combustion are let off into the atmosphere*".

Explanation: In actual practice, the combustion products are not cooled to room temperature, but simply let off into the atmosphere. Since this calorific value does not include the latent heat of steam, net calorific value is always lower than the gross calorific value.

$$\text{Net calorific value} = \text{Gross calorific value} - \text{Latent heat of steam}$$

$$\text{NCV} = \text{GCV} - \frac{9 \times \%\text{H}}{100} \times 587$$

Because 1 g hydrogen gives 9 g of water and the latent heat of steam is 587 cal. g^{-1}.

3.5 DETERMINATION OF CALORIFIC VALUE USING BOMB CALORIMETER

Calorific value of a solid or liquid fuel is determined by using a bomb calorimeter (Fig. 3.1).

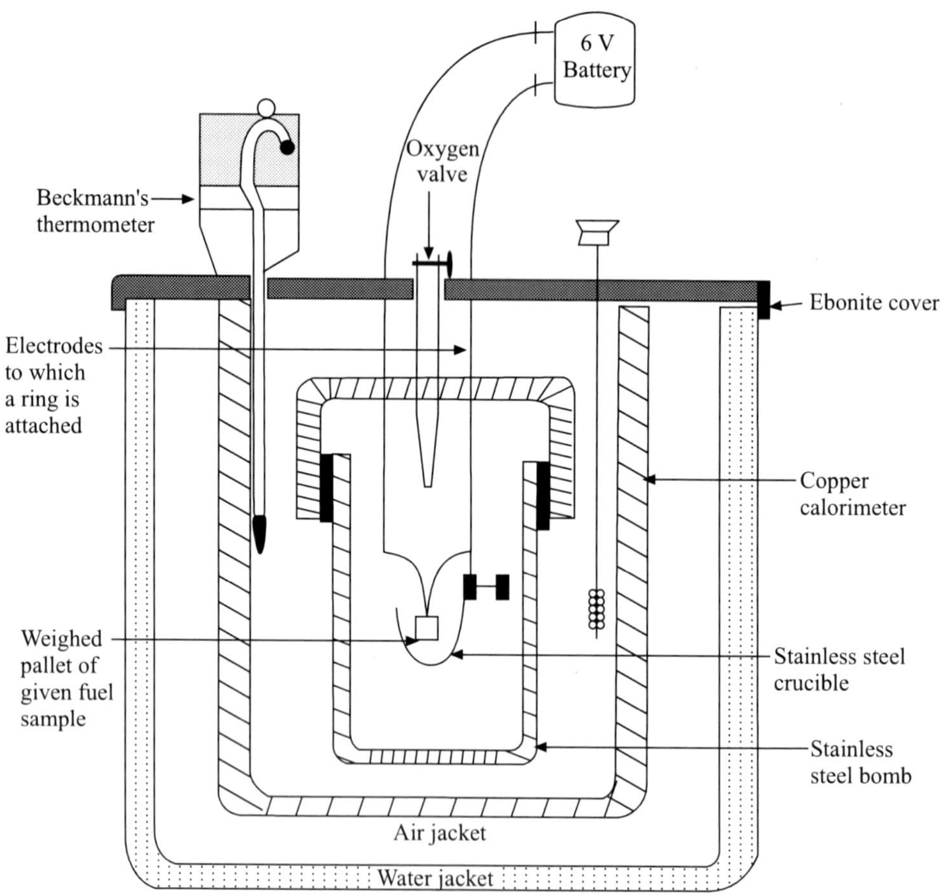

Fig. 3.1 *Bomb calorimeter*

3.5.1 Principle

A known weight of the sample fuel (solid or liquid) is burnt completely in excess of oxygen. The liberated heat is absorbed by the surrounding water and the calorimeter. Thus, the heat liberated during the combustion of fuel is equal to the heat absorbed by water and copper calorimeter. The higher calorific value of the fuel is calculated from the data.

3.5.2 Construction

It consists of a stainless steel airtight sealed cylindrical bomb. The bomb has an inlet valve for providing oxygen atmosphere inside the bomb and an electrical ignition coil for the initiation of combustion of fuel. The bomb is placed in a large, well insulated copper calorimeter. The copper calorimeter is equipped with a mechanical stirrer for dissipation of heat and a thermometer to read accurately the temperature rise.

3.5.3 Working

A known weight of fuel (solid or liquid) is placed in a small stainless steel crucible. The crucible containing the fuel is next placed inside the bomb. The bomb is sealed with airtight lid. The sealed bomb containing the sample is next placed in a large, well insulated copper calorimeter. It is covered by accurately measured quantity water. The water is kept in constant agitation by the mechanical stirrer. The initial temperature of the water is carefully measured. The bomb is filled with oxygen and the combustion of fuel is initiated by passing electric current through the ignition coil.

As the sample burns in the bomb, heat is liberated and is absorbed by the surrounding water and the copper calorimeter. The temperature of water gradually increases and attains the maximum value. The maximum temperature (final) is again carefully noted.

Water equivalent of the calorimeter is determined by burning a fuel of known calorific value.

3.5.4 Observation and Calculation

$$\text{Weight of the fuel} = m\,\text{g}$$

$$\text{Weight of water taken in the calorimeter} = W_1\ \text{g}$$

$$\text{Water equivalent of calorimeter} = W_2\,\text{g}$$

$$\text{Initial temperature of water} = t_1\ ^\circ\text{C}$$

$$\text{Final (maximum) temperature of water} = t_2\ ^\circ\text{C}$$

$$\begin{aligned}\text{Heat lost by } m \text{ g of fuel} &= \text{Heat gained by water} + \text{Heat gained by the calorimeter}\\ &= (W_1 + W_2)\,(t_2 - t_1)\ \text{cal}\end{aligned}$$

$$\text{Higher calorific value of the fuel} = \frac{(W_1 + W_2)\,(t_2 - t_1)}{m}\,\text{cal/g} \times \frac{4.184}{10^{-3}}\,\text{J kg}^{-1}$$

3.5.5 Calculation of Lower (Net) Calorific Value

Let the fuel contain H percentage of hydrogen.

$$2H + \tfrac{1}{2}\,O_2 \longrightarrow H_2O$$

2 g of hydrogen produces 18 g of water.

Water formed by 1 g of fuel = 18/2 × H/100 = 0.09 H g

Latent heat of steam = 587 cal/g

Latent heat of water formed = 0.09 H × 587 cal/g

Lower (net) calorific value = HCV – Latent heat of water formed

LCV = HCV – 0.09 H × 587 cal/g

WORKED PROBLEMS

Problem 3.1

Calculate the calorific value of a fuel sample of coal with the given data: Mass of coal 0.6 g. Water equivalent of calorimeter 2200 g. Specific heat of water 4.187 kJ kg^{-1} C^{-1}. Increase in temperature = 6.52 °C.

Solution: Here the percentage of hydrogen is not given. Hence, there is no need to calculate NCV. The water equivalent given may be taken as the total water equivalent mass.

Mass of fuel = 0.6 g = 0.6×10^{-3} kg

Water equivalent mass of calorimeter = 2200 g = 2200×10^{-3} kg

Specific heat of water = 4 187 kJ kg^{-1} C^{-1}

Increase in temperature = 6.52°C

$$CV = \frac{\text{Total water equivalence} \times \text{Specific heat of water} \times \text{Rise in temperature}}{\text{Mass of fuel}}$$

$$= \frac{2200 \times 10 \quad \times 4.187 \times 6.52}{0.6 \times 10}$$

CV = 100097.2 kJ kg^{-1}

Calorific Value 100097.2 kJ kg^{-1}.

Problem 3.2

Calculate the GVC and NCV of a fuel from the following data:

Mass of fuel burnt = 0.75 g

Water equivalent mass of calorimeter = 350 g

Mass of water taken = 1150 g

Increase in temperature = 3.02 °C

Percentage of hydrogen in the fuel = 2.8 °C

Solution:

$$GCV = \frac{\text{Total water equivalence} \times \text{Specific heat of water} \times \text{Rise in temperature}}{\text{Mass of fuel taken}}$$

$$= \frac{(350 + 1150) \times 10^{-3} \text{ kg} \times 4.187 \times 6.52}{0.6 \times 10^{-3}}$$

GCV = 25289 48 kJ kg^{-1}

Heat evolved by condensation of water vapour = $0.09x \times 587 \times 4.187$ kJ kg^{-1}

Where x is % of hydrogen higher = $0.09 \times 2.8 \times 587 \times 4.187$

= 619.357 kJ kg^{-1}

NCV = GCV – Heat of condensation of steam

= 25289.48 – 619.357

NCV = 24670.12 kJ kg^{-1}

GCV = 25289.48 kJ kg^{-1}, NCV = 24610.12 kJ kg^{-1}.

Problem 3.3

Calculate the gross calorific value of a coal sample from the following data:

Weight of coal sample taken = 5.5×10^{-3} kg

Weight of water taken in the calorimeter = 2.5 kg

Water equivalent of calorimeter = 0.5 kg

Initial temperature of water = 24 °C

Final temperature of water = 28 °C

Solution:

$$\text{GCV} = \frac{\text{Total water equivalence} \times \text{Specific heat of water} \times \text{Rise in temperature}}{\text{Mass of coal}}$$

$$\text{GCV} = \frac{[0.5 + 2.5] \times 4.187 \times 4}{5.5 \times 10^{-3}}$$

GCV = 9135.27 kJ kg^{-1}

GCV = 9135.27 kJ kg^{-1}.

Problem 3.4

0.6 g of a coal sample with 92%C, 5% hydrogen and 3% ash caused a rise in temperature of 2000 g water by 3.2°C in a bomb calorimeter experiment. Calculate the gross and net calorific values of coal.

Given, water equivalent = 200 g

Specific heat of water = 4.187 kJ kg^{-10} C^{-1}

Latent heat of steam = 580 calories / gm (1 cal = 4.18 J)

Mass coal taken = 0.60 g = 0.6×10^{-3}

Hydrogen in coal sample = 5 %

Rise in temperature of water = 3.2°C

Mass of water taken = 2000 g – 2.0 kg

Water equivalent of calorimeter = 200 g = 0.2 kg

Specific heat of water = 4.187 kJ kg^{-1} C^{-1}

Latent heat of steam = 587 calories/g = 587 x = 4.187 kJ kg^{-1}

Solution:

$$GCV = \frac{(W_1 + W_2) \times \text{Specific heat of water} \times \text{Rise in temperature}}{\text{Mass of coal}}$$

$$GCV = \frac{2.2 \times 4.187 \times 3.2}{0.6 \times 10^{-3}}$$

GCV = 49127.46 kJ kg^{-1}

Heat of condensation of steam = 0.09x × 4.187 kJ kg^{-1}

where x is the % of hydrogen contained in the sample

$= 0.09 \times 5 \times 587 \times 4.187 = 1104.147$ kJ kg^{-1}

Net calorific value = 49127.46 – 1104.147

NCV = 48021.47 kJ kg^{-1}

GCV = 49127.46 kJ kg^{-1}, NCV = 48021.47 kJ kg^{-1}.

Problem 3.5

The gross calorific value of a bituminous coal is 36,000 kJ kg^{-1}. In an experiment, 0.83 g of this coal is burnt under 1.2 kg of water in a bomb calorimeter. Due to combustion, the temperature of water rose by 3.92°C. Calculate the water equivalent of the calorimeter. Specific heat of water = 4.2 kJ kg^{-1}.

Gross calorific value of bituminous sample = 36,000 kJ kg^{-1}

Mass of coal taken = 0.83 g = 0.83 × 10^{-3} kg

Mass of water = 1.2 kg

Rise in temperature = 3.92 °C

Specific heat of water = 4.2 kJ kg^{-1} C^{-1}

Solution:

$$GCV = \frac{(W_1 + W_2) \times \text{Specific heat of water} \times \text{Rise in temperature}}{\text{Mass of coal}}$$

$$36{,}000 = \frac{(1.2 + W_2) \times 4.187 \times 3.92}{0.83 \times 10^{-3}}$$

$$36{,}000 = (1.2 \times 22.483) + (W_2 \times 22.483) = 26980 + (W_2 \times 22.483)$$

$$W_2 = \frac{36000 - 26980}{22.483}$$

$W_2 = 0.409$ kg

W_2 = 0.409 kg.

Problem 3.6

On burning 0.75×10^{-3} kg of a solid fuel in a bomb calorimeter, the temperature of 2.5 kg water is increased from 24°C to 28°C. The water equivalent of calorimeter and the latent heat of steam are 0.485 kg and 4.2×587 kJ/kg respectively. Specific heat of water is 4.187 kJ/kg/°C. If the fuel contains 2.5% hydrogen, calculate its gross and net calorific values.

Solution:

Mass of fuel = 0.75×10^{-3} kg

Mass of water taken = 2.5 kg

Water equivalent of calorimeter = 0.485 kg

Latent heat of steam = 4.2×587 kJ/kg^{-1}

Increase in temperature = 4°C

$$\text{Gross calorific value} = \frac{(W_1 + W_2) \times \text{Specific heat of water} \times \text{Rise in temperature}}{\text{Mass of coal}}$$

$$= \frac{[2.5 + 0.485] \times 4.187 \times 4}{0.75 \times 10^{-3}}$$

$$\text{GCV} = 66657.04 \text{ kJ/kg}^{-1}$$

$$\text{Heat of condensation of water vapour} = 0.09\, x \times 587 \times 4.187$$
$$= 0.09 \times 2.5 \times 587 \times 4.187$$
$$= 552.99 \text{ kJ/kg}^{-1}$$

$$\text{Therefore, the net calorific value} = \text{GCV} - \text{Heat of condensation of steam}$$
$$= 66657.0 - 553 = 66104.00 \text{ kJ/kg}^{-1}$$

GCV = 66657.04 kJ/kg^{-1}; NCV = 66104.00 kJ/kg^{-1}.

Problem 3.7

Calculate the gross and net calorific values of a coal sample from the following data obtained from bomb calorimeter experiment.

Weight of coal (m) = 0.73 g,

Weight of water taken in calorimeter (w) = 1500 g

Water equivalent of calorimeter (w_1) = 470 g

Initial temperature (t_1) = 25 °C

Final temperature (t_2) = 27.3 °C

Percentage of hydrogen in coal sample = 2.5%

Latent heat of steam = 587 cal/g

Solution:

$$\text{GCV} = \frac{(W_1 + W_2)\,(t_2 - t_1)}{m} \text{ cal/g}$$

$$= \frac{(1500 + 470)\,(27.3 - 25.0)}{0.73}$$

$$= 6206.8 \text{ cal/g}$$

$$\frac{6206.8 \text{ cal/g} \times 4.187 \text{ J/kg}}{10^{-3}} = 259693 \times 10^3 \text{ J/kg}$$

$$\text{NCV} = \text{HCV} - \text{Heat released by condensation of steam}$$

$$= 25969 \times 10^3 - \frac{0.09 \times 2.5 \times 587 \times 4.187}{10^{-3}}$$

$$= 25416.3 \times 10^3 \text{ J/kg}$$

GCV = 259693 × 10^3 J/kg, NCV = 25416.3 × 10^3 J/kg.

Problem 3.8

The following data is obtained from a bomb calorimeter experiment:

Weight of crucible = 3.649 gm
Weight of crucible + Fuel = 4.678 gm
Water equivalent of calorimeter = 570 gm
Water taken in calorimeter = 2200 gm
Observed rise in temperature = 23°C
Cooling correction = 0.047°C
Acid correction = 62.6 calories
Fuse wire correction = 3.8 calories
Cotton thread correction = 1.6 calories

Calculate the gross calorific value of the fuel sample. If the fuel contains 6.5% H, determine the net calorific value.

Solution:

$$\text{GCV} = \frac{(W_1 + W_2)(t_2 - t_1 + t_c) - (C_A + C_F + C_C)}{m}$$

$$= \frac{(2200 + 570)(2.3 + 0.047) - (62.6 + 3.8 + 1.6)}{(4.678 - 3.649)}$$

$$= \frac{(27702.347) - 68}{1.029} = 6261 \text{ cal/g}$$

$$\text{NCV} = (\text{GCV} - 0.09 \times \text{H} \times 587) \text{ cal/g}$$

$$= 6261 - 0.09 \times 6.5 \times 587$$

$$= 6261 - 343.4 = 5917.6 \text{ cal/g}$$

GCV = 6261 cal/g, NCV = 5917.6 cal/g.

3.6 KNOCKING AND ITS MECHANISM

The performance of a motor car measured in terms of kilometre per litre of petrol, depends on the quality of the fuel. An important method of obtaining more power from petrol is to increase the compression ratio of the engine. Increase of compression ratio not only increases the efficiency of the engine but also saves the fuel. So, it may be defined as:

"The production of a shock wave in an IC engine as a result of an explosive combustion of fuel – air mixture. Due to an increase in the compression ratio, beyond a certain value, leading to a rattling sound".

Knocking increases fuel consumption, reduces the efficiency of the engine and damages the engine parts and the life of the engine.

Mechanism

Under ideal conditions, there is slow oxidation of the fuel during which oxygen combines with a few molecules and activates them. The activated molecules combine with the hydrocarbon molecules and a chain reaction is set up resulting in smooth combustion.

$$C_2H_6 + 3\tfrac{1}{2}O_2 \longrightarrow 2CO_2 + 3H_2O \text{ (normal combustion)}$$

Knocking occurs because the chain reaction proceeds at a fast rate. Hydrocarbon molecules combine with oxygen to form peroxides. The unstable peroxides decompose readily to give a number of gaseous compounds. This give rise to pressure waves which knock against the engine walls.

$$C_2H_6 + O_2 \longrightarrow \underset{\text{Ethane peroxide}}{CH_3{-}O{-}O{-}CH_3} \text{ (explosive combustion)}$$

$$CH_3{-}O{-}O{-}CH_3 \longrightarrow CH_3\,CHO + H_2O$$

$$CH_3CHO + 1\tfrac{1}{2}O_2 \longrightarrow HCHO + CO_2 + H_2O$$

$$HCHO + O_2 \longrightarrow H_2O + CO_2$$

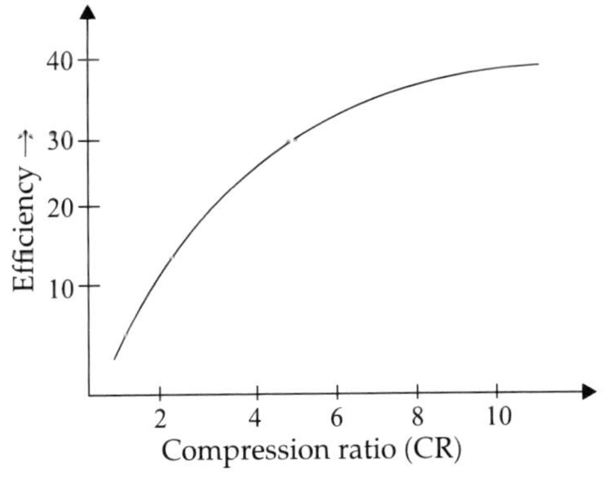

Fig. 3.2 *Variation of IC engine efficiency with compression ratio*

The following adverse effects are observed due to gasoline knock.

1. It produces undesirable rattling noise.
2. It increases the fuel consumption.
3. It results in decreased power output.
4. It causes mechanical damage due to overheating to engine parts such as spark plug, piston and engine walls.
5. Driving becomes unpleasant.

Remedial Measures (Prevention)

The knocking in IC engines can be controlled through the following measures:

1. By using high rating gasoline.
2. By using critical compression ratio.
3. By using anti-knocking agents.

3.7 UNLEADED PETROL

In order to enhance the octane rating of gasoline in IC engines, gasoline is normally mixed with 0.1 per cent TEL per gallon of gasoline. This increases octane rating by 10 – 15 points. Mixing of TEL with gasoline brought advantages together with disadvantages. In order to overcome the harmful effects of TEL, it is slowly being phased out from gasoline. *"Petrol whose octane number is increased by the addition of substances other than lead compounds is referred to as unleaded petrol"*.

The compelling reasons for using unleaded petrol are:

- Lead is toxic. The recognition of health hazard associated with its release into atmosphere from automobile exhaust leads to mandatory phasing out of TEL in gasoline.
- Further factor for reducing TEL is the requirement of vehicles to be fitted with catalytic converters because lead inactivates the catalysts used in converters.

Unleaded petrol rating can be enhanced by the addition of certain compounds in place of TEL. The octane rating enhancers such as methyl-*t*-butyl ether (MTBE), Ethyl-*t*-butyl ether (ETBE), methanol, ethanol which have octane rating above 100, are added to unleaded petrol in place of TEL.

3.8 POWER ALCOHOL

When ethyl alcohol is used as an additive to motor fuels to act as a fuel for internal combustion engines, it is called power alcohol. Blends containing up to 25% of alcohol with petrol are used.

The important raw materials for the manufacture of power alcohol or ethyl alcohol are saccharine materials (such as molasses, sugar beets , sugar cane, etc.) starchy materials (such as starch, potatoes, cereal grains, etc.) cellulose materials (such as sulphite liquor from paper mills) and hydrocarbon gases.

3.8.1 Ethyl Alcohol from Molasses

Molasses is a dark colour viscous liquid left after the crystallization of cane sugar from cane juice. It contains 50 – 55 % total sugar, of which 35 – 40% is cane sugar or sucrose. Molasses is converted into ethyl alcohol by means of yeast, which contains enzymes invertase and zymase responsible for fermentation.

$$C_{12}H_{22}O_{11} \xrightarrow{\text{Invertase}} \underset{\text{Glucose}}{C_6H_{12}O_6} + \underset{\text{Fructose}}{C_6H_{12}O_6}$$

$$C_6H_{12}O_6 \xrightarrow{\text{Zymase}} 2C_2H_5OH + 2CO_2$$

Before mixing with yeast, the molasses is diluted with water to bring down the concentration of sugar to 10–12 %. pH is kept between 4 and 5 to favour the function of enzymes, by using sulphuric acid. After fermentation for 48 – 60 hours, it is distilled to get ethyl alcohol which is mixed with water. Repeated distillation and condensation can raise the alcohol content to 97.6%.

Absolute alcohol can be obtained from the above by the azeotropic process in which benzene is used to form ternary mixture with water and alcohol. The ternary, constant boiling mixture boils at 65°C and pure dehydrated alcohol, which is absolute alcohol, boils at 78.5°C. So the vapours of ternary mixture go out at 65°C taking away all the water and leaving behind absolute alcohol.

3.8.2 Power Alcohol as a Fuel

Power alcohol is used as a fuel by blending with petrol in internal combustion engines as a motor spirit. Blends containing up to 25% of alcohol with petrol are used. Industrial alcohol containing 95% alcohol and 5% of water can also be blended with petrol but by using some blending agent such as benzene, ether, etc. Without the use of blending agent, industrial alcohol does not mix with petrol.

3.8.3 Advantages

The advantages of power alcohol as fuel are:

(a) The power output is good.

(b) Addition of ethanol to gasoline promotes efficient combustion of the hydrocarbons in gasoline, (it contains 35% oxygen by mass, i.e., twice the oxygen content of MTBE). This reduces the possibility of peroxide formation and hence has better antiknock property than MTBE (methyl-*t*-butyl ether).

(c) Due to better antiknock property, it can be used in engines with higher compression ratio.

(d) Addition of ethanol to gasoline also reduces the emission of carbon monoxide and volatile organic compounds.

(e) Petrol-alcohol blend has the same lubrication as petrol.

(f) Since ethanol is obtained from agricultural products, it has the potential to be a sustainable fuel.

3.9 BIODIESEL

Biodiesel is a renewable, oxygenated fuel made from a variety of agricultural resources such as soybeans, rapeseeds, cotton seeds, rice bran and rubber seed oil. It can be used as an alternative to diesel in automobiles.

The oils are triglycerides esters of fatty acids and their use in diesel engines requires extensive engine modifications. However, it is possible to convert these triglycerides into substances of diesel quality by transesterification. Transesterification is alcoholysis of the triglyceride oil in the presence of a base. The resulting mixture of monoalkyl esters of fatty acids is referred to as biodiesel. Thus, biodiesel is a mixture of monoalkyl esters of long chain fatty acids. The transesterification with methanol in presence of NaOH is given below.

$$
\begin{array}{llcll}
CH_2OCOR_1 & & & CH_3OCOR_1 & \quad CH_2OH \\
| & & & | & \quad | \\
CHOCOR_2 & + 3CH_3OH & \xrightarrow{NaOH} & CH_3OCOR_2 \quad + & \quad CHOH \\
| & & & | & \quad | \\
CH_2OCOR_3 & & & CH_3OCOR_3 & \quad CH_2OH \\
\text{Triglyceride oil} & & & \text{Biodiesel} & \quad \text{Glycerol}
\end{array}
$$

where R_1, R_2, R_3 are the long chain fatty acids in the oil. Glycerol is soluble in water and can be separated from the reaction mixture.

A variety of oils can be used to produce biodiesel. These include:

- Virgin oil feedstock: rapeseed and soybean oils are most commonly used, others can be jatropha and other crops such as mustard, flax, sunflower, palm oil, coconut, etc.
- Waste vegetable oil.
- Animal fats including tallow, lard, yellow grease, chicken fat, and the by-products of the production of fatty acids from fish oil.
- Algae, which can be grown using waste materials such as sewage and without displacing land currently used for food production.

Advantages

- It is made using renewable sources and feedstocks.
- Readily undergoes biodegradation.
- It has a higher flashpoint than diesel but lower ignition point.
- Biodiesel has a higher cetane number (48 – 60) compared to diesel (40 – 55).
- Use of biodiesel reduces greenhouse gases.
- It is non-toxic.
- Burns more efficiently than petroleum diesel.
- Substantially higher lubricity means it can reduce engine wear and hence prolonged engine life.
- Fewer harmful emissions compared to diesel (45% less particulate matter 65% less total HCs 41% less CO NO*x* remains the same). Diesel engines using petroleum sources however emit a number of toxic chemicals in the exhaust which include acetaldehyde ethyl benzene formaldehyde naphthalene and xylenes.

3.10 FUEL CELL

In a galvanic cell chemical energy is directly converted into electrical energy and the efficiency of conversion is high. In the case of combustion of fuels, the reaction taking place in a galvanic cell is also a redox reaction. But the only difference is that oxidation and reduction reactions in a cell take place at two different electrode surfaces separated from each other. The electrons liberated at the anode flow to the cathode, thereby generating electricity. Therefore, it was thought that if the oxidation of the fuel and reduction of oxidant, oxygen can be carried out separately at a distance between the two, then there can be identical situation as in the case of a galvanic cell. Electrons liberated during the oxidation of the fuel can flow to the reduction zone of oxygen. The attempt in this direction by several scientists over quite a long period of time lead to the invention of fuel cells in which chemical energy of a fuel is directly converted into electrical energy.

The principle of fuel cells were first demonstrated by **Sir William Grove** in 1839, who is acknowledged as the father of the Fuel Cell; they became popular only after NASA used them in their space ships.

3.10.1 Definition

A fuel cell is a galvanic cell which converts chemical energy of the fuels into electrical energy through catalyzed electrochemical process in which the fuel is oxidized at the anode.

The fuel cell has two electrodes and an electrolyte. However, the fuel and the oxidizing agents are continuously and separately supplied to the two electrodes of the cell, at which they undergo reactions. These cells are capable of supplying current as long as they are supplied with reactants.

The representation of the fuel cell is

Fuel/electrode/electrolyte/electrode/oxidant

At the anode, fuel undergoes oxidation

$$\text{Fuel} \longrightarrow \text{Oxidation product} + ne^-$$

At the cathode, the oxidant gets reduced

$$\text{Oxidant} + ne^- \longrightarrow \text{Reduction products}$$

The $_+e^\circ$ liberated from the oxidation process at the anode can perform useful work when they pass through the external circuit to the cathode.

3.10.2 Principle

The basic principle of the fuel cell is same as that of electrochemical cell. The only difference is that the fuel and oxidant are stored outside the cell. Fuel and oxidant are supplied continuously and separately to the electrodes at which they undergo redox reactions.

Fuel cells are capable of supplying current as long as reactants are restore.

$$\text{Fuel} + \text{Oxidant} \rightarrow \text{Oxidation products} + \text{Electricity}.$$

3.10.3 Advantages

1. High energy conversion
2. Ecofriendly (i.e., it converts a fuel to electrical energy with minimum emission of pollutants
3. No need of charging
4. Space required for fuel cell is less
5. Maintenance is less
6. High energy density
7. No moving parts and there is no wear and tear of cell parts
8. Silent operation
9. Absence of harmful waste products
10. It generates electricity as long as fuel is supplied.

3.10.4 Disadvantages

1. The power generated is moderate
2. Cost of power is high as a result of the cost of the electrodes
3. Fuels in the form of gases and oxygen need to be stored in tanks under high pressure.

3.11 COMPARISON BETWEEN FUEL CELL AND BATTERIES

Fuel Cell	Batteries
1. Reactants (fuel and oxidant) are not part of the cell they are supplied from outside	1. All reactants are integral parts of it
2. Do not store chemical energy	2. Store chemical energy
3. The electrodes are charged with catalyst	3. The electrodes are not helped with catalyst
4. The electrodes in some cases are not consumed during production of energy	4. The electrodes are consumed during production of energy
5. Products are continuously removed from a fuel cell	5. Products are not removed
6. Produce current on their own	6. Require recharging from external source of electric current

3.12 CLASSIFICATION OF FUEL CELL

Fuel cells are classified on the basis of temperature of operation.

1. Low temperature fuel cells (less than 100 °C): which use water based electrolyte.
2. Medium temperature fuel cell (100–250 °C): which use molten salt as electrolyte.
3. High temperature fuel cell (less than 500 °C): which use ceramic electrolyte.

3.13 TYPES OF FUEL CELL

The fuel cells can be classified into different types based on the fuel and electrolytes used.

1. Alkaline fuel cell – (AFC)

- Low temperature fuel cells (80 °C) using
- Porous carbon charged with Mi/Pd acts as anode
- Porous carbon charge with Ag-catalyst acts as cathode
- Hydrogen gas is fuel at anode
- Oxygen gas is fuel at cathode
- Aqueous KOH solution is used as electrolyte

2. Phosphoric acid fuel cell: (PAFC)

- Porous C + Sic + Teflon charge with Pt-catalyst + acts as anode
- Porous C + Sic + Teflon charge with Ag-catalyst acts cathode
- Pure H_2 gas is anodic fuel
- Pure O_2 gas is cathodic fuel
- Concentrated phosphoric acid is used as electrolyte.

3. Molten carbonate fuel cell: (MCFC)

- Anode is porous Ni/Ni – Cr alloy
- Cathode is porous NiO
- H_2 gas or CO gas is fuel at anode
- O_2 gas is fuel at cathode operates between 600 – 650 °C
- Fused carbonates (eutectic mixture of 32% $Li_2\,CO_3$ + 48% $NiAlCO_3$ + K_2CO_3) in porous inorganic material is used as electrolytic.

4. Polymer electrolyte fuel cell: (PEFC)

- Two different porous gas diffusion carbon electrodes charged with Pt acts as both anode and cathode.
- H_2 gas is used as anode fuel.
- O_2 gas is used as cathode fuel stable at 100 °C.
- Nafion membrane with 50% water as electrolyte is used as electrolyte.

  (Nafion, i.e., per fluorinated cation exchange polymer membrane)

5. Solid oxalate fuel cell (SOFC)

- The anode is CO–ZrO_2
- Cathode is Sr doped La MnO_3
- H_2 gas or reformed fuel (mixture of Co + H_2) is fuel at anode

- O_2 gas is used as fuel at cathode
- ZrO_2 doped with Y_2O_2 (migration O^{2-} ions takes place across the electrolyte) is used as electrolyte.

3.14 SUMMARY

Table 3.1 Five types of fuel cells

	PEFC	AFC		PAFC	MCFC	SOFC
Electrolyte	Nafion	KOH (40%)	KOH (85%)	Liquid H_3PO_4	Li/Na/K carbonate solution	Yttria stabilized ZrO_2
Temperature	50–100°C	90–120°C	250°C	220°C	600–700°C	1000°C
Charge carrier	H^+	OH^-		H^+	CO_3^{2-}	O^{2-}
Anode catalyst	Pt or Pt/C	Pt/Pd/Av/Ni		Pt/C	Ni	Ni
Fuel	Pure H_2, Methanol	Pure H_2		Pure H_2 Methanol reformed natural gas	H_2, natural gas, gasified coal	H_2, natural gas, gasified coal
Application	Back-up power, portable power transportation	Military, space		Distributed power generation	Large electricity distributed generation	Large electricity distributed generation

3.15 METHANOL OXYGEN FUEL CELL

Methanol-oxygen fuel cell is the simplest fuel cell in which methanol is used as the fuel and oxygen is used as an oxidant. The cell can be used either in alkaline or acidic medium. The most preferred electrolyte is the acidic medium. Methanol is also relatively cheeper and easily available fuel with less carbon content and has easily oxidizable OH group.

3.15.1 Construction

1. It consists of anodic and cathodic compartments.
2. Anode of the cell is made up of platinum and methanol along with sulphuric acid is continuously passed through anodic compartment as a fuel.
3. Cathode of the cell is also made up of platinum and oxygen is continuously supplied into cathodic compartment as oxidant.
4. Electrolyte used is sulphuric acid.

Note:

(a) KOH is not used as electrolyte as it reacts with CO_2 and gets converted into K_2CO_3. Thus, conductivity of the electrolyte decrease as well as the efficiency.

(b) The advantage of acid electrolyte is that the CO_2, a product of the reaction can be easily removed.

5. A membrane is provided which prevent the diffusion of methanol into the cathode. In the absence of the membrane, methanol crosses over to the cathode compartment and undergoes oxidation there itself.

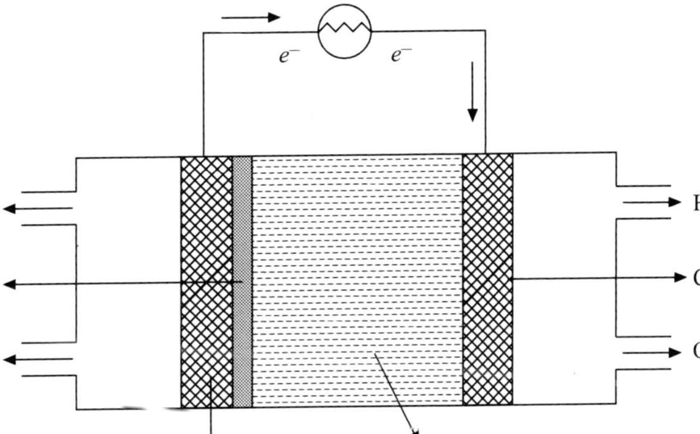

Fig. 3.3

3.15.2 Working

The electrode reactions are as follows:

Anode: $CH_3OH + H_2O \rightarrow CO_2 + 6H^+ + 6e^-$

Cathode: $\frac{3}{2}O_2 + 6H^+ + 6e^- \rightarrow 3H_2O$

Overall reaction: $CH_3OH + \frac{3}{2}O_2 \rightarrow CO_2 + 2H_2O$

The emf of the cell is 1.2 V.

Advantages

- Methanol has low carbon content OH group is easily oxidisable
- Methanol is highly soluble in water.

Applications

Used in military applications such as soldier carried equipment, power for test and training equipment, battery chargers, etc.

B. SOLAR ENERGY

3.16 INTRODUCTION

Solar energy is the energy obtained by capturing heat and light from the sun. Energy from the sun is referred to as solar energy. It is renewable and therefore a green source of energy. Using a range of ever-evolving technologies such as solar heating, photovoltaics, solar thermal energy, solar architecture and artificial photosynthesis provides a number of ways to utilize this abundant resource. The main advantages of using solar energy as an energy source are that it is readily available, free of cost and ecofriendly.

3.17 UTILISATION OF SOLAR ENERGY

The utilization of solar energy is of two types as explained below.

3.17.1 Direct Solar Power

It involves only one step transformation into a usable form.

Example: Photovoltaic cell.

Sunlight hits a photovoltaic cell and generates electricity.

3.17.2 Indirect Solar Power

It involves more than one transformation to reach a usable form.

Example: Where solar energy is converted to chemical energy which can be later used as energy source. Photosynthesis is a classical example for conversion of solar energy into chemical energy. It can be used for photo-splitting of water to produce H_2, which can be used as a fuel.

Advantages

1. Maintenance needed is minimum to keep solar cells running, there are no moving parts in a solar cell which makes it impossible to really damage them.
2. Solar power is becoming more and more economical as costs associated with production are decreasing and the technology is becoming more effective.
3. It causes no pollution.
4. It is a renewable energy source.
5. Solar energy is directly used for cooking in solar cookers, for heating rooms and for heating water.
6. Solar energy can be used for diverse purposes. You can generate electricity (photovoltaics) or heat (solar thermal). It can be used to produce electricity in areas without access to the energy grid, to distill water in regions with limited clean water supplies and to power satellites in space. It can also be integrated in materials used for building.

Disadvantages

1. Solar power is only practical in certain areas with a favourable climate and latitude. That is, areas near the tropics and which are relatively cloud free.
2. Solar power is not available at night.
3. Solar power decreases during cloudy atmosphere.
4. It must be converted into some other form of energy to be stored for times when conditions are prohibitive.

3.18 PHOTOVOLTAIC CELL

Photovoltaic cells are semiconductor devices which convert solar energy into electrical energy by the photovoltaic effect. The effect of creating a voltage is called photovoltaic effect.

3.18.1 Construction

Pv cell consists of a front contact which is made up of metallic grids. Below the front contact a *P* – *N* junction (semiconductor diode) of silion is placed. Bottom of the *P* – *N* junction is in contact with noble metal like silver.

3.18.2 Working

A metallic grid forms one of the electrical contacts of the diode and allows light to fall on the semiconductor between the grid lines.

An antireflective layer between the grid lines increases the amount of light transmitted to the semiconductor.

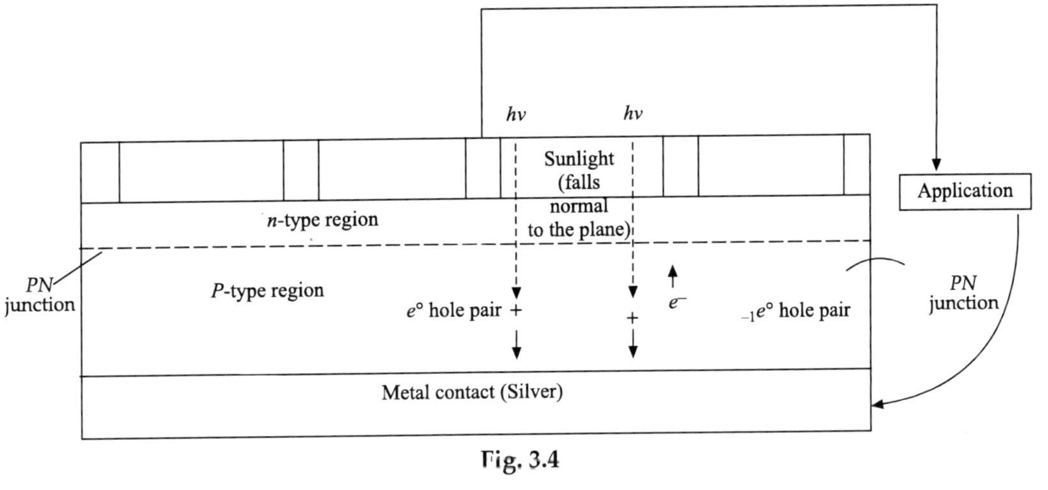

Fig. 3.4

The cell's other electrical contact is formed by a metallic layer on the back of the solar cell. When light radiation falls on the *P-n* junction diode, ${}_{-1}e^{\circ}$ hole pairs are generated by the absorption of radiation. The ${}_{-1}e^{\circ}$ are drifted to and collected at the *P*-type end. When these two ends are electrically connected through a conductor, there is a flow of current between the two ends through the external circuit. This photoelectric current is produced and available for use.

3.18.3 Importance

1. Photovoltaic cell produces no pollution so it is environment-friendly.
2. They can be conveniently used for a variety of applications like remote area lighting and other power required domestic activities, industrial power needs and irrigation, harvesting, spraying, etc.
3. It provides power for spacecraft and satellites.
4. Photovoltaic cell provides enormous amount of energy from the sun which is unlimited, inexhaustible and renewable.
5. Used in transportation system.
6. *PV* cell are used to improve remote area communication systems.

Advantages

1. Fuel source is infinite.
2. There is no moving parts in *PV* cells, hence no wear and tear.
3. Low operating cost (no fuel) corrosionless and serve for long duration.
4. High acceptance.
5. No recharging and produce electrical energy as long as sunlight is available.
6. Silicon based *PV* cell is available abundantly.
7. Generate electrical energy at ambient temperature.

Disadvantages

1. Energy can be produced only during daytime.
2. The installation cost is high.
3. It is relatively low density energy.
4. They require energy storage device, i.e., battery which require maintenance and replenishment from time to time.

3.19 DESIGN OF SOLAR CELLS

A solar cell is a device used to convert solar light incident on it into electrical energy. Each cell produces about 0.5 V when exposed to sunlight. Therefore, to obtain a potential of more than 12 V required to charge a battery, number of cells are connected.

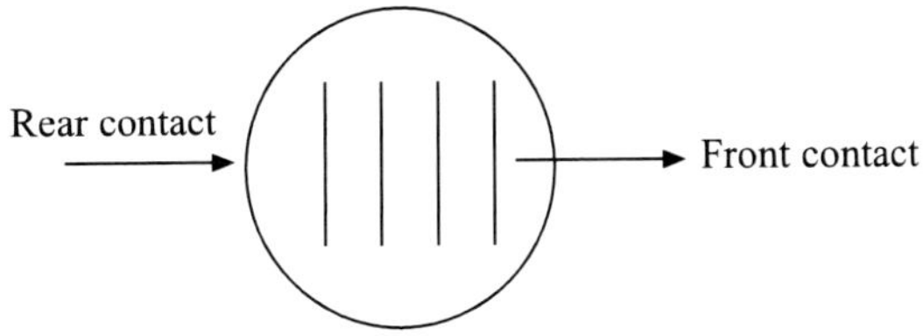

Fig. 3.5 *Solar cell*

3.19.1 Module

A number of solar cells electrically connected to each other and mounted in a support structure is called a photovoltaic module. A module can be used to charge a 12 V rechargeable battery.

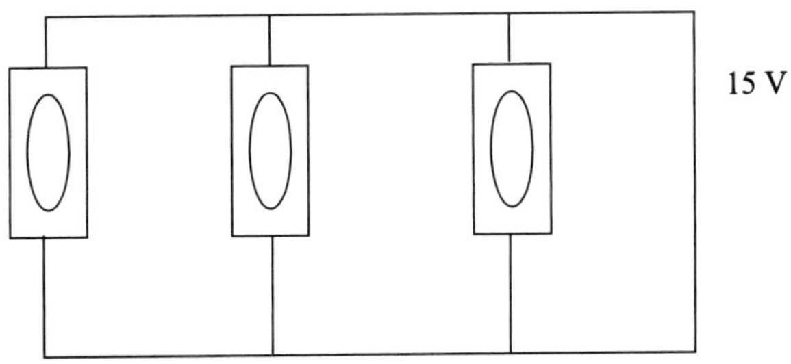

Fig. 3.6 *Three modules connected in parallel*

3.19.2 Panel

Multiple modules are interconnected to obtain a panel. In a panel also modules may be connected in parallel or in series to obtain the required voltage and current. In panels, modules are mechanically connected together and electrically wired and mounted on a metal. It is covered by transparent glass sheet with the help of aluminium frame. If two modules each giving 2.5 A at 16 V in bright sunlight are connected in series they give 2.5 A at 32 V. If they are connected in parallel, they give 5.0 A at 16 V. In both the cases power is the same 80 W.

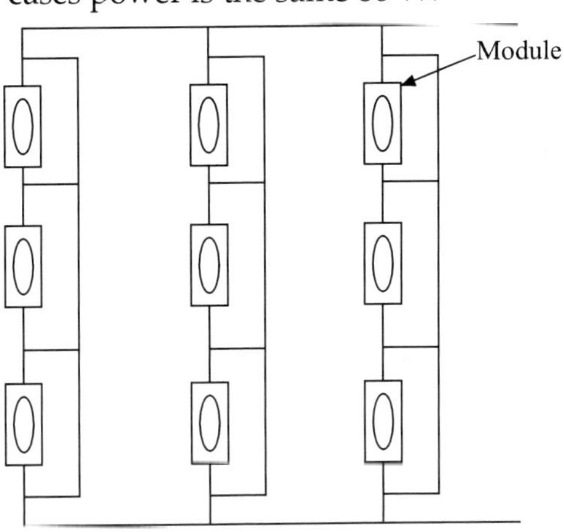

Fig. 3.7 *Series and parallel connection of modules in a panel*

3.19.3 Array

A number of panels are electrically interconnected to produce an array. In a *PV* array, panels may be wired together in a series and or parallel to deliver the voltage and current required for a particular system.

It is a ready to install, complete power-generating unit. An array can be as small as a single pair of panels or large enough to cover acres.

From cell to array

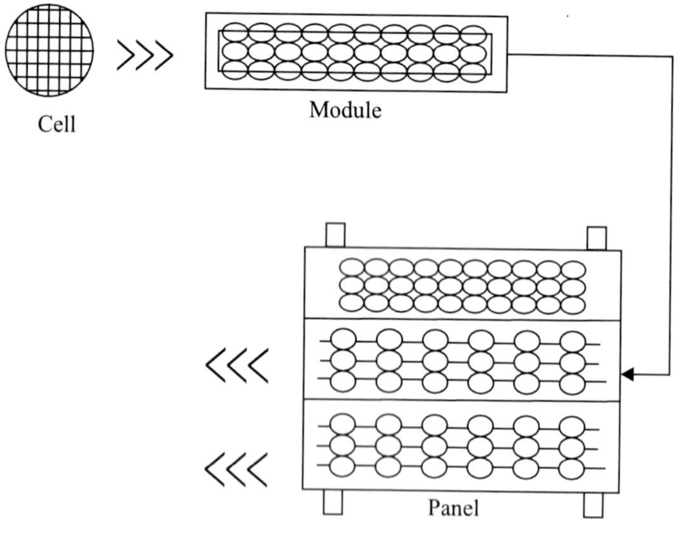

Fig. 3.8

3.20 PRODUCTION OF SOLAR GRADE SILICON

Solar grade silicon is metallurgical grade silicon which is about 97% pure which is then refined to over 99% purity. The grade of silicon is defined by its transition metal content.

It is commercially prepared by the reaction of high purity silica with carbon in an electric arc furnace using carbon electrodes. The mixture is heated to a high temperature and the following reaction takes place to give silicon.

$$SiO_2 + C \longrightarrow Si + CO_2$$
$$SiO_2 + 2C \longrightarrow Si + 2CO$$

It is also obtained by the strong heating of silanes.

$$SiH_4 \xrightarrow{\Delta} Si + 2H_2$$

Oxygen is passed through molten silicon mixed with a flux of silica. The elements such as Al, Ca and Mg present as impurities (gangue) are oxidized to corresponding oxides. These oxides, in turn, combine with silica forming a slag of silicates. The slag is removed and the refined melt is poured into moulds to solidify silicon. Silicon so obtained is called metallurgical grade silicon.

$$4Al + 3\,SiO_2 \longrightarrow 3Si + 2Al_2O_3$$

$$2Ca + SiO_2 \longrightarrow Si + 2CaO$$

$$2Ca + SiO_2 \longrightarrow Si + 2MgO$$

$$Si + O_2 \longrightarrow SiO_2$$

To get semiconductor grade silicon or polysilicon metallurgical grade silicon is processed in four stages.

1. Synthesis of silicon hydride
2. Purification of the hydride
3. Decomposition to elemental silicon
4. Removal of by-products.

The metallurgical grade silicon is treated with dry HCl gas at 300°C to form trichlorosilane and a small amount of tetrachlorosiliane.

$$Si + 3HCl \longrightarrow HSiCl_3 + H_2$$

$$Si + 4HCl \longrightarrow SiCl_4 + 2H_2$$

The mixture is distilled to get pure trichorosilane. Silicon tetrachloride is reduced to trichlorosilane by heating with hydrogen to about 1000°C in a reactor.

$$SiCl_4 + H_2 \longrightarrow HSiCl_3 + HCl$$

Trichlorosilane is passed through a fixed bed column containing quaternary ammonium ion exchange resin as catalyst to get silicon hydride.

$$2HSiCl_3 \longrightarrow H_2\,SiCl_2 + SiCl_4$$

$$3H_2\,SiCl_2 \longrightarrow SiH_4 + 2HSiCl_3$$

The products are separated by distillation. Tetrachlorosilane and trichlorosilane are recycled to the hydrogenation reactor and the ion exchange resins respectively.

Silane (silicon hydride) is further purified by distillation and passed into a reactor containing heated silicon rods. Silane undergoes pyrolysis to give semiconductor grade silicon or polysilicon

$$SiH_4 \longrightarrow Si + 2H_2$$

REVIEW QUESTIONS

1. What is calorific value of a fuel? Explain the bomb calorimeter method to determine calorific value of a solid fuel.
2. Calculate the gross and net calorific value of a coal sample from the following data:

   Weight of coal = 0.73 g, Weight of water taken in calorimeter = 1500 g

   Water equivalent of calorimeter = 470 g, Rise in temperature = 2.3°C

   Percentage of hydrogen in coal sample = 2.5%, Latent heat of steam = 587 cal/g.
3. 0.78 g of coal containing 1.9% hydrogen, when burnt in a bomb calorimeter, increased the temperature of 2.7 kg water from 27.2°C to 29.7°C. If the water equivalent of calorimeter is 1.2 kg. Calculate gross and net calorific value (specific heat of water = 4.187 kJ/kg/°C. Latent heat of steam = 2457 kJ/kg).
4. What is knocking in IC engines? Explain its mechanism with chemical reaction. Mention its ill effects.
5. Calculate the calorific value of a sample of coal from the following data:

   Mass of coal = 0.95 g, Mass of water in copper calorimeter = 2000 g

   Water equivalent of calorimeter = 700 g, Rise in temperature = 2.8°C

   Specific heat of water = 4.187 kJ/kg/°C.
6. What is biodiesel? Explain its production by transesterification of triglycerides.
7. On burning 0.96 × 10 kg of a solid fuel in a bomb calorimeter, the temperature of 3.5 kg of water was increased by 2.7°C. Water equivalent of the calorimeter and latent heat of steam are 0.385 kg and 2455 kJ/kg, respectively. If the fuel contains 5% hydrogen, calculate its gross and net calorific values.
8. On burning 1.15 g of a coal sample in a bomb calorimeter, the temperature of 3.5 kg of water in the calorimeter increased from 26.5°C to 28.5°C. Water equivalent of calorimeter is 325 g. Specific heat of water is 4.187 J/g/K and latent heat of steam is 2458 J/g. If the fuel contains 4% hydrogen, calculate its gross and net calorific values.
9. Discuss the following: (i) power alcohol (ii) biodiesel.

   On burning 0.85×10^{-3} kg of a solid fuel in a bomb calorimeter, the temperature of 2.1 kg water is raised from 24°C to 27.6°C. The water equivalent of calorimeter and latent heat of steam are 1.1 kg and 2454 kJ/kg, respectively. Specific heat of water is 4.187 kJ/kg. If the fuel contains 2 % hydrogen. Calculate its gross and net calorific values.
10. Calculate gross and net calorific values of a coal sample from the following data:

    Weight of coal sample = 0.98 g

    Weight of water taken in calorimeter = 2600 g

    Water equivalent of calorimeter = 368 g

    Latent heat of steam = 2454 J/g

Specific heat of water = 4.187 J/g/K
Rise in temperature = 2.8 K
Percentage of hydrogen in coal sample = 5.8

11. Calculate the gross and net calorific values of a sample of coke from the following data:
Mass of coke = 0.795×10^{-3} kg
Water equivalent of calorimeter = 2.5 kg
Mass of water = 1.3 kg
Specific heat of water = 4.187 kJ kg^{-1} K^{-1}
Rise in temperature = 1.8 K
% hydrogen in the fuel sample = 2.5
Latent heat of steam = 587 cal /g
12. What are chemical fuels? Give the classification of fuels with examples.
13. What is a fuel cell? How does a fuel cell differ from a conventional cell.
14. Explain the working of methanol-oxygen cell.
15. Explain the following fuel cells
(a) Molten carbonate (b) Solid polymer electrolyte
16. What are fuel cells? Give the difference between battery and fuel cell.
17. How does a fuel cell differ from a battery? Explain the construction and working of methanol-oxygen fuel cell with H_2SO_4 as an electrolyte? Indicate the advantage of H_2SO_4 as electrolyte over alkali electrolytes.
18. What are fuel cells? Describe, the construction and working of a methanol-oxygen fuel cell.
19. Describe the construction and working of methanol – oxygen fuel cell.
20. Give the characteristics of fuel cells.
21. What are fuel cells? Indicate the advantages of fuel cells. Explain methanol – oxygen fuel cells with the reactions involved.
22. What is a photovoltaic cell? Explain its construction and working.
23. List three important uses of photovoltaic cells.
24. How do you produce solar grade silicon by the union carbide process?
25. Distinguish between metallurigical grade siliocon and semiconductor grade silicon.
26. How do you purify silicon by the zone refining method?
27. Give the advantages and disadvantages of *PV* cells.
28. List three important physical properties of silicon relevant to solar cells.

EXERCISE

1. On burning 0.83 g of a solid fuel in a bomb calorimeter, the temperature of 3500 g of water increased from 26.5°C to 29.2°C. Water equivalent of calorimeter and latent heat of steam are 385.0 g and 587.0 cal/g, respectively. If the fuel contains 0.7% hydrogen, calculate its gross and net calorific values.

   [*Ans.* GCV = 12,638 cal /g , NCV = 12,601 cal/g]

2. 0.7 g of a fuel contains 80% carbon, when burnt in a bomb calorimeter, raised the temperature of water from 27.2°C to 29.2°C. If the calorimeter contains 250 g of water and its water equivalent is 150 g, calculate the HCV of the fuel.

   [*Ans.* HCV = 1142.8 cal/g]

3. 0.9 g coal, containing 93% C, 6%H, is burnt in a bomb calorimeter. It contains 550 g water and its equivalent of 2200 g. On combustion, temperature increases up to 2.45°C. If fuse wire and acid corrections are 10.0 and 50 cal, respectively, calculate the gross and net calorific values of coal. The latent heat of condensation of steam is 587 cal/g. **[*Ans.* GCV = 7419.4 cal /g, NCV = 7102 cal/g]**

4. A sample of coal contains C = 93%, H = 6% and ash = 1%. The following data were obtained when the above coal was tested in a bomb calorimeter.
   Weight of coal burnt = 0.92 g
   Weight of water taken = 2200 g
   Water equivalent of bomb calorimeter = 550 g
   Rise in temperature = 2.42°C
   Fuse wire correction = 10.0 cal
   Acid correction = 50.0 cal

   Calculate the gross and net calorific values of the coal, assuming the latent heat of condensation of steam is 580 cal/g. **[*Ans.* GCV = 7,168.5 cal/g, NCV = 6,855.3 cal/g]**

5. A coal has the following composition by weight: C = 90%, O = 3.0%, S = 0.5%, N = 0.5% and ash = 2.5%, let calorific value of the coal be 8,490.5 kcal /kg. Calculate the percentage of hydrogen and gross calorific value of coal.

   [*Ans.* % of H = 4.575, HCV = 8,731.8 kcal /kg]

6. 0.72 g of a fuel containing 80% carbon, when burnt in a bomb calorimeter, increased the temperature of water from 27.3° to 29.1°C. If the calorimeter contains 250 g of water and its water equivalent is 150 g. Calculate the HCV of the fuel in kJ/kg.

   [*Ans.* HCV = 1,000 kcal/kg]

7. A sample of coal contains: C = 83%, H = 6% and ash = 1%. The following data were obtained when the above coal was tested in a bomb calorimeter.
   Weight of coal burnt = 0.92 g
   Weight of water taken = 550 g
   Water equivalent of calorimeter = 2,200 g
   Rise in temperature = 2.42°C
   Fuse wire correction = 10.0 cal
   Acid correction = 50.0 cal

Calculate the GCV and NCV values of coal, assuming that the latent heat of condensation of steam is 580 cal/g. **[*Ans.* HCV = 7,168.5 cal /g , NCV = 6,855.3 cal/g]**

8. On burning 0.83 g of a solid fuel in a bomb calorimeter, the temperature of 3,500 g of water rises from 26.5°C to 29.2°C. Water equivalent of calorimeter and latent heat of steam are 385.0 g and 587.0 cal/g respectively. If the fuel contains 0.7% hydrogen, calculate its gross and net calorific values.
[*Ans.* HCV = 12,638 cal /g, NCV = 12,601 cal /g]

9. A sample of coal contains 89%C, 8% H, 3% ash. When this coal sample was tested in the laboratory for its calorific value in a bomb calorimeter, the following data were obtained.
Weight of coal burnt = 0.85 g
Weight of water taken = 650 g
Water equivalent of bomb and calorimeter = 2500 g
Rise in temperature = 2.5°C
Cooling correction = 0.03°C
Fuse wire correction = 10 cal
Acid correction = 50 cal **[*Ans.* GCV = 9205.2 cal/g, NCV = 8782.56 cal/g]**

10. In an experiment in a bomb calorimeter, a solid fuel of 0.90 g is burnt. It is observed that increased of temperature is 3.8°C of 4000 g of water. The fuel contains 1% of H. Calculate the HCV and LCV values.
(Equivalent weight of water = 385 g, Latent heat of steam = 587 Cal/g)
[*Ans.* HCV = 18514.5 cal/g, LCV = 18461.6 cal/g]

11. 0.72 g of a fuel containing 80% carbon, when burnt in a bomb calorimeter, increased the temperature of water from 27.3°C to 29.1°C. If the calorimeter contains 250 g of water equivalent is 150 g. Calculate the HCV of the fuel. Give your answer in kJ/kg. **[*Ans.* HCV = 4,200 cal/g]**

12. Calculate the GCV and NCV from the following data:
Mass of the fuel sample used for combustion = 0.85×10^{-3} kg
Mass of water in the copper calorimeter = 2.35 kg
Water equivalent of calorimeter = 0.45 kg
Specific heat of water = 4.187 kJ/kg/K
Increase in temperature of water = 3.2°C
Latent heat of condensation of steam = 2457 kJ/kg
Percentage of hydrogen in the fuel sample = 2.5
[*Ans.* GCV = 44,136 kJ/kg, NCV = 43,583 kJ kg^{-1}]

13. 0.945 g of fuel on complete combustion in excess of oxygen increased the temperature of water in a calorimeter from 13.25 to 19.2°C. The mass of water in calorimeter was found to be 1,458 g. Calculate the gross calorific value of the fuel, if the water equivalent of calorimeter is 144 g. **[*Ans.* 10,087 cal/g]**

14. A sample of coal containing 92% C, 5% H, 3% ash. When this coal was tested in the laboratory for its calorific value in the bomb calorimeter, the following data were obtained.

Weight of coal burnt = 0.95 g
Weight of water taken = 700 g
Water equivalent of bomb calorimeter = 2,000 g
Rise in temperature = 2.48°C
Cooling correction = 0.02°C
Fuse wire correction = 10.0 cal
Acid correction = 60.0 cal
Calculate the net and gross calorific values of the coal in cal/g. (Assume the latent heat of condensation of system is 580 cal/g)

[*Ans.* GCV = 7,031.6 cal /g, LCV = 6,770.6 cal/g]

15. A coal sample with 93%C, 5% H_2 and 2%ash is subjected to combustion in a bomb. Calculate the gross and net calorific values given that
Mass coal sample taken = 0.95 g
Mass of water in the calorimeter = 2000 g
Water equivalent of calorimeter = 700 g
Rise in temperature of water = 2.8°C
Latent heat of steam = 2457.2 J g^{-1} and specific heat of water = 4.2 $Jg^{-1} C^{-1}$.

[*Ans.* GCV = 33320 kJ/kg, NCV = 32214 kJ kg^{-1}]

4

Environmental Pollution and Water Chemistry

A. ENVIRONMENTAL POLLUTION

4.1 INTRODUCTION

Environmental pollution is the unfavourable alteration of our surroundings, wholly or largely as a byproduct of man's actions, through direct or indirect effects of the changes in the energy pattern, radiation levels, chemical and physical constitution and abundance of organisms.

Nowadays, the word environment is often used by almost all people around us, on television and in newspaper. Everyone is speaking about the protection and preservation of environment. Global summits are being held regularly to discuss environmental issues. During the last hundred years, mutual relationship among environment, social organization and culture has been discussed in sociology, anthropology and geography. All this shows the increasing importance of environment. Besides it is a fact that life is tied with the environment. Before understanding the linkages between the environment and the health, it is necessary to understand the meaning of environment and health.

4.2 AIR POLLUTION

Air pollution means any atmospheric condition in which certain substances such as dust, fumes, mist, odour, smoke or vapour are present in such concentrations that they can produce undesirable effects on man and his environment. It can either be natural or may be the result of various activities of man like industrial operations. The industrial contaminants can either be the by-product of external combustion like smoke, dust and sulphur dioxide or by-product of internal combustion like the reaction in petrol and diesel engine.

So, air pollution is the increase in concentration of undesired gases in the atmosphere.

or

The presence of materials in the air in such a concentration, which is harmful to man and the environment.

or

The occurrence or addition of foreign particles gases and other pollutants in the air, have adverse effects on human beings, animals, vegetation, buildings and other objects is called air pollution.

4.3 CLASSIFICATION

Air pollutants are classified according to:

1. origin,
2. chemical composition, and
3. state of matter.

1. According to Origin: According to their origin, the air pollutants are classified in two categories (Fig. 4.1).

1. Primary pollutants
2. Secondary pollutants

Primary pollutants: Primary pollutants are directly emitted from the sources into the atmosphere. Typical pollutants included under this category are:

(1) Inorganic gases such as SO_2, NOx, H_2S, CO, NH_3, CO_2, HF
(2) Particulate matter like ash, smoke, dust, fumes, mist, spray
(3) Olefinic and aromatic hydrocarbons
(4) Radioactive compounds

Secondary pollutants: Secondary pollutants are derived from the primary pollutants due to chemical or photochemical reactions in the atmosphere. **Example:** SO_3, NO_2, O_3, sulphate, nitrate, aldehydes and peroxyacetyl nitrate smog. These or usually formed by a photochemical reaction or by hydrolysis or oxidation reactions in the atmosphere.

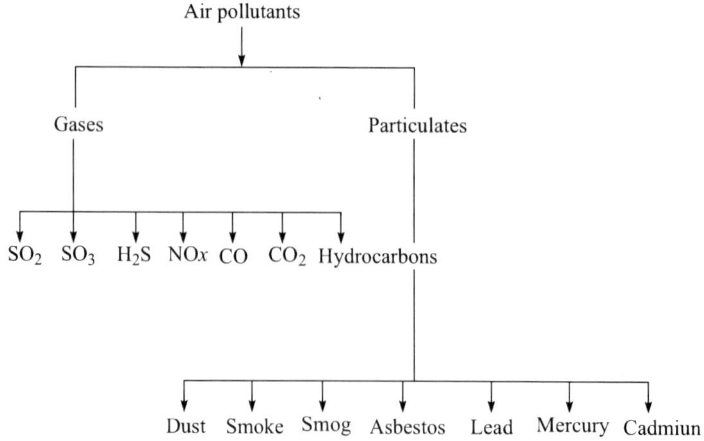

Fig. 4.1

2. According to Chemical Composition

(a) **Inorganic pollutants**

(1) Nitrogen compounds (NOx, NH_3)
(2) Sulphur compounds (H_2S, SO_2, SO_3, H_2SO_4)
(3) Carbon compounds (CO, carbonates)
(4) Halogen compounds (HF, HCl, fluorides)
(5) Oxidising agents like ozone
(6) Inorganic particles such as silica, asbestos, fly ash, dust, etc.

(b) **Organic pollutants**

(1) Hydrocarbons
(2) Ketones
(3) Alcohols

3. According to State of Matter: Air pollution can be classified into two main categories:

1. Gaseous air pollutants
2. Particulate air pollutants

4.3.1 Gaseous Air Pollutants

The pollutant which is mixed with air in the gaseous form and does not settle down is known as a gaseous air pollutant. **Example:** CO_2, nitrogen oxides, sulphur oxides hydrocarbons and photochemical oxidants. Important pollutant gases include the following:

1. Oxides of Carbon: These include CO and CO_2.

(A) Carbon monoxide (CO)

Nature:

1. It is colourless, odourless, tasteless gas which is chemically inert under normal conditions of temperature and pressure.
2. It has no effect on life and property at normal concentrations (0.1 ppm) but at higher concentration it seriously affects human metabolism.

Effects:

1. High CO level is potentially deadly and fatal to human life as CO is a very dangerous asphyxiant.
2. CO impairs a person's time interval discrimination, reduces vision and causes cardiovascular disorders.
3. It has a very high affinity for haemoglobin and on inhalation, combines selectively with haemoglobin of the blood (Hb) and forms carboxyhaemoglobin thereby reducing the oxygen carrying capacity of the blood. This affects the central nervous system causing giddiness, laziness and exhaustion, (Table 4.1).

Table 4.1

CO level (ppm)	Effects on Humans
10	Impairment of judgement and visual perception
100	Headache, dizziness, weariness
250	Loss of consciousness
750-1000	Death

Source:

1. Coal refuse (charcoal stoves)
2. Forest fires
3. Solid waste disposal
4. Transportation sources (traffic jams)
5. Gas heaters
6. Coal mines

Detection and Analysis: A large number of methods are available for both intermittent and continuous sampling of carbon monoxide. These methods use gravimetric, chemical, electrochemical and calorimetric process. Methods for monitoring ambient air quality standards for CO are gas chromatography, catalytic conversion and flame ionization detection.

Control: Adsorption, absorption, condensation and combustion are four basic technical control methods for CO and use of these methods can control almost all carbon monoxide emissions. Control of CO at the source is far more desirable than control by dilution in the ambient air.

(B) Carbon dioxide (CO_2)

CO_2 is comparatively less dangerous than CO and merely causes nausea and headache. Its major effect is on the climate of earth. CO_2 absorbs infrared terrestrial radiations and continuous increase in CO_2 concentrations could result in raising the temperature of the earth's surface so significantly that arctic ice would meet causing a rise in the oceanic level. As a consequence the coastal towns would be flooded.

Sources of CO_2:

1. Forestry
2. Agricultural practices (e.g., deforestation)
3. Jet planes use O_2 and release CO_2
4. Fossil fuel combustion.

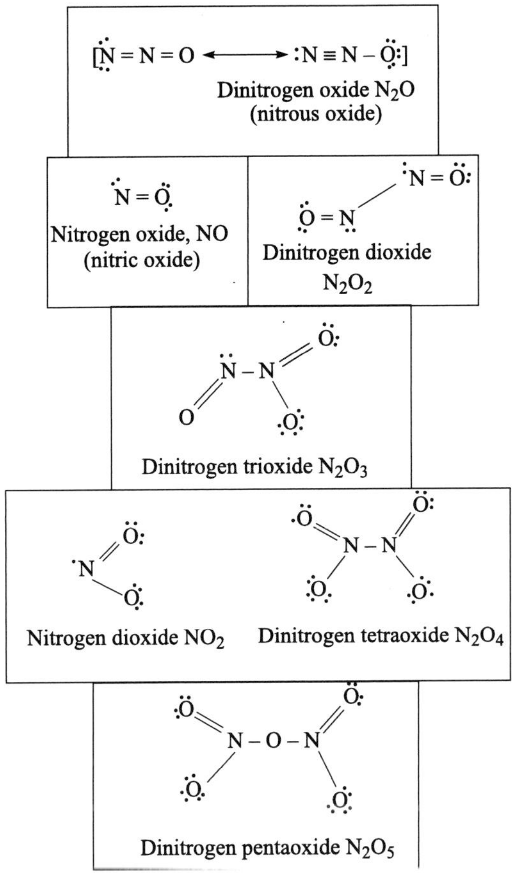

Fig. 4.2

2. Oxides of Nitrogen: These include nitrogen monoxide (NO) and nitrogen dioxide (NO_2).

NO is a colourless, odourless gas which is oxidised to NO_2 through secondary photochemical reactions. Being heavier than air, NO_2 is readily soluble in water and forms nitric acid, which falls out in the rain. NO is an inert gas but like CO, it can also combine with haemoglobin to reduce the oxygen carrying capacity of the blood. It is only moderately toxic as its concentration in the air is not high enough to become a health hazard. NO_2 is relatively more toxic as it irritates the alveoli of the lungs and high concentrations may even cause acute bronchitis (Fig. 4.2).

Sources of NO_2 pollution:

1. Lightning
2. Natural ionising radiations
3. Bacterial decomposition of organic matter
4. Forest fires

Detection and analysis: Montioring for NO requires that nitric oxide first be converted to NO_2 and its concentration determined indirectly. New methods for measuring NO and NO_2 have been developed including chemilominescense, ultraviolet spectrophotometry, electrical transducers and dispersive infrared spectrometry, pulverized ultraviolet fluroscene.

Control: Most control measures for NOx emission have been directed at modification of combustion conditions to decrease NOx production and at utilization of various devices to remove NOx from exhaust gas streams.

3. Oxides of Sulphur: These include sulphur dioxide (SO_2) and sulphur trioxide (SO_3).

Nature: SO_2 is a colourless gas having a characteristic sharp, pungent and suffocating odour. It is photochemically oxidised to SO_3. It is highly soluble in water and along with SO_3 forms sulphuric or sulphurous acid and is quickly washed out of the atmosphere by rain (acid rain).

Effects:

1. They are power pollutants and tend to irritate the mucous membranes of the respiratory tracts.
2. Higher concentrations cause bronchitis
3. Clothes, leather and paper are also affected.
4. They readily attack building materials especially, marble limestone and mortar. Plants are particularly sensitive to high concentration of sulphur dioxide and sulphur from chloresis (disappearance of chlorophyll) metabolic inhibition, plasmolysis and even death.

Source of SO_2 Pollution:

1. Transportation
2. Burning of solid and fossil fuels.

Detection and Analysis: Continuous and intermittent sampling are commonly used for SO_2 detection. The chlorimetric, conductometric and colorimetric methods are the most commonly used direct methods for continuous monitoring. Other direct monitoring techniques for SO_2 are spectroscopic, utilizing infrared or ultraviolet absorption as well as flame photometric methods. An indirect method used to approximate. SO_2 concentration involves using Huey sulphation plates.

Control: Methods for control of sulphur oxide emission include burning fuel with less sulphur, removing sulphur from fuel, converting coal by liquification or gasification substituting of another energy source, cleaning up the combustion products or dispersion by tall stacks.

4. Hydrocarbons: Gaseous and volatile hydrocarbons are responsible for causing air pollution, e.g., CH_4, C_2H_2, C_2H_4, terpenes.

Sources of hydrocarbons:

1. Coal fields
2. Forest fires
3. Agricultural burning
4. Industrial sources
5. Coal waste fires
6. Natural fires
7. Incomplete combustion from car engines

Detection and analysis: It is necessary to separate methane and non-methane components for the detection and analysis for hydrocarbons. Gas chromatography in combination with flame ionization is the most acceptable procedure capable of doing this, while gas chromatography provides the necessary sensitivity for hydrocarbons. It is still an expensive sophisticated procedure and at present is limited to research or short term quality studies.

Control: It is divided into five general categories

1. Incineration
2. Adsorption
3. Absorption
4. Condensation
5. Substitution of other materials

1. ***Incineration:*** It is the process for control of hydrocarbons are classified as either after-burners or catalytic after-burners. After-burners can be used to burn dilute organic vapours by adding fuel to the gas stream. The catalytic after-burner uses a solid active surface upon which combustion takes place.

2. ***Adsorption:*** Carbon adsorption of hydrocarbon is a proven method of reducing emissions with the possible economic advantage of recovering the material. The activated carbon can be adversely affected by particulate matter in the gas stream.

3. Absorption: Absorption of the hydrocarbons in a scrubbing solution is usually carried out in bubble plate columns, jet scrubbers packed towers, spray towers and venture scrubbers.

4. Condensation: The temperature of the gaseous hydrocarbons is lowered so that they turn into a liquid, which can be recovered for reuse. It is usually used only for high level hydrocarbon emissions.

5. Substitution of other materials: The substitution of less reactive hydrocarbons is utilized to reduce photochemical oxidant formation.

4.3.2 Particulate Air Pollutant

"Those pollutants which are mixed with air in liquid or solid state and after sometime settle down at the surface of the earth are called particulate pollutants". They differ widely in terms of particle density and particle size with sizes varying from 0.1 μm to 100 μm.

Particles larger than 50 μm can be seen with the naked eyes and those smaller than 0.005 μm can be seen only through an electron microscope. Larger particles like sand and water droplets quickly settle down in still air and smaller particles like dust remain in air for a long time whereas very fine particles like tobacco smoke do not settle down at all (Fig. 4.3).

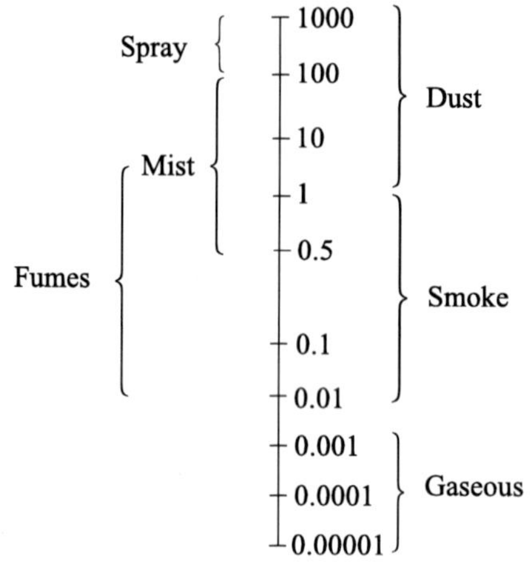

Fig. 4.3

They are categorized according to size, mode of formation or physical state.

1. ***Aerosols:*** These include all air-borne suspensions of solid or liquid particles smaller than 1 μm. **Example:** Dust, smoke, mist and fumes.
2. ***Dust:*** It consists of small solid particles (size 1 to 200 μm) which is formed by breaking up of larger masses of rock and soil either by natural disintegration or by mechanical processes of crushing, grinding, blasting, etc. They remain suspended in air temporarily but ultimately settle under the influence of gravity.
3. ***Smoke:*** It consists of fine solid particles (size 0.1 to 1 μm) resulting from the incomplete combustion of organic particles like coal, wood, tobacco or other chemical process.
4. ***Fumes:*** Fumes are formed by the condensation of vapours by distillation, sublimation, calcination and other chemical reactions. It is fine solid particles. The size of these particles is less than 1 micron. They are odourous vapours which may or may not be visible and are usually released from chemical or metallurgical processes.
5. ***Mist:*** It consists of liquid droplets and it is formed by the condensation of vapours in the atmosphere or released from industrial operations. The size of these particles is less than 10 μm.
6. ***Fog:*** If the mist is made up of water droplets whose concentration is high or dense enough to obscure vision, then the mist is called a fog.
7. ***Flyash:*** It consists of finely divided non-combustible particles present in the gases arising from fuel combustion. It contains inorganic, metallic or mineral substances released when the organic part of coal is burnt.

8. *Soot:* Incomplete combustion of carbon containing materials release carbon particles impregnated with tar. A collection of such particles is soot.
9. *Natural particulates:* Natural particulates are pollen grains, spores, bacterial viruses, protozoa, fungal spores and volcanic dust.
10. *Effect of particulates:* The greatest health hazard is constituted by trace element particulates such as Pb, Cd, Ni and Hg. Lead particulates are emitted by combustion of petroleum in automobiles, affect haemoglobin formation. Respiratory problems are caused when fly ash and soot are added to the air by the burning of coal. Metal dust containing Pb, Cr, Ni, Arsenic, Cd, zinc and mercury from metallurgical processes and cotton dust from textile milles also affect the respiratory system.
11. *Aerosols:* Fluorocarbons (NO_x) (SO_x) are released as emissions from the jet aeroplanes and supersonic aeroplanes. They deplete the ozone layer in the higher atmosphere thinning it considerably and hence permitting harmful ultraviolet light to reach the earth, smog has toxic effects on man, animals, plants and materials (cracks rubber) dust coating on leaves reduces photosynthesis and reduces plant growth. Animals feeding on particulate coated plants containing fluorides, arsenic or lead also suffer from toxic effects. Particulates damage materials by soiling textiles, corroding metals, discolouring and destroying painted surfaces.

4.3.3 Detection and Analysis

Settable particulates are usually measured by means of a dust fall bucket, while suspended particulate matter is often sampled by means of a high volume sampler opacity and pollen tests are also used to evaluate the amount of suspended particulates in the air. Other types of units for particulate sampling include centrifugal devices, impringers and impactors, scrubbers, electrostatic precipitators and thermal precipitators.

4.3.4 Control

Although the control of particulate matter can be undertaken either at the source or by dilution. "Dilution is the solution to pollution" is no longer applicable and cannot be considered a viable control method. Control at the source, the only acceptable method may depend upon principles of sedimentation, centrifugation, impaction, filtration or electrical charge.

4.4 SOURCES OF AIR POLLUTION

Air gets polluted largely due to the smoke produced by automobiles, power plants and kitchens. Due to large scale burning of fossil fuels such as coal, diesel, petrol, kerosene and so on.

1. Thermal power plants are coal based. The main pollutants are fly ash, soot and sulphur dioxide.
2. Fertilizer, plants produce, oxides of sulphur, particulate matter and fluorine. These pollutants come from sulphuric and phosphoric acid units, ammonia, nitrogen oxides.

3. The burning of fossil fuels produces carbon dioxide, carbon monoxide, sulphur dioxide, oxides of nitrogen, hydrocarbons, particulate matter and metallic traces. Coal produces a lot of smoke and dust while petrol and its products produce more sulphur dioxide.
4. The major pollutants from the textile industry are cotton dust, nitrogen oxides, chlorine, naphtha vapours, smoke and sulphur dioxide.
5. There are thousands of chemical and pesticide plants, which prepare caustic soda and produce chlorine gas.
6. Steel plants produce carbon monoxide, carbon dioxide, sulphur dioxide, fluorine particulate matter, phenol, cyanide, sludge slag, etc.
7. Automobiles contribute 60% of the air pollution by releasing compounds like carbon monoxides, oxides of nitrogen and hydrocarbons. Gases emitted during deceleration and acceleration are more harmful than those produced during constant speed. Incomplete combustion produces a hydrocarbon called 3-4 benzpyrene.
8. Decomposition of organic waste and municipal garbage produces foul smelling gases which pollute the air (Table 4.2).

Table 4.2

Pollutants	Major sources
1. Smoke, dust	Cement and gas works, iron and steel works
2. CO_2 and CO	Combustion of fossil fuels
3. SO_2	Oil refineries, iron and steel works, electricity generation, etc.
4. Oxides of nitrogen	Nitric acid industry, electricity generation, fertiliser plants, iron and steel industry
5. Ammonia	Fertilisers and ammonia industries
6. SO_3	Brickwork, H_2SO_4 industry
7. Chlorinated hydrocarbons	Drycleaning works
8. Mercaptans	Oil refineries, coke ovens
9. Freon (CCl_2F_2)	Refrigeration works
10. Cr, Mi, CO, etc.	Dyeing and printing industries
11. Sulphur and sulphides	Generating stations, rubber vulcanising, coke ovens, metal smelting, etc.

4.5 EFFECTS OF AIR POLLUTION

Our health depends on the quality of air we breathe in our immediate environment. Foul air is responsible for various health hazards and diseases. Airborne-spores, pollen grains, bacteria, fungi, hair, etc., cause various allergic reactions, bronchial asthma, tuberculosis and other infections. With rapid increase in industries and vehicular traffic, the air of our cities is becoming increasingly polluted. The various effects of air pollution are discussed below.

1. Effect of Particulate Matter: The important effects of particulate matter are:

(i) Smoke and dust particles which settle over the vegetation not only clog stomata, but also form a thin layer on the leaves to reduce light absorption and so photosynthesis.

(ii) It is found that cotton dust produced in fine mills, coarse and ginning mills produces pneumoconiosis or lung fibrosis known as byssinosis. Asbestos dust produces cancers of lung, stomach and intestine.

(iii) Particulate matter suspended in air scatters and also absorbs light, whereby the visibility is much reduced. It is hazardous to vehicle drivers.

(iv) Dust and smoke particles act as nuclei for condensation of water vapours and produce smog. The smog particles attract molecules of H_2S, NO_2, SO_2, etc. Thus, smog not only reduces visibility, but is harmful to mankind and plant life. Dust and smoke particles cause irritation of the respiratory tract and cause asthma, bronchitis and lung diseases.

2. Effect of Human Health: Air pollution can affect our health in many ways with both short-term and long-term effects. Different groups of individuals are affected by air pollution in different ways. Some individuals are more sensitive to pollutants than others. Young children and elderly people often suffer more from the effects of air pollution. People with health problems such as asthma, heart and lung diseases may also suffer more when the air is polluted. The extent to which an individual is harmed by air pollution depends on the total exposure to the damaging chemicals (Table 4.3).

Table 4.3

Gaseous pollutant	Source	Pathological effect on man
1. SO_2	Coal and oil combustion	Causes chest constriation, headeche, vomitting and death from respiratory system
2. NO_2	Soft coal, automobile exhaust	Inhibits cilia action, so that dust penetrate far into the lungs
3. H_2S	Refineries, chemical industries	Causes nausea, irritate eyes and throat
4. CO	Burning of coal, gasoline motor exhausts	Reduces the oxygen carrying capacity of blood
5. HCN	Blast furnace, fumigation chemical manufacturing metal plating, etc.	Interference with nerve cells, produce dry throat, headache
6. NH_3	Explosive, dry making fertilizers, plants and lacquers	Inflame upper respiratory passage
7. Aldehydes	Thermal decomposition of oil, fats and glycerol	Irritate nasal and respiratory tracts
8. Suspended particles (ash, soot, smoke, etc.)	Almost every manufacturing process	Cause emphysema, eye irritation and possibility of cancer

Short-term effects of air pollution include irritation to the eyes, nose and throat and upper respiratory infections such as bronchitis and pneumonia. Other symptoms include headache, nausea and allergic reactions. Short term air pollution can aggravate the medical conditions of individuals with asthma and emphysema.

Long-term effects include chronic respiratory disease, lung cancer, heart disease and even damage to the brain, nerves, liver or kidneys. Continuous exposure to air pollution affects the lungs of growing children and may aggravate or complicate medical conditions in the elderly people with heart or lung diseases also react more severely to polluted air.

3. Effect of photochemical oxidant: Some substances combine with other substances in the atmosphere to produce poisonous products. The tolerance limit of ozone is 0.3 ppm per hour when ozone combines with unburnt hydrocarbons of automobile exhausts, sunlight and aldehyde compounds known as *photochemical oxidants*. These causes irritation of eyes, nose and throat.

Another pollutant peroxyacetyl nitrate (PAN) is produced due to the combination of aldehydes, oxides of nitrogen, sunlight and ozone. It causes irritation of skin, eyes, nose and throat. It also produces cancer in human being.

4. Effect of agriculture: These effects have been classified as:

(a) ***Visible symptoms*** are the deviation from the healthy appearance to the leaves, chlorosis, glazing or silvering of leaf surfaces, stripping and decrease in yield.

(b) ***Non-visible symptoms*** involve, reduced plant growth, change in reproductive cycle, alteration in photosynthesis and metabolic activities, etc.

Major air pollutants which are photo-toxic to plants are O_3, SO_2, NO_2, fluorides, ethylene, PAN, acid mists, a number of organic oxidants, etc.

5. Effect on Forests: Forest as well as other natural systems, act as a sink for the removal of trace gases from the atmosphere. It consists of reduction in forest growth, change in forest species and susceptibility to forest pests.

6. Aesthetic loss: Air pollution has strong bearing on the aesthetic side of human life.

(i) Foul odour emitted by industries, automobiles, dirty drains and garbage heaps make urban life unpleasant.

(ii) Coal dust and their materials discharged from the industries settle down on the floor and other objects of houses and give dirty look.

(iii) A clear and transparent atmosphere is not only aesthetically pleasing but is necessary for clear vision. A dust haze or hanging smoke blurs our views.

7. Effect of Hydrocarbon

(a) ***On Man:***

(i) Aromatic hydrocarbons are toxic and cause irritation of the mucous membrane.

(ii) Secondary pollutants such as peroxyacetyl nitrate (PAN) and peroxybenzoyl nitrate, irritate nose, throat, eyes and cause chest constriction.

(iii) Higher levels of methane create narcotic effects on man.

(iv) Hydrocarbons like benzene, toluene and benzopyrene at 600 to 1000 ppm produce oedema, headache, drowsiness and symptoms akin to drunkenness.

(v) Cyclic hydrocarbons affect respiratory system, nervous system, dilate pupil's of eyes, cause loss of coordination, affect DNA and cell growth.

(b) ***On plants:*** Hydrocarbons and photochemical oxidants are injurious to plants. Ethylene at 0.1 ppm for several hours causes epinasty, leaf abscission flower dropping. Peroxyacetyl nitrate (0.01 to 0.05 for a few hours) cause glazing or bronzing of underside of leaf, damages sensitive plants, young leaves, necrosis of leaf tip. Grapes are particularly susceptible to attack.

(c) ***On material:*** Hydrocarbons include chemical alteration in paper, textile, rubber polymers making them more brittle and fragile.

8. Effect of Climate: Earth's climate depends on various factors including composition of atmosphere and balance of gases. Therefore, air pollution may bring about harmful effect on the climate.

- Aerosols and nitrogen oxides deplete ozone layers in the stratosphere which permits most of the harmful ultraviolet radiation to reach the earth.
- Heat produced by the industrial plants raises the temperature of the area.
- CO_2 content of the atmosphere is increasing due to destruction of forests and excessive burning of fossil fuels. It will lead to rise in global temperature by more than two or three degree due to greenhouse effect. A rise in global temperature may result in melting of glaciers and polar ice caps flooding of low lying coastal plains and submersion of islands. Rainfall pattern may also change thus affecting agricultural output.

9. Effect on Animals: The process by which the animals get poisoned is entirely different from that by which human beings exposed to air pollutants are poisoned. In case of animals, it is a two-step process.

(i) Accumulation of air pollutants in the vegetation and forage.

(ii) Subsequent poisoning of the animals when they eat the contaminated vegetation.

The pollutants mainly responsible for most livestock damage are:

Fluorine: Of all the farm animals, cattle and sheep are the most susceptible to fluorine toxicity. It is a cumulative poison under conditions of continuous exposure to sub-acute doses. Its effects are lack of appetite, rapid loss of weight, lameness, periodic diarrhoea, muscular weakness, wearing of teeth and death.

Lead: Chronic lead poisoning has been observed frequently in animals that have been grazing near smelters and lead mines. It causes paralysis and difficulty in breathing. There is complete loss of appetite, paralysis and diarrhoea.

Arsenic: In acute cases, it can cause severe salivation, thirst, vomiting, irregular pulse and respiration, abnormal body temperature and death in few hours. Chronic arsenic poisoning causes cough, diarrhoea, anaemia, abortion, paralysis and death.

4.6 LONG-TERM EFFECTS

1. Global warming: Warming on the global scale is expected to occur as a result of the increase of CO_2 and other greenhouse gases. The concentration of CO_2 has increased due to burning of fossil fuels and deforestation, which decreases the surface area available for photosynthesis. Other greenhouse gases such as methane and chlorofluorocarbon have increased by many times. The short lived constituents of the atmosphere such as NO_2, CO and non-methane hydrocarbon also play roles resulted to global warming because of their chemical reaction to the longer lived greenhouse gases.

4.7 FORMATION AND DEPLETION OF OZONE IN THE STRATOSPHERE

Ozone is an important chemical species present in the stratosphere. At a height of about 30 km, the concentration of ozone is about 10 ppm. The ozone layer present in the stratosphere acts as a protective shield for the life on the earth. It strongly absorbs ultraviolet radiations from the sun in the region 220 – 330 nm and thus protects the life on the earth from severe radiation damage, such as DNA mutation and skin cancer. Therefore, only a small fraction of ultraviolet radiations reaches the lower atmosphere and the earth's surface.

Formation of ozone: Three forms of oxygen are involved in the ozone-oxygen cycle. Oxygen atoms (O), oxygen molecule (O_2) and ozone (O_3). Ozone is formed in the stratosphere when oxygen molecule photodissociate after absorbing an ultraviolet photon whose wavelength is shorter than 240 nm. This produces two oxygen atoms. The atomic oxygen then combines with O_2 to create O_3. Ozone molecules strongly absorb UV light, between 310 and 200 nm, following which ozone splits into a molecule of O_2 and one oxygen atom. The oxygen atom then joins up with an oxygen molecules to regenerate ozone. This is a continuous process which terminates when an oxygen atom recombines with an ozone molecule to make two oxygen molecules.

Ozone is formed in the stratosphere by the following photochemical reactions:

$$O_2 + h\nu \xrightarrow{2A\,2\,\text{nm}} O + O$$

$$O + O_2 + M \longrightarrow O_3 + M$$

where, M is a third body, e.g., N_2.

4.7.1 Ozone Destruction

The third body absorbs the excess energy produced by the above reaction, whereby the ozone molecule is stabilized. Thus, ozone is constantly formed in the stratosphere. However, it is also destroyed by chlorine, released due to volcanic activity and also by reaction with nitric oxide, atomic oxygen and reactive hydroxyl radical. These substances are also present in the atmosphere, by the following reactions.

(a) $O_3 + NO \longrightarrow NO_2 + O_2$

(b) $O_3 + O \longrightarrow O_2 + O_2$

(c) $O_3 + HO^\bullet \longrightarrow HO_2 + HOO^\bullet$

$HOO^\bullet + O \longrightarrow HO^\bullet + O_2$

Manmade factors: Ozone, in the stratosphere is also found to be destroyed by manmade chlorofluorocarbons (CFCs) which are used as coolants in refrigerators, air conditioners, propellants in aerosol sprays and in plastic foams, such as thermocole or styrofoam. The CFC molecules, escaping into the atmosphere, decompose to release chlorine in the ozone layer (by photodissociation) and each atom of chlorine, thus, formed is capable of attacking several ozone molecules.

$$Cl + O_3 \longrightarrow ClO + O_2$$

This reaction is followed by

$$ClO + O \longrightarrow Cl + O_2$$

which regenerates chlorine atoms. So, a long chain process is invalid, which conserves chlorine atoms.

4.7.2 Ozone Hole

The ozone hole is an area of the stratosphere in which the recent ozone levels have dropped to as low as 33% of their pre -1975 value. The depletion of O_3 layer was found more severe in polar regions (North and South Poles). The overall cause of ozone depletion is the presence of chlorine containing source gases (primarily CFCs and related hydrocarbon). In the presence of UV light, these gases dissociate, releasing chlorine atoms which then go on to catalyze ozone destruction. The chlorine catalyzed ozone depletion can take place in the gas phase, but it is dramatically enhanced in the presence of polar stratospheric clouds.

Detection of the ozone hole over antarctica in 1985 attracted the attention of whole scientific community in the world. The USA immediately banned the use of CFCs in spray cans. Further in 1987, twenty-four nations of the world signed the Montreal Protocol which aims at 35% reduction in the global production of the CFCs by the year 1999. Efforts are still on to curb the use of CFCs. Simultaneously, efforts to produce chlorine-free substitutes have also started. In fact, synthesis of a product called HFC-134 has already been reported as an effective substitute for CFC.

4.7.3 Effects of Ozone Layer Depletion

With the depletion of atmospheric ozone layers, the extent of ultraviolet radiation over the earth's atmosphere increases. Ultraviolet radiations are harmful to man's life (Fig. 4.4). They cause skin cancer, swelling of skin, sun burns, burning sensation, skin ageing, leukaemia, breast cancer, cataract of eyes, oedema haemorrhage, lungs injury, lung cancer, dizziness, visual impairment, premature ageing, DNA breakage, inhibition and alteration of DNA's replication and formation of DNA adduct and may even lead to death (Table 4.4).

Table 4.4 Effect of ozone on human beings

Concentration (ppm)	0.2	0.3	0.8	1.0–3.0	9.0–10.0
Ill effects on health	Not several effects	Irritation of nose and throat	Lungs undergo genetic alteration within a week	Extreme fatigue after 2 hours	Chronic pulmonary oedema

4.7.4 Importance of Ozone Layer

The ozone layer acts as a protective shield for the existence of life on the earth. It absorbs most of the harmful UV-radiations coming from the sun in the region (220 – 330 nm) and thereby protects the life from radiation damage. In the absence of ozone layers, these ultraviolet radiations would cause:

(a) Diseases like skin cancer in men and animals, swelling of skins, burning sensation skin aging, leukaemia, breast cancer, cataracts of eyes, haemorrhage, lungs injury, premature aging, visual impairment, dizziness, inhibition and alteration of DNA's replication and formation of DNA adduct and may even lead to death.

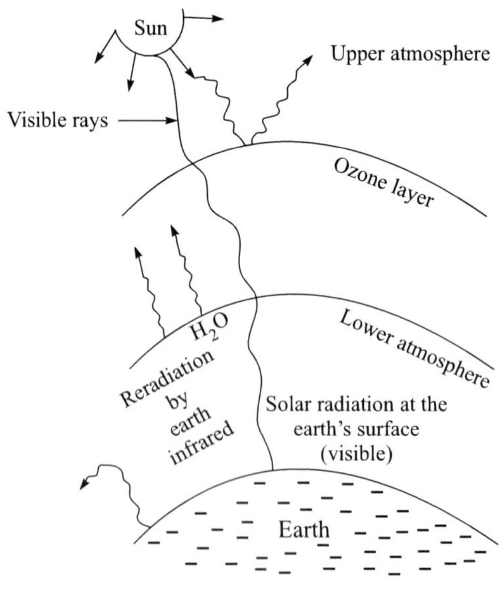

Fig. 4.4 *Importance of O_3 layer*

(b) DNA mutation
(c) Damage to plants
(d) Faster deterioration of paints, fabric's plastics, etc.

4.8 SMOG

The term 'smog' was first coined by Dr. Henry Antoine des Voeux in his 1905 paper "Fog and Smoke" for a meeting of the Public Health Congress, it may be defined as:

"The poisonous mixture of smoke, fog air and other chemicals is called smog"

or

Smog is a mixture of smoke (from coal combustion) and fog in suspended droplet form. There are two general types of smog.

1. Photochemical smog (Los Angeles smog)
2. Chemical smog (London smog)

4.8.1 Photochemical Smog (Los Angeles Smog)

It is an oxidising smog. Photochemical smog is formed in traffic congested metropolitan cities where warm conditions and intense solar radiation are present. It is composed mainly of ozone (O_3), peroxyacetyl nitrate (PAN) and NOx. Oxides of nitrogen and hydrocarbons are produced by the combustion of fuels in automobiles and in industries. Absorption of UV-rays by oxides of nitrogen initiates the process of smog formation. Nitrogen dioxide splits up into reactive species and starts a number of chain reactions in which the hydrocarbons are involved. This results in the formation of many secondary pollutants.

$$N_2(g) + O_2(g) \xrightarrow[\text{or furnaces}]{\text{In i.c. engines}} 2\,NO(g)$$

$$2NO(g) + O_2(g) \longrightarrow 2NO_2(g)$$

$$NO_2(g) \xrightarrow{\text{UV light}} NO(g) + [O]$$

$$O_2(g) + [O] \longrightarrow O_3(g)$$

Hydrocarbons + O_3, O_2[O], $NO_2 \longrightarrow$ Peroxide, Peroxyacetyl nitrate (PAN) Formaldehyde, ozone aldehyde, acrolein, etc. (oxidized hydrocarbons).

In presence of humidity, oxidized hydrocarbons and ozone cause "photochemical smog, which is characterised by the formation of aerosols that reduce visibility, generation of brown hazy fumes, which causes irritation to eyes and lungs and also causes extensive damage to vegetation and rubber goods.

This type of smog was observed in some parts of Los Angeles and Denver in USA and that is why it is sometimes referred to as Los Angeles smog (Table 4.5).

Table 4.5 Effects of photochemical smog

Components of photochemical smog	Characteristics and hazardous effects
1. Ozone (O_3)	(1) Pungent small, toxic (2) It can cause coughing, wheezing, bronchial constriction and irritation to the respiratory mucous system. Rubber has a high affinity for ozone and is cracked and aged by it. (3) O_3 also causes damage to vegetation and reduction in plant growth and crop productivity.
2. Peroxyacyl nitrates (PAN) and aldehydes	(1) Eye irritants (2) PAN has the highest toxicity to plants of all the components of smog. It attacks younger leaves and causes bronzing and glazing of their surfaces.
3. Nitrogen oxides (NO*x*)	NO_2 is extremely toxic to living tissues and harmful to paints, textiles and metals.

4.8.2 Chemical Smog (London Smog)

This type of smog is thought to be caused by sulphur dioxide. It is formed by combination of smoke, dust, fog and SO_2. It is usually formed around industrial and power plants. It is generally bad in the early morning hours and becomes worse after sunrise due to sunlight induced oxidation of SO_2 to SO_3 followed by reaction with humidity. *Sulphuric acid is formed in this type of smog which is very harmful to people suffering from bronchial diseases such as asthma or emphysema.*

4.9 CONTROLLING MEASURES

For controlling the formation of noxious photochemical pollutants in the atmosphere, the emission of primary pollutants, viz., level of hydrocarbons and NO_2 should be controlled by providing catalytic converters to modern automobiles. There are four approaches available for the control of emissions discharged into the atmosphere. They are:

(i) ***Dispersion of source locations:*** Air pollution can be controlled by dispersion of the sources of air pollutants through allocation of land, i.e., by proper planning and zoning of industrial areas.

(ii) ***Dilution:*** By using tall stacks for thermal plants, the pollutants can be discharged at a sufficient height from the ground, where the air movement, both horizontal and vertical is more and chances of downward movement of air are less. This helps in dispersion of pollutants over a larger area in less time and hence dilute the concentration of pollutants near the source.

(iii) ***Reduction at source by process changes:*** This can be achieved by:

(a) Substitution of raw materials: **Example:** smoke, soot.

(b) Substitution of fuel: **Example:** desulphurization and reduced emissions of SO_2 suspended particulate matter.

(c) Modification of the process: Saintary landfills can be used instead of incinerators.

(d) Modification of the process equipment or repair and maintenance of the exising equipment helps in reducing atmospheric pollution.

(iv) ***Reduction at source by using control equipment:*** Various control devices are used for controlling different types of pollutants emitted from different sources.

4.9.1 Control of Gaseous Pollutants

Gaseous pollutants can be controlled through the techniques of combustion, absorption, adsorption, and condensation.

Combustion: In this method, the gaseous emissions are burnt at high temperatures (600 to 1000°C). The combustion control of gaseous pollutants is used in petrochemical fertilizer paints and varnish industries.

Absorption: In absorption technique gaseous pollutants are absorbed in suitable adsorbent materials. The most widely used absorbers (or solvents) are magnesium oxide, calcium oxide and calcium carbonate (lime).

Adsorption: This method is used to remove toxic gases, vapours and inflammable compounds from the polluted air effluents by employing multiple fixed-bed adsorber where gaseous pollutants are adsorbed by adsorbents. The most commonly used adsorbents are activated carbon, activated alumina, bauxite bone char, silica gel, etc.

Condensation: The widest application of condensors is in control of hydrocarbons. Otherwise, they are considered pre-treatment devices for air pollution control measures and are thus used in conjunction with adsorber, absorber and combustion units.

4.9.2 Control of Automobile Exhaust

1. The load of lead in the exhaust can be reduced by using lead-free petrol.
2. The unburnt hydrocarbons in auto emissions can be reduced by the use of efficient engines.
3. Catalytic converter filters in the vehicles can convert NO to nitrogen and reduce the potential hazards of NO*x*.
4. By using good quality automobile fuels, the toxic contaminants in exhaust can be reduced drastically.
5. The toxic contaminates in exhaust can significantly be minimized by the use of automobile engines operated with compressed natural gas (CNG).

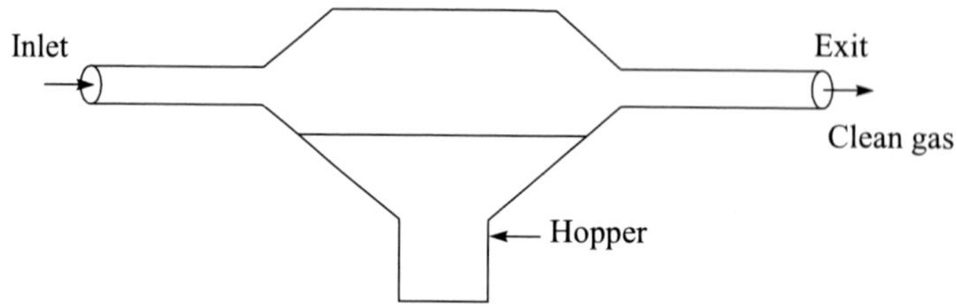

Fig. 4.5 *Settling chamber*

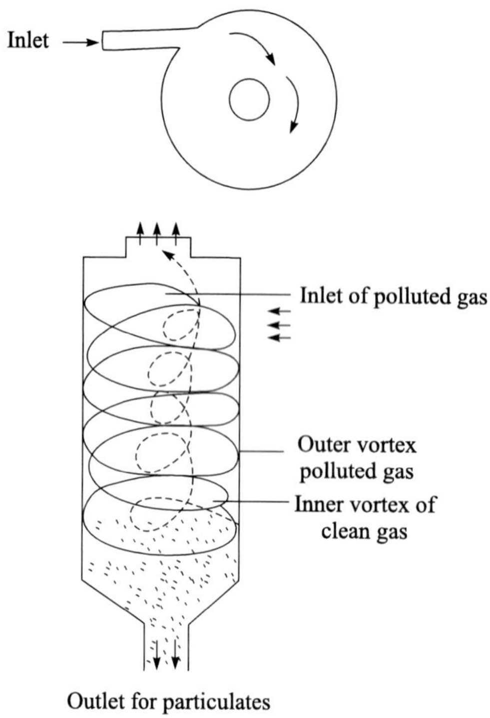

Fig. 4.6

4.9.3 Control of Particulate Matter

The various types of control devices or equipment used for the removal of particulate matter are briefly discussed as follows.

1. **Settling chamber:** It is the simplest type of equipment used for the collection of solid particles. It consists of a chamber in which the carrier gas velocity is reduced, so as to allow the particulates to settle out of the moving stream under the action of gravity at the base of the chamber. The gas velocity in the chamber is kept sufficiently low, so that the settling time of a particle entering the chamber at the top is same or less than the time taken by the flue gas to pars through the chamber (Fig. 4.6).

2. **Cyclone separators (Reverse flow cyclone):** A simple cyclone separator consists of a cylinder with a conical base. A tangential inlet discharging near the top and an outlet for discharging the particulates is present at the base of the cone (Fig. 4.7).

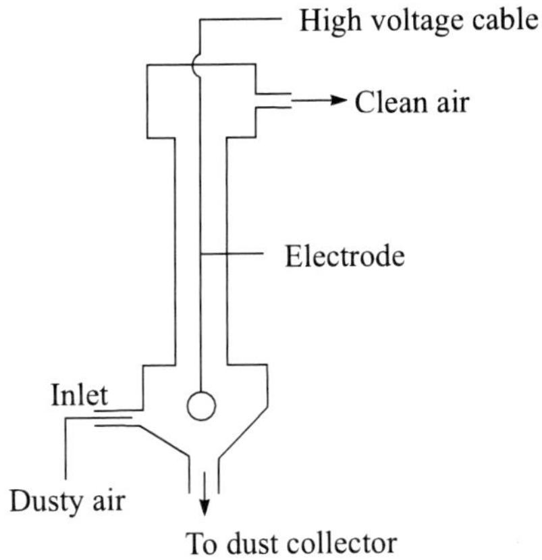

Fig. 4.7 *Electrostatic precipitators*

Mechanism of action: The dust laden gas enters tangentially, receives a rotating motion and generates a centrifugal force due to which the particulates are thrown to the cyclone walls as the gas spirals upwards inside the cone, i.e., flows reverse to form an inner vortex which leaves flow through the outlet. The particulates slide down the walls of the cone and are discharged from the outlet.

3. **Filters:** Particulate matter can be filtered, if the fumes containing them are forced to pass trough a filtering device. The particles are held, while the gases pass through the media. Cloth fabric or fibrous medium, like mats of wool, cellulose, etc., may be used as a separator. The most common type of fabric collector used in industries is bag filters.
4. **Electrostatic precipitators:** The electrostatic precipitator works on the principle of electrostatic precipitation, i.e., electrically charged particulates present in the polluted gas are separated from the gas stream under the influence of the electrical field (Fig. 4.8).

   A typical wire and pipe precipitator consist of:
   1. A positively charged collecting surface (grounded)
   2. A high voltage (50 kV) discharge electrode wire
   3. Insulator to suspend the electrode wire from the top
   4. A weight at the bottom of the electrode wire to keep the wire in position.

Mechanism: The polluted gas enters from the bottom, flows upwards. The high voltage in the wire ionises the gas. The negative ions migrate towards the grounded surface and pass on their negative charge to the dust particles also. Then these negatively charged dust particles are electrostatically drawn towards the positively charged collector surface, where they finally get deposited. The collecting surface is rapped or vibrated to periodically remove the collected dust particles so that the thickness of the dust layer deposited does not exceed 6 mm, otherwise the electrical attraction becomes weak and efficiency of the electrostatic precipitators gets reduced.

As the electrostatic precipitation has 99 + present efficiency and can be operated at high temperature (600°C) and pressure at less power requirement therefore, it is economical and simple to operate compared to other devices.

4.10 SOLID WASTE MANAGEMENT

It can be defined as "those wastes which have been rejected for further use and which can neither be transported by water into streams nor can readily escape into the atmosphere. Any unwanted or discarded material from residential, commercial, industrial, mining and agricultural activities that causes environmental problems may be termed solid waste".

Solid waste management comprises purposeful and systematic control of the generation, storage collection, separation, transportation, processing treatment recycling recovery and disposal of solid waste.

4.11 SOURCE AND CLASSIFICATION OF SOLID WASTE

Solid waste can be classified into the following categories:

1. Municipal solid waste: Municipal solid waste consists of household activities, restaurants, public places, institutions markets, street-sweepings, etc. This garbage is generated mainly from residential and commercial complexes. They may be categorised as:

> *Garbage:* It includes putrescible organic waste like the animal, fruit or vegetable residues resulting from the handling, preparation, cooking and eating of foods.
>
> *Rubbish:* It includes combustible and non-combustible solid waste, excluding food wastes or putrescible materials, combustible rubbish includes paper, cardboard textiles, plastic, rubber wood, garden-trimmings, etc. Non-combustible rubbish consists of glass, crockery, tin-cans, aluminium cans, metals, construction wastes, etc.

2. Domestic waste: These include waste from household preparation, cooking and serving, of food, waste paper, cloth rags, plastic, etc.

3. Industrial waste: These include chemicals, paints, sand, metal ore processing, fly ash, sewage treatment sludge, etc. Manufacturing industries produce wastes which are solid or semi-solid. This waste can be self-igniting, explosive, toxic or radioactive. Chemical process industries generate a variety of waste both organic and inorganic, which are mixtures with wide range of component concentration.

4. Agricultural waste: These wastes result from farms, feed lots and livestock yards including paddy husk passage from sugarcane tobacco and corn residues, slaughter house wastes, manures, etc.

5. Biodegradable waste: The solid waste material which can be degraded by microorganisms is called biodegradable waste. **Example:** dry leaves, peanut shells, egg shells, tea leaves, etc.

6. Non-biodegradable waste: The waste that cannot be degraded by microorganisms is called non-biodegradable waste, **Example:** polyethyene bags, scrap metal, glass bottles, etc.

7. Hazardous waste: A waste is said to be hazardous if it exhibits any of the following characteristics: viz., ignitability, corrosivity, reactivity or toxicity. Typical hazardous wastes are radioactive substances, chemicals, biological wastes, flammable, wastes, and explosives. The sources of hazardous wastes are industries, nuclear plants, hospitals research institutes, laboratories.

4.12 REASONS FOR SOLID WASTES

The reasons for the rapid growth in the quantity of solids wastes are over-population, affluence and technology.

Over-population: The main reason for increasing quantity of solid waste is the over-population. The over-population increases the amount of polluants. It is also true for solid waste pollution.

Affluence: There is a tendency to declare items as being in or out of fashion and promptly throw away the ones out of fashion. This result in solid waste pollution.

Technology: Rapidly growing technologies for most economic goods are leading to returnable packing to non-returnable packaging. For example: returnable glass bottles/ containers being replaced by non-returnable cans, plastic containers, etc.

4.13 EFFECT OF SOLID WASTES POLLUTION

The improper handling and transfer of the solid wastes results in various health and environmental hazards such as:

1. Improper handling of solid waste damages the environment. The environment damage caused by solid waste is mostly unaesthetic in nature.
2. Diseases like bacillary dysentery, diarrhoea and amoebic dysentery may result in humans from food and water contamination though flies, which breed on the refuse dump and solid waste.
3. Rats depending upon these solid wastes may also cause plague, salmonellosis, trichinosis, endemic typhus like diseases through direct bite.
4. The crops and water supply may also get contaminated and may result in large scale epidemic of cholera, jaundice, gastrointestinal diseases, hepatitis, etc.
5. There is also aesthetic danger to the surrounding environment, as the stray animals and scavengers invade the garbage dumps on roadsides.
6. There is danger of water pollution when the leachate from a solid waste dump enters surface water or groundwater resource.

4.14 TECHNIQUES OF SOLID WASTE MANAGEMENT

It is important to note that four techniques have to be followed for solid waste management.

1. **Refuse:** Instead of buying new containers from the market, the ones that are already in the house should be used.

2. **Reuse:** Do not throw away the soft drink cans or the bottles; cover them with home made paper or paint on them and use them as pencil stands or small vases. Use shopping bags made of cloth or jute, which can be used over and over again.
3. **Recycle:** Segregate your wastes to make sure that it is collected and taken for recycling.
4. **Reduce:** The generation of unnecessary waste should be reduced. **Example,** carrying one's own shopping bag while going to the market and put all your purchases directly into it.

4.15 METHODS OF SOLID WASTE MANAGEMENT

1. Sanitary landfilling
2. Composting
3. Incineration
4. Pyrolysis
5. Recycling

1. Sanitary landfilling: It is the cheapest method of refuse disposal. It may be defined as "A method of disposing refuse on land without creating nuisance or hazards to public health". This is a biological method of waste treatment and bacterial refuse digestion results in decomposition products like CO_2, CH_4, H_2S and H_2O which can be used as renewable sources of energy.

This is a modified form of open dumping. Waste is deposited in 0.9 – 4.5 m thick layers in depressions and then compacted and covered at least once a day by earth with bulldozers. The covering prevents breeding of flies, rats, etc. While selecting site for landfilling, it is necessary to examine, if any underground potable water source in the vicinity will be polluted and if the neighbourhood habitation would be affected by odour or fire.

This method does not cause environmental damages by creating nuisance or health hazards as the refuse is covered and prevents breeding of pests and disease vectors. Besides there is no danger of air pollution resulting from burning and no water pollution provided precaution is taken to avoid leachates of refuse from contaminating the surface or underground water source. This prevention can be taken by using a plastic membrane or water tight membrane on the base.

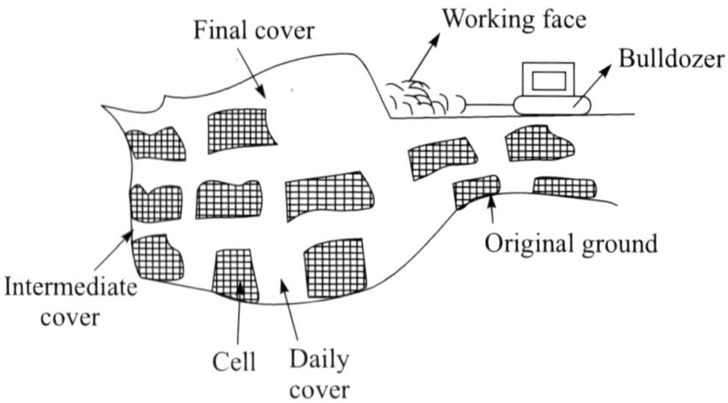

Fig. 4.8 *Sanitary landfill operation*

Advantages:

1. The process is completely sanitary.
2. Converts low-lying, marshy wasteland into useful area.
3. Natural resources are returned to soil and recycled.
4. Land can be safely overloaded without causing any problems.
5. Highly skilled personnel are not required.
6. Natural resources are returned to soil and recycled.
7. Fire hazards are minimum.
8. There is no air pollution from burning.

Disadvantages:

1. A large area is required.
2. Since land is available away from town, transportation cost is heavy.
3. Fire hazard due to formation of methane in wet weather.
4. Use of insecticides is required.
5. Covering good earth required for top layer may sometimes be difficult to obtain.
6. Continuous evolution of foul smell near the site of disposal.

2. Composting: It is a biological process in which microorganisms, mainly fungi and bacteria, convert degradable organic waste into humus like substances. This finished product, which looks like soil is high in carbon and nitrogen and is an excellent medium for growing plants. It recycles the nutrients and returns them to the soil as nutrients.

Method of composting: Composting may be carried out naturally under controlled conditions or in mechanized composting plants. The whole composting process involves:

1. *Waste preparation:* The waste is ground in hammer mills or wet pulpers to desired size mixed with nutrient source filler and water.
2. *Digestion:* The mixture is placed in windrows, decomposed and allowed to stabilize for 2 to 5 weeks.
3. *Product upgradation:* It is achieved by curing, grinding, screening, pelletising and bagging.

   Apart from being clean, cheap, safe composting can significantly reduce the amount of disposable garbage. The organic fertilizers can be used instead of chemical fertilizers. It increases the soil's ability to hold water and makes it easier to cultivate. It helps the soil retain more of the plant nutrients.

Advantages:

1. It aids in preventing soil erosion by keeping the soil covered.

2. It helps in controlling the growth of weeds in the garden.
3. It helps keep the soil cool in summer and warm in winter.
4. It makes soil easier to cultivate.
5. It helps reduce the adverse effects of excessive alkalinity, acidity or the excessive use of chemical fertilizers.
6. Compost allows the soil to retain more plant nutrients over a longer period.
7. It supplies part of the 16 essential elements needed by the plants.

3. Incineration: It is the process of burning waste in large furnaces. The wastes like plastics, carboard paper, wood scrap, food wastes, rubber, etc., are subjected to incineration. This process results in air pollution by producing fly ash, hydrogen chloride, corrosive fumes of sulphur dioxide, etc. This method is hygeinic as all the pathogens and insects are destroyed. This method can generate some revenue by raising steam power.

The final products will be ashes and clinkers. The ashes are dispersed by dumping in low lying areas, while the clinkers can be used as aggregate for low grade concrete or as road material.

Advantages:

1. This method is sanitary, as all the pathogens and insects are destroyed.
2. No odour and dust nuisance.
3. Some revenue can be generated by raising steam power and selling of the clinkers.
4. It requires very little space.
5. Cost of transportation is not high as incinerators are located within city limits.

Disadvantages:

1. Its capital and operating cost is high.
2. Needs skilled personnel
3. Air pollution may be caused.

4. Pyrolysis: It is an endothermic process and thus differs from the conventional incineration. In this disposal method the solid wastes are heated under anaerobic conditions, i.e., burning without oxygen. The organic components of the solid waste split up into gaseous liquid and gaseous fractions (CO, CO_2, CH_4, etc.).

5. Recycling: It is an integral part of solid waste management. The recovery of solid waste components for possible use as raw material is called recycling or salvaging. It involves separating materials (such as scrap metal, glass paper, plastic, etc.) from refuse and reprocessing them for reuse. It also makes economic sense. For stable economic growth, resources must be used carefully and technologies for recycling of waste are to be evolved. Available materials is for future generations and will also save energy and environment.

The various recycling techniques are as under:

1. Melting plastic domestic toys and moulding them into new ones.
2. Melting blast furnace slag for making artificial jewellery, etc.
3. Use of incineration heat.
4. Food sources from organic wastes for livestock.
5. Crushing of materials such as discarded furniture, appliances, etc.
6. Converting waste into solid fuel (wood chips, raw dust, etc.).
7. Methane fermentation and residual sludge.
8. Thermal decomposition of waste organics in the form of gas and oil.
9. Selection for magnetic, non-magnetic and specific gravity based materials.
10. Composting garbage and using it as manure.
11. Alcohol fermentation

4.16 REUSE OF WASTE MATERIALS

The refillable containers which are discarded after use can be reused. Villagers make casseroles and silos from waste paper and other waste materials. Making rubber rings from the discarded cycle tubes which are used by the newspaper vendors instead of rubber bands reduces the waste generation during manufacturing of rubber bands. Because of financial constraints poor people reuse their materials to the maximum.

B. WATER CHEMISTRY

4.17 INTRODUCTION

Water is God's greatest gift to mankind. It is vital for all known forms of life. It covers nearly 72% of the earth's surface. Without food human being can survive for a number of days but without water no body can survive. It provides a medium for all biochemical reactions that take place inside our body. It is a universal solvent and the versatile nature of water is due to some of its interesting properties. From an engineer's point of view, water is no less important. It is required in boilers for production of steam, which acts as a source of energy, as a coolant in many power and chemical plants in steel, paper, textile, ice and many other such industries. Despite its incredible importance, water is receiving least attention and is getting contaminated in ways beyond description.

Water is obtained from nature in various forms such as rainwater, and river water. In addition to the inherent impurities water gets contaminated due to a number of contaminants from sewage, industries, etc. during usage. Water is not only essential for the lives of animals and plants but also occupies a unique position in industry. Surface and groundwater is normally used for industrial and domestic purposes. Each industry has its own specifications for water and so the treatment of water depends upon the purposes for which it is to be used. Some of the major parameters which are generally checked for assessing the water quality for a particular use are determination of hardness, alkalinity, turbidity, total dissolved solids, suspended solids, dissolved oxygen (DO), biological oxygen demand (BOD), chemical oxygen demand (COD), total residual chlorine pH, temperature, etc. Hence, a complete analysis with respect to these is desired for determining the suitability of water for a particular industrial use.

Depending upon the nature of impurities present in water, it has to be treated. In this chapter, we will study the determination of some of these parameters. We will also study the main sources of water.

4.18 SOURCES OF WATER

The main sources of water are:

1. Rainwater
2. Surface water
3. Groundwater

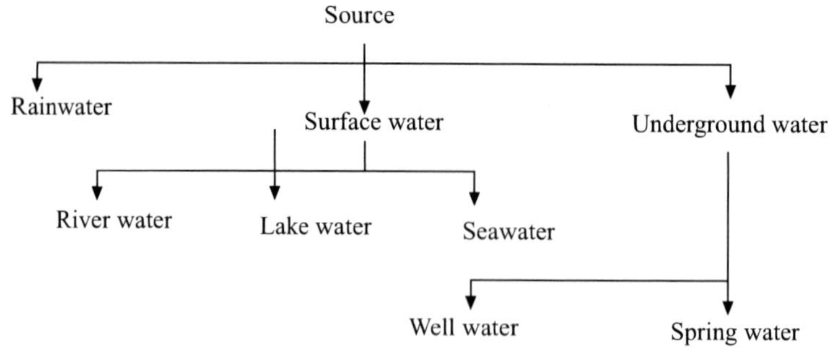

1. Rainwater: It is obtained from evaporation of surface water. It is the purest form of natural water. But rainwater during its downward path to the earth dissolves a considerable amount of gases, for example: CO_2, SO_2, NO, etc., and suspended solid particles present in the atmosphere. So it becomes polluted.

2. Surface Water: The water which comes to the surface through rain is of three types:
(i) River water (ii) Lake water (iii) Seawater

(i) River Water: Springs and rainwater are the sources of river water. This water is less pure than rainwater. River water contains dissolved inorganic salts such as chlorides. In addition to the dissolved impurities obtained from the soil, river water also contains organic materials from the soil. River water also contains the organic products (from decomposition of plants and animals) and small particles of sand and rock as suspended impurities.

(ii) Lake Water: In this type of water the main impurities are that of organic matter. It contains less of dissolved impurities compared to river water and thus, it is much purer than river water.

(iii) Seawater: This is the most impure form of natural water. This source of water is also fed by rain and spring waters. It contains 3.6% of dissolved salts of which 2.6% is NaCl. Other salts present are sulphates, bicarbonates, bromides of sodium, potassium, magnesium, etc. In addition, rivers also carry some other impurities into seawater.

3. Groundwater: Various forms of underground water sources are:

(i) Spring Water: The quality of water from these water sources depend on formation of earth. Some spring waters possess minerals of medicinal value such as colloidal sulphur, hydrogen sulphide and borax, etc.

(ii) Well Water: The sources are rich in mineral content. It is formed by digging earth deep down or near the underground source.

4.19 IMPURITIES IN WATER

Water, from all the natural sources, has got impurities in dissolved, suspended and colloidal forms.

Impurities in Water

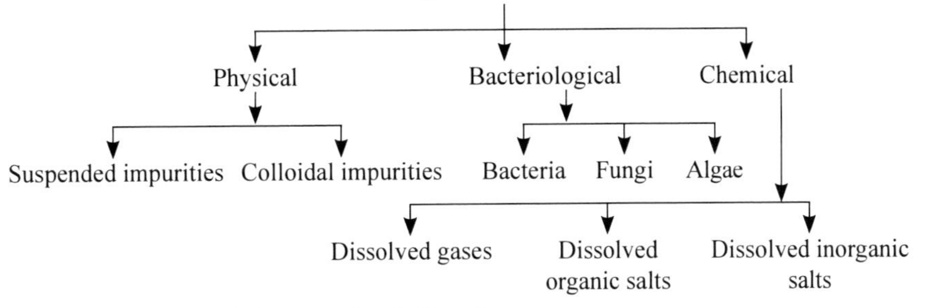

The sources of these impurities are the following.

- Water takes up impurities from the ground, or soil with which it comes into contact.
- When water comes with industrial waste it may be contaminated with acids, bases, heavy metal ions and other harmful chemicals.
- Water from well or tubewell may contain dissolved minerals such Ca^{2+}, Mg^{2+}, F^-, Cl^- ions, etc.
- Organic impurities in water, it may be contaminated with microorganisms like, bacteria, algae, viruses, etc.

On the basis of the above discussions, the impurities present in water are classified into the following types:

1. Dissolved impurities
2. Suspended impurities
3. Biological impurities
4. Colloidal impurities

Table 4.6

Impurities	Organic/Inorganic water
Dissolved impurities	(a) Salts: carbonates, bicarbonates, sulphates and chlorides of calcium, magnesium, iron, sodium and potassium (b) Gases: CO_2, O_2, N_2, H_2S, NH_3, etc.
Suspended impurities	(a) Inorganic – sand, clay (b) Organic – vegetable and animal matter
Colloidal impurities	Finely divided silica and clay, organic waste products, colouring matter, etc.
Biological impurities	Bacteria, algae, fungi, virus, etc.

4.19.1 Dissolved Impurities

These impurities are present in the dissolved form in water. These are mostly inorganic salts and dissolved gases. For example, salts, such as bicarbonates, chlorides, sulphates of calcium, magnesium, iron, sodium potassium and gases such as O_2, CO_2, N_2, H_2S, NH_3, etc. Gases can be removed by deaeration and by certain chemical treatments. Some inorganic salts like chlorides, bicarbonates and sulphates of calcium and magnesium increase hardness of water. Inorganic salts can be removed by ion exchange, reverse osmosis, electrodialysis, etc.

4.19.2 Suspended Impurities

These impurities are present in both organic form such as vegetable and animal material and inorganic form such as sand, clay, silt, rust powder, etc. This type of impurities are heavy and settle down to bottom. These can be removed either by filtration or by coagulation.

4.19.3 Biological Impurities

These consist mainly microorganisms like amoeba, bacteria, algae, fungi, protozoans, etc., act as biological impurities. They cause certain diseases like typhoid, hepatitis, cholera, etc. They also impart, colour, odour and turbidity to water. These can be removed from water by boiling and chlorination.

4.19.4 Colloidal Impurities

The solid impurities which may be present in water in the finely divided and in suspended state constitute colloidal impurities. These do not settle easily and impart colour and turbidity to water. Inorganic impurities like iron and aluminium hydroxide may be removed by coagulation and filtration. Organic suspended impurities like oil globules, animal matter, etc., require specific chemical treatments.

4.20 HARDNESS OF WATER

Hardness is the characteristics of water by which water does not form lather with soap. Hardness of a water sample is a measure of its capacity to precipitate soap or its soap consuming capacity, on the other hand, the water which forms copius lather with soap is termed soft water.

4.20.1 Cause of Hardness

Hardness of water is due to the presence of Ca and Mg salts. Generally, soaps consist of sodium salts of higher fatty acids such as stearic, oleic, palmitic, etc., which are highly soluble in water and thus exert their cleansing action. If calcium and magnesium salts are present in water, these react with soluble sodium soap to form insoluble salts of calcium and magnesium.

For example:

$$2C_{17}H_{35}COONa + CaSO_4 \longrightarrow (C_{17}H_{35}COO)_2\,Ca + Na_2SO_4$$
(Sodium sterate) (Insoluble salt)

$$2\,C_{17}H_{35}COONa + MgCl_2 \longrightarrow (C_{17}H_{35}COO)_2\,Mg + 2NaCl$$

Thus, a white scum or precipitate of insoluble soap of calcium and magnesium is formed. Lather is not produced until the cations, Ca^{2+} and Mg^{2+} are completely precipitated in the form of insoluble salts. As a result a large quantity of soap is required to produce lather with hard water.

Thus, hardness is soap consuming capacity of water.

4.20.2 Types of Hardness

Hardness of water is of two types:

1. Temporary hardness
2. Permanent hardness

1. Temporary hardness (carbonate hardness) or alkaline hardness

It is caused due to the presence of bicarbonates of calcium and magnesium. It is called temporary hardness because it can be removed easily by boiling.

The presence of bicarbonates in water is due to dissolution of insoluble carbonates in the presence of atmospheric CO_2.

$$CaCO_3 + CO_2 + H_2O \rightarrow Ca(HCO_3)_2$$

On boiling bicarbonates of calcium and magnesium get decomposed to give insoluble carbonates or hydroxides. These carbonates or hydroxides being insoluble are removed in the form of precipitates and thus temporary hardness is removed.

$$Ca(HCO_3)_2 \xrightarrow{\Delta} CaCO_3\downarrow + H_2O + CO_2\uparrow$$
$$Mg(HCO_3)_2 \xrightarrow{\Delta} Mg(OH)_2\downarrow + 2CO_2\uparrow$$

2. Permanent hardness (Non-carbonate hardness or Non-alkaline hardness)

It is caused due to the presence of chlorides and sulphates of calcium and magnesium. It is called permanent hardness, because it cannot be removed by boiling.

However, it can be removed by using various chemical agents.

4.20.3 Units of Hardness

The results of the analysis of hard water are expressed in one of the following units.

1. **Parts per million (ppm):** ppm is the parts of calcium carbonate equivalent hardness per 10^6 parts of water.
2. **Milligrams per litre (mg/L):** It is the number of milligrams of $CaCO_3$ equivalent hardness present per litre of water.

1 mg/L = 1 mg of $CaCO_3$ equivalent hardness per L of H_2O

But 1 L of water weighs = 1000 g = 1000 × 1000 mg

1 mg/L = 1 mg/10^6 mg = 1 ppm

3. **Clarke's degree:** It is the number of grains (1 grain = $\frac{1}{7,000}$ lb) of $CaCO_3$ equivalent hardness per gallon (10 lb) of water or simply it is the part of $CaCO_3$ equivalent hardness per 70,000 parts of water. 1°Clarke = 1 part of $CaCO_3$ equivalent hardness per 70,000 parts of water.
4. **Degree French (°Fr):** It is the parts of $CaCO_3$ equivalent hardness per 10^5 parts of water.

Relationship between various units of hardness

$$1 \text{ ppm} = 1 \text{ mg/L} = 0.1°\text{Fr} = 0.07°\text{Cl}$$
$$1°\text{Fr} = 10 \text{ ppm} = 10 \text{ mg/L} = 0.7°\text{Cl}$$
$$1°\text{Cl} = 14.3 \text{ ppm} = 14.3 \text{ mg/L} = 1.43°\text{Fr}$$

4.20.4 Determination of Hardness

Hardness of water can be estimated by a number of methods, but the most important among them is discussed below.

Determination of hardness using EDTA

Principle: Ethylene diamine tetraacetic acid (EDTA) is a hexadentate ligand and forms complexes with Ca^{2+} and Mg^{2+} ions. EDTA, commonly represented as H_4Y, has two replaceable hydrogen ions in its molecule. It has the following structure.

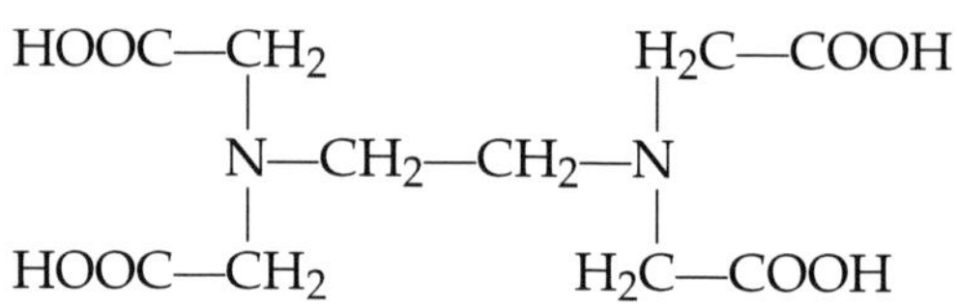

The ionization of EDTA in solution is represented as

$$H_4Y \rightarrow H_2Y^{2-} + 2H^+$$

The anion forms complexes with metal ions, M^{2+}

$$M^{2+} + H_2Y^{2-} \rightarrow MY^{2-} + 2H^+$$

where M^{2+} is Ca^{2+} or Mg^{2+}

The total hardness (temporary + permanent) of water is determined by titrating a known volume of water against standard EDTA solution at pH 10 using eriochrome black-T indicator. The colour of the free indicator at pH 10 is blue. Eriochrome black-T forms a wind red complex with M^{2+} ions. On adding EDTA, it forms a complex first with free M^{2+} ions and near the equivalence point with M^{2+} indicator complex releasing the free indicator. Hence, a colour change from wine red to blue is seen at the equivalence point. Since the reaction involves the release of H^+ ions, a buffer mixture ($NH_3 - NH_4Cl$) is used to maintain a pH of 10.

To determine the temporary hardness, same volume of another portion of the water sample is boiled. This converts soluble bicarbonates to insoluble carbonates. The precipitate of calcium carbonate is allowed to settle and then filtered. The filtrate is titrated against EDTA. This gives the permanent hardness. The difference b/w the total hardness and the permanent hardness gives the temporary hardness.

Procedure

Total hardness: Pipette out 50 mL of the sample of water into a clean titration flask, add 1 mL of $NH_3 - NH_4Cl$ buffer solution and 3-4 drops of eriochrome black-T indicator. Titrate against 0.01 M EDTA till the colour changes from wine red to clear blue. Let the volume of EDTA required be V_1 mL.

Permanent hardness: Transfer 50 mL of the sample of water into a clean 500 mL beaker and boil gently for 20-30 minutes, cool and filter it directly into a 250 mL conical flask. Add 1 mL of buffer solution followed by 3-4 drops of eriochrome black-T indicator. Titrate against standard 0.01 M EDTA. Let the volume of EDTA required be V_2 mL.

Calculation

(i) Total hardness

1000 mL of 1 M EDTA = 100 g $CaCO_3$ (molecular mass of $CaCO_3$ = 100 g)

$$1 \text{ mL of } 1 \text{ M EDTA} = \frac{100}{1000} \text{ g of } CaCO_3$$

$$V_1 \text{ mL of } 0.01 \text{ M EDTA} = \frac{V_1 \times 0.01 \times 100}{1000} \text{ g of } CaCO_3$$

50 mL (parts) of water sample

$$= \frac{V_1 \times 0.01 \times 100}{1000} \text{ g of } CaCO_3$$

10^6 (1 million parts) mL of water sample contains

$$= \frac{V_1 \times 0.01 \times 100 \times 10^6}{1000 \times 50}$$

$$= 20 \times V_1 \text{ g of } CaCO_3$$

or total hardness of water

$$= 20 \times V_1 \text{ ppm } CaCO_3 \text{ equivalent}$$

(ii) Permanent hardness

$$\text{Permanent hardness} = \frac{V_2 \times 0.01 \times 100 \times 10^6}{1000 \times 50}$$

$$= 20 \times V_2 \text{ g of } CaCO_3$$

(iii) Temporary hardness

$$\text{Temporary hardness} = 20\ (V_1 - V_2)\ \text{ppm}\ CaCO_3\ \text{equivalent}$$

NUMERICAL PROBLEMS

1. A water sample contains 180 mg of $MgSO_4$ per litre. Calculate the hardness in terms of $CaCO_3$ equivalent

$$\text{Mass of } MgSO_4\ (W) = 180\ \text{mg/L}$$
$$\text{Equivalent mass of } MgSO_4(E) = 60$$
$$\text{Equivalent of } CaCO_3 = \frac{50 \times W}{E} = \frac{50 \times 180}{60}$$
$$= 150\ \text{Mg/L} = 150\ \text{ppm}$$

2. Calculate temporary hardness and permanent hardness of a sample of water which on analysis is found to contain the following:

$Ca(HCO_3)_2 = 16.2$ mg/L $Mg(HCO_3)_2 = 7.3$ mg/L
$MgCl_2 = 9.5$ mg/L $CaSO_4 = 13.6$ mg/L

Substance	**Mass of substance**	**Equivalent mass of substance (E)**	**$CaCO_3$ equivalent** $= \left(\frac{W \times 50}{E}\right)$
$Ca(HCO_3)_2$	16.2 mg/L	81	10 mg/L
$Mg(HCO_3)_2$	7.3 mg/L	73	5 mg/L
$MgCl_2$	9.5 mg/L	47.5	10 mg/L
$CaSO_4$	13.6 mg/L	68	10 mg/L

Temporary hardness = Hardness due to bicarbonate ions = 15 mg/L
Permanent hardness = Hardness due to Cl^-, SO_4^{2-} = 20 mg/L

3. 100 mL of a sample of water required 18 mL of 0.01 M EDTA for titration using eriochrome black-T indicator. In another experiment, 100 mL of the same sample of water was gently boiled and the precipitate was removed by filtration. The filtrate required 9.0 mL of 0.01 M EDTA using eriochrome black-T indicator. Calculate (i) the total hardness, (ii) permanent hardness, and (iii) carbonate hardness.

(i) Total hardness

$$1000\ \text{mL of 1 M EDTA} \equiv 100\ \text{g of } CaCO_3$$
$$18\ \text{mL of 0.01 M EDTA} \equiv \frac{100 \times 18 \times 0.01}{1000}\ \text{g of } CaCO_3$$
$$= 0.018\ \text{g of } CaCO_3$$

100 mL of water sample contains 0.018 g of $CaCO_3$

10^6 mL of water sample contains 180 ppm of $CaCO_3$

Total hardness of the water sample = 180 ppm of $CaCO_3$

(ii) Permanent hardness

$$1000 \text{ mL of 1 M EDTA} = 100 \text{ g } CaCO_3$$

$$1 \text{ mL of 1 M EDTA} = \frac{100}{1000} \text{ g of } CaCO_3$$

$$9.0 \text{ mL of 0.01 M EDTA} = \frac{9.0 \times 0.01 \times 100}{1000} \text{ g of } CaCO_3$$

$$= 0.009 \text{ g of } CaCO_3$$

100 mL of water sample contains = 0.009 g of $CaCO_3$

10^6 (1 million) mL of water sample contains

$$= 0.009 \times 10^6 \text{ g of } CaCO_3$$

$$= 90 \text{ g of } CaCO_3$$

$$\text{Permanent hardness} = 90 \text{ ppm of } CaCO_3$$

(iii) Carbonate hardness

$$\text{Carbonate hardness} = 180 - 90 = 90 \text{ ppm of } CaCO_3$$

4. If 50 mL of a sample of hard water consumed 15 mL of 0.01 M EDTA. What is the hardness of water?

Let us first calculate the molarity of hard water sample consuming 15 mL of 0.01 M EDTA by applying molecularity equation

$$M_1V_1 = M_2 V_2$$

$$50 \text{ mL} \times \text{molarity of hard water} = 15 \text{ mL} \times 0.01 \text{ M EDTA}$$

$$\text{Molarity of hard water} = \frac{15 \times 0.01}{50} \; 0.003 \text{ M}$$

$$= 0.003 \times 2\text{N} = 0.006 \text{ N}$$

We know strength in g/L = Molarity × mol. wt. In order to express hardness in terms of $CaCO_3$ equivalents, we should take the molecular weight of $CaCO_3$, i.e., 100.

$$\text{Strength of hard water} = 0.003 \times 100 \text{ g/L}$$

$$= 0.3 \text{ g/L}$$

$$= 300 \text{ mg/L}$$

$$= 300 \text{ ppm}$$

4.21 BOILER FEED WATER

Water being a universal solvent is used for many industrial processes directly as well as indirectly. One of the most important applications of water is, it absorbs large quantities of heat for a given rise in temperature. Then it evaporates to form steam at atmospheric pressure and the steam so formed carries large amount of heat. Due to these wonderful properties, water becomes an ideal raw material for heating and power generating processes.

A large quantity of water is used in boilers for generating steam. Boilers play an important role for generation of steam in industries.

A boiler is a closed vessel which operates under different pressures. Water heated in a boiler under pressure is transformed to steam. The water used in these types of boilers is called boiler feed water. Hard water creates a number of problems like corrosion, scale and sludge formation, caustic embrittlement, priming foaming, etc. It is dangerous because it causes the explosions of boilers due to high pressure. Natural water is not directly used in boilers because it contains hardness producing salts. Hence, water should be properly softened and sufficiently pure before feeding into the boiler. Thus, the boiler feed water must be pure and should have specific characteristic properties.

4.21.1 Characteristics of Boiler Feed Water

Water available in nature contains different amounts of dissolved, suspended substances and dissolved gases. If concentration of these substances is above the tolerance limits, they cause problems in boilers. Therefore, if water does not satisfy the following characteristics, it has to be pretreated to remove impurities. The possible impurities and their specifications for boiler feed water are given in Table 4.7.

Table 4.7

Parameters	**Specification**
Hardness	< 0.2 ppm
Caustic alkalinity	0.15 – 0.45 ppm
Soda alkalinity	0.15 – 1.0 ppm
Soda ash	0.3 – 0.5 ppm
pH	7.5 – 10
Dissolved oxygen	0.02 ppm
Iron	< 0.3 mg/L
Silica	< 5.0 mg/L
Cu	< 0.01 mg/L

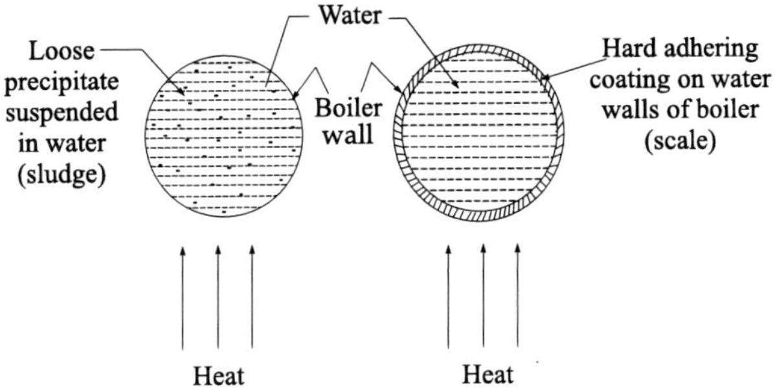

Fig. 4.9 *Sludge and scale formation in boilers*

4.21.2 Boiler Problems

When the natural water without any pretreatment is directly fed into the boilers, various physical as well as chemical reactions in boiler water take place with the action of heat. This results in the following defects.

1. Scale and sludge formation
2. Priming and foaming
3. Boiler corrosion

4.22 SCALE AND SLUDGE FORMATION

Water is heated under high pressures to high temperature inside the boilers which causes the water to evaporate in the form of steam. Due to this, the impurities present in the water get progressively concentrated. When the impurities reach a saturation point, they get precipitated out. The precipitate so formed is in the form of hard deposit which is coated to the walls of the boiler is called *scale*. It is very difficult to remove.

If it is in the form of loose, silky, non-adherent precipitate then it is called *sludge*.

4.22.1 Scale

Scales are hard deposits which stick very firmly to the inner walls of the boiler. They are difficult to remove even with a chisel and hammer, scales mostly contain $CaCO_3$, $Ca(OH)_2$, $Mg(OH)_2$, $CaSO_4$, $CaSiO_3$ and $MgSiO_3$ formed by the impurities in water.

(*i*) *Decomposition of bicarbonates:* It is decomposed into calcium carbonate on boiling with low pressure

$$Ca(HCO_3)_2 \xrightarrow{\Delta} \underset{\text{(Scale)}}{CaCO_3} \downarrow + H_2O + CO_2 \uparrow$$

In high pressure boilers calcium carbonate is soluble which forms $Ca(OH)_2$ whose solubility decreases with the increase of temperature, hence, it is also deposited as scale.

$$CaCO_3 \xrightarrow{\Delta} CaO + CO_2 \uparrow$$

$$CaO + H_2O \longrightarrow \underset{\text{Soluble}}{Ca(OH)_2}$$

(*ii*) *Presence of calcium sulphate:* The solubility of calcium sulphate decreases with increase in temperature. Therefore, at high temperature, calcium sulphate gets precipitated to form a very hard and adhering scale.

(*iii*) *Hydrolysis of magnesium salt:* Dissolved magnesium salts undergo hydrolysis at high temperatures to give magnesium hydroxide precipitates which form a soft type scale.

$$MgCl_2 + 2H_2O \longrightarrow \underset{\text{Soft scale}}{Mg(OH)_2} \downarrow + 2HCl$$

(*iv*) *Presence of silica:* Salts like calcium silicate and magnesium silicate are sparingly soluble in cold water, but are completely insoluble in hot water. At high temperature of boiler, they precipitate and deposit as scale.

4.22.2 Disadvantages of Scale Formation

(*i*) *Wastage of fuel:* They are bad conductors of heat so a lot amount of heat is wasted.

(*ii*) *Decrease in efficiency:* Due to deposition of scales in boilers, valve and pipes, it may be choked and hence efficiency of the boiler is decreased.

(*iii*) *Danger of explosion:* Sometimes at high temperature, the scales crack and water suddenly comes in contact with overheated iron plates. This results in formation of large amount of steam suddenly which may cause explosion.

(*iv*) *Expenses of cleaning:* Scale must be removed regularly and this cleaning process is expensive.

(*v*) *Corrosion:* Due to scale formation excessive heating of boiler material takes place. At such high temperatures, there is greater corrosion of boiler tubes.

4.22.3 Methods of Removing Scales

1. Soft scales are removed by scrapper or wire brush.
2. Hard scales can be removed by thermal shocks, i.e., alternate heating and cooling that makes the scales brittle.
3. Scales can be removed by dissolving in some suitable chemicals, $CaCO_3$ scales can be dissolved in 5–10% diluted HCl. Similarly, calcium ion can be removed by EDTA solution.

4.22.4 Sludge

Sludge is a soft, loose and slimy precipitate formed inside the boiler. These are formed by substances which have greater solubilities in hot water than in cold water.

It is formed at comparatively colder portions of the boiler and collects in the area where flow rate is slow. These can be easily removed by scrapping with a wire brush. If sludge is not removed frequently, it settles down to form scales. It can be removed by blow down operation.

The main compositions of sludge are $MgCO_3$, $MgCl_2$, $MgSO_4$, $CaCl_2$, etc.

4.22.5 Disadvantages of Sludge Formation

1. Sludges are conductors of heat so they tend to waste a lot of heat and the efficiency of the boiler is decreased.
2. Sludge needs to be removed regularly and this cleaning process is expensive.
3. Since sludge settles in areas of poor water circulation such as joints, bends, therefore, choking of pipes takes place.
4. It reduces the efficiency of the boiler.

4.22.6 Prevention of Sludge Formation

1. It can be prevented by using soft water for boiler feed purposes.
2. Removing concentrated salty water from time to time prevents deposition of sludge.
3. It can easily be removed by scrapping off with a brush.

4.23 COMPARISON BETWEEN SCALES AND SLUDGES

Scales	Sludges
1. They form an adherent coating within the boiler which cannot be removed even by mechanical means.	They form slimy or less adherent coating within the boiler which can be removed by mechanical means.
2. Scales are harder and more permeable.	Sludges are soft and less permeable.
3. Scales are bad conductors of heat.	Sludges are poor conductors of heat.
4. These are formed throughout the surface in contact with water.	These are formed at comparatively colder portions of the boiler.
5. They are formed by substances like $Mg(OH)_2$ and $CaSO_4$.	They are formed by substances like $CaCl_2$, $MgCl_2$, $MgSO_4$, and $MgCO_3$.
6. They decrease the efficiency of the boiler and are more dangerous because of the chances of explosion.	They decrease the efficiency of the boiler but are less dangerous.

4.24 PRIMING AND FOAMING

When water is boiled in a boiler, it produces steam which evaporates from the surface of water. If steam carries small droplets of water with it then it is called *wet steam*. Steam associated with liquid water droplets is called *priming*. These tiny droplets moving along with steam can carry some suspended and dissolved impurities present in the boiler water. This process of carrying of small droplets of water along with impurities by steam is called *carry-over*. This carry-over is mainly due to priming and foaming.

4.24.1 Priming

"When water is boiled rapidly in a boiler. The bubbles and drops of liquid water along with steam in the form of spray into the steam outlet."

OR

The process of carrying water droplets along with steam is called priming.

4.24.2 Causes of Priming

It is caused due to the presence of

- High velocity of steam.
- Water level is very high.
- Presence of suspended and dissolved impurities in water.
- Faulty boiler design.

4.24.3 Prevention

It is prevented by

- Proper designing of the boiler for proper evaporation of water.
- Keeping the low water level.
- Using soft and filtered water.
- Good boiler design fitted with mechanical steam purifiers.

4.24.4 Foaming

"It is the formation of small but persistent bubbles on the surface of the boiled water".

It is caused by alkalis, oils, fats, certain organic matters and suspended solids in boiler feed water.

4.24.5 Causes of Foaming

It is caused due to the presence of:

- Some impurities like oil, fat, etc., which produce foams by reducing the surface tension of water.
- Oil or grease in water.
- The difference in concentration of dissolved salts between layer surfaces and bulk of the water in boiler.

4.24.6 Prevention

It is prevented by.

- Using suitable coagulants like ferrous sulphate, so that the oil or grease present in water may be removed.
- Addition of anti-foaming agents like castor oil, polyamides, etc.
- Removal of clay and suspended matter using coagulating agents like $Al(OH)_3$.
- Oil and greases are removed by using sodium aluminate.

4.25 BOILER CORROSION

Decay of boiler material by chemicals or electrochemicals of its environment is called boiler corrosion.

OR

The decay of boiler material due to the presence of impurities in boiler feed water is called boiler corrosion.

The dissolved gases like oxygen and carbon dioxide present in the boiler feed water also cause boiler corrosion.

4.25.1 Factors Causing Boiler Corrosion

1. *Presence of dissolved oxygen*: Generally, natural water contains 8-9 ppm of dissolved oxygen, oxygen levels of more than 7 ppm cause boiler corrosion. When water is heated the dissolved oxygen is liberated and attacks the boiler material.

$$2Fe + 2H_2O + O_2 \longrightarrow 2Fe(OH)_2$$

$$4Fe(OH)_2 + O_2 \longrightarrow 2[Fe_2O_3.2H_2O]$$

2. *Presence of dissolved CO_2*: It is present in the boiler water either from air or due to the presence of residual temporary hardness.

$$Ca(HCO_3)_2 \longrightarrow CaCO_3 + CO_2 + H_2O$$

$$Mg(HCO_3)_2 \longrightarrow Mg(OH)_2 + 2CO_2$$

This carbon dioxide dissolved in water forms a weak carbonic acid which is slightly acidic in nature and hence causes corrosion.

$$H_2O + CO_2 \longrightarrow H_2CO_3$$

3. *Presence of acid forming salts:* Some inorganic salts like $MgCl_2$, $CaCl_2$, etc., present in water gets hydrolysed to produce acids which corrode the boiler material.

$$MgCl_2 + 2H_2O \longrightarrow Mg(OH)_2 + 2HCl$$

$$CaCl_2 + 2H_2O \longrightarrow Ca(OH)_2 + 2HCl$$

The resultant acid then attacks the boiler to form rust.

$$Fe + 2HCl \longrightarrow FeCl_2 + H_2$$

$$FeCl_2 + H_2O \longrightarrow Fe(OH)_2 + 2HCl$$

4.25.2 Prevention

It may be prevented by the following methods.

1. By removing dissolved oxygen, i.e., adding a calculated amount of sodium sulphite or hydrazine or sodium sulphide.

$$2Na_2SO_3 + O_2 \longrightarrow 2Na2SO_4$$

$$\underset{\text{Hydrazine}}{N_2H_4} + O_2 \longrightarrow N_2 + 2H_2O$$

$$Na_2S + 2O_2 \longrightarrow Na_2SO_4$$

2. By removing carbon dioxide, i.e., by mechanical deaeration method. Chemical removal of carbon dioxide can be done by treating with lime or ammonium hydroxide.

$$Ca(OH)_2 + CO_2 \longrightarrow CaCO_3 + H_2O$$

$$2NH_4OH + CO_2 \longrightarrow (NH_4)_2CO_3 + H_2O$$

3. By removing acidic impurities, i.e., by treating with alkaline agents like ammonium hydroxide.

4.26 DISSOLVED OXYGEN

Oxygen in water is measured as dissolved oxygen (DO). One unit of measure of dissolved oxygen in water is parts per million (ppm), which is the number of oxygen (O_2) molecules per million total molecules in a sample. It is also defined as "the number of moles of molecular oxygen (O_2) dissolved in one litre of water at a temperature, expressed as mg O_2/L".

4.26.1 Determination of Dissolved Oxygen

The dissolved oxygen can be determined by the Winkler or iodometric method. The method involves the estimation of equivalent amount of iodine which is liberated by dissolved oxygen content of water by standard sodium thiosulphate.

Principle: The dissolved oxygen in water is in molecular form which is not capable of oxidizing KI to I_2. Therefore, it requires oxygen carrier such as $Mn(OH)_2$, which is produced in water sample by the reaction of $MnSO_4$ with KOH.

$$MnSO_4 + 2KOH \longrightarrow Mn(OH)_2 + K_2SO_4$$

This $Mn(OH)_2$ (divalent manganese) reacts with the dissolved oxygen to form a brown precipitate of manganic oxide (tetravalent manganese)

$$2Mn(OH)_2 + O_2 \longrightarrow \underset{\text{Manganic oxide}}{2MnO(OH)_2}$$

The manganic oxide which is treated with concentrated H_2SO_4 and to liberate nascent oxygen.

$$MnO\,(OH)_2 + H_2SO_4 \longrightarrow MnSO_4 + 2H_2O + [O]$$

Under acidic condition (in the presence of H_2SO_4), nascent oxygen quantitatively oxidizes KI to I_2.

$$2KI + H_2SO_4 + [O] \longrightarrow K_2SO_4 + H_2O + I_2$$

The liberated iodine is estimated by titrating against standard sodium thiosulphate solution using starch as an indicator towards the endpoint.

$$I_2 + 2Na_2S_2O_3 \longrightarrow Na_2S_4O_6 + 2NaI$$

Interference: Water sample may contain nitrites and sulphides which can oxidise KI to I_2. Therefore, these are removed by using sodium azide.

4.26.2 Procedure

Pipette out 300 cm^3 of water sample into a clean glass stoppered bottle. Add 3 cm^3 of manganous sulphate solution dipping the pipette below the surface of water. Add 3 cm^3 of alkaline potassium iodide solution. The bottle is corked immediately and shaken well for 5–10 minutes to get the brown precipitate. Allow to stand for 3–5 minutes for the development of precipitate now add 1 cm^3 of concentrated sulphuric acid slowly and shake well until the dissolution of precipitate.

Pipette out 100 cm^3 of this solution into a conical flask and slowly titrate against 0.02 N sodium thiosulphate solution using 2 cm^3 starch indicator till the disappearance of blue colour. Record the volume of sodium thiosulphate solution used in titration.

From the above procedure, the following data are recorded. Let,

Volume of water sample taken for titration = V_1 mL

Normality of water sample taken for titration = N_1

Volume thiosulphate used for titration = V_2 mL

Normality of thiosulphate used for titration = N_2.

$$\underset{\text{Water sample}}{M_1V_1} = \underset{\text{Hypo}}{N_2V_2}$$

$$\text{Normality of water sample, } N_1 = \frac{N_2V_2}{V_1}$$

Strength of dissolved oxygen in water sample

$$= \text{Normality} \times \text{Equivalent weight of } O_2$$

$$= \frac{N_2V_2}{V_1} \times 8 \text{ gm/lit}$$

$$= \frac{N_2V_2}{V_1} \times 8 \times 1000 \text{ mg/L}$$

$$= \frac{N_2V_2}{V_1} \times 8000 \text{ ppm}$$

The amount of oxidizable impurities present in polluted water is measured in two ways:

1. Biological oxygen demand (BOD)
2. Chemical oxygen demand (COD).

4.27 BIOLOGICAL OXYGEN DEMAND (BOD)

BOD is an important parameter of water quality. It is a measure of the amount of oxygen needed (in milligrams per litre or ppm) by bacteria and other microorganisms to oxidize the organic matter present in a water sample over a period of 5 days.

"BOD is the amount of dissolved oxygen needed for the biological oxidation of organic matter under aerobic conditions present in a given water sample at a certain temperature (20°C) over a specific time period, i.e., 5 days.

4.27.1 Characteristics of BOD Parameter

1. The unit of BOD is mg/dm^3 or ppm.
2. It is empirical and semi-quantitative.
3. It is the only biodegradable organic load in sewage. Strict aerobic conditions are needed.

4. Determination is slow and a time consuming method. It indicates the amount of decomposable organic matter in the sewage. It is an expression of how much oxygen is needed by microbes to oxidize the organic matter in the sewage (water sample).

4.27.2 Determination of BOD Principle

BOD of water sample is determined by measuring the dissolved oxygen (DO) content before and after 5 days of incubation period by the iodometric method.

This method is based on the fact that in an alkaline medium. DO oxidises Mn^{2+} to Mn^{4+}, which gets precipitated as MnO_2. H_2O or $H_2\ MnO_3$

$$MnSO_4 + 2KOH + \frac{1}{2} O_2 \longrightarrow H_2MnO_3 + K_2SO_4$$

On acidification, Mn^{4+} oxidises iodide to free iodine

$$H_2MnO_3 + 2H_2SO_4 + 2KI \longrightarrow MnSO_4 + K_2SO_4 + 3H_2O + I_2$$

The liberated iodine is then titrated with standard sodium thiosulfate solution using starch as indicator.

$$I_2 + 2Na_2S_2O_3 \longrightarrow Na_2S_4O_6 + 2NaI$$

From the volume of sodium thiosulphate consumed, DO content of water sample can be determined.

4.27.3 Procedure

A definite volume of water sample is diluted to a known volume with distilled water which contains nutrients for bacteria and sufficient free oxygen. Equal volume of the diluted water sample is filled and kept for 5 days for incubation and another is for immediate use. DO determination is called *blank titration.*

1. Blank Titration: DO content in one of the bottles is determined immediately –2 mL of $MnSO_4$ solution + 3 mL of alkaline KI solution are added into the bottle, stopper the bottle and shake well for 5–10 min to get brown precipitate and allow to stand for 2–3 minutes for the complete formation of precipitate. Now 1 mL of conc. H_2SO_4 is added, shake well to dissolve the MnO_2 precipitate which gives deep brown colour solution due to the liberation of iodine.

A known volume of this solution is titrated with standard sodium thiosulphate solution using starch as indicator till the discharge of violet colour.

2. Sample (Back) Titration: The second bottle which was kept for 5 days at 20°C is taken for DO determination by the above mentioned method and DO is noted as D_2 mg/L. Thus, BOD of water sample is given by,

$$\text{BOD} = \frac{D_1 - D_2}{A} \times B \text{ mg/dm}$$

where
$D_1 \rightarrow$ DO in mg/dm in the solution
$D_2 \rightarrow$ DO in mg/dm in the solution after 5 days of incubation
$A \rightarrow$ Volume of sample in mL before dilution
$B \rightarrow$ Volume of sample in mL after dilution

WORKED PROBLEMS

Problem 4.1

30 mL of water sample was diluted to 600 mL and equal volumes were filled in two BOD bottles. DO in one bottle was determined immediately and 100 mL of diluted wastewater when titrated immediately required 6.1 mL of 0.02 N $Na_2S_2O_3$. 100 mL of the incubated sample after 5 days required 3.6 mL of same $Na_2S_2O_3$. Calculate BOD of the waste water.

Solution: Volume of the sample before dilution A = 30 mL

Volume of the sample after dilution B = 600 mL

Volume of dilute water used for titration = 100 mL

Blank titre value = 6.1 mL 0.02 N $Na_2S_2O_3$

Titre value after 5 days = 3.6 mL 0.002 N $Na_2S_2O_3$

(a) Dissolved oxygen of blank titration

$$\underset{\text{Water sample}}{N_1Y_1} = \underset{Na_2S_2O_3}{N_2Y_2}$$

$$N_1 \times 100 = 0.02 \times 6.1$$

$$\text{Normality of DO, } N_1 = \frac{0.02 \times 6.1}{100} = 1.22 \times 10^{-3}\text{ N}$$

$\therefore$ DO in blank titration, D_1 = Normality × Equal weight of O_2

$$= 1.22 \times 10^{-3}\ 8\text{ g/L}$$
$$= 9.76 \times 10^{-3}\text{ g/L}$$
$$= 9.76 \times 1000 \times 10^{-3}$$
$$= 9.76\text{ mg/L}$$

(b) DO of sample titration,

$$\underset{\text{Water sample}}{N_1V_1} = \underset{\text{Hypo}}{N_2V_2}$$

$$N_1 \times 100 = 0.02 \times 3.6$$

$$N_1 = \frac{0.02 \times 3.6}{100} = 7.2 \times 10^{-4}\text{ N}$$

DO in sample titration,

$$D_2 = \text{Normality} \times \text{Equal weight of } O_2$$
$$= 7.2 \times 10^{-4} \times 8\text{ g/l}$$
$$= 5.76 \times 10^{-3} \times 1000\text{ mg/L}$$

Now,
$$\text{BOD} = \frac{D_1 - D_2}{A}\text{ mg/L}$$

$$\frac{9.76 - 5.76}{30} \times 600 = 80\text{ mg/L}$$

Problem 4.2

Calculate the BOD value of a sewage sample having an organic compound, CH_2O containing 12.0 mg/L.

Solution: Molecular weight of CH_2O = 30

$$\underset{30g}{CH_2O} + \underset{32g}{O_2} \longrightarrow CO_2 + H_2O$$

30g of CH_2O requires 32 g of O_2 for complete oxidation

$$12.0 \text{ mg/l of } CH_2O \text{ requires } = \frac{12 \times 32 \times 10^3}{30 \times 10^3} \text{ mg/L of } O_2$$

$$\text{BOD of the effluent } = 12.8 \text{ mg/L}$$

Problem 4.3

Calculate the BOD value of a sewage sample containing 20.0 mg/L of an organic matter $C_6H_{12}O_6$ calculate BOD.

Solution: Molecular weight = 180g

$$\underset{180\text{ g}}{C_6H_{12}O_6} + \underset{6 \times 32 = 192\text{ g}}{6O_2} \longrightarrow 6CO_2 + 6H_2O$$

180 g of $C_6H_{12}O_6$ requires 192g of O_2 for complete oxidation

$$20 \text{ mg/l of } C_6H_{12}O_6 \text{ requires } = \frac{20 \times 192 \times 10^3}{180 \times 10^3} \text{ mg/L}$$

$$= 21.3 \text{ mg/L}$$

Problem 4.4

An effluent water sample emerging out from a chemical factory is found to contain only an organic compound of molecular formula $H_2C_2O_4$. If the wastewater sample contain 60 mg/litre of the above compound, what would be the amount of oxygen required to oxidise by microorganisms under experimental conditions of BOD estimation. [C = 12, H = 1 and O = 16]

Solution: Molecular weight of $H_2C_2O_4$ = 90

$$\underset{2 \times 90}{2H_2C_2O_4} + \underset{32\text{ g}}{O_2} \longrightarrow 4C_2 + 2H_2O$$

Therefore, 180 mg/L of $H_2C_2O_4$ in wastewater requires 32 mg of oxygen to oxidise into carbon dioxide and water

$$60 \text{ g/L of } H_2C_2O_4 \text{ requires } = \frac{60 \times 32}{180} = 10.67 \text{ mg/L.}$$

4.28 CHEMICAL OXYGEN DEMAND (COD)

BOD value takes into account only for biologically oxidisable impurities and very slowly oxidisable impurities. Therefore, COD value is introduced to measure both biologically oxidisable and non-oxidisable but chemically oxidisable organic matter present in the water sample.

COD is defined as "the amount of oxygen consumed in the oxidation all oxidizable impurities organic, inorganic, biologically and chemically oxidizable present in 1 litre of wastewater using a strong oxidizing agent such as acidified $K_2Cr_2O_7$.

It is expressed in mg/L or ppm.

4.28.1 Characteristics

1. It is preferable to BOD as the results are reliable.
2. Rapidly measurable parameter and needs about 3 hours for completion.
3. It is a satisfactory quantitative method for measuring total organic load.
4. When used along with BOD test, it gives biologically resistant organic matter.
5. In general COD > BOD test, it gives biologically resistant and non-biodegradable organic load are completely oxidised.

4.28.2 Determination of COD

Principle: The principle of the method involves the oxidation of organic matter using chemical oxidizing agents such as acidified potassium dichromate in the presence of a silver sulphate as catalyst, which catalyzes the oxidation of organic matter. Chloride ions interfere in the oxidation titration. Ag_2SO_4 is used as catalyst for the oxidation straight chain aliphatic compounds. However, Ag^+ ions become ineffective in the presence of chloride ions (present in wastewater) due to their precipitation as silver halide. This difficulty is overcome by treating the wastewater with mercuric sulphate. Mercuric sulphate binds the halide ions and makes them unavailable.

$$3CH_2O + 16H^+ + 2Cr_2O_1^{2-} \xrightarrow[HgSO_4]{Ag_2SO_4} 4Cr^{3+} + 3CO_2 + H_2O$$

$$Hg^{2+} + 4Cl^- \longrightarrow [HgCl_4]^{2-}$$

The unreacted $K_2Cr_2O_7$ is estimated by titrating against standard ferrous ammonium sulphate (FAS) using ferrion indicator.

4.28.3 Procedure

1. Sample (Back) Titration: Pipette out a known volume of the standard $K_2Cr_2O_7$ into a 250 mL conical flask. Further, the mixture is acidified with 1 : 1 sulphuric acid (containing silver sulphate and silver sulphate and mercuric sulphate). Reflux the mixture for half an hour and cool to room temperature. Add 2-3 drops of ferroin indicator and titrate against the standard ferrous ammonium sulphate solution until the colour changes from bluish-green to reddish-brown.

2. Blank Titration: The above procedure is repeated by taking same volume of distilled water in place of water sample.

4.28.4 Calculations

From the above procedure the following data are recorded.

Volume of water sample taken for titration = V mL

Volume of FAS used in sample titration = A V mL

Volume of FAS used in blank titration = B V mL

Normality of FAS solution = N

Volume of $K_2Cr_2O_7$ consumed in satisfying COD in terms of FAS solution = (B – A) mL

$$N_1V_1 = N_2V_2$$

(Water sample) (FAS = Ferrous ammonium sulphate)

$$\text{Normality of COD sample, } N_1 = \frac{N(B-A)}{V}$$

$$\therefore \quad \text{COD of the sample} = \text{Normality} \times \text{Equivalent weight of } O_2 \text{ g/L}$$

$$= \frac{N(B-A)}{V} \times 8 \text{ g/L}$$

$$= \frac{N(B-A)}{V} \times 8 \times 1000 \text{ mg/L}$$

4.29 COMPARISON BETWEEN BOD AND COD

1. BOD is a slow process. It takes more than five days for estimation of BOD value of water sample. Whereas COD value can be estimated in about 3 hours.
2. Both BOD and COD are expressed in mg/of O_2/dm^3 of wastewater.
3. For any water sample COD value is always greater than its BOD value. Because BOD is the measure of only biologically oxidizable impurities present in water whereas COD is the measure of all the oxidizable impurities.

NUMERICAL PROBLEMS

1. $MK_2Cr_2O_7$ = 294 g of $K_2Cr_2O_7$ in 1 litre of solution; 1 N $K_2Cr_2O_7 \equiv$ 49 g of $K_2Cr_2O_7$ in 1 litre of the solution

1 equivalent of $K_2Cr_2O_7 \equiv$ 1 equivalent O = 8 g of O

49 g of $K_2Cr_2O_7 \equiv$ 8 g of O

294 g of $K_2Cr_2O_7 \equiv$ 48 g of O

Problem 4.5

20 mL of sewage water sample for COD is reacted with 40 mL of $K_2Cr_2O_7$ solution and the untreated $K_2Cr_2O_7$ required 18.0 mL of 0.25 N FAS solution. Under similar conditions in blank titration 30.0 mL of FAS solution was used. Calculate the COD of sewage water sample.

Solution: Volume of sewage water sample, V = 20 mL

Volume of FAS used in sample titration, A = 18 mL

Volume of FAS used in blank titration, B = 30 mL

Normality of FAS solution, N = 0.25 N

Volume of $K_2Cr_2O_7$ in terms of FAS = $V_2 = (B - A)$

$$\underset{\text{Water sample}}{N_1V_1} = \underset{\text{FAS}}{N_2 V_2}$$

$$\text{Normality of COD sample, } N_1 = \frac{N \times (B - A)}{V}$$

$$= \frac{0.25 \times (30 - 18)}{20}$$

$$= 0.15 \text{ N}$$

$\therefore$ COD of the sewage water sample= Normally × Equivalent weight of O_2

$$= 0.15 \times 8 \text{ g/L}$$

$$= 1.2 \text{ g/L}$$

$$= 1.2 \times 1000 \text{ mg/L}$$

$$= 1200 \text{ mg/L}$$

Problem 4.6

50 mL of an industrial sewage requires 11.5 mL of 0.4N $K_2Cr_2O_7$ solution for complete oxidation. Calculate COD of the industrial sewage, assuming that the effluent contains only oxalic acid. Also calculate the amount of oxalic acid present in 1 dm^3 (equivalent mass of oxalic acid as 45).

1 equivalent of $K_2Cr_2O_7$ = 1 equivalent of oxalic acid

1000 mL of 1N $K_2Cr_2O_7$ = 5 g of oxalic acid

$$11.5 \text{ mL of } 0.4\ K_2Cr_2O_7 = \frac{11.5 \times 0.4 \times 45}{1000} \text{ g}$$

$$= 0.207 \text{ g of oxalic acid}$$

Amount of oxalic acid present in one litre of the effluent

$$= \frac{0.207 \times 1000}{1000}$$

$$= 4.14 \text{ g}$$

For complete oxidation,

1 equivalent of oxalic acid ≡ 1 equivalent of oxygen

50 g of oxalic acid ≡ 8 g oxygen

$$4.14 \text{ g of oxalic acid} = \frac{4.14 \times 81}{50}$$

$$= 0.6624 \text{ g}$$

$$= 662.4 \text{ mg of oxygen}$$

$$\text{COD of effluent sample} = 662.4 \text{ mg/L}$$

$$\text{Amount of oxalic acid present in 1 litre} = 4.14 \text{ g}$$

Another method

$$(N \times V) \text{ of effluent} = (N{\times}V)\ K_2Cr_2O_7$$

$$\text{N of effluent} = \frac{(N \times V)K_2Cr_2O_7}{\text{V of effluent}} = \frac{0.4 \times 115}{50}$$

$$\text{Weight of oxalic acid in 1 litre} = 0.092 \times 45 = 4.14 \text{ g} = 0.092$$

Problem 4.7

25 mL of an effluent sample consumes 8.3 mL of 0.001 M $K_2Cr_2O_7$ solution for oxidation. Calculate COD of the effluent sample.

Solution: Method I:

$$1000\ 1 \text{ mL N } K_2Cr_2O_7 \equiv 8 \text{ g of oxygen}$$

$$1 \text{ mL } 1 \text{ N } K_2Cr_2O_7 \equiv 8 \text{ mg of oxygen}$$

$$\text{Molar mass of } K_2Cr_2O_7 \equiv 294$$

$$\text{Equivalent mass of } K_2Cr_2O_7 \equiv 49$$

$$1 \text{ mole} \equiv 6 \text{ equivalence } 1 \text{ M} = 6\text{N}$$

$$1 \text{ mL } 1 \text{ mol} = 6 \text{ mL } 1 \text{ N } K_2Cr_2O_7$$

$$\text{COD} = \frac{(B - A) \times M \times 8000}{V} \text{ mg/L}$$

$$6\text{N } K_2Cr_2O_7 = 1 \text{ M } K_2Cr_2O_7$$

$$\text{COD} = \frac{8.3 \times 6 \times 0.001\text{N} \times 8000}{25} = 15.976 \text{ mg/L}$$

Method II:

$$1000 \text{ cc of } 1 \text{ M } K_2Cr_2O_7 = 294\text{g of } K_2Cr_2O_7$$

$$8.3 \text{ cc of } 0.11\ 1 \text{ M } K_2Cr_2O_7 = \frac{8.3 \times 0.001 \times 294}{1000}$$

$$= 2.4402 \times 10^{-3} \text{ g of } K_2Cr_2O_7$$

$$K_2Cr_2O_7 + 4H_2SO_4 \longrightarrow K_2SO_4 + Cr_2(SO_4)_3 + 4H_2O + 3[O]$$

i.e., $$294\text{g of } K_2Cr_2O_7 = 48\text{g of } O_2\ (3 \times 16)$$

$\therefore$ $$2.4402 \times 10^{-3} \text{ g of } K_2Cr_2O_7 = \frac{2.4402 \times 10^{-3} \times 48}{294} = 3.984 \times 10\text{g of } O_2$$

$$= 3.984 \times 10^{-4} \times 1000 \text{ mg of } O_2$$

$$= 0.3984 \text{ mg of } O_2$$

$$\therefore \quad \text{COD of the sample} = \frac{0.3984 \times 1000}{25} \text{ mg/L}$$
$$= 15.936 \text{ mg/L.}$$

4.30 DETERMINATION OF SULPHATE IONS IN WATER BY GRAVIMETRIC METHOD

4.30.1 Introduction

Gravimetric analysis is based on the measurement of the mass of a substance of known composition that is chemically related to the analyte. There are two ways to perform gravimetric analysis.

1. Precipitation method
2. Volatilization method

Precipitation method involves the transformation of the analyte into precipitates that can be filtered. This process requires the container usually a crucible to be weighed to constant weight. The voltalization method on the other hand volatilizes the analyte to suitable temperature and the volatile products are then weighed. This particular experiment that determines sulphate from a sample uses the precipitation method where an unknown sample that contains sulphate was mixed with a reagent $BaCl_2$ to form a precipitate that was filtered and weighed. The weight is then used in computation for the present yield of sulfate in the unknown.

Principle: The sulphate ions in the sample are precipitated by addition of barium chloride solution to water sample acidified with hydrochloric acid and kept near the boiling point.

$$SO^{2-}_{4(aq)} + Ba^{2+}_{(aq)} \longrightarrow BaSO_4(s)$$

The solution is first acidified with HCl to prevent the precipitation of $BaCO_3$ and $Ba(OH)_2$ and it also aids the larger precipitate crystals.

4.30.2 Materials

1. Porcelain crucible
2. 1 g sample containing sulfate
3. 200 mL distilled water
4. Ashlers filter paper
5. Conc. HCl
6. $BaCl_2$ solution

4.30.3 Procedure

Take 200 mL sample solution, add 2 mL conc. HCl in it. Heat the solution on hot plate to near boiling add 100 mL of 0.05 M $BaCl_2$ to the hot sample while stirring vigorously. Now filter the ppt to ashless filter paper like Wattman 42 or 44. Wash the filter paper with a little amount of deionized water.

Weigh the crucible and note down its weight. Put moist filter paper in it and place in the furnace, check the crucible till the filter paper completely burns. Cool the crucible in the desiccators. Now reweigh the content.

4.30.4 Calculation

Weight of crucible + filter paper

$$x + y$$

Weight of crucible + residue after ignition + filter paper

$$x + z + y$$

$$\text{Weight of } BaSO_4 \text{ ppt} = (x + z + y) - (x + y)g$$

$$\text{Weight of } BaSO_4 \text{ in mg} = (x + z + y) - (x + y) \times 1000 \text{ mg}$$

$$\text{Molecular weight of } BaSO_4 = 233.4 \text{ g}$$

$$\text{Molecular weight of } SO_4 = 96.06 \text{ g}$$

$$SO_4 \text{ in mg/l} = \frac{\text{mg of } BaSO_4 \times 96.06 \times 1000}{\text{Quantity of sample} \times 233.4}$$

4.30.5 Result

Given sample solution contains SO_4 by using gravimetric analysis.

4.30.6 Impacts of Sulphate Ions

1. **Environmental impacts:** Sulphates increase the acidity of the atmosphere and hence results in acid rain. It causes the scattering of light and effectively increasing the earth's albedo.
2. **Impacts on human:** If some one drinks water that has above 500 ppm of sulphates then after a short time following symptoms will start appearing.
   (i) Diarrhoea and intestinal pain (especially in babies)
   (ii) Dehydration due to diarrhoea
   (iii) A slight decrease in normal stomach acidity
   (iv) Catharsis in young males

   When sulphates are taken in by breathing then it may cause lung irritation but there are no long-term human health effects (WHO, 1996).

4.31 DETERMINATION OF FLUORIDE BY SPADNS METHOD

4.31.1 Principle

Under the acidic conditions fluorides react with zirconium SPADNS solution and colour of Spadans reagent is bleached with increase in concentration of fluoride ions.

4.31.2 Procedure

Take five 25 mL volumetric flasks. Add 2, 4, 6, 8 cm^3 of NaF solutions and 10 mL of zirconium SPADANS reagent. And dilute the solution by adding water. Also for the given water sample in which fluoride ion to be determined add 10 mL of SPADANS reagent and dilute up to 25 mL mark. The absorbance of these samples is measured at 570 nm. Plot the graph of absorbance with volume of NaF. From the calibration curve, the volume of fluoride ions in water sample can be determined.

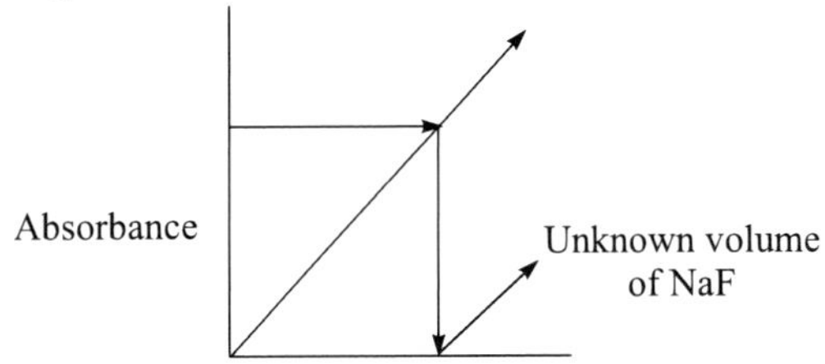

Fig. 4.10

4.32 SEWAGE

Sewage is the liquid waste from human and household wastewaters, industrial activities, ground wastes, farm activities, etc., that is disposed of via sewage.

Fresh domestic sewage is yellowish-green in and gradually turns dark brown due to decomposition which produces bad/offensive smell due to the release of ammonia, hydrogen sulphide, phosphine, etc.

Sewage besides about 99.9% water, contains organic and inorganic matters in dissolved, suspended and colloidal states. It also contains pathogenic and non-pathogenic bacteria.

Aerobic bacteria: The bacteria which live and grow on dissolved or free oxygen and decompose organic impurities of sewage to nitries, nitrates, sulphates, carbon dioxide, water, etc., are called aerobic bacteria. This decomposition does not produce bad smell.

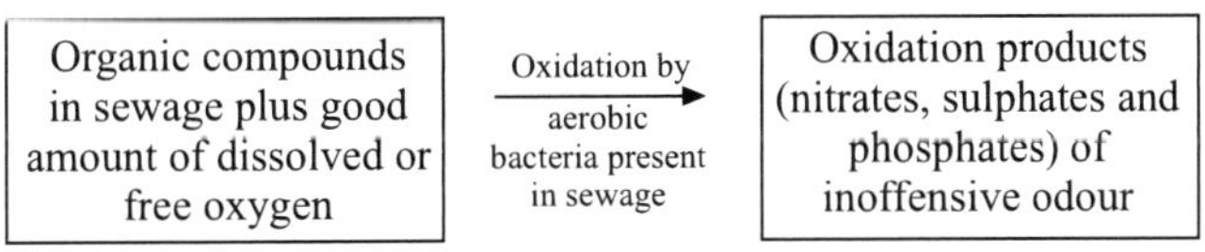

Fig. 4.11 *Aerobic oxidation*

Anaerobic bacteria: The bacteria which live and grow in the absence of free or dissolved oxygen are called anaerobic bacteria. These bacteria start decomposition of sewage when the DO content of sewage falls below a particular limit and produces H_2S, methane, ammonia, etc.

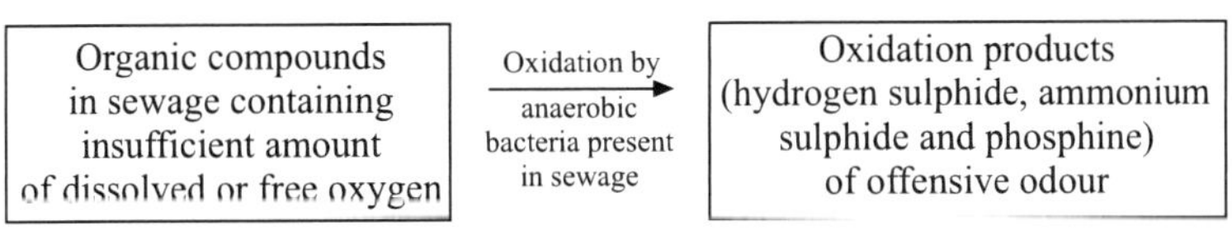

Fig. 4.12 *Anaerobic oxidation*

Sewage Treatment: Sewage treatment is removal of impurities. It can be treated in three different stages to make it suitable for domestic use:

1. Primary treatment (physical and chemical)
2. Secondary treatment (biological)
3. Tertiary treatment (chemical)

4.32.1 Primary Treatment

It involves:

(a) Screening
(b) Silt and grit removal
(c) Oil and gas removal
(d) Sedimentation

(a) Screening: It is a physical process which removes large suspended or floating matters in the sewage. This is capable by using bar screens and mesh screens. In this process, all the suspended impurities are left behind and sewage with colloidal and dissolved impurities moves for further treatment.

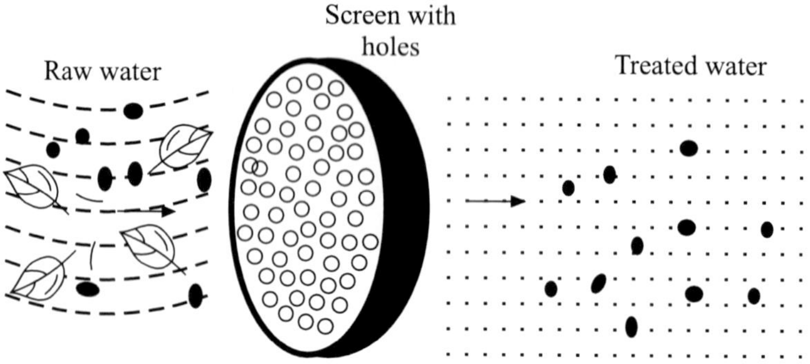

Fig. 4.13 *Screening for removal of floating matter*

(b) Silt and Grit Removal: Grit, i.e., small pieces of materials, heavy sand and mud are removed by slow passing of sewage through grit chambers, in which the velocity of the flow of sewage is reduced. This process allows the grit to settle down to bottom by the action of gravity.

(c) Oil and Grease Removal: Oil, grease, etc., are removed in skimming tanks. They are converted into a soapy mixture by blowing compressed air through sewage water in skimming tanks and lifted to the surface in the form of a foam. The floating substance is skimmed off.

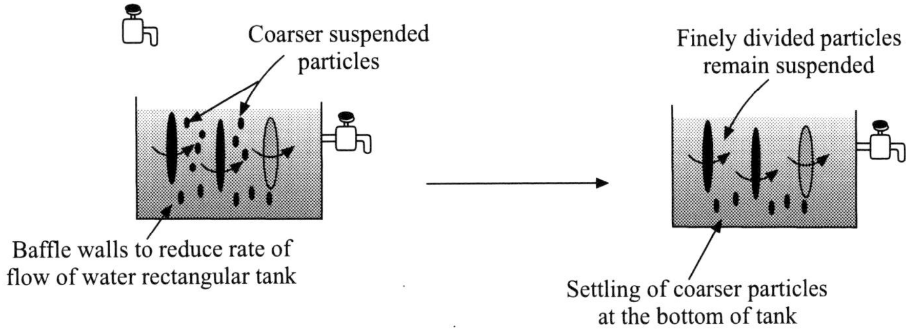

Fig. 4.14 *Oil and grease removal*

(d) Sedimentation Process: Colloidal impurities are removed from water by this process. The water is allowed to enter a sedimentation tank or by sedimentation with coagulation. The coagulants used is alum for settlement of the finely suspended particles.

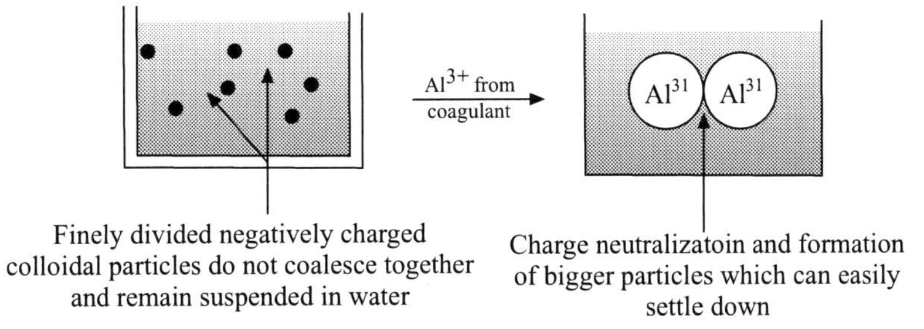

Fig. 4.15 *Sedimentation with coagulation*

4.32.2 Secondary Treatment

It involves aerobic biochemical oxidation or aeration. During this aerobic oxidation process, carbon of the organic matter is converted into CO_2, nitrogen into NH_3 and finally into nitrites and nitrates. The dissolved bases, present in the sewage water, then form salts such as ammonium nitrite, ammonium nitrate, and calcium nitrate.

$$NH_2COH_2 \xrightarrow[\text{aerobic}]{\text{Oxidation}} CO_2 + 2NH_3$$

$$NH_3 \xrightarrow{\text{Oxidation}} HNO_3, HNO_2, \text{etc.}$$

$$HN_3, HNO_2 \xrightarrow[\text{of soils}]{\text{+Bases}} NH_4NO_2, NH_4NO_3, Ca(NO_3)_2, KNO_2, \text{etc.}$$

This process is generally done by either the trickling filter method or activated sludge process.

Trickling Filter Method: It consists of a rectangular or circular vessel with a filter bed made up of broken rock, or broken bricks. The sewage is allowed to trickle over the filtering medium by means of rotating arm distributor. As the trickle sewage percolates downward through the filter bed, microorganisms grow on the surface of aggregates, using organic material of the sewage as food. Aerobic conditions always maintain and purified sewage is removed from the bottom. This process generally removes 90% of biologically oxidisable impurities.

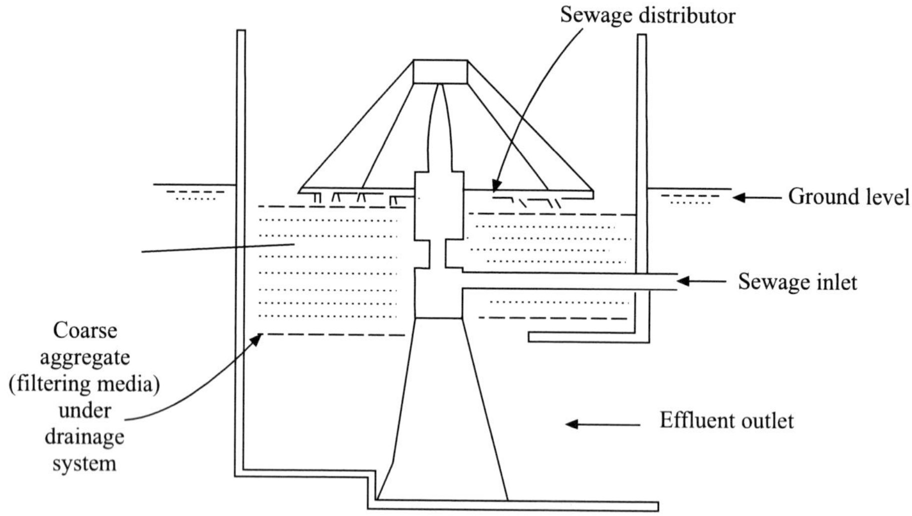

Fig. 4.16 *Trickling filter*

Activated Sludge Process: In this process for complete aerobic oxidation, adequate amount of oxygen or air is passed through sewage containing aerobes. However, if this aeration is carried out in the presence of a part of sludge from the previous oxidation process, the oxidation process is much quicker. This so-added sludge from the previous process is known as activated sludge, since it contains organic matters inhabited by numerous bacteria, etc.

Wastewater after sedimentation is mixed with the required quantity of activated sludge (containing microorganisms and aerobic bacteria) in a aeration tank. The mixture is aerated by blowing air through it for several hours. Because of the aerobic conditions the organic matter in the sewage is fully oxidised. The purified water plus sludge is sent to a tank where sludge settles down. A part of the sludge is used for purification of fresh batch of sewage while the rest is pumped to sludge disposal tank. Purified water is pumped out and collected separately.

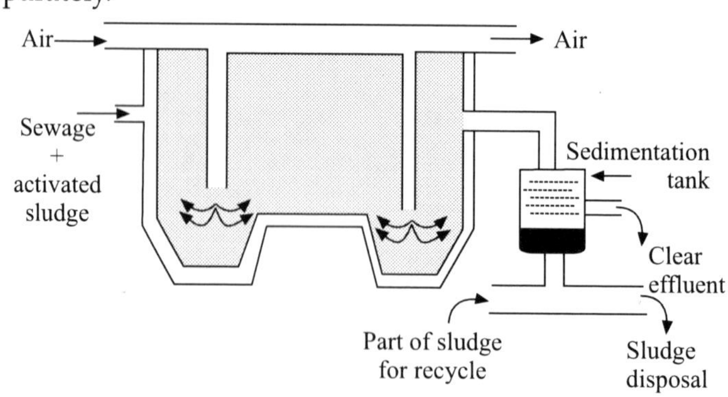

Fig. 4.17 *Activated sludge process for secondary (biological) treatment of sewage*

4.32.3 Tertiary Treatment

The aim of tertiary treatment is further purification wastewater as well as its recycling. Treatment in this stage depends on the type of impurity present in water. It generally involves the following:

1. Removal of Phosphate: It is removed by adding lime at pH above 11. At this pH, ammonium salts are converted into ammonia.

$$3Ca(OH)_2 + 2PO_4^{3-} \longrightarrow Ca_3(PO_4)_2 + 6OH^-$$
$$Ca(OH)_2 + NH_4^+ \longrightarrow Ca^{2+} + NH_3 + H_2O + OH^-$$

2. Coagulation and Sedimentation: Fine suspended particles are coagulated in a sedimentation tank by adding a coagulating agent, i.e., alum, $FeSO_4$, etc. These compounds produce a gelatinous precipitate of their metal hydroxides. This gelatinous precipitate adsorbs fine suspended particles on its surface and settles at the bottom of the sedimentation tank.

3. Filtration: The last traces of suspended matter are removed through conventional filter beds filled with gravels, coarse sand and fine sand.

4. Degasification: Removal of dissolved gases like NH_3, CO_2, H_2S, etc., by passing it through a degasifier. The degasifier consists of a large tower fitted with a number of perforated plates. Hot water trickels through perforated plates provided on the tower of the degasifier. High temperature and the exposure to large surface area lead to stripping out of gases from water.

5. Disinfection: Harmful pathogenic bacteria are removed by disinfection. Among many disinfectants chlorine is cheap and effective.

$$Cl_2 + H_2O \longrightarrow HOCl + H^+ Cl^- \text{ (pH 6.5)}$$

Unionized HOCl formed by the above reaction kills all the pathogenic bacteria present in water.

Water obtained after tertiary treatment is free from odour, colour, pathogenic bacteria, particulate matter, has low BOD and hence can be used for domestic purposes.

4.33 ION EXCHANGE PROCESS OR DEMINERALIZATION PROCESS OF SOFTENING OF WATER

It is the process of complete removal of all ions (cations and anions) present in water by using ion-exchange resins.

It is an insoluble crosslinked long chain organic polymer. It consists of microporous structure with functional groups for exchange of ions. Generally, polystyrene divinyl benzene, polymenthacrylic acid – divinyl benzene or phenol – formaldehyde polymer is used as the ion exchange resin. The functional groups of resin may be acidic or basic on the basis of functional group ion exchange resins are of two types:

1. Cation exchange resins
2. Anion exchange resins

4.33.1 Cation Exchange Resins

These resins contain acidic groups like –COOH or –SO_3H. The resin styrene divinyl benzene on sulphonation becomes cation exchange resin and is capable of exchanging its H^+ ions with cations of water sample. It is generally represented as RH^+.

A cation exchange may be represented as

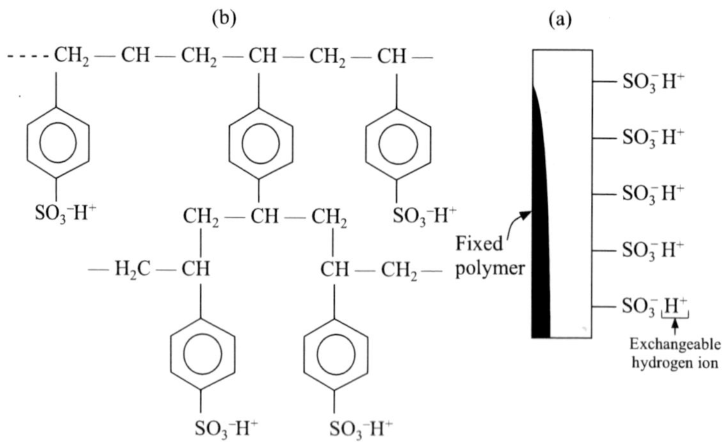

Fig. 4.18

4.33.2 Anion Exchange Resins

These resins contain basic functional group with hydroxide salts. The resin styrene divinyl benzene complexes on treatment with NaOH becomes an anion exchange resin and capable of exchanging its OH^- ions with anions of water sample. It is represented as ROH^-.

An anion exchange resin is capable of exchanging its OH^- ions with anions of water sample. It is represented as shown in Fig. 4.19.

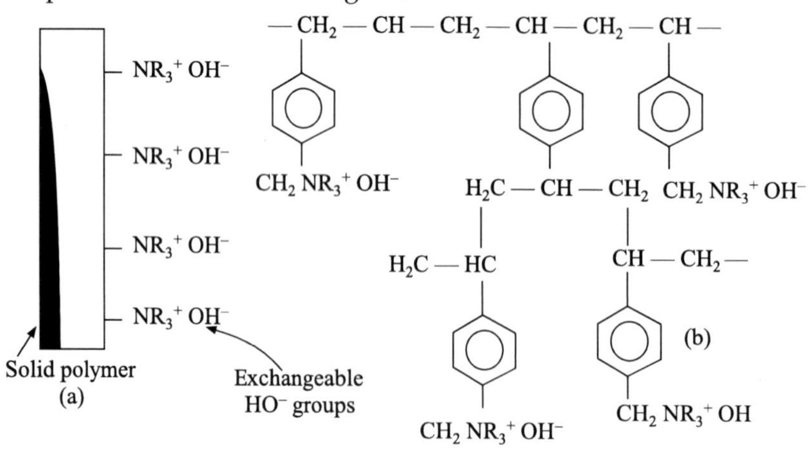

Fig. 4.19 *Representation of exchange anion*

4.33.3 Working

Cation exchanger and anion exchanger both are interconnected with a pipe.

The high minerals water sample is first allowed to pass through cation exchange resin, which exchanges its H^+ ions with Ca^{2+} and Mg^{2+} ions of high in mineral water.

$$2RH^+ + Ca^{2+} \longrightarrow R_2Ca^{2+} + 2H^+$$

$$2RH^+ + Mg^{2+} \longrightarrow R_2Mg^{2+} + 2H^+$$

Now, the water sample after passing through cation ion exchange resin is free from cations, i.e., Mg^{2+} and Ca^{2+}.

This water sample is further allowed to pass through anion exchanger resin columns where anions (SO_4^{2-}, Cl^-) present in the hard water are replaced by OH^- ions. Following reactions are takes place.

$$ROH^- + Cl^- \longrightarrow RCl^- + OH^-$$

$$2RH^- + SO_4^{2-} \longrightarrow R_2SO_4^{2-} + 2OH^-$$

$$2RH^- + CO_3^{2-} \longrightarrow R_2CO_3^{2-} + 2OH^-$$

The released H^+ from cation exchanger treatment and OH^- from anion exchange treatment get combined to produce water.

$$H^+ + OH^- \longrightarrow H_2O$$

It is called deionized water because it is free from cations and anions.

4.33.4 Regeneration of Resins

When the resins are exhausted or inactivated they lose their capacity to exchange ions and are regenerated (i.e., when capacities of cation and anion exchangers they are then said to be exhausted).

Cation exchange resins are regenerated by passing a solution of dilute HCl or H_2SO_4 and then washing with deionized water.

$$R_2Ca + 2H^+ \longrightarrow 2RH + Ca^{2+} \text{ (in washing)}$$

$$R_2\,Mg + 2H^+ \longrightarrow 2RH + Mg^{2+} \text{ (in washing)}$$

Similarly, anion exchange resin is washed with ammonium or sodium hydroxide solution

$$R_2SO_4 + 2OH^- \longrightarrow 2ROH + SO_4^{2-} \text{ (in washing)}$$

Then washed with deionized water.

4.33.5 Advantages

1. Easy to operate and control
2. Highly acidic or highly alkaline water also can be softened.
3. By this process water of very low hardness (~ 2 ppm) is produced.

4.33.6 Disadvantages

1. Equipment and process is costly.

2. Turbid water decreases the efficiency of resins.
3. Regeneration of resins is also costly.

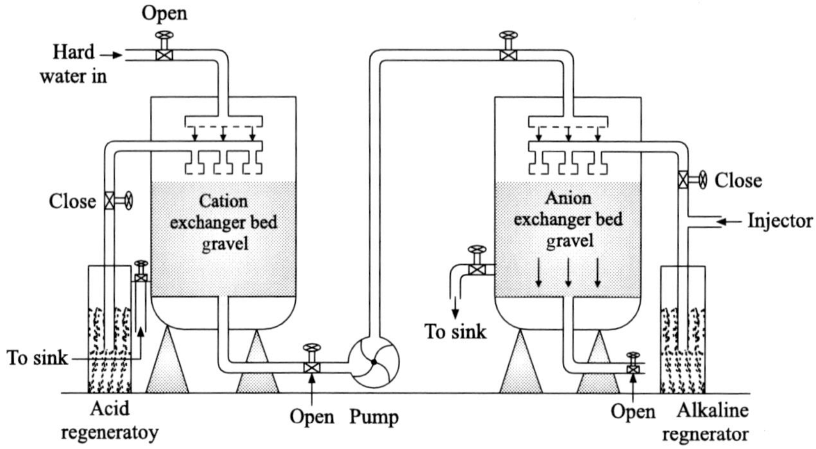

Fig. 4.20 *Softener for demineralization*

4.34 POTABLE WATER

It is defined as water that is suitable for human consumption (i.e., water can be used for drinking or cooking). It is generally obtained from rivers and lakes. Water obtained from these sources are never pure and as such it is unfit for drinking. The drinking water should be of right quality. The principal cause of many diseases is water. Therefore, the water used for domestic purposes should be potable (safe) and palatable (pleasant).

Following are the characteristics of quality drinking water.

- It should have a high degree of clarity.
- It should have a good taste and should be free from odour.
- It should be reasonably soft.
- It should be free from any pathogenic organisms.

4.34.1 Water Treatment for Domestic Use

The type of treatment given to water largely depends on the quality of raw water and also upon the specified standards. In general, water treatment for municipal supply or domestic use consists of the following stages–

1. Screening
2. Sedimentation
3. Coagulation
4. Filtration
5. Disinfection

1. **Screening:** The raw water obtained from rivers, reservoirs and lakes is passed through steel screens, having large number of holes. The screening removes solid floating

materials such as dead fish and animals, bits of wood, leaf, weeds and other debris present in water. These materials are simply retained by the screen.

2. **Sedimentation:** The process of removing suspended clay and other colloidal matter present in water by the process of settling is known as sedimentation.

3. **Sedimentation with coagulation:** In this method sedimentation is carried out in the presence of a coagulant which assists in removing the finely divided and colloidal suspended particles from water. It is a more effective and rapid method compared to plain sedimentation.

   The common coagulants used in most of the water treatment plants are alum, sodium aluminate, ferrous sulphate, ferric sulphate, etc. A calculated amount of coagulant solution is added to water, mixed thoroughly and allowed to stand in a large sedimentation tank for a few hours. The particles settle down as sludge.

4. **Filtration:** It is a process of removal of suspended matter from water by passing it through the porous medium or bed. When water precolates through the pores of filter bed, the suspended particles are retained by the bed. The filtered water acquires a high level of clarity. There are two types of filters: gravity filters and pressure filters.

5. **Disinfection:** Contaminated water is the main cause of many diseases like cholera, typhoid, etc. Therefore, as a final assurance against the presence of harmful bacteria and viruses, drinking water is treated with certain chemicals. The chemicals which are used to destroy the bacteria and viruses are called disinfectants. The process of destruction of bacteria and viruses is called disinfection.

The important methods of disinfection are:

1. Boiling
2. Chlorination
3. Addition of chloramines
4. Ozone treatment

1. **Boiling:** Boiling of water for 15-20 minutes kills all types of bacteria present in water. This method is commonly used in hospitals and for household purpose. Since the method is expensive, it is not used for municipal water supply.

2. **Chlorination:** Among the common disinfectants, chlorine is the most widely used disinfectant in municipal water treatment. The high popularity of chlorine is due to the following reasons:

   - It is readily available either in the liquid or in the gaseous form.
   - It has powerful bactericide property.
   - It does not introduce any impurities in water.
   - It is economical.

   Mechanism of chlorination: When chlorine is added to water, it produces two species, namely, ionized hydrochloric acid and unionized hypochlorous acid at lower pH value of 6.5.

$$Cl_2 + H_2O \rightarrow HOCl + H^+ + Cl^- \text{ (pH 6.5)}$$

Hypochlorous acid is a powerful bactericidal. It is the unionized hypochlorous acid that destroys the pathogenic bacteria in water.

At higher pH value of 8, hypochlorous acid undergoes appreciable ionization and exists as hypochlorite ion (OCl^-) and H^+ ion. Hypochlorite ion is a weak bactericidal. Chlorine is a powerful desinfectant at lower pH values.

3. **Addition of chloramines:** Chloramines are chlorine containing derivatives of ammonia. These include monochloramine (NH_2Cl), dichloramine ($NHCl_2$) and trichloramine (NCl_3). Among these monochloramine is found to be a powerful bactericidal. When chloramine is added to water, it slowly decomposes to produce unionized active hypochlorous acid which destroys the pathogenic bacteria.

$$NH_2Cl + HOH \rightarrow HOCl + NH_3$$

Chloramine acts as a reserve for chlorine. Available residual chlorine in chloramines inactivates any contamination that may be introduced after the initial disinfection and thus provides adequate protection.

***Advantage*:**

- Chloramine is a powerful bactericidal.
- It imparts good taste to water.
- High residual chloramines in treated water do not cause any health hazard.

4. **Ozone treatment:** Ozone is an excellent and efficient desinfectant. It destroys all types of bacteria present in water and makes the water absolutely safe for consumption. It leaves behind no odour. Ozone is an unstable allotropic form of oxygen when added to water, ozone breaks up into oxygen molecule and nascent oxygen

$$O_3 \rightarrow O_2 + [O]$$

The nascent oxygen so produced being a very strong oxidizing agent, destroys the pathogenic germs. It also removes both colour and odour present in water through the oxidation of organic matter. It makes the treated water safer, tasty and pleasant. Ozone treatment of water is too expensive to be used for large quantity of water supply.

4.35 DESALINATION OF WATER

Water containing dissolved salts and having very salty taste is called brackish water, like the seawater and it is not suitable for drinking. "The process of removal of common salt from this brackish water is called desalination". It can be done by the following methods.

1. Electrodialysis: Electrodialysis is the process of migration of ions present in the saltwater towards the oppositely charged electrodes under the influence of applied field.

This process is used to transport salt ions from one solution through ion-exchange membranes to another solution under the influence of an applied electric potential difference.

It consists of two electrodes and ion selective membrane which are permeable to either cation or anion. Cathode is placed near cation permeable membrane. A cation selective membrane has negatively charged functional group like $RCOO^-$ and allows only positively charged ions to pass through it. Similarly, the anion selective membrane has positively charged functional groups such as R_4N^+ and therefore allows negatively charged ions to pass through them.

It is an important method for desalination of brackish water and may be applied at room temperature. It is based on the fact that when direct electric current is applied to saline (brackish) water by using electrodes and ion selective permeable membranes, the ions present in saline water migrates towards oppositely charged electrodes and water gets purified.

When direct electric current is passed through saline water, the sodium ions (Na^+) start moving towards negative pole (cathode), while the chloride ions (Cl^-) start moving towards the positive pole (anode) through the membrane. Thus, each alternate compartment gets concentrated with anions and cations. Meanwhile, other adjacent alternate compartments get filled with desalinated water. Finally concentrated water and desalinated water are removed separately as shown in Fig. 4.21.

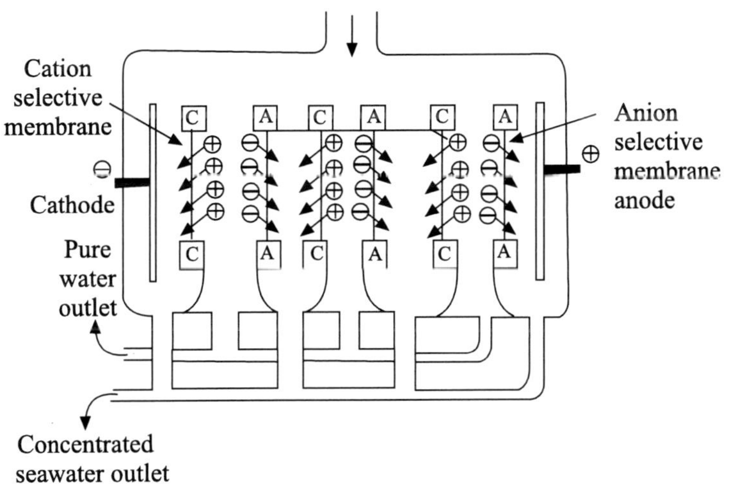

Fig. 4.21 *Schematic diagram of electrodialysis cell: anion and cation permeable membrane*

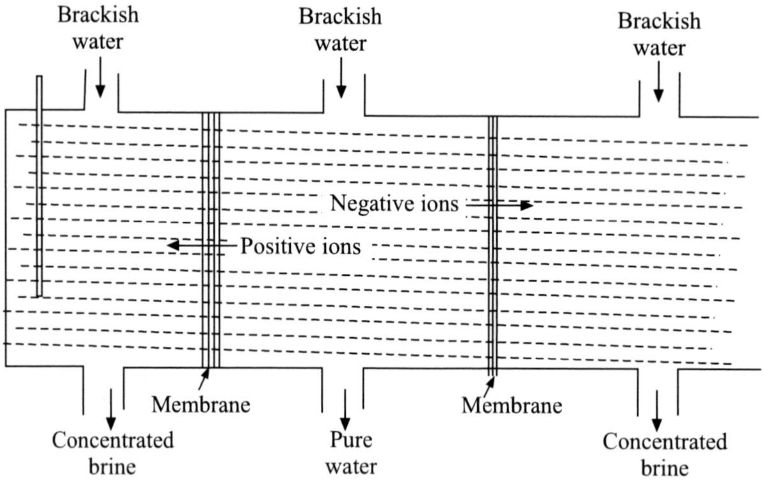

Fig. 4.22 *Line diagram of electrodes*

2. Reverse Osmosis: When two solutions having different concentrations are separated by a semipermeable membrane. The flow of solvent molecules is from the region of lower concentration to that of higher concentration. The phenomenon is called reverse osmosis. The driving force in this phenomenon is called osmotic pressure.

If the pressure is applied on solution side, the solvent will flow in the reverse direction.

That is called reverse osmosis. The membrane chosen allows only water to pass through it and not the salt.

Method

In this process, pressure of 15–40 kg/cm is applied to the seawater/impure water (to be treated) to force its pure water out through the semipermeable membrane. This membrane consists of very thin films of cellulose acetate affixed to either side of a perforated tube. It is a result of reverse osmosis, fresh water moves down and can be collected from outlet at the bottom of the cell.

Applications

1. The dialysis used by people with kidney failure is based upon this technique.
2. It is used for the production of deionized water.
3. It is used in the wine industry for the concentration of wine.
4. It is used to clean effluent and brackish groundwater.
5. It is more economical for concentrating liquids like tomato juice and fruit juices like orange juice.

Advantages

1. It has long life and the membrane is easy to replace.
2. Maintenance cost is also economical and depends upon the membrane.
3. It is compact and simple.

Disadvantages

1. It does not remove bacteria, virus, etc., present in water.
2. It does not remove the odour of water.

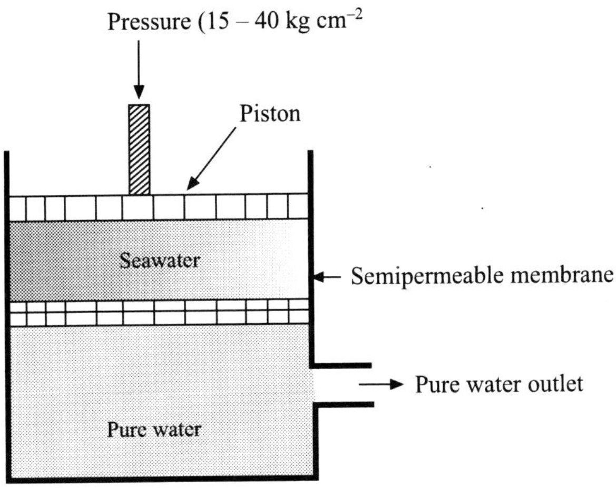

Fig. 4.23 *Reverse osmosis cell*

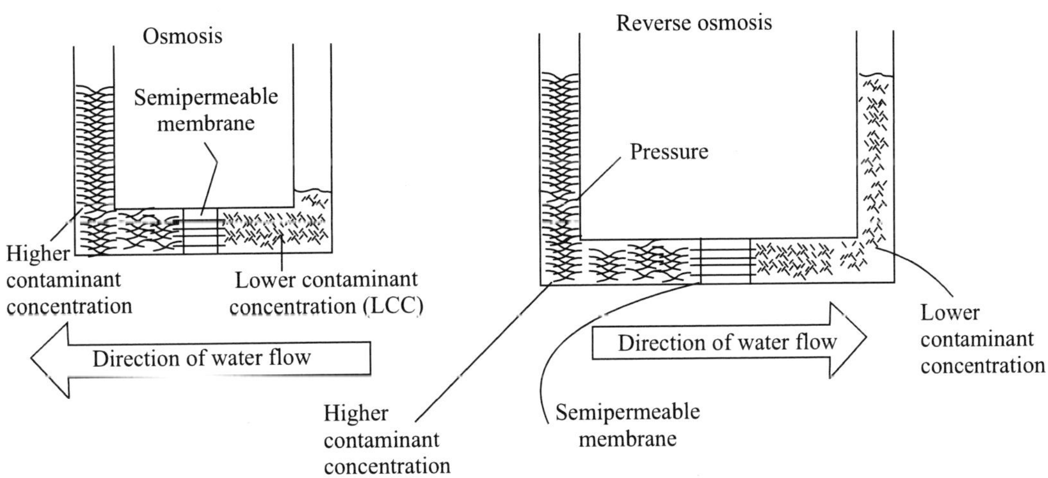

Fig. 4.24 *Diagrammatic representation of osmosis and reverse osmosis processes*

Review Questions

1. Define air pollution? Give examples of air pollution.
2. List the natural sources of air pollution.
3. Briefly discuss the anthropogenic sources of air pollution.

4. Differentiate between the following:
   (a) Primary and secondary air pollutants
   (b) Mist and fog
   (c) Smoke and fumes
5. Discuss the air pollution caused by oxides of nitrogen and sulphur.
6. Define particulate pollutants. Discuss the effect of particulate pollutants.
7. Discuss the effect of air pollution on the following:
   (a) Human health
   (b) Plants
   (c) Animals
   (d) Materials
   (e) Climate
   (f) Aesthetic beauty
8. Discuss the source correction methods for controlling air pollution.
9. What is the effect of pollution on human beings, animals and plants? Explain.
10. Discuss the sources and correction methods for controlling air pollution.
11. Explain the various methods commonly employed for the disposal of solid wastes, with their advantages and disadvantages.
12. What are solid wastes? Discuss their types, effects and name the various methods used to dispose of solid wastes. Explain any one of them with its merits and demerits.
13. Briefly describe the idea of solid waste management stating suitable examples. Distinguish between the incineration and composting methods of solid waste disposal.
14. What do you mean by *E*-waste? Explain the factors effect. The process of composting and various classification of the composting process. In your opinion, which one will be ideal for Indian conditions.
15. What is recycling? Discuss its importance by giving suitable examples, in social waste management also give the advantages of recycling and waste utilization.
16. Explain the various methods commonly employed for the disposal of solid waste, with their advantages and disadvantages.
17. Classify solid wastes and discuss their health effects.
18. What is a solid waste? Discuss the sources of solid wastes.
19. What adverse effects can solid waste cause? How can the solid waste be managed?
20. Classify solid waste. What are the sources of urban and industrial solid wastes?
21. Briefly explain the techniques of wastes utilization by giving examples.
22. Define solid waste management. Give classification of solid wastes.

23. Enumerate the causes of solid waste pollution.
24. Discuss the health effects of solid waste pollution.
25. Describe the negative impacts of solid waste pollution.
26. Describe control measures of urban and industrial solid wastes.
27. What are different types of impurities present in water?
28. What are the characteristics of potable water?
29. What is brackish water?
30. Write short notes on:
    (a) Sedimentation
    (b) Coagulation.
31. What is desalination process? Explain its significance?
32. What is the principle of revere osmosis for desalination of water?
33. Explain the main advantage of reverse osmosis process over ion exchange.
34. How can scale formation be prevented by colloidal conditioning?
35. Write the chemical reaction involved in water softening by lime-soda process.
36. Give equations to explain the following:
    (a) $Al_2(SO_4)_3$ requires equivalent amount of lime for softening.
    (b) Mg $(HCO_3)_2$ requires double amount of lime for softening.
37. Explain the following:
    (a) Disadvantages of zeolite process
    (b) Regeneration of zeolite
38. What are ion exchange resins and how are they classified? Write the chemical reactions involved in ion exchange process.
39. How can exhausted cation and anion exchange resins be regenerated?
40. Explain BOD and its significance.
41. Explain COD and its significance.
42. How do you estimate dissolved oxygen in waste?
43. What are the steps involved in tertiary of sewage?
44. Write a brief note on secondary treatment sewage.
45. What are the disadvantages of using hard water in a boiler operation?
46. What are the requirements of boiler feed water?
47. What are the boiler problems caused by the presence of silica in water?
48. Distinguish between scale and sludge.

49. Write a short note on scale and sludge formation.
50. Why should the presence of CO_2 be avoided in boiler feed water? How is dissolved CO_2 removed from water by chemical means?
51. What are the disadvantages of scale formation and the prevention methods for it?
52. Compare between priming and foaming.
53. Compare between external and internal treatments of boiler feed water.
54. What is sewage? Mention its characteristics.
55. What is desalination? Explain briefly the methods for desalination of brackish water.
56. With the help of a neat diagram, explain the reverse osmosis method for desalination of brackish water.
57. Explain the following methods for desalination of water:
    (a) Reverse osmosis
    (b) Electrodialysis
58. What is potable water? Discuss the purification of water by reverse osmosis process.
59. Describe the Winkler method for the determination of dissolved oxygen. Give the reactions involved.
60. What are scales? How are they formed? What are the disadvantages and the methods of preventing scale formation?
61. Describe the process of scale and sludge formation in boilers and methods for their prevention.
62. Describe the factors responsible for the corrosion of a boiler. Discuss the measures for its prevention.
63. Write short notes on:
    (a) Priming and foaming
    (b) Boiler feed water.
64. Explain the lime-soda process for water softening. Give the chemistry involved during the softening.
65. Describe the ion-exchange process for the purification of water with a neat diagram. Discuss the merits and demerits of this process.
66. Explain the ion-exchange method for the removal of dissolved impurities and its advantages over lime-soda process.
67. What are ion-exchange resins? How will you purify water by using the resins? What are the advantages of this method over other methods?
68. Explain the principles and processes involved in determination of:
    (a) Biological oxygen demand.
    (b) Chemical oxygen demand.

69. Explain the activated sludge process used in biological (secondary) treatment of sewage.
70. Justify the following statement:
    (a) COD is greater than BOD
    (b) Sedimentation with coagulation is a better process than sedimentation alone.
71. Explain the terms BOD and COD. What are the steps involved in the tertiary treatment of sewage?

EXERCISE

1. In a COD experiment, 29.5 cm^3 and 20 cm^3 of 0.0225 N Mohr's salt solution were required for blank and sample titrations respectively. The volume of test sample used was 25 cm^3. Calculate the COD of the sample solutions. **[*Ans.* 76 mg/L]**
2. 25 mL of wastewater was mixed with 25 mL of $K_2Cr_2O_7$, acidified and refluxed. The unreacted $K_2Cr_2O_7$ acidified required 8.2 mL of FAS. In a blank titration 25 mL of $K_2Cr_2O_7$ acidified required 16.4 mL of same 0.2 N FAS. Calculate COD of wastewater. **[*Ans.* 524.8 mg/L]**
3. Calculate the COD of the effluent sample when 2 cm^3 of the effluent requires 10.5 cm^3 of 0.005 M $K_2Cr_2O_7$ for complete oxidation. **[*Ans.* 1260 mg/L]**
4. In a COD experiment, 30 cm^3 of an effluent sample required 9.8 cm^3 of 0.001 M $K_2Cr_2O_7$ for oxidation, calculate the COD of the sample. **[*Ans.* 2.613 mg/L]**
5. A 50 mL of sample contains 840 ppm of DO after 5 days. The DO value becomes 230 ppm after the sample has been diluted 80 mL. Calculate the BOD of the sample. **[*Ans.* 976 mg/L]**
6. An effluent water sample emerging out from a chemical factory is found to contain only an organic compound of molecular formula $H_2C_2O_4$. If the wastewater sample contains 15.0 mg/L of the above compound, what would be the amount of oxygen required to oxidise by microorganisms under experimental contains of BOD estimation. [C=12, H=1 and 0 =16] **[*Ans.* 2.67 mg/L]**
7. 50 mL of an industrial effluent is allowed to react with 100 mL of $K_2Cr_2O_7$ solution. The unreacted $K_2Cr_2O_7$ consumed 25.0 mL of 0.05N FAS solution. In blank titration 35.0 mL of same FAS is consumed. Calculate COD of effluent. **[*Ans.* 80 mg/L]**
8. In COD determination, 25 cm^3 of an effluent required 8.9 cm^3 of 0.001 M $K_2Cr_2O_7$ for complete oxidation, calculate COD of effluent sample. **[*Ans.* 17.088 mg/L]**
9. Calculate BOD, when 1 litre of effluent from sugar industry containing 150 mg of glucose was completely oxidized into CO_2 and H_2O. **[*Ans.* 160 mg/L]**
10. 20 mL of a sample of COD analysis was reacted with 10 mL of 0.25 N $K_2Cr_2O_7$ and the unreacted dichromate required 6.5 mL of 0.1 N FAS. 10 mL of same $K_2Cr_2O_7$ and 20 mL of distilled water under the same conditions as the sample required 26.0 mL of 0.10 N FAS. What is the COD of sample? **[*Ans.* 780 mg/L]**

5

Instrumental Methods of Analysis and Nanomaterial

5.1 INTRODUCTION

Instrumental methods of chemical analysis have become the principal means of obtaining information in diverse areas of science and technology. These include analysis of geochemical composition, new chemical substances, pollutants, engine oils, body fluids in humans and animals for diagnosis of disease, etc., involve instrumentation. Information based on the data obtained from instrumental analysis is useful in solving problems in various fields, owing to their accuracy and the advantage that even a small quantity of sample is sufficient for analysis. Most instrumental techniques fall into three principal areas: electroanalytical methods, spectroscopic techniques and chromatography.

5.2 FLAME PHOTOMETER

In 1980, Bowling Barnes, David Richardson, John Berry and Roberthood developed an instrument to measure the low concentrations of sodium and potassium in a solution. They named this instrument *flame photometer*.

It is one of the branches of atomic absorption spectroscopy. It is also known as flame emission spectroscopy. Nowadays it has become a necessary tool in the field of analytical chemistry. It can be used to determine the concentration of certain metal ions like sodium, potassium, lithium, calcium, cesium, etc. In flame photometer spectra the metal ions are used in the form of atoms.

5.2.1 Principle

The principle of flame photometer is based on the measurement of the emitted light intensity when a metal is introduced into the flame. The wavelength of the colour gives information about the element and the colour of the flame gives information about the amount of the element present in the sample. Characteristic colours imparted to the flame by the emission of radiation from the alkali metals excited in flame is the basis of flame emission spectroscopy, for example: the compounds of the alkali and alkaline earth metals dissociate into atoms when introduced into the flame. Some of these atoms further get excited to even higher levels. But these atoms are not stable at higher levels. Hence, these atoms emit radiations when returning back to the ground state. These radiations generally lie in the visible region of the spectrum. Each of the alkali and alkaline earth metal has a specific wavelength (Table 5.1).

Table 5.1

Element	Emitted wavelength	Flame colour
Sodium	589 nm	Yellow
Potassium	766 nm	Violet
Barium	554 nm	Lime green
Calcium	622 nm	Orange
Lithium	670 nm	Red

For certain concentration ranges: The intensity of the emission is directly proportional to the number of atoms returning to the ground state. And the light emitted is, in turn, proportional to the concentration of the sample (Table 5.2).

Table 5.2

Fuel-oxidant mixture	Temperature (°C)
Natural gas – Air	1700
Propane – Air	1800
Hydrogen – Air	2000
Hydrogen – Oxygen	2650
Acetylene – Air	2300
Acetylene – Oxygen	3200
Acetylene – Nitrous oxide	2700
Cyanogen – Oxygen	4800

5.2.2 Instrumentation

It is based on 4 parts:

1. Burner
2. Monochromators
3. Detectors
4. Recorder

1. Burner: A burner in the flame photometer is the source of flame. It can be maintained in at a constant temperature. It is a part which produces excited atoms. Here the sample solution is sprayed into fuel and oxidant combination. A homogeneous flame of stable intensity is produced. Fuel and oxidant are required to produce the flame such that the sample converts to neutral atoms and gets excited by heat energy. The temperature of flame should be stable and also ideal. If the temperature is high, the elements in sample convert into ions instead of neutral atoms. If it is too low, atoms may not go to excited state. So a combination of fuel and oxidant is used such that there is desired temperature (Fig. 5.1).

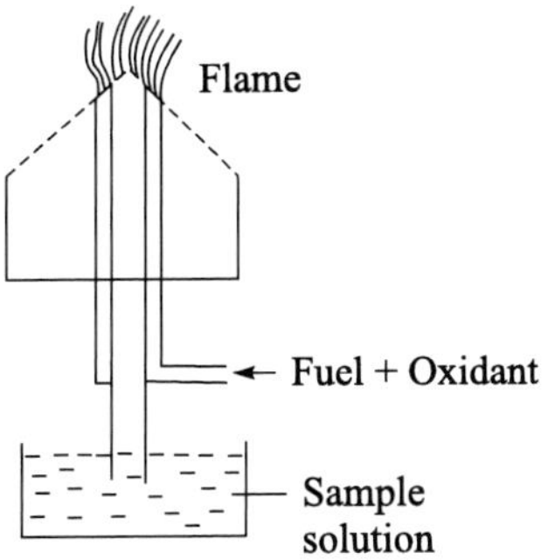

Fig. 5.1

2. ***Monochromators:*** Filters and monochromators are needed to isolate the light of specific wavelength from the remaining light of the flame. For this simple filters are sufficient as we study only a few elements like Ca, Na, K and Li. So a filter wheel with filter for each element is taken. When a particular element is analyzed, the particular filter is used so that it filters all other wavelengths.

3. ***Detector:*** Flame photometric detector is similar to that used in spectrophotometry. The emitted radiation is in the visible region, i.e., 400 nm to 700 nm. Further the radiation is specific for each element, so simple detectors are sufficient for the purpose of photovoltaic cells, phototubes, etc.

4. ***Recorders:*** These are the devises to read out the recording from detectors (Fig. 5.2).

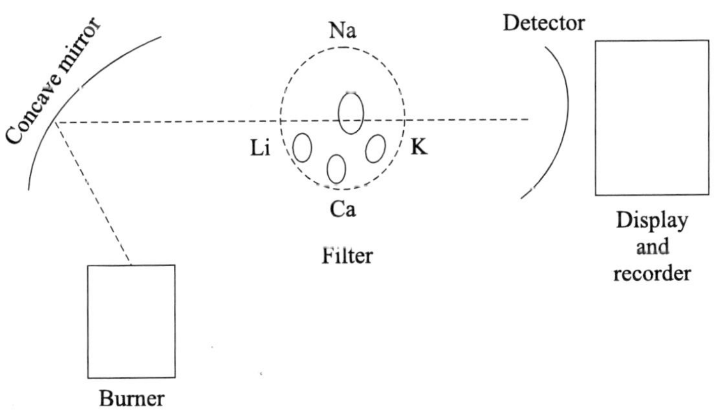

Fig. 5.2 *Schematic diagram of flame photometry*

5.2.3 Working

1. Both the standard stock solution and sample solution are prepared in fresh distilled water.

2. The flame of the photometer is calibrated by adjusting the air and gas. Then the flame is allowed to stabilize for about 5 min.
3. Now the instrument is switched on and the lids of the filter chambers are opened to insert appropriate colour filters.
4. The readings of the galvanometer are adjusted to zero by spraying distilled water into the flame.
5. The sensitivity is adjusted by spraying the most concentrated standard working solution into the flame. Now the full scale defection of the galvanometer is recorded.
6. Again distilled water is sprayed into the flame to attain constant readings of galvanometer. Then the galvanometer is readjusted to zero.
7. Now each of the standard working solutions is sprayed into the flame for three times and the readings of the galvanometer are recorded. After each spray, the apparatus must be thoroughly washed.
8. Finally, sample solution is sprayed into the flame for three times and the readings of galvanometer are recorded. After each spray, the apparatus must be thoroughly washed.
9. Calculate the mean of the galvanometer readings.
10. Now plot the graph of concentration against the galvanometer reading to find out the concentration of the element in the sample (Fig. 5.3).

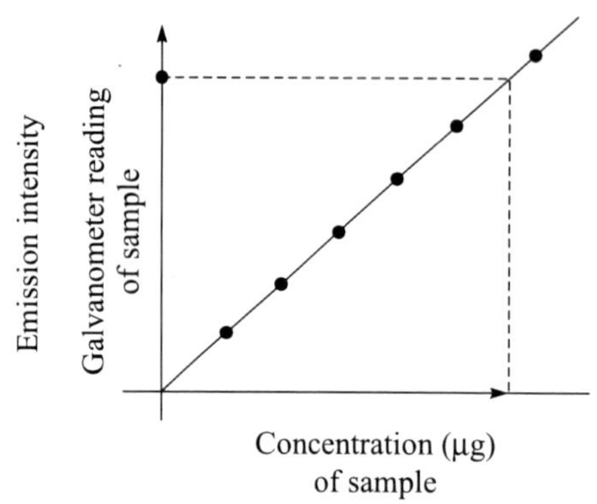

Fig. 5.3

5.2.4 Illustration of Steps in Flame Photometry

1. Sodium atom in ground state.
2. Energy is absorbed promoting an electron to an excited energy level.
3. Na atom in excited state.
4. Energy is released as light and the $_{-1}e^0$ drops back to the ground state.
5. Sodium atom in ground state (Fig. 5.4).

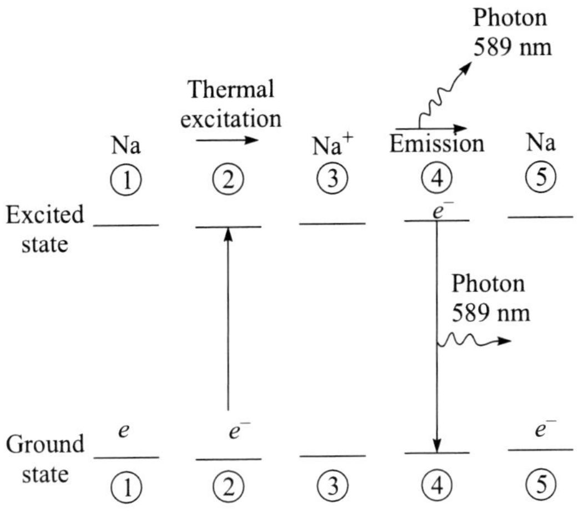

Fig. 5.4 *Illustration of excitation and emission processes in flame photometry*

The processes occurring during flame photometer analysis are summarized below.

1. Desolvation: It involves drying a sample in a solution. The metal particles in the solvent are dehydrated by the flame and thus solvent is evaporated.

2. Vaporization: The metal particles in the sample are also dehydrated. This also leads to the evaporation of the solvent.

3. Atomization: It is the separation of all atoms in a chemical substance. The metal ions in the sample are reduced to metal atoms by the flame.

4. Excitation: The electrostatic force of attraction between the electrons and nucleus of the atom helps them to absorb a particular amount of energy. The atoms then jump to the higher energy state when excited.

5. Emission: Since the higher energy state is not stable the atoms jump back to the ground state or low energy state to gain stability. This jumping of atoms emits radiation with characteristic wavelength. The radiation is measured by the photodetector.

5.2.5 Application

1. It can be applied both for quantitative and qualitative analyses of elements. For qualitative analysis of samples by comparison of spectral emission wavelengths with that of standards and for quantitative analysis to determine the concentration of group I A and II A elements. For example:

(a) Concentration of calcium in hard water
(b) Concentration of sodium, potassium in urine
(c) Concentration of Ca and other elements in bio-glass and ceramic material.

2. It is used in agriculture field for the analysis of water, soil and plant materials.
3. It is used in medical science for the testing of urine and blood samples.
4. It is used in industries for the detection of elements in fuel, glass, cement, etc.

5.2.6 Advantages

1. This method can be used to estimate elements which are rarely analysed.
2. It is both qualitative and quantitative in nature.
3. The method of analysis is very simple and economical.
4. It is quick, convenient, selective and sensitive analysis.
5. Even very low concentrations (parts per million/ppm to parts per billion/ppb range) of metals in the sample can be determined.
6. This method compensates for any unexpected interfering material present in the sample solution.

5.2.7 Disadvantages

1. It cannot be used for the direct determination of each and every metal atom. A number of metal atoms cannot be analysed by this method. The elements such as C, hydrogen and halides cannot be detected due to their non-radiating nature.
2. It cannot directly detect and determine the presence of inert gases.
3. Only liquid samples may be used. Also sample preparation becomes lengthy in some cases.
4. The accurate concentration of the metal ion in the solution cannot be measured.

5.3 ATOMIC ABSORPTION SPECTROSCOPY

The phenomenon of atomic absorption was first observed in 1802 with the discovery of the Fraunhofer lines in the sun's spectrum. It is a spectroanalytical procedure for the quantitative determination of chemical elements using the absorption of optical radiation (light) by free atoms in the vapour state. In this method, the radiation is absorbed by non-excited atoms in the vapour state. A much longer number of the gaseous metal atoms will normally remain in an unexcited state or in other words in the ground state (Fig. 5.5). These ground state atoms are capable of absorbing radiant energy of their own specific resonance wavelength, which in general is the wavelength of the radiation that the atom would emit if excited from the ground state. Hence, if light of the resonance wavelength is passed through a flame containing the atoms in question then part of the light will be absorbed and the extent of absorption will be proportional to the number of ground state atoms present in the flame. This is the basic principle of atomic absorption spectroscopy.

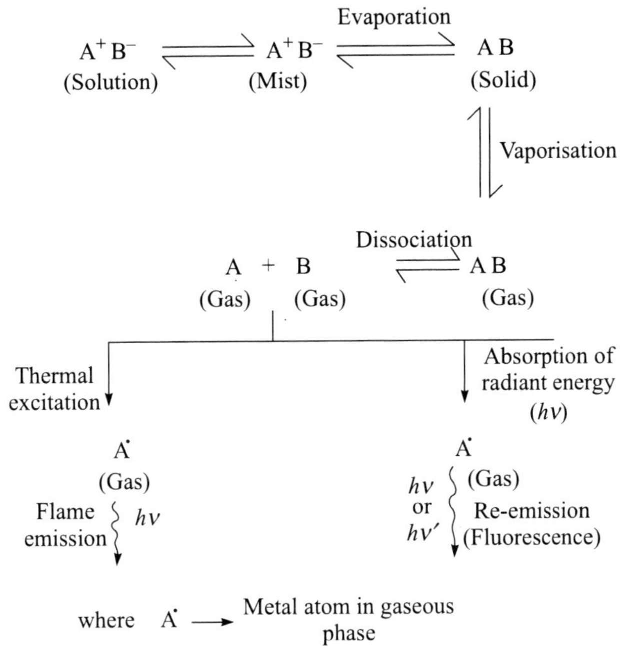

Fig. 5.5

Almost all analytical applications of the atomic absorption method involve spraying a solution of the sample into the flame. For this reason the technique is also known as absorption flame photometry.

5.3.1 Procedure

When a solution containing a suitable compound of the metal to be investigated is aspirated into a flame, the following facts occur:

1. The solvent is evaporated leaving behind a solid residue.
2. The solid residue vaporises and dissociates into its constituent atoms. The atoms will be in the ground state.
3. There are a few atoms which are excited by the thermal energy of the flame to higher energy levels. After a certain limit, they radiate energy.

5.3.2 Principle

Consider that the atom is in ground state (ε_0) and excited to energy level, ε_1, ε_2 or higher energy levels. The transitions between two quantized energy levels, say from ε_0 to ε_1, correspond to the absorption of radiant energy and the amount of energy absorbed ($\Delta\varepsilon$) is determined by the equation:

$$\Delta\varepsilon = \varepsilon_1 - \varepsilon_0$$
$$= h\nu = hc/\lambda$$

$c \rightarrow$ Velocity of light

$h \rightarrow$ Planck's constant

$\nu \rightarrow$ Frequency

$\lambda \rightarrow$ Wavelength of radiation absorbed

The absorption spectrum of a given element is usually associated with transitions from the ground state to higher energy state and is consequently simpler in character. The relationship between the ground state and excited state population is given by

$$\frac{N}{N_0} = \left(\frac{\mu_1}{\mu_0}\right) e^{-\Delta E/kT}$$

$N_0 \rightarrow$ Number of atoms in ground state

$N \rightarrow$ Number of atoms in excited state

$\mu_1/\mu_0 \rightarrow$ Ratio of statistical weights for ground and excited states

$\Delta E \rightarrow$ Energy of excitation

$K \rightarrow$ Boltzmann's constant

$T \rightarrow$ Temperature in kelvin

The atomic absorption spectra of most elements are simple in character as compared with the emission spectra. It follows that the absorption spectroscopy is less prone to inter-element interference. It should be noted that in atomic absorption spectroscopy, as with molecular absorption, the absorbance A is given by the logarithmic ratio of the intensity of the incident light signal, I_0, to that of the transmitted light, I_t

i.e.,
$$A = \log I_0/I_t = KLN_0$$

$N_0 \rightarrow$ Concentration of atoms in the flame

$L \rightarrow$ Path length through the flame

$K \rightarrow$ Absorption coefficient

5.3.3 Instrumentation

A large number of commercially available instruments are available in the market. These are based either on a single or a double beam design. The important instrument facilities of a modern atomic absorption spectrophotometer (AAS) are shown in Fig. 5.6.

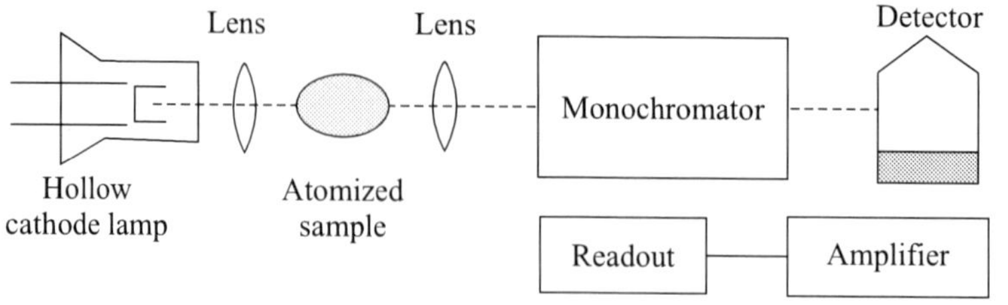

Fig. 5.6 *Schematic diagram of AAS*

1. ***Light source:*** The light source is usually a hollow cathode lamp of the element that is being measured. Lasers are also used in this instrument because lasers are intense enough to excite atoms to higher energy levels.
2. ***Atomizer:*** Atomic absorption spectroscopy requires that the analytic atoms be in the gas phase ions or atoms in a sample must undergo desolvation and vaporization in a high-temperature source such as a flame or graphite furnace. Flame atomic absorption can only analyze solutions, while graphite furnace atomic absorption can accept solutions, slurries or solid samples.
   Flame atomic absorption uses a slot type burner to increase the path length and therefore increases the total absorbance. Sample solutions are usually aspirated with the gas flow into a nebulizing/mixing chamber to form small droplets before entering the flame.
   The graphite furnace has several advantages over a flame. It is a much more efficient atomizer than a flame and it can directly accept very small absolute quantities of sample.
3. ***Light separation and detection:*** Atomic absorption spectrometers use monochromators and detectors for UV and visible light. The main purpose of the monochromator is to isolate the absorption line from background light due to interference.

5.3.4 Application

1. Atomic absorption is also used as an ideal method for the analysis of many ores and minerals.
2. Determination of trace lead in a ferrous alloy.
3. Determination of trace elements in contaminated soil.
4. A number of elements can also be determined in biological fluids by using atomic absorption method.
5. It can be used to determine some sixty metals.
6. This is an ideal method for analysis of nodular cast iron for magnesium.
7. It is widely used in the fields of metallurgy and general inorganic analysis.
8. Determination of vanadium in lubricating oil.

5.4 POTENTIOMETER

Potentiometer is an instrument used for the measurement of electrical potential (voltage) difference. It is based on the Nernst equation.

5.4.1 Principle

The principle involved in the potentiometry is when the pair of electrodes is placed in the sample solution it shows the potential difference by the addition of the titrant or by change in the concentration of ions (Fig. 5.7).

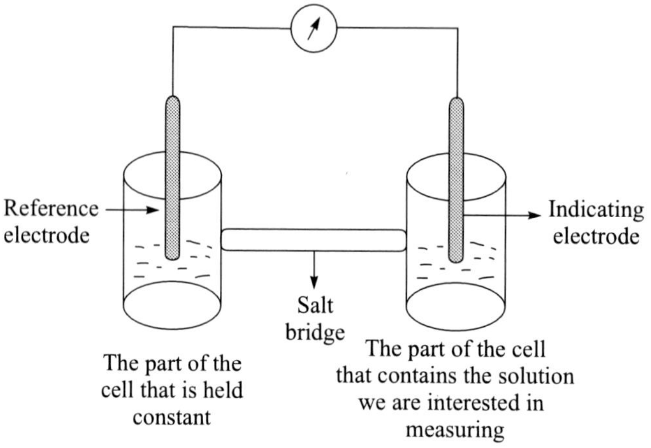

Fig. 5.7

Reference electrode is the electrode which contains its own potential value and it is stable when dipped into the sample solution.

The indicator electrode is the electrode which responds to change in the potential of analyte solution.

The salt bridge is used to prevent the interference of the analyte solution with that of reference solution.

Here analyte solution is the solution whose potential is to be measured.

For example: Redox titrations can be carried out potentiometrically using platinum-calomel electrode combination in a manner similar to acid-base neutralizations.

For the reaction:

$$\text{Reduced form} \rightarrow \text{Oxidised form} + n \text{ electrons}$$

The potential is given by the Nernst equation

$$\varepsilon = \varepsilon^0 + \frac{0.0591}{n} \log \frac{(\text{Oxidised form})}{(\text{Reduecd form})}$$

$$\varepsilon = \varepsilon^0 + \frac{0.0591}{n} \log \frac{(Fe^{3+})}{(Fe^{2+})}$$

where $\varepsilon^0 \rightarrow$ standard potential of the system.

The potential of the system is thus controlled by the ratio of concentration of the oxidized to that of the reduced species present. As the reactions proceed the ratio increases and hence the potential also increases more rapidly in the vicinity of the endpoint of titration. A plot of change in potential against volume is characterized by a sudden change of potential at the equivalent point. At the endpoint potential is determined by a large increase in the potential at the endpoint.

5.4.2 Procedure

Pipette out 25.0 cm^3 of ferrous ammonium sulphate solution into a beaker. Add a test tube containing dilute sulphuric acid. Immerse the platinum-calomel electrode assembly into the beaker containig the solution (Fig. 5.8). Connect the electrodes to the potentiometer. Note down the emf of the cell before the addition of potassium dichromate. Now add 0.5 mL potassium dichromate from the semi-microburette. Allow the solution to mix well and measure the potential. Continue the procedure till a sudden rise in emf of the cell is observed. Take about 5 – 6 more readings.

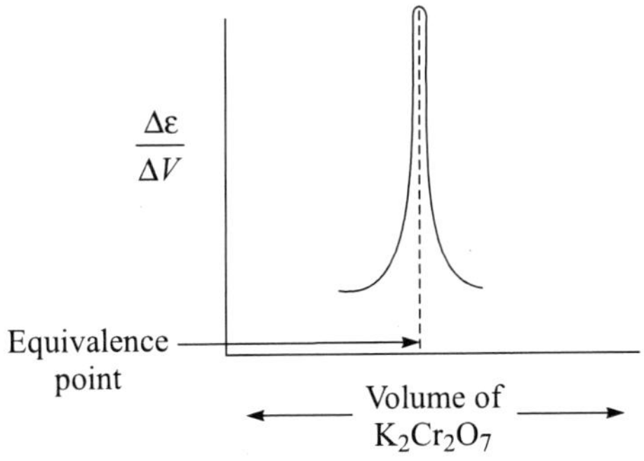

Fig. 5.8

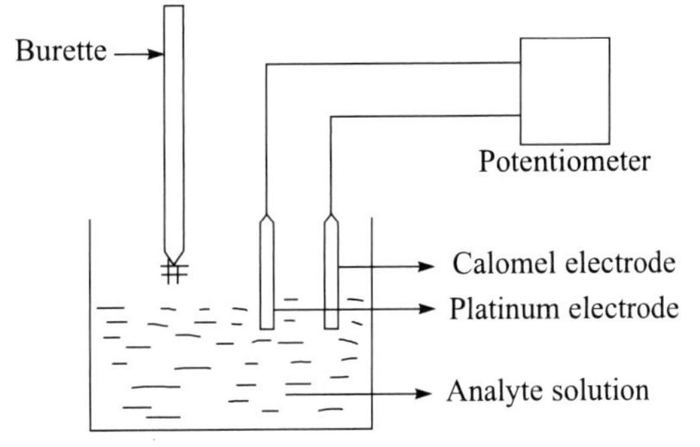

Fig. 5.9

5.4.3 Instrumentation

A potentiometer consists of an indicator electrode (e.g., platinum), a standard reference electrode (e.g., calomel electrode) and potentiometer to read the values directly as change in potential.

Vol. of $K_2Cr_2O_7$ (V) mL	Emf ε (mv)	ΔV $(V_2 - V_1)$ mL	ΔE $(\varepsilon_2 - \varepsilon_1)$	$\Delta\varepsilon/\Delta V$ (mv/mL)

Determine the endpoint by plotting $\Delta\varepsilon/\Delta V$ against the volume as shown in Fig. 5.8.

$$N_{FAS} = \frac{(NV)_{K_2Cr_2O_7}}{V_{FAS}}$$

Amount of FAS per litre = N_{FAS} x eq. wt of FAS

= ———— g(say '*a*')

392 g of FAS contains = 55.85 g Fe

$$a \text{ g of FAS contains} = \frac{55.85 \times a}{392}$$

= ———— g

Result: Normality of the given FAS solution

= ———— N

Weight of FAS/litre = ———— g/litre

Weight of Fe present in one litre = ———— g/litre

5.4.4 Application

1. Coloured solution can also be titrated.
2. Acid-base titration can also be done in this method.
3. Precipitation reactions can also be carried out potentiometrically.
4. In this method oxidation-reduction titrations can also be carried out.
5. Use a pH electrode to monitor the change in pH during the titration.

5.5 COLORIMETRY

Colorimetry is a technique used to determine the concentration of coloured compound in solution. It is frequently used in biochemical investigations. This involves the quantitative estimation of colours. This means that if you want to measure the quantity of a substance in a mixture, you could use the technique of colorimetry by allowing the substance to bind with colour forming chromogens. The difference in colour resulting in the difference in the absorption of light is called colorimetry.

5.5.1 Apparatus

It comprises the following parts (Fig. 5.10).

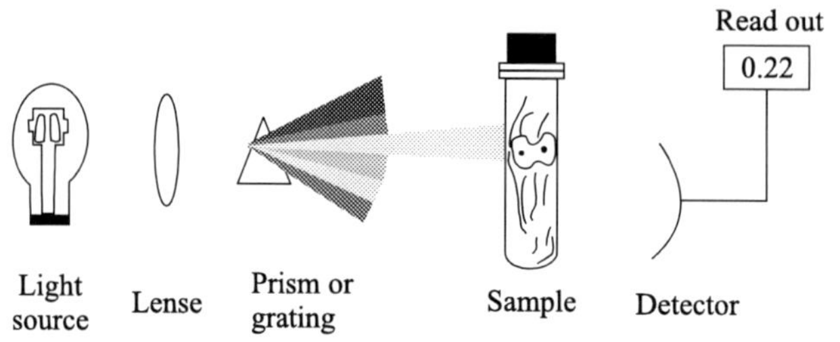

Fig. 5.10

1. Light source
2. Filter (the device that selects the desired wavelength). Three types of filters are available – blue, green and red – with corresponding light wavelength transmission rays from 470 – 490 nm, 500 – 530 nm and 620 – 680 nm, respectively.
3. Cuvette chamber (the transmitted light passes through compartment wherein the solution containing the coloured solution is kept in cuvette made of glass or disposable plastic).
4. Detector (this is a photosensitive element that converts light into electrical signals).
5. Galvanometer (measures electrical signal quantitatively).

5.5.2 Steps for operating the colorimeter

1. Choose the glass filter recommended in the procedure and insert it in the filter (Table 5.3).
2. Fill two of the cuvettes with blank solution to about three-fourths and place them in the cuvette slot.
3. Switch on the instrument and allow it to warm up for 4 – 5 minutes.
4. Adjust to zero optical density.
5. Take the test solution in another cuvette and read optical density.
6. Take the standard solution in varying concentrations and note down the optical density as S_1, S_2, S_3, S_4, S_5 and so on.
7. A graph is plotted taking concentration of standard solution versus the optical density.
8. From the graph concentration of the test solution or the unknown solution can be calculated.

Table 5.3 For choosing the wavelength of absorption

S. No.	Colour of the solution	Colour Absorbed	Wavelength of Absorption
1.	Yellow to green	Violet	400 nm – 435 nm
2.	Yellow to orange	Blue	435 nm – 490 nm
3.	Red	Blue to green	490 nm – 500 nm
4.	Purple	Green	500 nm – 560 nm
5.	Violet	Yellow to green	560 nm – 580 nm
6.	Blue to green	Yellow to orange	580 nm – 650 nm
7.	Bluish green	Red	650 nm – 700 nm

5.5.3 Principle

When a monochromatic light of intensity *Io* is incident on a transparent medium, a part I_a of it is absorbed, a part I_r is reflected and the remaining part is transmitted

$$I_o = I_a + I_r + I_t$$

For a glass air interface I_r is negligible and therefore,

$$I_o = I_a + I_t$$

$$I/I_o = T$$

$I_o \rightarrow$ Intensity of incident radiation

$I \rightarrow$ Intensity of transmitted radiation

$$A = \varepsilon = \log_{10}(1/T) = \log_{10}(I_o/I)$$

It is called the absorbance or optical density.

The relation between absorbance *A*, concentration *C* (expressed in mol/dm^3) and path length, *t* is expressed in cm

From Beer-Lambert's law

$$A - \log_{10}/l'I') - \in Cl$$

$\in \rightarrow$ Molar extinction coefficient

$C \rightarrow$ Concentration of the solution

l – Length that the light passes through (also known as the mean free path). Aside from this if there is a continual changing of the solution, i.e., it is a reaction then % of transmittance against time is generally used. To measure concentrations, the amount of light absorbed is dependent upon the amount of solute in the solution—a higher concentration of dissolved solute means that more light will be absorbed and vice versa hence, the concentration can be backed out from the absorption of specific wavelengths. A plot of aborbance against concentration gives a straight line.

For example: Colorimetric estimation of copper.

Procedure: A series of standard solution of copper sulphate pentahydrate is treated with ammonia to get blue cuprammonium complex and is diluted to a definite volume. The absorbance of each of these solutions is measured at 590 nm since the complex shows maximum absorbance at this wavelength. The absorbance values are plotted against concentration to get a calibration curve. The given test solution is treated with strong ammonia and diluted to the same volume as above. The absorbance of this solution at 620 nm is measured and its concentration is determined from the calibration curve.

Observation:

Vol. of Cu SO_4	Vol. of NH_3	Wt. of Cu in mg	Optical density
Blank	5		
2	5		
4	5		
6	5		
8	5		
10	5		

Draw a calibration curve (Fig. 5.11) by plotting absorbance against volume of copper sulphate solution. Using the calibration curve, find out the volume of copper sulphate solution given, i.e., the volume of test solution and calculate the amount of copper in the given solution.

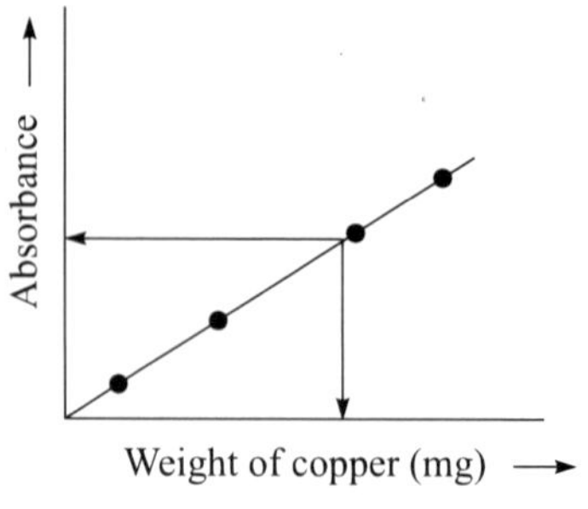

Fig. 5.11

Calculation:

From the graph.

Volume of $CuSO_4$ in test solution = ——— mL (say V mL)

Working solution concentration = 4 mg/cc.

249.6 mg of $CuSO_4 \cdot 5H_2O$ contains 63.55 mg of CO^{2+}

$$\text{Therefore, 4 mg of } CuSO_4 \cdot 5H_2O \text{ contains} = \frac{63.55 \times 4}{249.6}$$

$$= 1.02 \text{ mg}$$

1 mL of Cu $SO_4 \cdot 5H_2O$ contains 1.02 mg of Cu_2^+

V mL of Cu $SO_4 \cdot 5H_2O$ contains $V \times 1.02$ mg of Cu_2^+

Since the stock solution is diluted to 50 times.

$$Cu^{2+} \text{ present in test solution } \frac{V \times a}{50}$$

$$= \text{———} \text{ mg}$$

Result: Concentration of Cu^{2+} in test solution = ——— mg

5.5.4 Applications

1. Determination of the composition of coloured solution.
2. Photometric titration, i.e., equivalence point can also be determined.
3. In quantitative analysis: large number of metal ions, anions and cations compounds can be determined by this method.

Advantages:

1. Used for lower concentrations
2. It is a simple method
3. It can determine the concentration of the coloured solution
4. Colorimetery gives most accurate results.

5.6 CONDUCTOMETRY

Conductance implies the ease with which electric current can flow through a conductor or its is the reciprocal of resistance, $C = 1/R$

Units: ohms, ohms^{-1}, mhos

In SI system, the unit of conductance is Siemen, S.

5.6.1 Theory

The conductance of the solution is explained by considering Ohm's law. According to Ohm's law the current flowing through the conductor is directly proportional to voltage and inversely proportional to the resistance

$$I = \varepsilon/R \quad \varepsilon = IR$$

The resistance of any conductor is directly proportional to the length of inversely proportional to the area of cross section of the conductor.

Therefore, $$R = S\,(l/a)$$

where $S \rightarrow$ Specific resistance

Therefore, $$C = 1/R = 1/S(a/l)\; K(a/l)$$

where $K \rightarrow$ Specific conductance

It is defined as the conductance of the solution which is placed between two electrodes of area 1 cm^3 and 1 cm apart. The conductance of the solution depends on mobility and number of the ions.

5.6.2 Types of Conductance

There are three types, namely, specific conductance, equivalence conductance, and molar conductance.

Specific conductance (*K*) is the conductance of the solution which is placed between two electrodes of area 1 cm^2 and at 1 cm apart.

$$K = 1/R(l/a) = 1/\text{ohm cm} = \text{ohm}^{-1}\,\text{cm}^{-1} = \text{mho/cm}$$

Equivalence conductance (λ) is the conductance of the solution when 1 g equivalent weight of the solution is placed between two electrodes of area 1 cm^2 at 1 cm apart.

Let C be the concentration of the solution in gram equivalent per litre, then the concentration per cm^3 is $C/100$. Thus, the volume containing one equivalent of the solute is $1000/C$.

$$\lambda_{eq} = K \times \frac{1000}{C} = \frac{1000\,K}{C}\,\text{S cm}^2\text{eq}^{-1}$$

Molar conductance is the conductance of the solution when 1 g molecular weight of solute is placed between two electrodes of area 1 cm^2 at 1 cm apart.

It gives the conducting power of the ions produced by 1 mole of an electrolyte at any particular concentration. Let C be the molar concentration of given electrolyte, then

$$\lambda_m = \frac{1000\,K}{C}\,\text{S cm}^2\text{mol}^{-1}$$

Relationship between equivalent conductivity and molar conductivity: Let $v_+ z_+$ (or $v_- z_-$) = Total charge of the cations (or anions) of a single molecular of an electron, it follows that

$$1 \text{ mole} = (v_+ z_+)\,(1 \text{ equivalent})$$

Hence, $$\lambda_m = \frac{1000\,K}{C} \text{ and } \lambda_{eq} = \frac{1000\,K}{(v_+ z_+)C}$$

$$= \lambda_{eq} = \frac{\lambda_m}{v_+ z_+}$$

For example, the total charge of the cations for different electrolytes is

Electrolyte	$v_+ z_+$
KCl	1
Zn SO_4	2

5.6.3 Conductometric Titration

There are different types.

1. Titration of strong acid (like HCl) with strong base like NaOH: Let us consider the titration of strong acid (like HCl) with strong base (like NaOH). The HCl solution is taken in a beaker, conductivity cell is dipped into it and initial conductance is measured after appropriate calibration of the conductivity meter. NaOH is taken in the burette and gradually added to HCl solution. Conductance is measured after every addition. The titration is carried out at constant temperature and when the conductivity is plotted against the volume of NaOH is obtained.

At the start of the titration, the acid solution has a high conductivity (represented by point x in Fig. 5.12) due to highly mobile hydrogen ions and little contribution from chloride ions. When NaOH is added to HCl, the highly mobile H^+ ions are replaced by less mobile Na^+ ions. This will result in decrease in conductivity (along xy) rapidly.

$$\underset{\text{Strong acid}}{H^+Cl^-} + \underset{\text{Base}}{Na^+OH^-} \rightarrow \underset{\text{Salt}}{Na^+Cl^-} + \underset{\text{Water}}{H_2O}$$

The solution at equivalence point (y) contains only Na^+ and Cl^- ions and has minimum conductivity. After the equivalence point, conductivity again increases (along yz) as more NaOH is added because OH^- ions are also highly mobile Fig. 5.12.

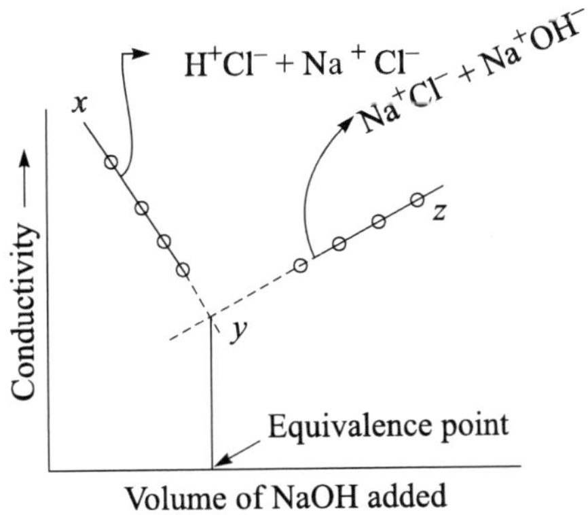

Fig. 5.12 *Conductometric titrations of strong acid and strong base*

2. Titration of strong acid (like HCl) with weak base (like NH_4OH): When ammonium hydroxide (NH_4OH) is added to hydrochloric acid (HCl), the conductivity decreases along xy because of the replacement of the fast moving H^+ ions by slow moving NH_4^+ ions.

$$\underset{\text{Strong acid}}{H^+Cl^-} + \underset{\text{Base}}{NH_4^+OH^-} \rightarrow \underset{\text{Salt}}{NH_4^+Cl^-} + \underset{\text{Water}}{H_2O}$$

This trend is continued till the equivalence point (y) is reached. The addition of NH_4OH after the equivalent point does not cause any appreciable change in the conductance. This is because NH_4OH is poorly ionized and has a very small conductivity compared with that of the acid or its salt (Fig. 5.13).

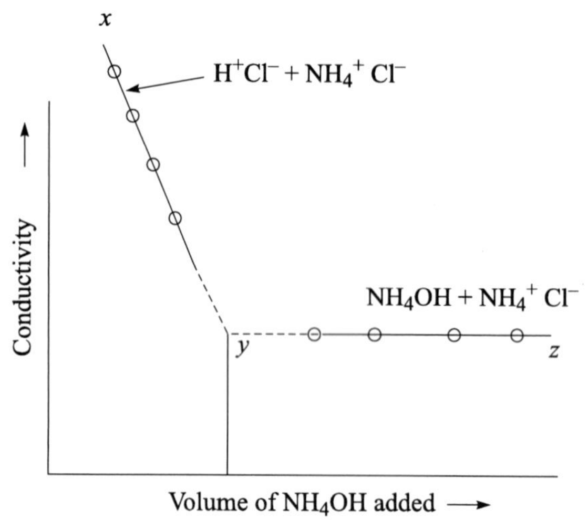

Fig. 5.13

3. Titration of weak acid (like CH_3COOH) with strong base (like NaOH): The conductance of the acid will be initially low since acetic acid is a weak electrolyte when NaOH is added to the acid the salt formed is highly ionized and the conductance increases. On complete neutralization of the acid, further addition of base leads to an increase in the number of mobile OH^- ions. Hence, the conductance increases sharply.

$$\underset{\substack{\text{Weak acid}\\ \text{(Poorly conducting)}}}{CH_3COOH} + \underset{\substack{\text{Strong}\\ \text{base}}}{NaOH} \rightarrow \underset{\text{Salt}}{CH_3COONa} + \underset{\text{Water}}{H_2O}$$

A plot of conductance against the volume of base added is shown in Fig. 5.14. The point of intersection of two curves gives the neutralization point (Fig. 5.14).

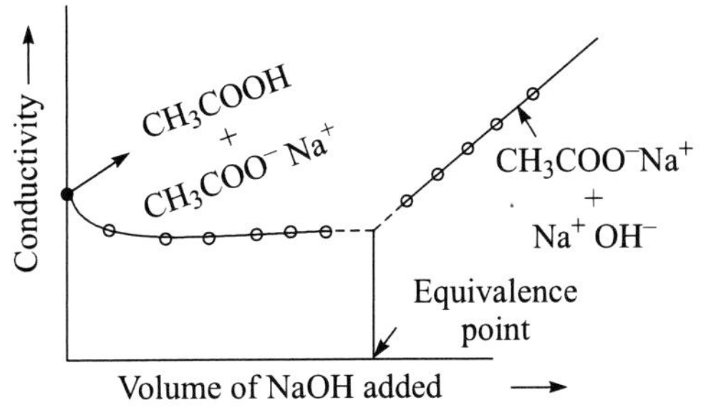

Fig. 5.14

4. Titration of weak acid (like CH_3COOH) with weak base (like NH_4OH): Consider the titration of aectic acid against NH_4OH. The conductance of acid will be initially low since aectic acid is a weak electrolyte. When NH_4OH is added to acid salt is formed and the conductance of the solution increases. On complete neutralization of acid conductance does not change rapidly since NH_4OH is a weak base.

$$CH_3COOH + NH_4OH \rightarrow CH_3COONH_4 + H_2O$$

A plot of conductance against the volume of base added is shown in Fig. 5.15. The point of intersection of two curves gives the neutralization point (Fig. 5.15).

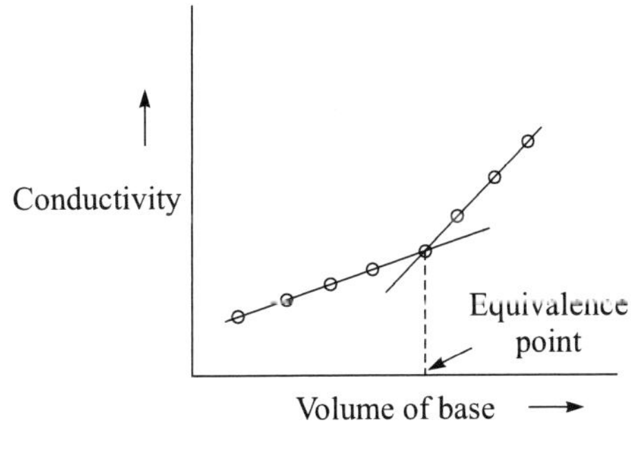

Fig. 5.15

5. Titration of a mixture of a strong acid (like HCl) and a weak acid (like CH_3COOH) with strong base (like NaOH) (HCl + CH_3COOH V/S NaOH): The strong acid like HCl is titrated against a strong base such as NaOH, the conductance first decreases due to replacement of fast moving H^+ ions by slow moving Na^+ ions.

$$HCl + NaOH \rightarrow NaCl + H_2O$$

The weak acids do not get neutralized initially because of the common ion effect. In the presence of excess of H^+ ions, the ionization of the weak acid is suppressed and hence weak acid like CH_3COOH ionizes gradually after the first endpoint and the available H^+ ions are neutralized giving the second endpoint. Because of common ion effect dissolution of acetic acid is suppressed. Hence, it does not provide H^+ ions which is required for neutralization. After the neutralization point of HCl, CH_3COOH, conductivity rapidly rises with further addition of NaOH because of continuous addition of fast moving OH^- ions.

A plot of conductance against the volume of base added is shown in Fig. 5.16. The point of intersection of two curves gives the neutralization point.

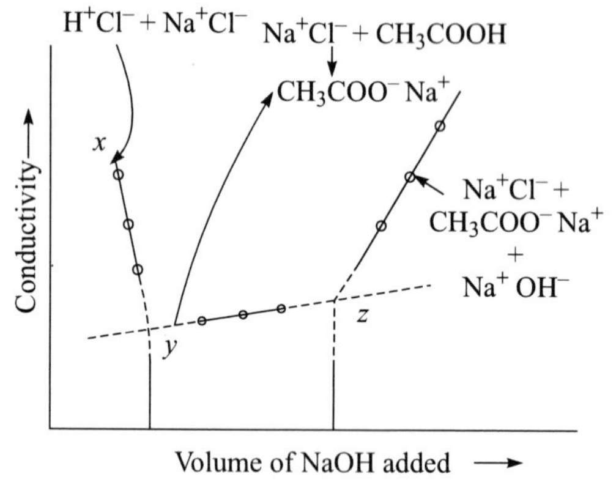

Fig. 5.16

5.7 TITRIMETRIC ANALYSIS

Introduction: Analytical chemistry provides the information about the chemical composition of a substance.

Chemical analysis is carried out in two ways:

- Qualitative analysis
- Quantitative analysis

Qualitative analysis: It is the detection of ions or radical in an inorganic salt or it involves a method to find out what the substance is made up of.

Quantitative analysis: It involves the determination of quantity of a particular component present in a substance or it involves a method to find out how much is present. It can be carried out in two ways:

- Gravimetric analysis
- Volumetric analysis

When the analysis is based on the measurement of weight then it is called *gravimetric analysis*, on the other hand, when it is based on measuring the volume of the solution of a substance then it is called **volumetric analysis.**

Thus, volumetric analysis is a technique of finding the volume of one solution which will exactly react with a definite volume of another solution of known concentration.

In this method, a known volume of one solution is taken in any glass vessel (conical flask) and the solution of other substance is added very slowly to it till the reaction is complete. The completion of the reaction is indicated by certain substances known as indicators, which change colour on completion of the reaction. This process is known as titration. The stage during which the reaction is just complete is known as endpoint (since the indicator changes colour when the reaction is very close to equivalence point and not at the equivalence point, it is called endpoint). From the volume of unknown solution which is required to completely react with definite volume of the other solution of known strength. Strength of the unknown solution can be calculated using the **law of equivalence or normality equation**.

5.7.1 Normality Equation

According to the law of equivalence, the substances react with each other in a definite ratio by mass.

For example: If an acid is titrated against a base, then at the end point according to normality equation.

Number of gram equivalents of the acid = Number of gram equivalents of the base

$$\boxed{N_1V_1 = N_2V_2}$$

N_1 and N_2 are the normalities of acid and base

V_1 and V_2 are the volumes of acid and base respectively.

If N_1, V_1 and V_2 are known then N_2, the normality of the base and hence its strength can be calculated

$$N_1 V_1 = N_2 V_2$$

$$N_2 = \frac{N_1V_1}{V_2} = x$$

Strength = Normality × Equivalent weight

Strength = x × equivalent weight

5.8 STANDARD SOLUTION

A solution of known concentration is known as standard solution. There are two types of substances which are generally employed for preparing solution as explained below.

5.8.1 Primary Standard Substances

Any substance whose standard solution can be prepared directly by dissolving an accurately weighed amount of it in water and making up the solution to a known volume. A primary standard substance should possess the following characteristics:

- It should be easily available in highly pure state. The impurities, if present must be inert with respect to the substance with which it is brought to react.
- It should have a high equivalent weight or molar mass so that error due to weighing is minimised.
- It should be readily soluble in water.
- It should neither be hygroscopic nor efflorescent.
- It should be stable at oven temperature used for drying. The hydrated substances are seldom used as primary standards.
- It should be stable and unaffected by air.
- Its composition should not change on standing during storage.

5.8.2 Secondary Standard Substances

The substances which do not fulfil the above requirements and their standard solutions are not prepared directly by weighing are called secondary standard substances.

To prepare the standard solution of secondary standards, an approximate weight of the substance is dissolved in a known volume of the solvent (water). Then the exact strength of the solution is found by titrating it against some primary standard. The process is called standardization.

The common secondary standards are:

- Alkali hydroxides like NaOH, KOH, etc., as they are hygroscopic and their weights keep on changing during weighing.
- $KMnO_4$, as it is not available in pure form and is associated with manganese dioxide which catalyses the autodecomposition of permanganate solution on standing.
- Inorganic acids HCl, H_2SO_4, etc., because their concentrations are known only approximately.

5.9 REQUIREMENTS OF VOLUMETRIC ANALYSIS

Following are the requirement in the volumetric analysis.

(1) For carrying out the reaction, a suitable standard solution must be available as titrant.

(2) To mark the endpoint of titration, there must be a suitable indicator.

(3) In performing titration, less time should be lost. This requires that the reaction must be rapid.

(4) The reaction should proceed according to definite weight relation between reactants to form a definite product (there should be no side reaction) under given conditions. Hence, the reaction must be well defined.

(5) When equivalent amounts of titrant and titrand are brought together, the reaction must be complete. On this basis, calculations are made from volumetric data.

5.10 TERMS USED IN VOLUMETRIC ANALYSIS

1. **Titration:** The process of finding out the volume of one of the solution required to react completely with the definite volume of the other solution is known as titration.
2. **Titrant:** The solution of known strength is called titrant.
3. **Titrate:** The solution which contains a substance to be estimated or whose strength is to be estimated.
4. **Endpoint and equivalence point:** The point at which the completion of reaction occurs is called equivalence point whereas an endpoint is an observable physical change that occurs near the equivalence point.
5. **Indicator:** The substance which indicates the endpoint of titration is called indicator. The indicator indicates the completion of reaction by change in colour at the endpoint.

Strength of solution: It indicates the amount of solute present in definite amount of solvent. When small quantity of solute is present in solvent it is called dilute solution.

Example: When 0.1 g of sugar is present in 100 mL of water it is dilute solution.

When large quantity of solute is present in solvent, it is called concentrated solution.

Example: When 10 g of sugar is present in 100 mL of water, it is called concentrated solution.

There are many ways to express the concentration of solution.

(a) Normality (N): It is the number of gram equivalent weight of a substance present in 1000 cm^3 of solution.

$$\text{Normality} = \frac{\text{Weight of substance per } dm^3}{\text{Gram equivalent weight}}$$

If one equivalent weight of substance is present in 1 cm^3 then it is called 1 normal (1 N) solution

Problem:

(a) 1.53 g of sodium carbonate is dissolved in 250 cm^3 of water. Calculate the normality of the solution.

$$\text{Normality} = \frac{\text{Weight of substance per } dm^3}{\text{Gram equivalent weight}}$$

$$\text{Normality} = \frac{1.53 \times 4}{53} = 0.1155$$

(b) Calculate the normality of oxalic acid solution containing 1.35 g of oxalic acid in 100 cm^3 of water (molecular mass of oxalic acid = 126)

$$\text{Normality} = \frac{\text{Weight of substance per } dm^3}{\text{Gram equivalent weight}}$$

$$\text{Normality} = \frac{1.35 \times 10}{53} = 0.2547$$

(c) Calculate the amount of Mohr's salt required to prepare 500 cm^3 of 0.5 N solution (molecular mass of Mohr's salt = 392)

$$\text{Normality} = \frac{\text{Weight of substance per } dm^3}{\text{Gram equivalent weight}}$$

$$\text{Weight of Mohr's salt} = \frac{\text{Normality } \times \text{Equivalent weight}}{2}$$

$$= \frac{0.5 \times 392}{2} = 98.0 \text{ g}$$

(b) Molarity (M): It is the number of moles of substance present in 1000 cm^3 of solution.

$$\text{Molarity} = \frac{\text{Weight of substance per } dm^3}{\text{Gram molecular mass}}$$

If one molecular mass of substance is present in 1 dm^3 then it is called 1 molar solution.

Problems:

(a) 6.0 g of glucose is dissolved in 100 cm^3 of water. Calculate the molarity of the solution.

$$\text{Molarity} = \frac{6.0 \times 10}{180} = 0.3333$$

(b) Calculate the molarity of NaOH solution containing 4.0 g NaOH in 750 g of water.

$$\text{Molarity} = \frac{\text{Weight of substance per } dm^3}{\text{Gram molecular mass}}$$

$$= \frac{4.0 \times 750}{1000 \times 180} = 0.0167$$

(c) Calculate the amount of potassium biphthalate required to prepare 500 cm^3 of 0.3 M solution (molecular mass of potassium biphthalate = 204).

$$\text{Molarity} = \frac{\text{Weight of substance per } dm^3}{\text{Gram molecular mass}}$$

$$\text{Weight of potassium biphthalate} = \frac{\text{Molarity} \times \text{Molecular weight}}{2}$$

$$= \frac{0.3 \times 204}{2} = 30.6 \text{ g}$$

(c) Molality of a solution: It is the number of moles of solute dissolved per 1000 g (or 1 kg) of the solvent. It is denoted by m.

$$\text{Molality (m)} = \frac{\text{Moles of solute}}{\text{Weight of solvent in kg}}$$

or

$$= \frac{\text{Moles of solute}}{\text{Mass of solvent in gram}} \times 1000$$

units → moles per kilogram

Problems

Calculate the molality of a solution containing 20.7 g of potassium carbonate dissolved in 500 mL of solution (assume density of solution = 1 g mL^{-1}).

Mass of K_2CO_3 = 20.7 g

Molar mass of K_2CO_3 = 138

$$\text{Moles of } K_2CO_3 = \frac{20.7}{138} = 0.15$$

$$\text{Mass of solution} = 500 \text{ mL} \times 1\text{g mL}^{-1} = 500 \text{ g}$$

$$\text{Amount of water} = 500 - 20.7 = 479.3 \text{ g}$$

$$\text{Molality} = \frac{\text{Moles of solute}}{\text{Mass of solvent in gram}} \times 1000$$

$$= \frac{0.15}{479.3} \times 1000 = 0.313 \text{ m}$$

(d) Mole fraction: It is the ratio of number of moles of one component to the total number of moles (solute and solvent) present in the solution. It is denoted by x. Let us suppose that a solution contains n_A moles of solute and n_B moles of the solvent, then,

$$\text{Mole fraction of solute } (x_A) = \frac{n_A}{n_A + n_B}$$

$$\text{Mole fraction of solvent } (x_B) = \frac{n_B}{n_A + n_B}$$

The sum of mole fractions of all the components in solution is always equal to one as shown below:

$$x_A + x_B = \frac{n_A}{n_A + n_B} + \frac{n_B}{n_A + n_B} = 1$$

Thus, if the mole fraction of one component of a binary solution is known, that of the other can be calculated.

For example, the mole fraction x_A is releated to x_B as

$$x_A = 1 - x_B \qquad x_B = 1 - x_A$$

It may be noted that the mole fraction is independent of temperature.

Problems

A solution is prepared by adding 60 g of methyl alcohol to 120 g of water. Calculate the mole fraction of methanol and water.

$$\text{Mass of methanol} = 60 \text{ g}$$

$$\text{Moles of methanol} = \frac{60}{32} = 1.875$$

(molar mass = 32)

$$\text{Mass of water} = 120 \text{ g}$$

$$\text{Moles of water} = \frac{120}{18} = 6.667$$

(molar mass = 18)

$$\text{Total number of moles} = 1.875 + 6.667$$

$$= 8.542$$

$$\text{Moles fraction of methanol} = \frac{1.875}{8.542} = 0.220$$

$$\text{Mole fraction of water} = \frac{6.667}{8.542} = 0.780$$

(e) Parts per million: When a solute is present in very minute amounts (trace quantities), the concentration is expressed in parts per million abbreviated as ppm. It is the parts of a component per million parts of the solution.

$$\text{ppm} = \frac{\text{Mass of component } A}{\text{Total mass of solution}} \times 10^6$$

Problem

(a) One litre of public supply water contains about 3×10^{-3} g of chlorine. What is the mass percentage of chlorine

$$\text{Mass percentage of chlorine} = \frac{3.0 \times 10^{-3}}{1000} \times 100$$
$$= 3 \times 10^{-4}$$

$$\text{ppm of chlorine} = \frac{3 \times 10^{-3} \times 10^6}{1000} = 3$$

(b) If 11 g of oxalic acid are dissolved in 500 mL of solution (density = 1.1 g mL^{-1}). What is the mass percentage of oxalic acid in solution.

Density of solution = 1.1 g mL^{-1}
Mass of solution = 500 mL × 1.1 g mL^{-1}
= 550 g
Mass of oxalic acid = 11 g

$$\text{Mass \% of oxalic acid} = \frac{11}{550} \times 100 = 2\%$$

(f) Equivalent weight: The equivalent weight of a substance is the number of parts by weight of it that combine with or displace one part by weight of hydrogen or eight parts by weight of oxygen or 35.5 parts by weight of chlorine. It is expressed in grams and it is called gram equivalent weight.

Equivalent weight of an acid

$$= \frac{\text{Molecular weight of acid}}{\text{Number of replaceable hydrogen atoms in one molecule of acid}}$$

$$= \frac{\text{Molecular weight of acid}}{\text{Basicity of acid}}$$

(Basicity = Number of replaceable hydrogen atoms in an acid. Example: HCl, H_2SO_4, H_3PO_4 are monobasic, dibasic and tribasic respectively as they contain 1, 2, and 3 replaceable hydrogen atom respectively).

Equivalent weight of a base

$$= \frac{\text{Molecular weight of base}}{\text{Number of replaceable OH groups present in one molecule of a base}}$$

$$= \frac{\text{Molecular weight of base}}{\text{Acidity of base}}$$

(Acidity = Number of replaceable OH groups in one molecule of a base. Example: NaOH, $Ba(OH)_2$ and $Al(OH)_3$ are mono-, di- and triacid bases as they contain 1, 2 and 3 replaceable OH groups respectively.

Equivalent weight of an oxidising/reducing agent

$$= \frac{\text{Molecular mass of the substance}}{\text{Number of electrons gained/lost by one molecule in the redox reaction}}$$

Table 5.4 Equivalent weight of a few oxidising/reducing agents

Substance	**Molecular mass**	**Balanced ionic equation**	**No. of $_{-1}e^0$ lost/ gained**	**Eq. wt = $\frac{\text{Mol.wt}}{n}$ (g)**
Ferrous sulphate $FeSO_4 \cdot 7H_2O$	278	$Fe^{2+} \rightarrow Fe^{3+} + e^-$	1	278/1 = 278
Ferrous ammonium sulphate ($FeSO_4$ $(NH_4)_2$ SO_4 6 H_2O)	392	$Fe^{2+} \rightarrow Fe^{3+} + e^-$	1	392/1 = 392
Sodium thiosulphate $Na_2S_2O_3$ $5H_2O$	248	$2S_2O_3^{2-} \rightarrow 1/2\ S_2O_6^{2-} + 2e^-$	1	248/1 = 248
Iodine (I_2)	254	$I_2 + 2e^- \rightleftharpoons 2I^-$	2	254/2 = 127
Potassium dichromate ($K_2Cr_2O_7$)	294	$Cr_2O_7^{2-} + 14\ H^+ + 6e^- \rightarrow 2Cr^{3+} + 7H_2O$	6	$\frac{294}{6} = 49$
Potassium permanganate $KMnO_4$	158	$MnO_2^{2-} + 8H^+ + 5e^- \rightarrow Mn^{2+} + 4\ H_2O$ (acidic medium) $MnO_4^- + 2H_2O + 3e^- \rightarrow MnO_2 + 4OH^-$ (neutral medium) $MNO_4^- + e^- \rightleftharpoons MnO_4^{2-}$ (alkaline medium)	5 3 1	$\frac{158}{5} = 31.6$ $\frac{158}{3} = 52.67$ $\frac{158}{1} = 158$
Oxalic acid ($H_2C_2O_4$ $2H_2O$	126	$C_2O_4^{2-} \rightarrow 2CO_2 + 2e^-$	2	$\frac{126}{2} = 63$

5.11 ADVANTAGES OF TITRIMETRIC ANALYSIS

The methods of titrimetric analysis have very important role in modern analytical chemistry due to the following advantage.

1. The precision (0.1%) is better than most instrumental methods.
2. The equipment in titrimetric analysis does not require constant recalibration, unlike instrumental methods.
3. Titrimetic methods are relatively economical with low unit costs per determination.
4. Titrimetric analysis is often used to calibrate and/or validate routine analysis using instruments.
5. The methods in titrimetric analysis can be automated.

5.12 DISADVANTAGES OF TITRIMETRIC ANALYSIS

There are few disadvantages to classical titrimetric procedures such as:

1. Normally titrimetric procedures are less sensitive and frequently less selective than instrumental methods.
2. Titrimetric methods are labour-intensive, slower and often costlier than the instrumental methods, specially when a large number of similar determinations are required.

5.13 SPECTROSCOPY

It is the most powerful tool for the study of structures of atoms and molecules. It is the analysis of electromagnetic radiation scattered, absorbed or emitted by molecules. It deals with the transition that a molecule undergoes between its energy levels upon absorption of suitable radiations determined by quantum mechanical selection rules.

5.13.1 Definition

It is the study of the interaction between electromagnetic radiation with matter as a function of wavelength λ.

5.13.2 Spectrum

A plot of the response as a function of wavelength or more commonly frequency is referred to as a spectrum.

5.13.3 Spectrometry

It is the measurement of these responses and an instrument which performs such measurements is a spectrometer or spectrograph although these terms are more limited in use to the original field of optics from which the concept sprang.

5.13.4 Advantages

Spectroscopy has a number of advantages. These are as follows:

1. These methods are highly reliable establishing the identity of two compounds.

2. These methods require less time and less amount of sample (1 mg)
3. Substance generally remains unaffected or unchanged during spectroscopic examination and can be used again for other test.
4. Sample recovery is possible after spectroscopic analysis, because sample is not chemically affected. (*Mass spectroscopy is an exception here*).

5.13.5 Uses

It has a wide range of applications in the study of atomic and molecular structures. It is an important tool in the hands of analytical chemists for the identification and quantitative analysis of chemical substances. It is also heavily used in astronomy and remote sensing. Most large telescopes have spectrometers which are used either to measure the chemical composition and physical properties of astronomical objects or to measure their velocities from the doppler shift of their spectral lines. In medical diagnosis, various techniques employed (such as MRI, ECG, X-ray) are based on spectroscopy.

5.14 ELECTROMAGNETIC RADIATION

Electromagnetic radiation is a form of all different kinds of energy that is transmitted through space, viz., radio waves, microwaves, X-rays and gamma rays. Sunlight is also a form of electromagnetic energy but visible light is only a small portion of the electromagnetic spectrum which contains a broad range of electromagnetic wavelengths. The common example of electromagnetic radiation is visible light. Quantum mechanics suggests that electromagnetic radiation has a dual character, i.e., exhibiting wave as well as particle behaviour.

According to wave nature of radiations, the electromagnetic waves consist of oscillating electric and magnetic fields. The vectors of electric and magnetic fields are perpendicular to the direction of propagation of radiations and are coplanar (Fig. 5.17).

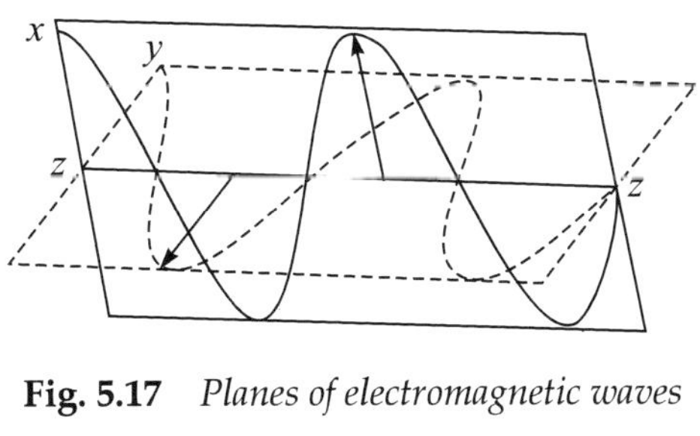

Fig. 5.17 *Planes of electromagnetic waves*

The wavelengths of visible light range from 400 nm (violet) to 800 nm (red). The visible region however, is a very small part of the entire electromagnetic spectrum wavelength slightly shorter than those of the visible fall into ultraviolet region, while slightly longer wavelengths fall into infrared region. Some components of the electromagnetic spectrum are listed in Table 5.5.

Table 5.5 Some components of electromagnetic spectrum

Type of spectra	Wavelength	Energy in kcal/mol	Effect on the molecule
Ultraviolet	200 – 400 nm	143 – 83	Changes in electronic energy levels within the molecule
Visible	400 – 800 nm	82 – 36	Changes in electronic energy levels within the molecule
Infrared	2 – 16 μm	14.3 – 1.8	Changes in vibrational and rotational movement of the molecule
Nuclear magnetic resonance	–	–	Induces changes in the magnetic energy levels of certain nuclei

The electromagnetic radiations are characterized by different parameters such as:

1. ***Wavelength:*** It is the distance between the two adjacent crests (point C – C in Fig. 5.18) or troughs (point *T* – *T*) in a particular wave. It is denoted by the letter "λ" (lambda).

$$\text{Mathematically, } \lambda = \frac{C}{\lambda}$$

$C \rightarrow$ Velocity of light and is equal to 3×10^{10} cm/sec
$\nu \rightarrow$ Frequency
It is expressed in a number of units such as cm, nm, Å, etc.

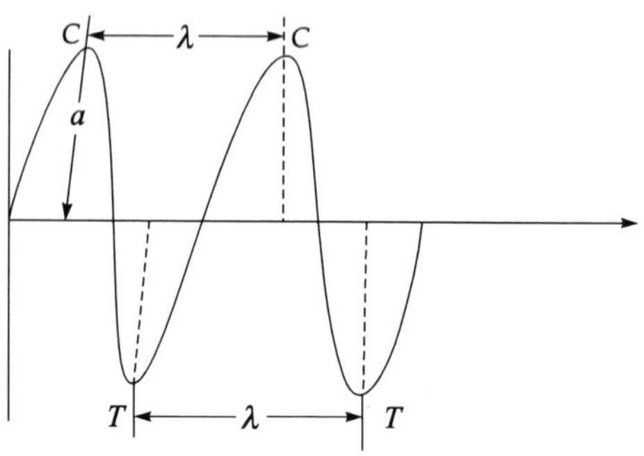

Fig. 5.18 *Wave motion*

2. ***Frequency:*** It is the number of waves which can pass through a given point in one second. It is denoted as 'ν'. It is expressed in cycles per second or in hertz

where 1 Hz = 1 cycle sec^{-1}

$$\text{Mathematically, } \nu = \frac{C}{\lambda}$$

$C \rightarrow$ Velocity of light
$\lambda \rightarrow$ Wavelength
As λ and ν are inversely related to each other. Shorter the wavelength, higher the frequency and vice versa.

3. ***Wave number:*** Wave number is the total number of waves which can pass through a space of one cm. It is the reciprocal of the wavelength expressed in centimetres. It is denoted as $\bar{v}$.

$$\text{Mathematically, } \bar{v} = 1/\lambda \text{ in cm}$$

4. ***Energy:*** Energy of a wave of the particular radiations can also be calculated by applying the relation.

$$\varepsilon = hv \quad v = C/\lambda$$

Put the value of λ

$$\varepsilon = hC/\lambda$$

$h \rightarrow$ Planck's constant and its value is 6.626×10^{-27} erg sec
$v \rightarrow$ Frequency of radiation in cycles sec^{-1}
$C \rightarrow$ Velocity of electromagnetic radiation
$\lambda \rightarrow$ Wavelength in centimetres

The energy of the light radiation can be calculated in ergs which can also be converted into mol^{-1}.

For example: Calculate the energy associated with a radiation having wavelength 4000 Å. Give answer in kcal mol^{-1} and also in kilojoules $mole^{-1}$.

$$\lambda = 4000 \text{ Å} = 4000 \times 10^{-8} \text{ cm}$$

$$\varepsilon = \frac{h.C}{\lambda}$$

$$= \frac{6.628 \times 10^{-7} \text{ erg sec} \times 2.998 \times 10^{10} \text{ cm sec}^{-1}}{4000 \times 10^{-8} \text{ cm}}$$

$$= 4.968 \times 10^{-12} \text{ ergs}$$

$$= \frac{4.968 \times 10^{-12} \times 6.023 \times 10^{23}}{4.18 \times 10^{10}}$$

$$= 71.6 \text{ kcal mole}^{-1}$$

Also 1 kcal = 4.184 kJ

$$\varepsilon = 71.6 \times 4.184 = 299.5 \text{ kJ/mole.}$$

5.15 ELECTROMAGNETIC SPECTRUM

"The arrangement of all types of electromagnetic radiations in order of their increasing wavelengths or decreasing frequencies is known as complete electromagnetic spectrum". In other words, electromagnetic spectrum is the arrangement of electromagnetic radiations in order of their increasing wavelengths or decreasing frequencies. The visible spectrum (from violet to red through rainbow colours) represents only a small portion of the electromagnetic spectrum. If we arrange all types of electromagnetic radiations in order of their increasing wavelengths, then the portion above the visible region is called infrared while that below it is ultraviolet region. Infrared radiations have longer wavelength and are thus less energetic; cosmic rays carry high energy while radio waves are most energetic. Microwaves have larger wavelengths and are used in telephone transmission. Table 5.6 gives an idea about the approximate wavelengths, frequencies, and energies of different types of radiations.

Table 5.6 The electromagnetic spectral regions

Spectral region	Wavelength	Frequency, Hz	Energy J/Mole	Type of quantum change
Y-rays	100 – 1 pm	$3 \times 10^{8} - 3 \times 10^{20}$	$10^{9} - 10^{11}$	Change of nuclear configuration
X-rays	10 nm – 100 pm	$3 \times 10^{16} - 3 \times 10^{18}$	$10^{7} - 10^{9}$	Change of $_{-1}e^{0}$ distribution
UV and visible	1000 – 10 nm	$3 \times 10^{14} - 3 \times 10^{16}$	$10^{5} - 10^{7}$	Change of $_{-1}e^{0}$ distribution
IR	100 μm – 1000 cm	$3 \times 10^{12} - 3 \times 10^{14}$	$10^{3} - 10^{5}$	Change of configuration
Microwave	1 cm – 100 μm	$3 \times 10^{10} - 3 \times 10^{12}$	$10^{-1} - 10^{3}$	Change of orientation
ESR	100 cm – 1 cm	$3 \times 10^{8} - 3 \times 10^{10}$	$10^{-1} - 10$	Change of spin
NMR	10 m – 100 cm	$3 \times 10^{6} - 3 \times 10^{8}$	$10^{-3} - 10^{-1}$	Change of spin

5.16 ORIGIN OF ELECTRONIC SPECTRA

A molecule can absorb radiation of certain frequency, if there exists within a molecule an energy difference of magnitude ($\Delta\varepsilon$) which is equal to $h\nu$. Suppose E_1 and E_2 are the two possible energy states of a system. The energy difference between the two states is given by $\Delta\varepsilon$.

$$\Delta\varepsilon = \varepsilon_2 - \varepsilon_1$$

The energy transition from ε_1 to ε_2 or from ε_2 to ε_1 corresponds to the absorption or emission of energy exactly equivalent to the energy of the wavelength absorbed or emitted. The absorbed or emitted energy is the form of electromagnetic radiation.

$$\nu = \Delta\varepsilon / n$$

Thus, spectra so obtained can be broadly classified into two categories.

1. Emission Spectrum

It is produced by heating a substance directly on a flame or heating it electrically and then passing the emitted radiation through a prism. When there is a transition from a state of high energy to a stage of lower energy, the excess energy is emitted as a photon of energy $h\nu$ provided, $\Delta\varepsilon = h\nu$. This is called emission spectrum (Fig. 5.19(b)).

2. Absorption Spectrum

When the light from a source emitting a continuous spectrum is first passed through an absorbing substance, and then observed through a spectroscope is called absorption spectrum. It is found that certain wavelengths of light (γ) absorbed by the substance leave dark lines or bands at their place giving rise to absorption spectrum (Fig. 5.19(a)).

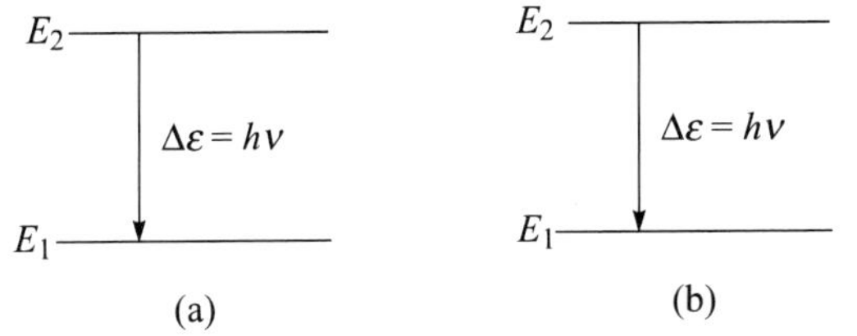

Fig. 5.19 *Energy transition for*
(a) absorption, and
(b) emission of electromagnetic radiation

When there is a transition from a state of lower energy to a state of higher energy the energy absorbed will be exactly equal to the energy difference, $\Delta\varepsilon$. The spectrum is called absorption spectrum.

It may be classified as atomic spectra and molecular spectra.

(a) Atomic Spectra

It is the study of the interaction of electromagnetic radiation with atoms which are commonly in their ground stage. In this spectra, the transition of electrons takes place from one electronic energy state to another, i.e., between the atomic energy levels.

(b) Molecular Spectra

It is the study of the interaction of electromagnetic radiation with molecules. The spectra by molecules are quite different from the atomic spectra because it involves the transition between rotational and vibrational energy levels in addition to electronic transitions.

5.17 TYPES OF ENERGY CHANGES

The total energy of the molecules is the sum of the translational energy, ε_t, rotational energy, ε_r, vibrational energy, ε_v, and electronic energy, ε_e.

$$\varepsilon_{\text{total}} = \varepsilon_t + \varepsilon_r + \varepsilon_v + \varepsilon_r$$

The rotational and vibrational energies are also quantized as is the case with electronic energy. So,

$$E_e >> E_v >> E_r > E_t$$

The translational energy which is very small is fast living as heat and is not involved in the molecular spectra.

The energy levels fall into widely spaced groups which are considered to correspond to different electronic states of the molecule. Associated with each electronic state are a series of vibrational energy levels designated by the vibrational quantum number, v. Moreover, each vibrational energy level has a set of rotational sub-level designated by the rotational quantum number J. From Fig. 5.20 electronic energy is associated with the transition of an electron from the ground state energy level to an excited state energy level of the molecule due to the absorption of a photon of suitable frequency.

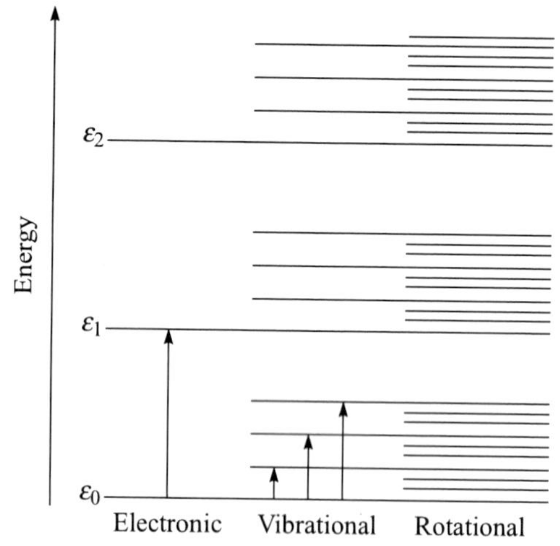

Fig. 5.20 *Energy level diagram showing the various molecular energies*

Vibrational energy is due to the to-and-fro motion of the molecules such that centre of gravity does not change.

Rotational energy arises when the molecule rotates about an axis perpendicular to the internuclear axis and passes through the centre of gravity of molecules.

Thus, energy required for transition will be least for rotational and higher for vibrational and still higher for electronic transitions.

From the above discussion, it is clear that absorption of energy can cause three changes in a molecule.

1. If the exciting energy is small so as to produce only rotational transition within a vibrational energy level, pure rotation spectra will result. The rotational spectra of a molecule are observed in the microwave region or far infrared region (> 20 μm).
2. If the energy required for excitation are sufficiently large to cause vibrational transitions within the same electric level, the energy changes observed correspond to the change in vibrational quantum numbers. But change in vibrational levels involves change in rotational level also thus the resulting spectrum is called vibrational rotational spectrum. Such type of spectra are generally observed in the near infrared region (< 20 μm).
3. If the exciting energies are still higher, the electronic transitions take place. The electronic transitions are accompained by a number of vibrational transitions and each vibrational transition, in turn, is associated with the number of rotational transitions. Thus, the resulting spectrum will be complex, consisting of lines due to electronic,

vibrational and rotational transitions. The set of rotational lies corresponding to a particular vibrational transition is called a band each vibrational transition gives rise to a series of such bands. The entire set of bands for a given electronic transition is known as band group. Such type of spectrum is called electronic band spectra. These types of transitions take place in visible and ultraviolet spectral region since considerably higher energies are involved.

5.17.1 Types of Molecular Spectra

Molecular spectra are classified according to the region of electromagnetic spectrum in which they are observed or according to the transition that take place (Table 5.7).

Table 5.7

Spectral region	**Wavelength involved**	**Transition**	**Spectroscopy**
Microwave	$1 - 100\ cm^{-1}$	Between rotational energy levels	Microwave spectroscopy (Rotational spectroscopy)
Infrared	$500 - 400\ cm^{-1}$	Between vibrational energy levels	Infrared spectroscopy (Vibrational and vibrational rotational spectroscopy)
Visible and ultraviolet	$12500 - 25000\ cm^{-1}$ (visible) 25000-70,00 cm^{-1} (UV)	Between energy levels	Ultraviolet-visible spectroscopy (Electronic spectroscopy)
Radio frequency $< 1\ cm^{-1}$ region		Between the nuclear spin energy of the molecules in applied magnetic field	Nuclear magnetic resonance spectroscopy
Microwave region	$1 - 100\ cm^{-1}$ (2 – 9.6 GHz)	Between the electron spin energy of the molecules in applied magnetic field	Electron spin resonance spectroscopy

5.17.2 Lambert Beer or Beer's Law

Lambert's law states that when a monochromatic light is passed through a solution, the decrease in the intensity of light with the thickness of the solution is directly proportional to the intensity of incident light.

$$I_t = I_o e^{-kx} \qquad \ldots(1)$$

I_t → Intensity of the transmitted light
I_o → Intensity of the incident light
x → Thickness of the medium or length of the solution
k → Absorption coefficient

Beer's law, an extension of Lambert's law states that "when a monochromatic light is passed through a solution, the decrease in the intensity of light with the thickness of the solution is directly proportional to the intensity of incident light as well as the concentration of the solution.

$$I = I_o e^{-\varepsilon Cx} \qquad \ldots(2)$$

where $\varepsilon \rightarrow$ Molar absorption coefficient
$c \rightarrow$ Concentration of solution

Equation (2) may be written as

$$\frac{I}{I_o} = e^{-\varepsilon Cx}$$

$$\log \frac{I}{I_o} = -\varepsilon Cx$$

where I/I_o = transmittance, T
$-\log I/I_o = -\log T$ is called absorbance
$-\log I/I_o = -\log T = A$

$$A = \varepsilon Cx$$

$$\varepsilon = \frac{A}{Cx}$$

$\varepsilon \rightarrow$ Molar absorbing coefficient
$C \rightarrow$ Molar concentration
$x \rightarrow$ Path length in centimetres

The magnitude of the molar absorption coefficient is directly related to the probability of a particular transition.

$$\varepsilon_{max} = 0.87 \times 10^{20}\, P.a.$$

$P \rightarrow$ Transition probability with values ranging from 0 to 1
$a \rightarrow$ Target area of the absorbing system.

Generally, transition with ε_{max} value more than 10^4 are called *allowed transitions* and ε_{max} value less than 10^4 are called *forbidden transitions*.

5.18 GENERAL INSTRUMENTATION

The instrument which is used to measure the absorption or emission of electromagnetic radiation by the substance under examination is called spectrometer. Spectrometers, used in various regions of the spectrum, differ widely in their construction but the basic features remain the same (Fig. 5.21).

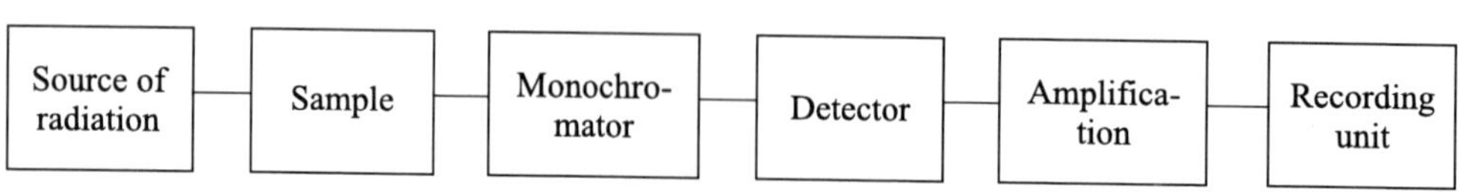

Fig. 5.21 *Block diagram of spectrometer*

The absorption spectrometer consists of the following parts.

1. *Source of electromagnetic radiation:* The incident light of required wavelength is provided by different sources for different spectroscopic techniques. The sources for different spectroscopic techniques are tabulated in Table 5.8.

Table 5.8 Source for different spectroscopic techniques

Spectral region	Source	Dispersing elements	Detector
Ultraviolet spectroscopy	Hydrogen discharge lamp Deuterium discharge lamp	Quartz prism	Photomultiplier tube
Visible spectroscopy	Tungsten lamp	Glass prism	Photomultiplier
Infrared spectroscopy	Electrically heated (~ 1500°C) rods of → Nerst glower (sintered mixtures of oxides of Zr, Y, Er, etc.) → Globar (silicon carbide) → Ceramic elements	Rock salt prism	Thermocouple
Microwave spectroscopy	Klystron tube (A specially developed electron tube in which electrons generate electromagnetic radiations	–	Quartz crystal detector

2. *Monochromator:* It is used for conversion of polychromatic radiation from the source into narrow bands or monochromatic radiations.
3. *The sample container:* The sample is placed in the sample container which are cuvettes or cells. The windows of the sample container must be chemically resistant to the sample. In the visible-ultraviolet infrared region, NaCl, KBr, glass and quartz are used. For the study of aqueous solution, AgCl is necessary.
4. *Detector unit:* The main function of any detector is to absorb the energy of photons and convert it into measurable quantity such as electric current or thermal changes. Generally, the detectors generate an electrical signal which activates some type of meter or recorder. In UV-visible spectroscopy, photo cells or photomultipliers are used as detectors. In the infrared regions, thermocouples are used whereas in microwave region, the detector is a crystal rectifier.
5. *Amplification and recording unit:* An amplifier takes an input signal from the circuit and through electronic operations, produces on output signal which is several orders of magnitude than input. The signal is recorded on the graph or on the computer monitor. The main function of a amplifier is to amplify a signal so that it can be measured. The absorbance (A) or optical density is given by

$$A - \log\frac{I_0}{I}$$

I_o → Intensity of incident light
I → Intensity of transmitted light

5.19 ULTRAVIOLET AND VISIBLE SPECTROSCOPY

It is also referred to as electronic spectroscopy because it involves the promotion of electrons (σ, π, n^* electrons) from ground state to the higher energy state when energy is absorbed by any substance. It is very useful to measure the number of conjugated double bonds and also aromatic conjugation within the various molecules. The electromagnetic radiations in UV-visible region extends from 200 – 750 nm (1 nm = 10^{-7} cm).

When a molecule absorbs ultraviolet radiation of frequency (ν), the electrons in that molecule may undergo transition from a lower to higher energy level.

The energy difference between the two states is given by

$$\Delta\varepsilon = h\nu$$

h → Planck's constant

$$\Delta\varepsilon = \varepsilon_2 - \varepsilon_1$$

ε_1 → Energy of the ground state
ε_2 → Energy of the excited state

We know that the total energy of the molecule is equal to the sum of the electronic, vibration and rotational energies. Therefore,

$$\varepsilon_{\text{total}} = \varepsilon_{\text{electronic}} + \varepsilon_{\text{vibrational}} + \varepsilon_{\text{rotational}}$$

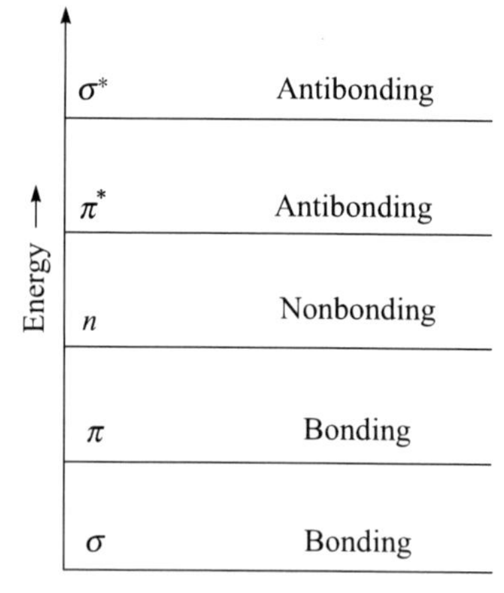

Fig. 5.22

The absorption in the UV region involves single electronic transition from lower energy state to the higher energy state, so the spectrum must contain a single discrete line. But in actual practice a broad absorption band appears. This is because the electronic transitions are also accompained by vibrational and rotational transitions. As the energy required for these transition is much less so the broad band appears. In UV range the absorbed energy produces changes in the valence electrons of the molecule (Fig. 5.22).

5.19.1 Types of Electronic Transition

The electronic transitions involved in the UV and visible region are of the following types:

(i) $\sigma \rightarrow \sigma^*$ transition
(ii) $n \rightarrow \sigma^*$ transition
(iii) $n \rightarrow \pi^*$ transition
(iv) $\pi \rightarrow \pi^*$ transition

$\sigma \rightarrow \sigma^*$ transition

1. It is a high energy process or transition since σ bonds are in general very strong bonds.
2. These transitions require high energy and radiations of very short wavelength.
3. These transitions occur by the promotion of sigma electrons from the bonding to anti-bonding (σ^*) orbitals.
4. The organic compound in which all the valence shell electrons are involved. Information of σ-bond shows this type of transition like saturated hydrocarbon-methane, ethane, etc. **Example:** consider $\sigma \rightarrow \sigma^*$ transition in a saturated hydrocarbon.

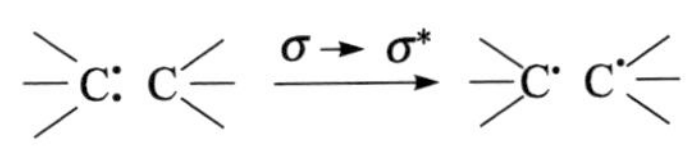

Fig. 5.23

$n \rightarrow \sigma^*$ transition

1. It is a low energy process or transition compared to $\sigma \rightarrow \sigma^*$ transition.
2. The excitation of an electron in an unshared pair on an atom to an antibonding σ orbital.
3. This type of transitions are seen in saturated compounds, containing heteroatom in addition to $\sigma \rightarrow \sigma^*$ transitions.
   For example, CH_3OH absorbs at ordinary UV region and λ_{max} is 183 nm, for methyl chloride λ_{max} is 173 nm and water absorbs at 167 nm.

$n \rightarrow \pi^*$ transition

1. The electrons from n-orbitals are transferred to π^* orbitals and require less amount of energy and occur at longer wavelength.
2. In this transition, an electron of unshared electron pair on a hetero atom is excited to π^* anti-bonding orbital.
3. These transitions occur with compound containing double bonds involving hetero atoms having unshared pair of electrons (**Example:** C = O, C = S and N = O, etc.)
   Example: Aldehydes and ketones exhibit two bands at 180 – 200 nm ($\pi - \pi^*$ transition) and 280 nm ($n - \pi^*$ transition).
4. It is always less intense because the electrons in the n-orbital are situated perpendicular to the plane of the π bond and hence the probability of the jump of an electron from n to π^* is less.

Example: >C=Ö:→C=Ȯ:

$\pi - \pi^*$ transition

1. These transitions require less energy and occur at higher wavelengths.

   Example: >C = Ö:→C ⁻⁻ Ö:
2. The electron transfer occurs from the bonding π-orbital to the anti-bonding π^*-orbital.
3. This type of transition takes place in the compounds containing double or triple bonds.

   Example: Unconjugated alkenes show an absorption band around 170 nm to 190 nm and saturated aldehydes and lactones absorb at about 180 nm (Fig. 5.24).

The relative energies required for the various transitions obey the order

$$\sigma - \sigma^* > n - \sigma^* > \pi - \pi^* > n - \pi^*$$

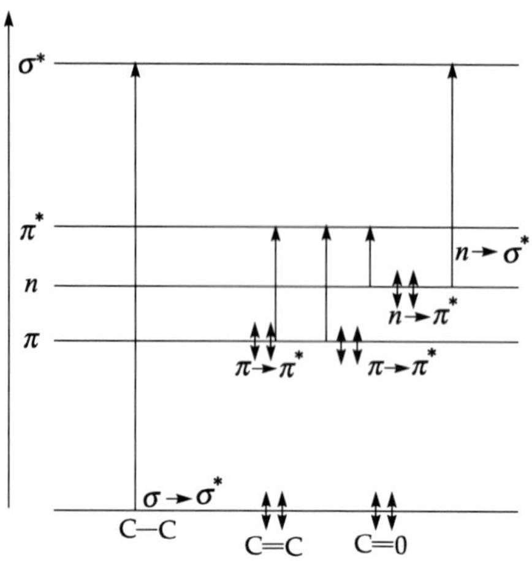

Fig. 5.24 *Various transitions involved in electronic spectroscopy*

5.19.2 Transition Probability

Depending upon the value of extinction coefficient the transitions can be classified as

(*a*) – allowed transitions (*b*) – forbidden transitions

The transitions with values of ε_{max} more than 10^4 are called allowed transitions. It is due to $\pi - \pi^*$ transition.

The transitions with values of ε_{max} below 10^4 are called forbidden transitions. It is due to $n - \pi^*$ transitions.

Example: Benzophenone

The two types of transitions involved are:

(i) 252 nm ε_{max} 20,000 (allowed)

(ii) 325 nm ε_{max} 180 (forbidden)

5.19.3 Principle

It follows the Beer-Lambert law. This law states. "Whenever a beam of monochromatic light is passed through a solution with an absorbing substance. The decreasing rate of radiation intensity along with the thickness of the absorbing solution is proportional to the concentration of the solution and the incident radiation". This law is expressed through this equation

$$A = \log I_o/I = \varepsilon Cl$$

$A \rightarrow$ Absorbance
$I_o \rightarrow$ Intensity of light upon a sample cell
$I \rightarrow$ Intensity of light departing the sample cell
$C \rightarrow$ Concentration of the solute
$l \rightarrow$ Length of the sample cell
$\varepsilon \rightarrow$ Molar absorptivity

The term transmittance (T) is the ratio of the intensity of the light transmitted by a sample (I) to the intensity of light incident on the sample (I_o).

Thus, $T = I/I_o$

Moreover % $T = T \times 100 = I/I_o \times 100$

The logarithm of the reciprocal of the transmittance [log (I_o/I) or log (I/T)] is termed absorbance (A).

$$A = \log 1/T$$

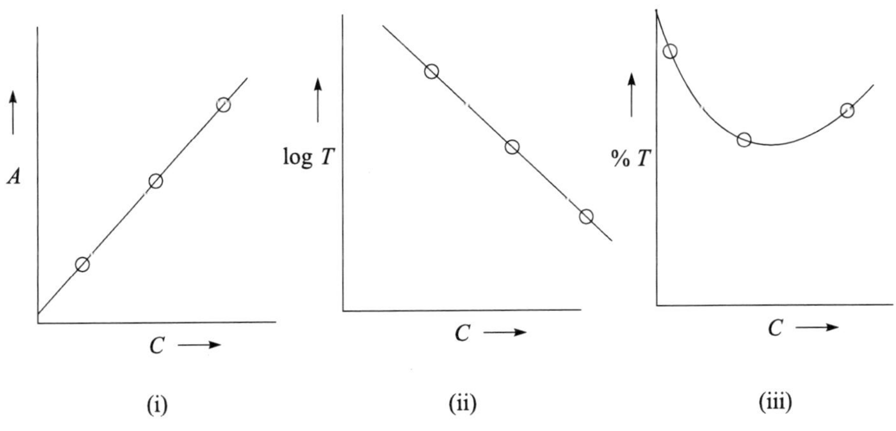

Fig. 5.25 *(i) Plots of A*
(ii) Plots of log T
(iii) Plots of % T vs concentration

If multiple species that absorb light at a given wavelength is present in a sample, the total absorbance at that wavelength is the sum due to all absorbers.

$$A = (\varepsilon_1 \times C_1 \times L) + (\varepsilon_2 \times C_2 \times L) + \ldots$$

where the subscripts refer to the molar absorptivity and concentration of the different absorbing species present.

5.19.4 UV-bands

Following types of bands originate as a result of the possible transition in a compound.

1. *B-bands:* The bands due to $\pi - \pi^*$ transitions are called *B*-bands. These bands occurs in aromatic or hetero-aromatic molecule. The *B*-bands are observed at longer wavelength than the more intense *K*-bands. They occurs in the region from 250 – 255 nm.
2. *R-bands:* These bands are due to the $n - \pi^*$ transition of a single chromophoric group and having at least one lone pair of electrons on the heteroatom. They have low molar absorptivities and are called forbidden bands.
3. *K-bands:* They originate from a compound containing a conjugated (π) system. These bands occur due to the $\pi - \pi^*$ transitions.
   Example: Dienes, polyenes, enones, etc.
   The ε_{max} is usually more than 10^4.
4. *ε-bands:* These bands are originate due to the electronic transitions in the benzenoid system of three ethylenic bonds which are in closed cyclic conjugation. These are further characterized as ε_1 and ε_2 bands. For example: ε_1 and ε_2 bands of benzene appear at 184 and 204 mμ, respectively. ε_1 band which appears at lower wavelength is usually more intense than the ε_2-band for the same compound which appears at longer wavelength.

5.19.5 Chromophores and Auxochromes

It is defined as a covalently bonded unsaturated group that shows a characteristic absorption in UV or visible region. The absorption occurs irrespective of the fact whether the colour is produced or not.

Originally, a chromophore was considered any system which is responsible for imparting colour to the compound.

Example: Benzene does not display colour as it does not have a chromophore, but nitrobenzene is pale yellow in colour because of the presence of a nitro group ($-NO_2$) which acts as a chromophore (Table 5.9).

Table 5.9 Simple unconjugated chromophores

Chromophore	Transition	Absorption max. (mμ)	Emax	Solvent
$>C=C<$	$\pi - \pi^*$	~ 175 ~ 175 196	15000 ~ 10000 ~ 2000	Vapour Hexane Vapour
$-C \equiv C-$	$n \rightarrow \sigma^*$	220 160	~ 150 18000	Hexane Hexane

$>C=0$	$\pi \rightarrow \pi^*$ $n \rightarrow \pi^*$ $\pi \rightarrow \pi^*$	180 285* ~ 200	10000 15 5000	Hexane Hexane Methanol
$R-NO_2$	$n \rightarrow \pi^*$	~ 274	15	
–C = O | OH	$n \rightarrow \pi^*$	204	60	Methanol
– N = N	$n \rightarrow \pi^*$	338	~ 5	Ethanol
$-CONH_2$	$n \rightarrow \pi^*$ $n \rightarrow \pi^*$	178 220	9500 63	Hexane Hexane

Types: There are two known types of chromophore.

1. Chromophore which contain π electrons and undergo $\pi \rightarrow \pi^*$ transitions.
   Example: C = O, C ≡ C, etc.
2. Chromophores which contain both π electron as well as non-bonding electrons and undergo $\pi - \pi^*$ and $n - \pi^*$ transitions.
   Example: C = O, C ≡ N, etc.

5.19.6 Identification of Chromophores

There is no set rule for the identification of a chromophore. The change in position as well as the intensity of the absorption depends on a large number of factors. Following points may be useful.

1. Spectrum having a band near 300 mμ may possess two or three conjugated units.
2. Absorption bands near 270 – 350 mμ with very low intensity ε_{max} 10 – 100 are because of $n \quad \pi^*$ transitions of carbonyl group.
3. Simple conjugated chromophores likes dienes or β-unsaturated ketones have ε_{max} values from 10,000 to 20,000.
4. The absorption with ε_{max} value between 1000-10,000 reveals the presence of an aromatic system. If aromatic nucleus is substituted with groups which can extend the chromophore, absorption takes place at still higher value of extinction coefficients.

5.19.7 Auxochrome

It is derived from two Greek words *Auxano* = increase, *chrome* = colour. It is basically a colour enhancing group. It is a group which does not itself act as a chromophore. It shifts the absorption band towards the longer wavelength along with an increase in the intensity of absorption.

Example: $-OH$, $-OR$, $-NH_2$, $-NR_2$, etc.

When the auxochrome $-NH_2$ group is attached to benzene ring, its absorption changes from λ_{max} 225 (ε_{max} 203) to λ_{max} 280 (ε_{max} 1430). All auxochromes have one or more non-bonding pairs of electrons. If an auxochrome is attached to a chromophore, it helps in extending the conjugation by sharing of non-bonding pair of electrons as shown below.

$$CH_2 = CH - NR_2 \rightarrow CH_2 - CH - NH_2$$

The extended conjugation is responsible for bathochromic effect of auxochromes.

1. **Bathochromic shift:** It is derived from two Greek words: *bathys* = deep and *chroma* = colour: It is a change of spectral band position in the absorption reflectance transmittance or emission spectrum of a molecule to a longer wavelength. It may be due to the presence of an auxochrome or change in the polarity of the solvent.

Table 5.10 Terminology for absorption shift

Nature of shift	Descriptive term
To longer wavelength	Bathochromic
To shorter wavelength	Hypsochromic
To greater absorbance	Hyperchromic
To lower absorbance	Hypochromic

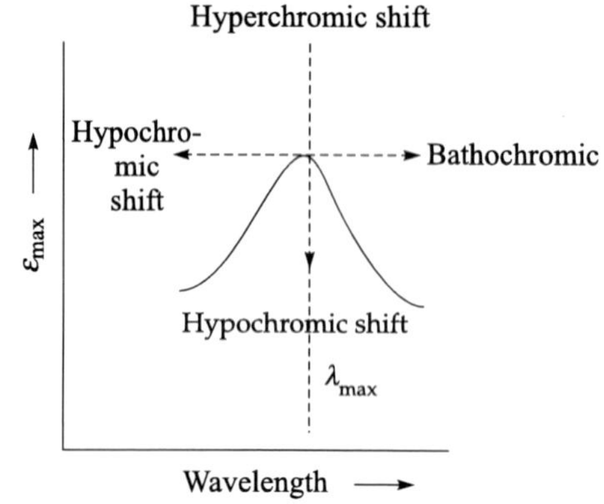

Fig. 5.26 *The various shifts in ε_{max} and λ_{max}*

Example: The $n - \pi^*$ transition for carbonyl compounds experiences bathochromic shift when the polarity of the solvent is decreased.
Conversion of phenol to phenolate anion causes bathochromic shift because the non-bonding electrons become available for the interaction with the π-electron system of the ring.

2. **Hypsochromic shift (blue shift):** Shifting of λ_{max} to lower value due to substitution, solvent pH, etc., is called hypsochromic shift. It may be due to increase in the polarity of the solvent or the removal of conjugation.

Example: Aniline absorption maximum occurs at 280 mμ because the pair of electrons on nitrogen atom is in conjugation with the π bond system of the benzene ring. In its acidic solution, a blue shift is caused and absorption occurs at shorter wavelength

(~ 203 mμ). 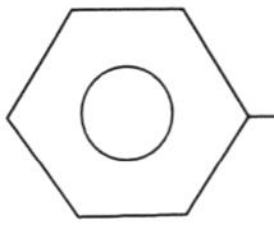 —N^+H_3 ion formed in acidic solution, the electron pair is no longer present and hence conjugation is removed.

3. **Hyperchromic effect:** It is defined as an effect due to which the intensity of absorption maximum increase, i.e., ε_{max} increases.

   For example: Biphenyl absorps at 250 mμ ε_{max} 19000 whereas 2-methyl biphenyl absorbs at 237 mμ ε_{max} 10250 (ε_{max} decreases). Introduction of methyl group distorts the geometry of the molecules thus causes hypochromic shift.

5.20 SOLVENT EFFECT

Choice of Solvent

1. It should not itself absorb radiation in the region under investigation.
2. It should be less polar so that it has minimum interaction with the solute molecule.
3. Most commonly 95% ethanol.
4. Cheap, good, dissolving power does not absorb radiation above 210 nm.
5. The position and intensity of absorption bands is greatly affected by the polarity of the solvent used.

Following shifts are observed on changing the polarity of the solvents.

1. Conjugated dienes and aromatic hydrocarbons experience very little shifts on changing the polarity of solvents.
2. The polar compounds show a change in absorption maximum on changing the polarity of the solvent. α-β-unsaturated carbonyl compounds show two different shifts on increasing the polarity of the solvent.
   (a) The $n - \pi^*$ transition (less intense): The absorption band moves to shorter wavelength (blue shift) by increasing the polarity of the solvent. This is because the ground state is more polar as compared to the excited state. This stabilization is more with hydrogen-bonding solvents (H_2O, ethanol). Thus, the net energy of n orbital is lowered thus increasing the energy gap between ground and excited states. Hence, absorption band moves to longer wavelength (Fig. 5.27).

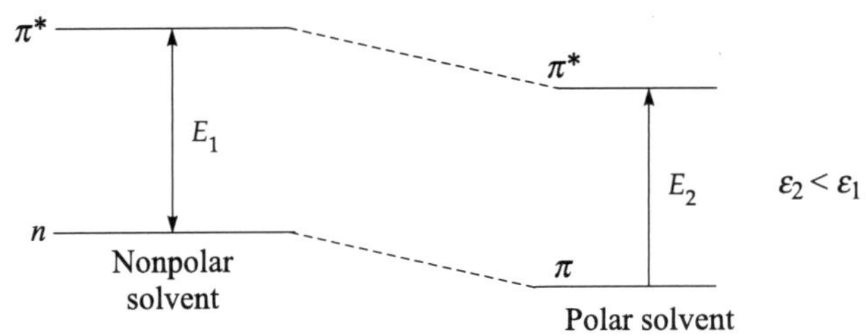

Fig. 5.27 *Absorption shift with change in polarity in $n - \pi^*$ band*

Example: Absorption maximum of acetone is at 279 nm in hexane as compared to 264 nm in water.

(b) $\pi \rightarrow \pi^*$ **transition (less intense):** The band moves to longer wavelength by increasing the polarity of the solvent. This is because the dipole-dipole interaction with the solvent molecules lowers the energy of excited state more than that of the ground state. In other words π^* orbitals are more stabilized by hydrogen bonding with polar solvents because of greater polarity of π^* compared to π orbital. Thus, the energy required for $\pi - \pi^*$ transition is less and absorption moves to longer wavelength (Fig. 5.28).

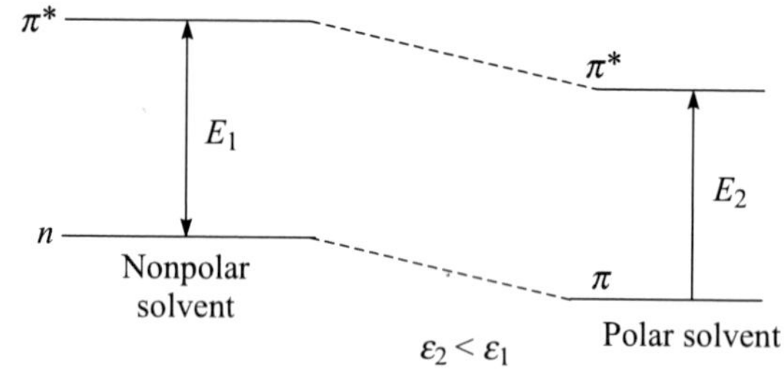

Fig. 5.28 *Absorption shift with change in polarity in $\pi - \pi^*$ band*

5.20.1 Woodward Fieser Rules

It predicted the absorption maxima for a conjugated or substituted diene. According to these rules, each type of diene has certain fixed values and the value of λ_{max} depends upon the following points:

1. The number of alkyl substituents or ring residues on double bond.
2. The number of double bonds with extended conjugation.
3. The presence of polar groups such as – Cl, Br, –OR, –SR, etc.

The various rules are summarized in Table 5.11.

Table 5.11 Woodward Fieser rules for diene and trienes

Base value	λ_{max}
Parent acyclic diene Homoannular diene (double bond present in the same ring) Heteroannular diene (double bonds are not present in the same ring)	217 nm 253 nm 214 nm
Addition for each substituent	
– *R*(alkyl) or ring residue – (alkoxy) Exocyclic double bond (to one ring) Exocyclic double bond (to two rings) Double bond extending conjugation	5 nm 6 nm 5 nm 10 nm
Auxochrome	
–OR	6 nm
–SR	30 nm
–Cl, –Br	5 nm

$-NR_2$	60 nm
$-OCOCH_3$	0 nm

$$\lambda_{max}\text{(calculated)} = \text{Base value} + \text{Substituent contribution}$$

EXAMPLES

Example 1: Calculate the λ_{max} value in the following compounds.

(a) 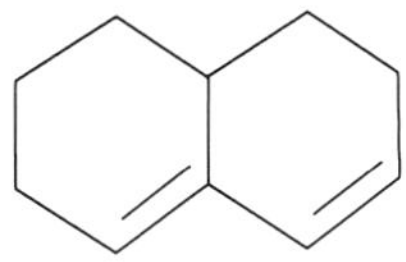

Solution: Base value of heteroannular diene = 214 nm

Substituents or ring residues = $3 \times 5 = 15$ nm

Exocyclic double bond = $1 \times 5 = 5$ nm

Calculate λ_{max} = 234 nm

(b) 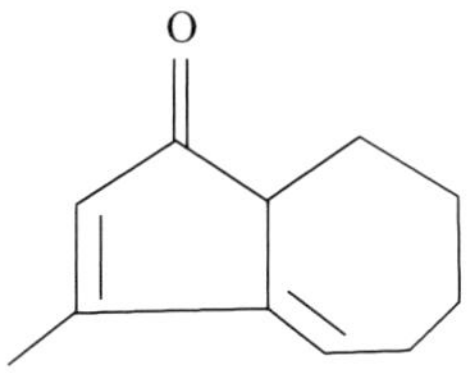

Solution: It is an α-β unsaturated cyclopentenone system.

Basic value = 202 mμ

β-alkyl substitution = 12 mμ

1 Exocyclic double bond = 5 mμ

1 Double bond extending conjugation = 30 mμ

1 γ- ring residue = 18 mμ

1 δ - ring residue = 18 mμ

Calculated λ_{max} = 285 mμ

(c) 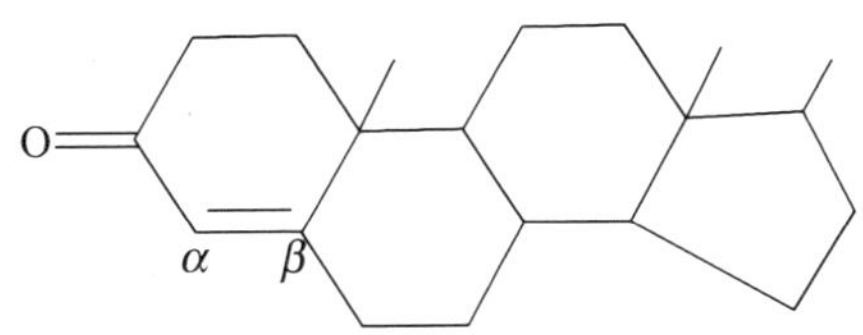

Solution: Basic value = 215 mμ

2 β-ring residue = 24 mμ

1 Exocyclic double bond = 5 mμ

Calculated λ_{max} = 244 mμ

(d) 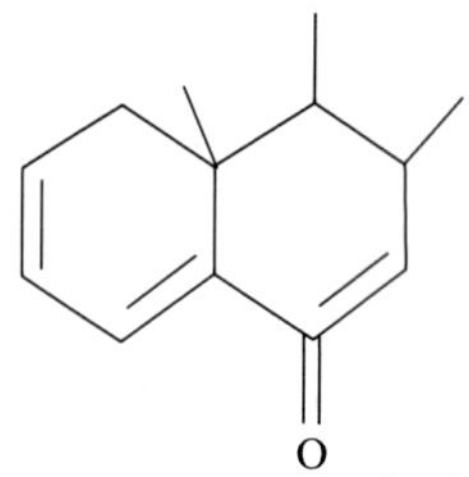

Solution: Basic value = 215 mμ
α-ring residue = 10 mμ
δ-ring residue = 18 mμ
1 Exocyclic double bond = 5 mμ
Homoannular conjugated diene = 39 mμ
1 double bond extending conjugation = 30 mμ

317 mμ

(e) $CH_3 - CH = CH - CH = CH - CH_3$.

Solution: Basis unit is CH = CH – CH = CH, i.e., butadiene
Substituents are two alkyl groups
Basic value = 217 nm
2 Alkyl substituents (2 × 5 nm) = 10 nm
Calculated λ_{max} 227 nm

(f) 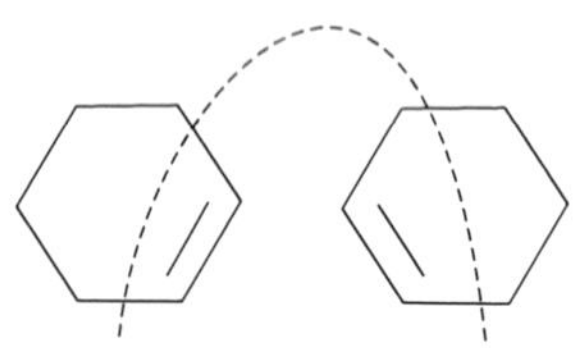

Solution: Base value of heteroannular diene
= 214 nm
Ring residue (4 × 5) = 20 nm
Calculated λ_{max} 234 nm

(g) 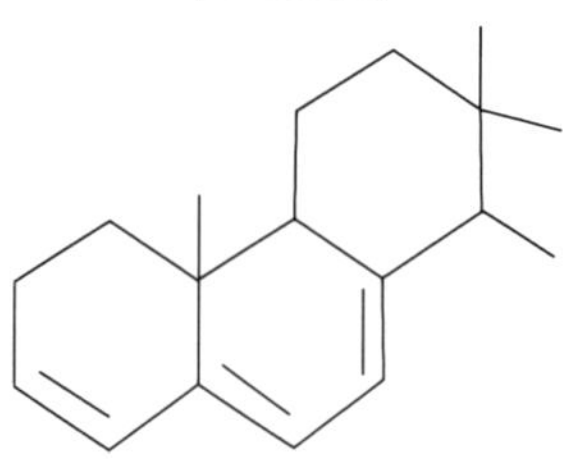

Solution: It is a homoannular conjugated diene system.
Basic value = 253 mμ
4 ring residues = 4 × 5 = 20 mμ
2-exocyclic double bonds = 2 × 5 = 10 mμ
1 double bond extending conjugation = 30 mμ
Calculated value 313 mμ.

In case there is a cross-conjugation in a compound, i.e., the carbonyl group has $\alpha\beta$-unsaturation on either side, then the value of absorption maximum is estimated by considering the most highly substituted conjugated system. Consider the following example.

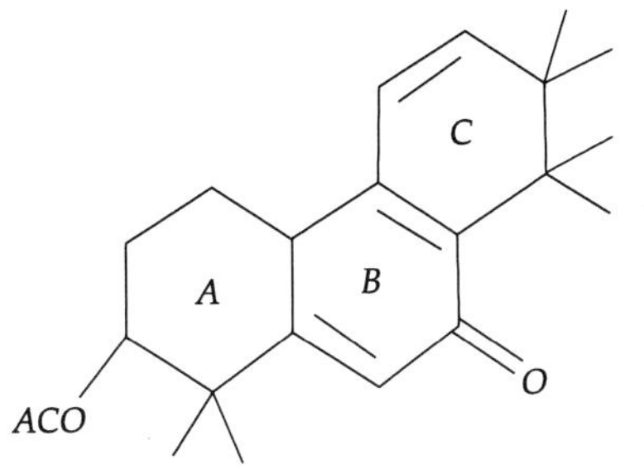

In ring *B*, the carbonyl group is conjugated on either side but conjugated system is highly substituted towards ring *C*. Thus, the maximum value of absorption is calculated as follows:

Basic value	= 215 mμ
1 α ring residue	= 10 mμ
1 β ring residue	= 12 mμ
1 δ ring residue	= 18 mμ
1 double bond extending conjugation	= 30 mμ
Homoannular conjugation diene	= 39 mμ
	324 mμ

5.20.2 Bicyclic Compounds

If a diene system is present in a bicyclic compound, then due to the strain in the bicycle compound, 15 mμ is added as the ring strain correction in the calculated value of absorption maximum.

Example 2: Calculate the absorption maximum in the ultraviolet spectrum of 2, 3 dimethylene bicyclic [2, 2, 1] heptane.

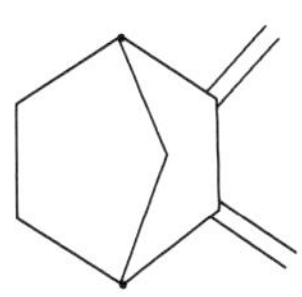

Solution:

Basic value	= 217 mμ
2-Ring residues	= 2 × 5 = 10 mμ
2 Exocyclic	= bonds = 10 mμ
1 - Bicyclic system (strain correction)	= 15 mμ
Calculated value	= 252 mμ

5.20.3 Rules for Calculating Absorption Maximum for Derivatives of Acyl Benzene

Like Woodward Fieser rules, Scott devised a set of rules for calculating the absorption maximum for the derivatives of acyl benzene. These rules helps in estimating the position of absorption maximum in ethanol in a number of monosubstituted aromatic ketones, aldehydes, acids and esters (Table 5.12).

Table 5.12 Auxochrome acting as a substituent

Auxochrome	Increment in mm positive of the substituent		
	Ortho-	**Meta-**	**Pera-**
Alkyl	+3	+3	+10
OH, OR	+7	+7	+25
Cl	0	0	+10
Br	+2	+2	+15
NH_2	+13	+13	+58
NHAC	+20	+20	+45
NR_2	+20	+20	+85
O^-	+11	+20	+75

For example:

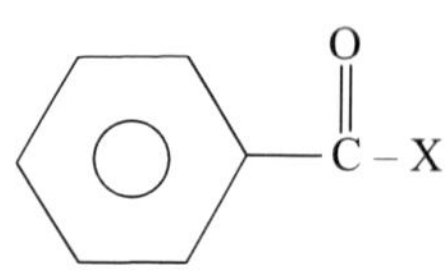

1. The basic value is 246 mμ. If *X* is an alkyl group or alicyclic residue.
2. If *X* is hydrogen atom, the basic value becomes 250 mμ.
3. The basic value is 230 mμ. If *X* is OH or OR. The structural increments in mμ for further substitution on the aromatic ring or the meta and para positions are given in Table 5.12.

   For example: Calculate the absorption maxima in ethanol for ρ-chloroacetophenone

   Basic value = 246 mμ

   Cl - substitution at para position = 10 mμ

   256 mμ

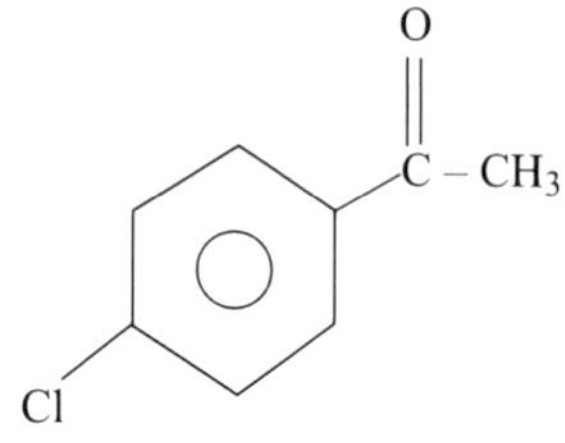

Example 3: Calculate the absorption maxima for the following compound.

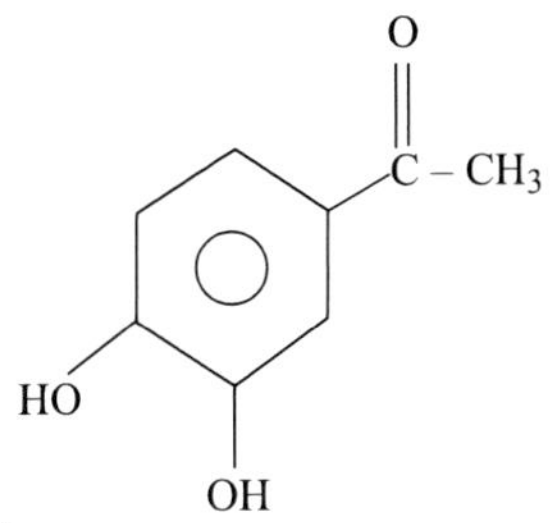

Basic value = 246 mμ
OH subsitution at meta position = 7 mμ
OH substitution at para position = 25 mμ
278 mμ

Example 4: Calculate the absorption maxima for the following compound.

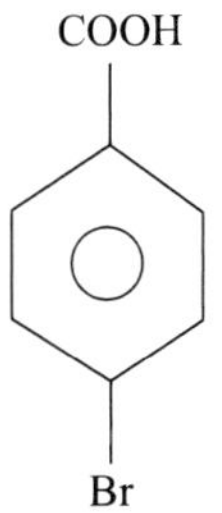

Basic value = 230 mμ
Br substitution at para = 15 mμ
245 mμ

5.20.4 Applications of UV Spectroscopy

Ultraviolet spectroscopy is mainly applied for the detection of functional groups (chromophore), the extent of conjugation, detection of polynuclear compounds by comparison, etc. There are some more important applications of ultraviolet spectroscopy are as follows:

1. *Extent of conjugation:* A conjugated system requires lower energy for the $\pi - \pi^*$ transition than an unconjugated system. Longer the conjugation, longer the absorption wavelength.
   For example: Ethylene has one double bond and absorbs at 170 nm due to $\pi - \pi^*$ transition whereas butadiene having two double bonds in conjugation absorbs at 217 nm.
2. *Determination of geometrical isomers:* It has been found that *cis*-alkenes absorb at different wavelengths as compared to their corresponding *trans* isomers.

Coplanarity is needed for the most effective π orbital which eases the $\pi - \pi^*$ transition. The steric strain in *cis*-isomer prevents complete coplanarity. Thus, in *cis*-isomer, λ_{max} arises at shorter wavelength.

Example: *cis*-stilbene absorbs at 274 and *trans*-stilbene absorbs at λ_{max} 294 nm because the steric hindrance in *cis*-isomer prevents complete coplanarity required for full conjugation and hence it absorbs at shorter wavelength than *trans*-isomers.

3. *Detection of impurities:* In ethanol, benzene is present as an impurity. Because of toxic nature of benzene, it is important that ethanol for internal use should be freed from benzene. The presence of even minute amounts of benzene can be confirmed by UV spectrum of sample at 280 nm. At this wavelength ethanol is transparent but benzene exhibits an absorption band with $\varepsilon_{max} = 230$.
4. *Identification of an unknown compound:* An unknown compound can be identified by comparing its spectrum with the known spectra. If the two spectra coincide, then the two compounds must be identical. If the two spectra do not coincide, then the expected structure is different from the known compounds.
5. *Distinction between conjugated and non-conjugated compounds*

For example:

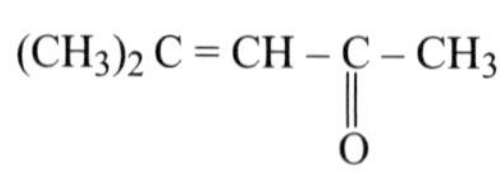

(a)

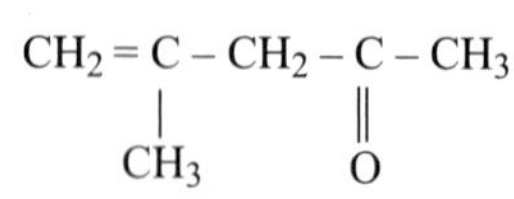

(b)

The compound (a) will have longer λ_{max} value because C = O group is in conjugation with double bond.

6. *Detection of functional group:* The technique is applied to detect the presence or absence of the chromophore. The absence of a band at a particular wavelength may be regarded as an evidence for the absence of a particular group in the compound. If the spectrum is transparent above 200 mμ. It shows the absence of:
   (i) Conjugation
   (ii) A carbonyl group (aldehydes and ketones)
   (iii) Benzene or aromatic compounds
   (iv) Bromo or iodo atoms.

   An isolated double bond or some other atoms may be present. It means that no definite conclusions can be drawn if the molecule absorbs below 200 mμ.
7. *Quantitative analysis:* It is based on:
   Beer-Lambert law, $A = \varepsilon Cl$
   A : Absorbance at a particular wavelength
   C : Concentration of the sample
   ε : Constant known as molar absorptivity or extinction coefficient.
   The absorbance of standard solution (A_1) and that of the sample is measured (A_2) in the same sample cell (l = constant), concentration of standard solution, knowing the (C_1) the concentration of the sample (C_2) can be determined as

$$A_1/A_2 = C_1/C_2$$

For example: For a solution of camphor in hexane in a 10 cm cell, the absorbance A was found to be 2.52 at 295 nm with $\varepsilon_{max} = 14$. What is the concentration of camphor.

$$A = \varepsilon Cl$$

$$A = 2.52, \varepsilon = 14\ l = 10 \text{ cm}$$

$$2.52 = 14 \times C \times 10$$

$$C = \frac{2.52}{14 \times 10} = 1.8 \times 10^{-2} \text{ moles/litre}$$

8. *Determination of strength of hydrogen bonding:* On the basis of shifts absorbed in polar solvents, it is possible to deduce the strength of hydrogen bonds.

5.21 FLUORESCENCE AND PHOSPHORESCENCE

Fluorescence is the light of comparatively longer wavelength emitted from a molecule after it has absorbed light of different and short wavelength.

Anthracene is a colourless substance and its electronic absorption spectrum lies in the ultraviolet region. But it is found that pure samples of anthracene when viewed in the ultraviolet light give off a blue – visible light. This light is called fluorescence. The emission of light of fluorescence stops at once when the irradiating light is removed. On the other hand, the phenomenon of phosphorescence is said to occur when the substance continues to emit the radiation of longer wavelength even when the irradiating light is removed.

For example: One electron excitation process.

When the molecule in the ground state absorbs light ($h\nu$), its one electron gets promoted from the ground state molecular orbital to one of the vibrational sub-levels of the first excited state molecular orbital (ε_1) or to the higher excited states ε_2, ε_3, etc. When the electron returns from ε_1 to ε_0 it emits radiation which is of longer wavelength or lower energy than $h\nu$. It is called $h\nu_f$ fluorescent radiation. In another mode, the energy may be lost from ε_1 to T_1 and then from T_1 to the ground state ε_0. The main difference between ε_1 and T_1 is the electron spin orientation. The spins of two electrons must be anti-parallel which originally occupy the ground state molecular orbital. The original promotion of electron to first excited state (ε_1) followed by its return to the lowest vibrational sub-level (ε_1) does not change the spin of the electron. But the transition from ε_1 to T_1 level does change the spin. Energy states containing only spin paired electrons are called *singlet states* while those with parallel spin electrons are called *triplet states*. In the exited triplet state electrons are farther apart in space and thus electron – electron repulsion is minimised. Thus, triplet states are more stable than singlet states and stay longer. Thus, this state may survive after the irradiating light is removed. After this electron jumps to ε_0 state by emitting the radiation of energy, $h\nu_p$. It is called *phosphorescent radiation*.

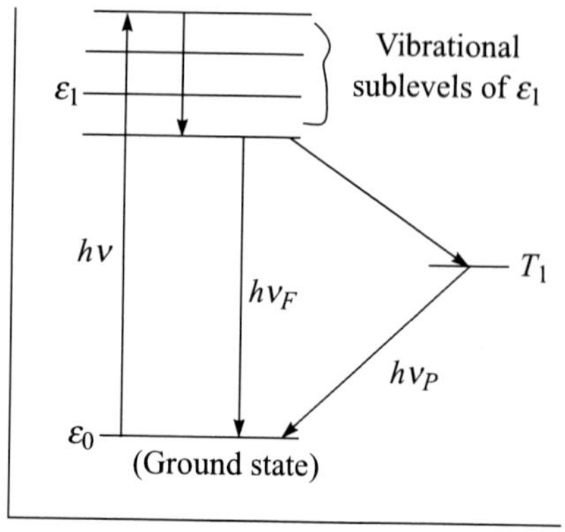

Fig. 5.29 *Mechanism of fluorescence and phosphorescence*

The fluorescent spectrum for anthracene consists of four maxima with λ_{max} 380, 400, 420 and 450 nm. Important uses of the fluorescence phenomenon are shown by the molecules of fluorescein and optical brightner derivative of *A, A′* diaminostiblene.

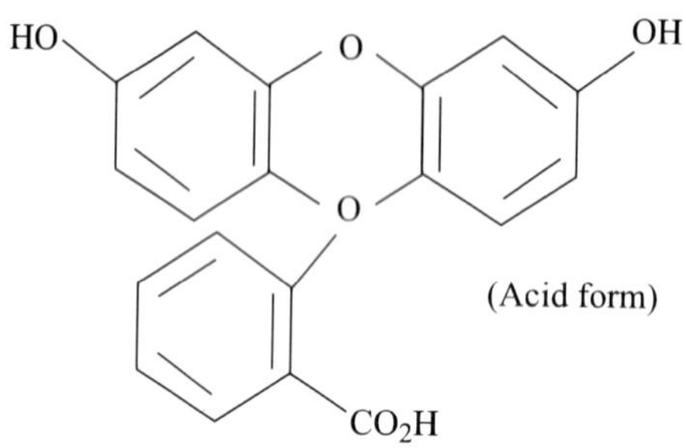

Fig. 5.30 *Fluorescein*

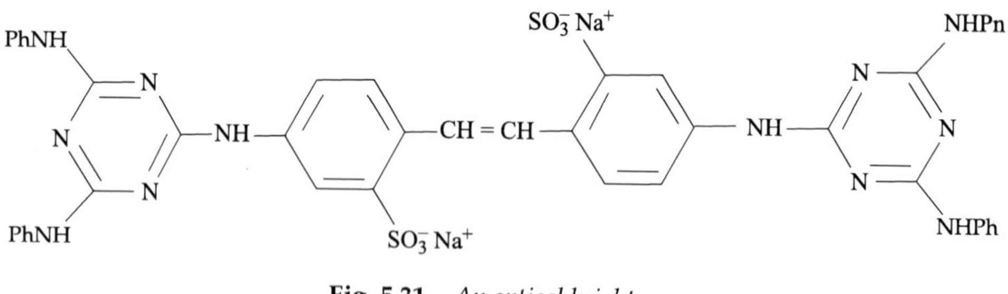

Fig. 5.31 *An optical brightner*

The intense green fluorescence of aqueous fluorescein solution makes it an excellent material for leak detection and an excellent marker for sea rescue operations, etc. The property of fluorescence is applied in polymer chemistry to detect and identify plasticizers and in the study of impurities. Biological applications include the study of the three-dimensional tertiary structures of proteins by measuring the proximity of known fluorescent groups within the protein. These fluorescent groups are generally aromatic amino acids. Phosphorescence seen at night in the sea is due to several species of marine microorganism which when agitated by an oar splash undergo an enzymatic alarm reaction which liberates energy in the form of green light. This phenomenon is called *bioluminescence.*

5.22 INFRARED SPECTROSCOPY

IR spectroscopy is used for detection of function groups in pure compounds and mixtures.

It is also known as *vibrational spectroscopy*, involves the transition between the vibrational energy levels or more precisely, the transition between vibrational and rotational energy levels of a molecule on absorption of radiation falling in the spectral range of 650 – 4000 cm^{-1} (infrared region). Infrared radiations do not have sufficient energy to cause the excitation of electrons. However, it causes atoms and groups of atoms of organic compounds to vibrate faster about the covalent bonds connecting them. As these vibrations are quantized and hence the compound absorbs the infrared energy in particular regions of the spectrum. In other words, an infrared spectrum is obtained when the frequency of molecular vibration corresponds to the frequency of the infrared radiation absorbed.

5.22.1 Principle

Absorption in the infrared region is due to the changes in the vibration and rotational levels. It has been studied that a single vibrational energy change is accompained by a large number of rotational energy changes. Thus, the vibrational spectra appear as vibrational, rotational bands. In the absorbed infrared spectroscopy, the absorbed energy brings about predominant changes in the vibrational energy — this causes the various band in molecule to stretch and bend with respect to one another. The vibrational energy depends upon:

- Mass of the atoms present in a molecules.
- Strength of the bonds.
- The arrangement of atoms within a molecule.

5.22.2 Units Used and Range in Infrared Radiation

Infrared spectra are generally expressed in terms of the frequencies of radiations absorbed rather than wavelengths. This is due to the reason that frequency is directly proportional to the energy of radiation, whereas wavelength is inversely proportional to energy. The most common unit used in infrared study is wave number ($\overline{v}$) which shows the number of vibrations of radiation per centimetre. Wavelength in μm can be converted into wave number. According to the relation

$$\overline{v} = 10^4/\lambda \quad (\because 1\ \mu m = 10^{-6}\ m = 10^{-4}\ cm)$$

The infrared radiation lies between the visible and microwave regions of the electromagnetic spectrum. This region may be divided into the following types:

(1) *Near infrared region:* It is the region which extends from 0.8 to 2.5 μm (12,500 to 4000 cm^{-1}). It is also known as *vibration rotation region.*

(2) *Far infrared region:* It is the region which extends from 15 to 200 μm (667 to 50 cm^{-1}). It is also known as *rotation region*.

(3) *Ordinary infrared region:* It is the region which extends from 2.5 to 1.5 μm (4000 to 667 cm^{-1}).

5.22.3 Molecular Vibrations

There are two types of molecular vibrations:

1. Stretching
2. Bending

1. Stretching: In this type of vibration, the distance between the two atoms increases or decreases but the centre atom remains in the same bond axis.

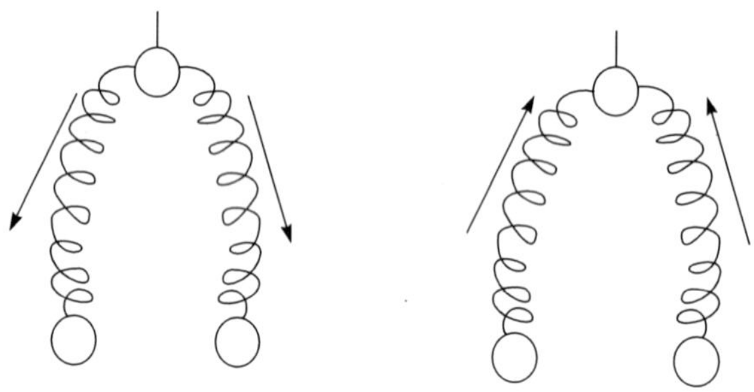

Fig. 5.32

Stretching vibrations are of two types:

(a) Symmetric stretching

(b) Asymmetric stretching

(a) Symmetric stretching: In this type of vibrations the movement of the atoms, with respect to a particular atom in a molecule is in the same direction.

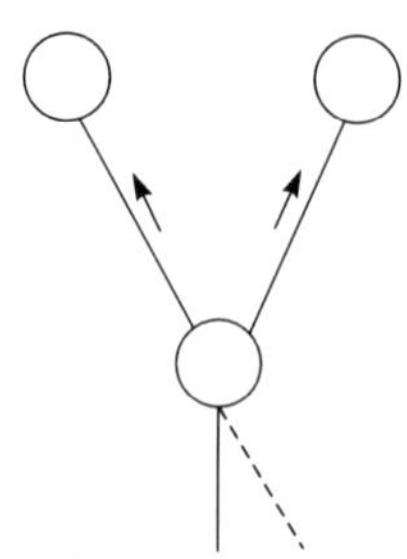

Fig. 5.33 *Symmetric mode of vibration*

(b) Asymmetric stretching: In this type of vibrations one atom approaches the centre atom, while the other departs from it.

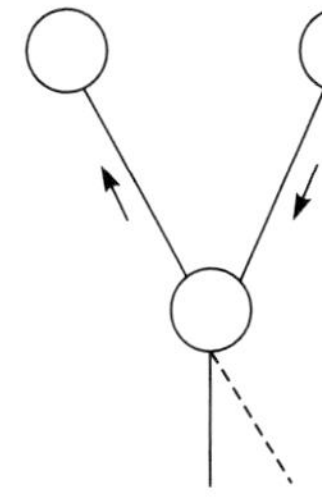

Fig. 5.34 *Asymmetric mode of vibration*

2. Bending: In this type of vibrations, the position of the atoms changes with respect to the original bond axis.

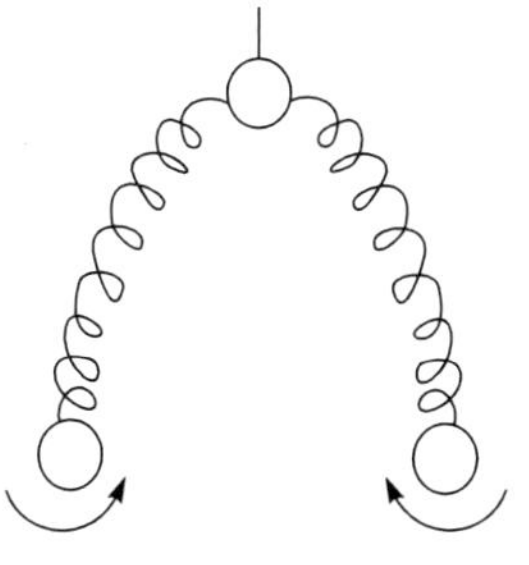

Fig. 5.35

Bending vibrations are of four types:

1. Scissoring
2. Rocking
3. Wagging
4. Twisting

1. Scissoring

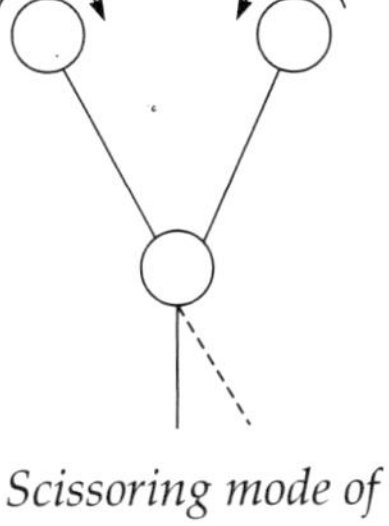

Fig. 5.36 *Scissoring mode of vibration*

In this type of vibrations, two atoms approach each other with respect to the central atom and the bond angle changes.

2. Rocking

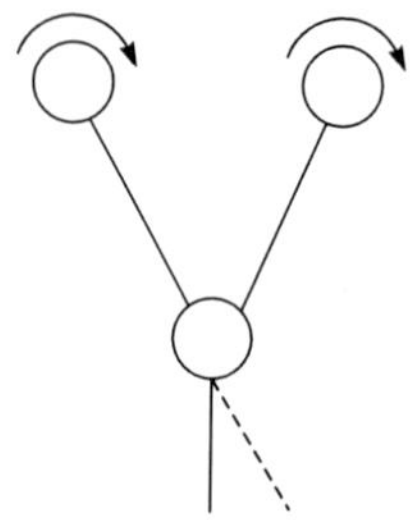

Fig. 5.37 *Rocking mode of vibration*

In this type of vibration, the two atoms joining a central atom, move back and forth in the plane of the molecule.

*3. **Wagging:*** In this kind of vibration, the structural unit moves back and forth out of the plane of the molecule.

In Figure 5.38 wagging mode of vibration (+) sign means vibrations perpendicular to the plane of paper.

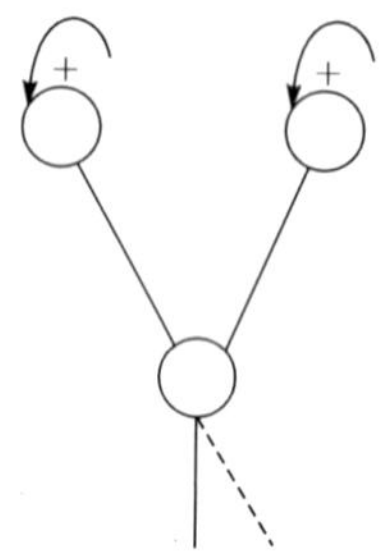

Fig. 5.38

4. Twisting: In this deformation, the structural unit rotates about the bond, which joins it to the rest of the molecule.

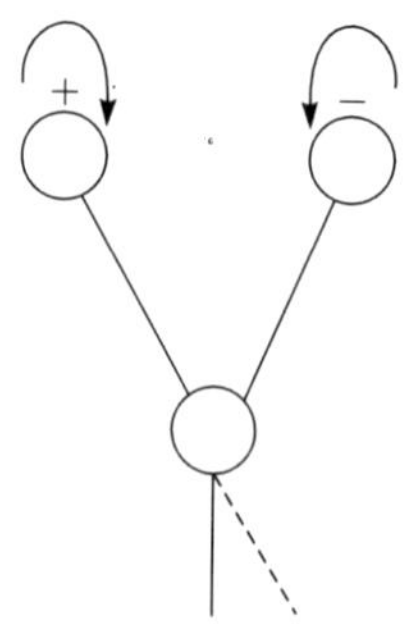

Fig. 5.39 *Twisting mode of vibration (+) or (–) sign signifies vibrations perpendicular to the plane of paper*

5.22.4 Number of Fundamental Vibrations

When atoms combine to from a molecule, no degrees of freedom are lost, i.e., the total number of degrees of freedom of a molecule will be equal to $3n$, where n is the number of atoms in a molecule.

There are three types of degree of freedom:

1. Translational degree of freedom
2. Rotational degree of freedom
3. Vibrational degree of freedom

Thus, $3N$ = Translational + Rotational + Vibrational degrees of freedom

Since *IR* spectroscopy is concerned with vibrational degree of freedom only then

Vibrational = $3N$ – Translational – Rotational

For linear molecule (for example, CO_2)

Number of atoms (n) = 3

Total degrees of freedom = $3n = 3 \times 3 = 9$

Translational = 3

Rotational = 2

Vibrational degrees of freedom = $9 - 3 - 2 = 4$

Hence, for CO_2 molecule, the theoretical number of fundamental bonds should be equal to four.

For non-linear molecule:

$$\begin{aligned} \text{Total degrees of freedom} &= 3n \\ \text{Translational degrees of freedom} &= 3 \\ \text{Rotational degrees of freedom} &= 3 \\ \text{Vibrational degrees of freedom} &= 3n - 3 - 3 \\ &= 3n - 6 \end{aligned}$$

Example: C_6H_6.

$$\begin{aligned} \text{Number of atoms } (n) &= 12 \\ \text{Total degrees of freedom} &= 3 \times 12 = 36 \\ \text{Translational} &= 3 \\ \text{Rotational} &= 3 \\ \text{Vibrational degrees of freedom} &= 36 - 3 - 3 \\ &= 30 \end{aligned}$$

So, theoretically there should be 30 fundamental bands in the infrared spectrum of benzene.

5.22.5 Conditions for the Infrared Absorption

The necessary condition for the absorption of infrared radiation can be summarised as below:

1. The natural frequency of vibration of the molecule must be equal to the frequency of the incident radiation.
2. Changes in vibration must stimulate changes in the dipole moment of the molecules.
3. The frequency of the vibration must satisfy the equation, $\varepsilon = h\nu$.
4. The intensity of absorption must be proportional to the square of the rate of change of dipole.

5.22.6 Characteristics of Infrared Spectrum

Maximal absorption of an infrared radiation or a frequency characteristics of a bond or group would show the presence of that particular bond or group in the substance. In case the compound does not absorb maximally that particular radiation it would mean that the particular bond or group is absent. So, we get characteristic infrared bands for different groups or bonds as shown in Table 5.7. It gives only the ranges of different bonds. The exact location of characteristic bond depends on several factors like structural environment of that bond or group in a specific molecule.

A strong band ~1700 cm^{-1} indicates the presence of a carbonyl group, but these are characteristic shifts in the position of the carbonyl band depending on the structural environment as shown.

$CH_3COCH_2CH_3$
Methyl ethyl ketone
(1710 cm^{-1})

$CH_3CO\ CH_2 = CH_2$
Methyl vinyl ketone
(1680 cm^{-1})

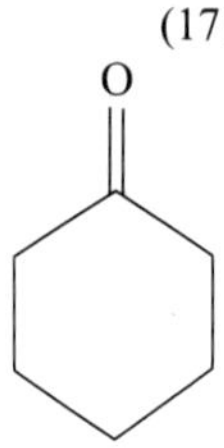

Cyclohexanone
(1710 cm^{-1})

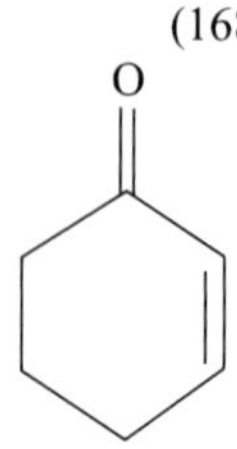

2-Cyclohexanone
(1680 cm^{-1})

Functional Graph	Stretching Frequency
–OH (alcohols)	3200 – 3600 cm^{-1}
–OD	2100 cm^{-1}
–OH (carboxylic acid)	3350 – 3500 cm^{-1}
>NH	3350 – 3500 cm^{-1}
$-C_SP - H$	3310 – 3320 cm^{-1}
–CHO	2715 – 2820 cm^{-1}
$-C_SP^2H$	3000 – 3100 cm^{-1}
= CO	1200 cm^{-1}
$C_SP^3 - O$	1025 – 1200 cm^{-1}
–SH	2500 cm^{-1}
C = C	1620 – 1680 cm^{-1}
C = O	1710 – 1750 cm^{-1}
> C = O (carboxylic acid)	1700 – 1725 cm^{-1}
> C = O (Acid anhydride)	1800 – 1850 cm^{-1}
> C = O (Ester)	1730 – 1750 cm^{-1}
> C = O (Amide)	1680 – 1700 cm^{-1}

$-C \equiv C-$	$2100 - 2200\ cm^{-1}$
$-C \equiv N$	$2240 - 2280\ cm^{-1}$
Aromatic CH stretching	$3030\ cm^{-1}$

C – H stretching vibrations	
Groups	**Infrared absorption wavelength (cm^{-1})**
Alkanes	$2800 - 3000\ cm^{-1}$
Alkenes and aromatics	$3000 - 3100\ cm^{-1}$
Alkanes	$3300\ cm^{-1}$

C – H bending vibrations		
Groups	**Symbol**	**Infrared Absorption**
Alkane	CH_3	$1420 - 1470\ cm^{-1}$
	$CH(CH_3)_2$	$1370, 1385\ cm^{-1}$
	$C(CH_3)_3$	$1370, 1395\ cm^{-1}$
Alkene	$RCH = CH_2$	$910 - 920\ cm^{-1}$
	$R_2C = CH_2$	$800 - 900\ cm^{-1}$
	RCH = CHR (cis)	$675 - 730\ cm^{-1}$
	RCH = CHR (trans)	$965 - 975\ cm^{-1}$
Aromatic	Mono-substituted	$690 - 710\ cm^{-1}$ and $730 - 770\ cm^{-1}$
	Di - substituted	
	ortho	$735 - 770\ cm^{-1}$
	meta	$690 - 710\ cm^{-1}$
	para	$810 - 830\ cm^{-1}$

5.22.7 Finger Print Region

This region ($4000 - 1650\ cm^{-1}$) is very useful for confirming the identification of two compounds and is called the finger print region. Two different substances howsoever closely related cannot possess identical or superimposability of the infrared spectra.

The region below $1500\ cm^{-1}$ is rich in many absorptions which are caused by bending vibration and those resulting from the stretching vibrations of C — C, C — O and C – N bonds. In a spectrum, the number of bending vibrations is usually more than the number of stretching vibrations. The said region is usually rich in absorption bonds and shoulders. It is called finger print region.

It can be subdivided into four regions as follows:

(1) 4000 – 1500 cm^{-1}
(2) 1500 – 1350 cm^{-1}
(3) 1350 – 1000 cm^{-1}
(4) Below 1000 cm^{-1}

Region 4000 – 1500 cm^{-1}: The region from 4000 – 1500 cm^{-1} (high frequency part) to the left in an infrared spectrum is useful for the identification of functional groups. This region shows absorption arising from stretching modes.

Region 1500 – 1350 cm^{-1}: The appearance of a doublet near 1380 cm^{-1} (m) and 1365 cm^{-1} (s) shows the presence of tertiary butyl group in the compound.

Region 1350 – 1000 cm^{-1}: All classes of compounds, viz., alcohols, esters, lactones, acid anhydrides show characteristic absorptions (strong brands) in this region due to C – O stretching. Primary alcohols from two strong bands at 1350 – 1260 cm^{-1} and near 1050 cm^{-1}. Phenols absorb near 1200 cm^{-1}. Esters show two strong bands between 1380 – 1050 cm^{-1}.

Region below 1000 cm^{-1}:

H > C = C – H deformation at ~700 cm^{-1} (s) and
|

that at 970 – 960 cm^{-1} (s) distinguishes between *cis* and *trans* alkenes. The higher value indicates that the hydrogen atoms in the alkene are *trans* with respect to each other.

5.22.8 Applications of IR Spectroscopy

(1) *Qualitative analysis:* The qualitative analysis can easily be done by comparing the spectrum of unknown sample with the standard one.

(2) *Quantitative analysis:* The quantitative determination is based on the determination of concentration of one of the functional groups of compound being estimated. Suppose there is a mixture of hexane and hexanol and then concentration of hexanol can be determined by measuring absorption of OH bond. The following formula is used for calculating the concentration,

$$A = -\log I/I_0 = abc$$

$A \rightarrow$ Absorbance
$I \rightarrow$ Intensity of radiation after leaving the sample
$I_0 \rightarrow$ Intensity of radiation before entering the sample
$a \rightarrow$ Absorptivity of cell
$b \rightarrow$ Initial path length of sample cell
$c \rightarrow$ Concentration of solution

If a and b are constants then $A \propto C$.

Hence, C can be measured by knowing A.

The different values of A are plotted against respective concentration to get a calibration curve from which the concentration of unknown solution is evaluated.

(3) *Determination of purity:* As impurity reduces the sharpness of individual bands and causes appearance of extra bands and blurring of spectrum hence, purity can easily be estimated.

(4) *Presence of water in a sample:* If the sample contains water, it gives bands at 3600 to 200 cm^{-1} region, in 1650 cm^{-1} region and in 600 – 300 cm^{-1} region. If anyhow, water is coordinated to a metal ion, an additional band can be seen in the range of 880 – 650 cm^{-1} region.

(5) *Shape and symmetry of a molecule:* Infrared spectroscopy has proved very successful in deducing the shape or symmetry of molecules. For example, if NO_2 is linear, only two bands will be active in *IR* spectrum and if bent then three bands will be found as in H_2O.

(6) *In the study of reaction kinetics:* The technique is helpful in the study of reaction kinetics as intermediate compounds can easily be detected and estimated. **Example:** Progress of oxidation of a secondary alcohol to ketone is studied by getting *IR* spectra of test portions at different times. In this case, we expect disappearance of O – H stretching ~2.8 μm (~3570 cm^{-1}) and appearance of C = O stretching ~5.8 μm (~1725 cm^{-1}).

(7) Given Tables 5.13 and 5.14 from which different functional groups can easily be detected.

Table 5.13 Saturated C – H and C – C

Group	Band, cm^{-1}	Remarks
(i) $>CH_2$, $-CH_3$	2960 – 2850(s)	Two or three bands usually
(ii) –CH	2890 – 2880 (w)	–C – H stretching
(iii) $>CH_2$, $-CH_3$	1470 – 1430 (m)	–C – H deformations
(iv) $-CH_3$	1390 – 1370 (m)	$-CH_3$ symmetrical deformation
(v) $> CH_2$	~720 (w)	$> CH_2$ rocking

Table 5.14 Alkene and aromatic C – H

Group	Bond cm–1	Remarks
(i) – C ≡ C – H	~3300 (s)	
(ii) $>C=C<$ (H, H)	3095 – 3070 (m)	C – H stretching
(iii) $>C=C<$ (H)	3040 – 3010 (m)	
(iv) Aryl – H	3040 – 3010 (w)	Often obscured
(v) R, II $>C=C<$ H, II	970 – 960 (s)	C – H out of plane deformation when (=) is conjugated with a C = O group, this band is shifted towards 990 cm^{-1}
(vi) RCII – CH_2	995 – 985 (s) and 940 – 900 (s)	
(vii) $R_2C = CH_2$	895 – 885 (s)	

(viii) $R_2C = C\begin{matrix} H \\ R \end{matrix}$	840 – 790 (m)	
(ix) $\begin{matrix} H \\ R \end{matrix}C = C\begin{matrix} H \\ R \end{matrix}$	730 – 675 (m)	

(8) The characteristic series of bands produced by –O – H in alcohols and phenols are given in Table 5.15.
It is also seen that if hydrogen bond is stronger then there will be longer O – H bond. Besides this, if the vibration frequency is lower, we will get broader and more intense absorption bonds. The study on the subject has shown that pure liquids, solids and many solutions exhibit the broad polymeric band in 3600 – 3200 cm^{-1}.

Table 5.15 Alcohol and phenol –O – H

Group	Band, cm^{-1}	Remarks
When H_2O in solution Free –OH H bonded, –OH (liquid and dilute solution, solid)	3710 3650 – 3590 (v) 3600 – 3200 (s)	Sharp, O – H stretching Generally, broad but it can be sharp for some intramolecular single bridge H-bonds, the lower the frequency the stronger the H-bond
Intramolecular H-bonded –OH is chelated	3200 – 2500 (v)	Broad, the lower the frequency the stronger the H-bond. Sometimes the band is so broad that it is to be overlooked
–OH	1410 – 1260 (s)	O – H bending
$\rightarrow$C – OH	1150 – 1040 (s)	C – O stretching
Water of crystallization	3600 – 3100 (w)	Generally, a weak band at 1640 – 1615 cm^{-1}

(9) The recent studies have revealed that N – H absorption is usually weaker (weak bands), overtones of strong carbonyl absorption ~1800 – 1600 cm^{-1} region also appear ~3600 – 3200 cm^{-1} region. The stretching and bending frequencies of N – H bonds are given in Tables 5.16 and 5.17.

Table 5.16 Amine, imine ammonium and amide N – H (N – H stretching)

Group	Band cm^{-1}	Remarks
(1) Amine and Imine > N – H = N – H	3500 – 3300 (m)	It is seen that primary amines exhibit two bands in this range. The unsymmetrical and symmetrical stretching secondary amines absorb weakly

$-NH_3^+$ Amino acids Amino salts	3130 – 3030 (m) ~3000 (m)	Values for solid state, broad bands also, near 2500 and 2000 cm^{-1}
$7NH_2^+, -NH^+ = {}_1NH^+$	2700 – 2250 (m)	Values for solid state; broad due to presence of overtone bands.
Primary amide $-CONH_2$	~3500 (m) ~3400 (m)	Lowered ~150 cm^{-1} in solid state and H-bonding
Secondary amide $-CONH^-$	3460 – 3400 (m)	Two bands, lowered on H-bonding and in solid state
	3100 – 3070 (w)	A weak extra band with bonded and solid state samples

Table 5.17 N – H Bending

Group	Band cm^{-1}	Remarks
> NH	1650 – 1560 (m)	
> NH	1580 – 1490 (w)	Very weak to be noticed
$-NH_3^+$	1600 (s) 1500 (s)	It is found that secondary amine salts have 1600 cm^{-1} band

(10) The bands due to cumulated double bonds are given in Table 5.18. The unusually high double bond frequencies encountered in X = Y = Z systems arise from strong coupling of two separate stretching frequencies becoming widely separated as shown in Table 5.18.

Table 5.18 Cumulated double bond

Group	Band cm^{-1}	Remarks
Isocyanates – N = C = O	2275 – 2250 (s)	Very high intensity, positive unaffected by conjugation
Carbon dioxide O = C = O	2349 (s)	This appears in many a spectra due to inequalities in path length
Ketenes > C = C = O	~2150 (s)	–
Azides $-NH_3$	2160 – 1220 (s)	–
Carbodi-imides	2155 – 2130 (s)	Very high intensity split into an unsymmetrical doublet by conjugation with aryl group
Diazo alkanes $R_2C = N^+ = N^-$	~2100 (s)	–
Alkenes C = C = C	~1950 (m)	Two bands when bonded to electron attracting group. **Example:** $-CO_2H$

5.22.9 Interpretation of Infrared Spectra

There are no rigid rules for interpreting an infrared spectrum. However, certain general observations may be of much help.

1. The limits of published range of group frequencies are defined by the extreme cases of electronic effects. A substance with no such effect is expected to absorb towards the centre of the range quoted.
2. We seldom require total interpretation of the spectrum. The much valuable structural information can be had from the relatively few bands. Only 10 – 20% bands can usually be assigned with certainty.
3. We should place more reliance on negative evidence, i.e., the absence of absorption in a particular region and not on positive evidence, i.e., the presence of a band which may exhibit several possible origins, **Example:** no absorption band in the 2500 – 3600 cm^{-1} region shows the absence of –OH group of alcohols, phenols, and carboxylic acid.
4. We should always start from the high frequency end of the spectrum and the finger print region should be used only for confirmation.

For example, a strong band ~1718 cm^{-1} shows almost certain presence of a carbonyl group (> C = O). It could be –CHO, > C = O, an ester, an amide or an acid.

1. Infrared spectrum of hydrocarbons: All hydrocarbons give absorption peaks in the 2800 – 3300 cm^{-1} region that are associated with carbon-hydrogen stretching vibrations.

The order of bond strength is

$$SP > SP^2 > SP^3$$

This too is the order of the bond stiffness. The C – H stretching peaks of hydrogen atoms attached to *SP*-hybridized carbon atoms occur at highest frequencies about 2200 cm^{-1}. Thus ≡ C – H groups of terminal alkynes give peaks in this region. We can see the absorption of the acetylenic C – H bond of 1-hexyne at 3320 cm^{-1}.

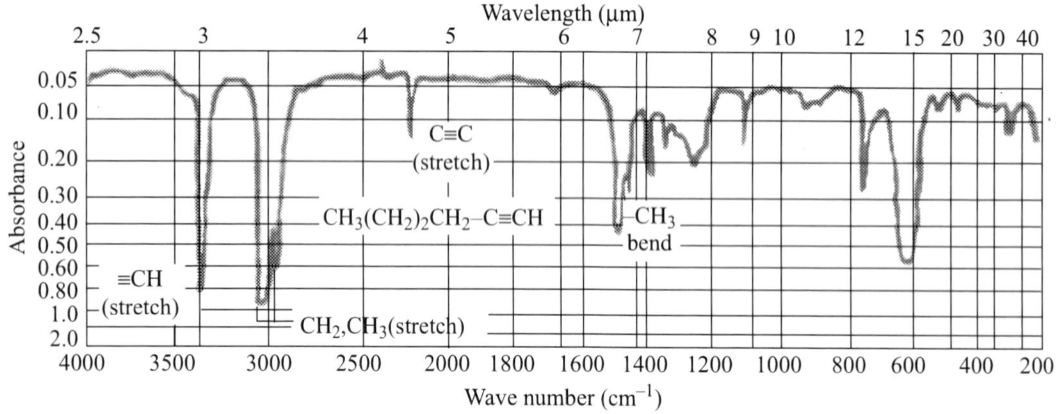

Fig. 5.40

The C-hydrogen stretching peaks of hydrogen atoms attached to SP^2-hybridized carbon atoms occur in the 3000 – 3100 cm^{-1} region. Thus, alkenyl C – H bond and the C – H group of aromatic rings give absorption peak in this region.

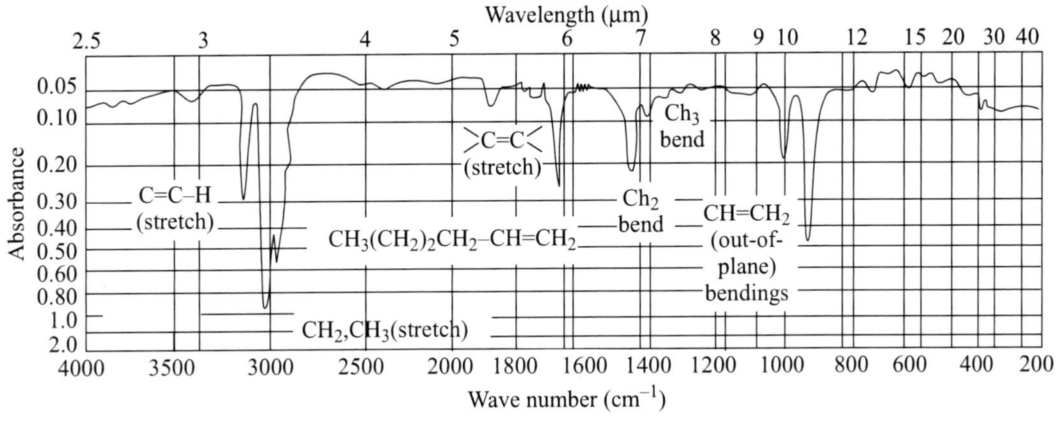

Fig. 5.41

Hydrocarbons also give absorption peaks in their *IR* spectra that result from carbon-carbon bond stretching. Carbon-carbon single bonds normally give rise to very weak peaks that are usually of little use in assigning structures. More useful peaks arise from multiple C – C bonds.

$$C = C \rightarrow 1620 - 1680 \text{ cm}^{-1}$$

$$C = C \rightarrow 2100 - 2260 \text{ cm}^{-1}$$

These absorptions are not usually strong ones and they are not present at all if the (=) or (≡) bond is symmetrically substitued. The stretching of the C – C bonds of benzene rings usually give a set of characteristic sharp peaks in the 1450 – 1600 cm^{-1} region. Absorption arising from carbon-hydrogen bending vibrations of alkenes occur in the 600 – 1000 cm^{-1} region. The exact location of these peaks can often be used to determine the substituted pattern of the double bond and its configuration.

Mono substituted alkenes give two strong peaks in 905 – 920 cm^{-1} and 985 – 1000 cm^{-1} regions. Disubstituted alkenes of the type $R_2C = CH_2$ give a strong peak in the 880 – 990 cm^{-1} range. *cis*-alkenes give an absorption peak in the 675 – 730 cm^{-1} region and *trans*-alkenes give a peak between 960 and 975 cm^{-1}. These ranges for the carbon-hydrogen bending vibrations can be used with fair reliability for alkynes that do not have an electron-releasing or electron withdrawing substituents on one of the carbon atoms of the double bond. When $_{-1}e^0$ releasing or $_{-1}e^0$ withdrawing substituents are present on a double-bond carbon, the bending absorption peaks may be shifted out of the regions, we have given.

2. Infrared spectrum of acetic acid.

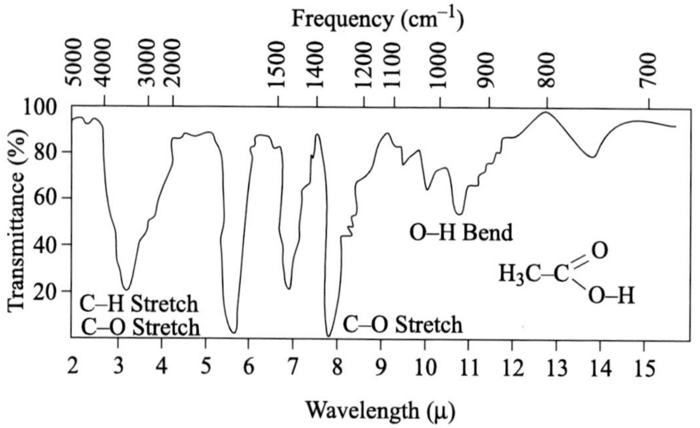

Fig. 5.42

(a) A broad band at ~3040 cm^{-1} is due to hydrogen bonded O – H stretching of a carboxylic acid.

(b) A band at ~1730 cm^{-1} is due to C = O stretching of a saturated chain carboxylic acid.

(c) A band at ~1280 cm^{-1} is due to C – O stretching of a carboxylic group.

(d) A band at ~950 cm^{-1} is due to O – H bending of a carboxylic acid.

3. Infrared spectrum of *n*-octane, $CH_3(CH_2)_6CH_3$

(a) A bond at ~2960 cm^{-1} is due to C – H stretch in CH_3 and a bond at ~2900 cm^{-1} due to C – H stretch in CH_2 show the presence of CH_3 and CH_2 group, respectively in the molecule.

(b) A bond at ~1475 cm^{-1} is due to C – H bending in the CH_2 group.

(c) A band at ~1375 cm^{-1} is due to C – H bending in the CH_3 group.

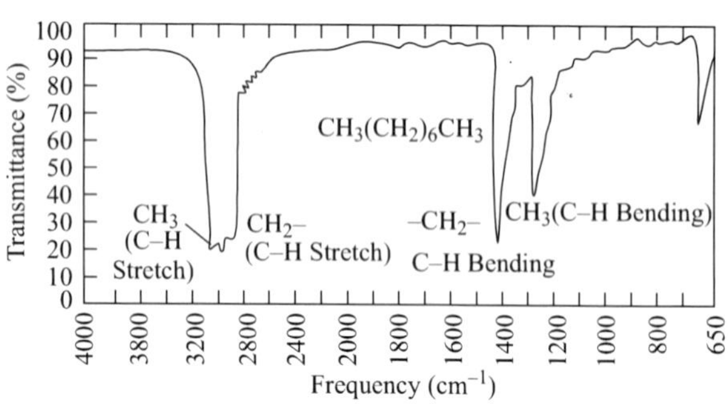

Fig. 5.43

4. Infrared spectrum of benzaldehyde $C_6H_5C{\overset{H}{\underset{O}{\lessgtr}}}$

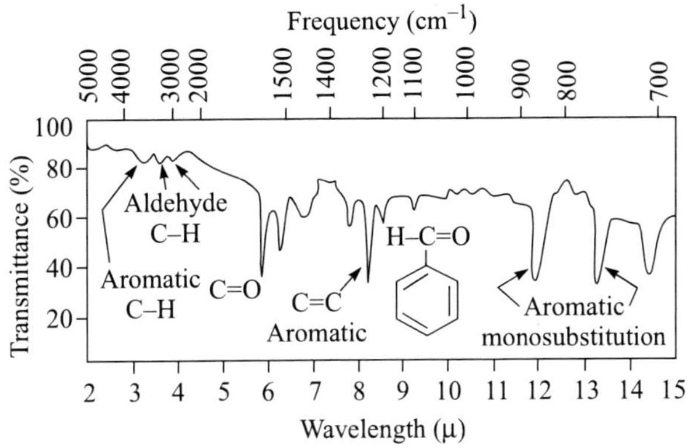

Fig. 5.44

(a) A band at ~3080 cm^{-1} is due to C – H stretching of an aromatic ring.

(b) Bands at ~2680 cm^{-1} and 2780 cm^{-1} are due to C – H stretching of an aldehyde group.

(c) A bond at ~1700 cm^{-1} is due to C = O stretching of an aromatic aldehyde group.

(d) A bond at ~1590 cm^{-1} is due to C = C stretching of an aromatic ring.

(e) Bonds at ~745 cm^{-1} and 685 cm^{-1} are due to C – H bending of monosubstituted benzene.

5. Infrared spectrum of other functional group: *IR* spectroscopy gives us an invaluable method for recognizing quickly and simply the presence of certain functional groups in a molecule. One important functional group that gives a prominent absorption peak in *IR* spectra is the carbonyl group > C = O. This group is present in aldehydes, ketones, esters, carboxylic acids, amides and so forth. The C-oxygen double bond stretching frequency of all these groups gives a strong peak between 1630 and 1780 cm^{-1}. The exact location of peak depends on whether it arises from an aldehyde, ketone, ester and so forth.

$R-\overset{O}{\overset{\|\|}{C}}-H$	$R-\overset{O}{\overset{\|\|}{C}}-R$	$R-\overset{O}{\overset{\|\|}{C}}-OR$
Aldehyde	Ketone	Ester
1690 – 1740 cm^{-1}	1690 – 1750 cm^{-1}	1735 – 1750 cm^{-1}

$R-\overset{O}{\overset{\|\|}{C}}-OH$	$R-\overset{O}{\overset{\|\|}{C}}-NH_2$
Carboxylic acid	Amide
1710 – 1780 cm^{-1}	1630 – 1690 cm^{-1}

The hydroxyl groups of alcohols and phenol are also easy to recognize in *IR* spectra by their O – H stretching absorptions. These bonds also give us direct evidence for hydrogen bonding. If an alcohol or phenol is present as a very dilute solution in CCl_4. O – H absorption occurs as a very sharp peak in the 3590 – 3650 cm^{-1} region. In very dilute solution or in the gas phase, formation of intermolecular hydrogen bonds does not take place because the molecules are too widely separated. The sharp peak in the 3590 – 3650 cm^{-1} region, therefore is attributed to free (unassociated) hydroxyl groups. Increasing the concentration of alcohol or phenol causes the sharp peak to be replaced by a broad band in the 3200 – 3550 cm^{-1} region. This absorption is attributed to OH groups that are associated through intermolecular hydrogen bonding. Very dilute solutions of 1° and 2° amines also give sharp peak in the 3300 – 3500 cm^{-1} region arising from free N – H stretching vibration. Primary amines give two sharp peaks, secondary amines give only one. Tertiary amines, because they have no N – H bond, do not absorb in this region.

RNH_2	R_2NH
1° Amine	2° Amine
Two peaks in 3300 – 3500 cm^{-1} region	One peak in 3300 – 3500 cm^{-1} region

Hydrogen bonding causes these peaks to broaden. The NH groups of amides give similar absorption peaks.

EXAMPLES

Example 1: Write down the approximate position of the characteristic *IR* bands in the following compounds.

(i) CH_3CH_2OH

(ii) CH_3COCH_3

(iii) $CH_2 = CHCOCH_3$

(iv) $C_6H_5COCH_3$

(v) CF_3COCF_3

(vi) CH_3OCH_3

Solution:

Compound	Characteristic IR bands
(i) CH_3CH_2OH	A strong and broad hydrogen bonded O – H stretching band at 3400 cm^{-1}, ~3200 cm^{-1} (s)
(ii) CH_3COCH_3	An intense C = O stretching band at ~1710 cm^{-1}
(iii) $CH_2 = CHCOCH_3$	An intense C = O stretching band of an α, β-unsaturated carbonyl group at ~1680 cm^{-1}

Contd...

(iv) $C_6H_5COCH_3$	An intense C = O stretching band at ~ 1680 cm^{-1} due to benzene ring
(v) CF_3COCF_3	An intense C = O stretching band at ~ 1790 cm^{-1} due to the presence of electron withdrawing groups
(vi) CH_3OCH_3	A characteristic C – O stretching band for methoxy group

Example 2: How would you differentiate between the following pairs of compounds using *IR* spectra.

(i) CH_3CH_2CHO and $CH_2 = CH - CH_2OH$

(ii)

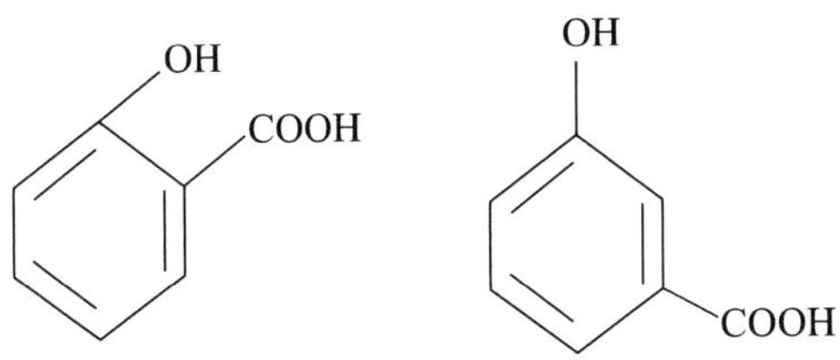

Solution:

(i) The compound CH_3CH_2CHO will show a strong characteristic band at ~ 1740 cm^{-1} to 1720 cm^{-1} for > C = O group. The compound $CH_2 = CH - CH_2 - OH$ will show a strong and broad hydrogen bonded OH stretching band at 3400 cm^{-1} to 3200 cm^{-1}, ~ 1050 cm^{-1} due to C – O stretching and bending, 3100 cm^{-1} to 3000 cm^{-1} for C = C bond and ~ 905 cm^{-1} for vinyl group $CH_2 = CH$.

(ii) In case of 2-hydroxy benzoic acid –OH and –COOH absorption peaks are obtained at 3550 cm^{-1} to 3450 cm^{-1} (due to intramolecular hydrogen bonding) and 2300 cm^{-1} to 2500 cm^{-1} respectively. A band ~ 755 cm^{-1} is obtained for C – H stretching and 1 : 2 disubstitution.

In the case of 3-hydroxy benzoic acid –OH and –COOH absorption peaks are obtained at ~ 3610 cm^{-1} and 2300 cm^{-1} to 2500 cm^{-1}, respectively. Along with these peaks at ~700 cm^{-1}, ~780 cm^{-1} and 880 cm^{-1} also appeared for C – H stretching.

Example 3: How would you identify the following pairs of compounds with the help of *IR* spectra?

(i) Ethanol and propanone
(ii) Methanol and ethanoic acid
(iii) Ethyl methyl ketone and methyl vinyl ketone
(iv) Cyclobutanone and cyclohexanone

Solution:

(i) Ethanol (CH_3CH_2OH) would show a broad hydrogen bonded OH stretching band at 3400 cm^{-1} to 3200 cm^{-1}, while propanone (CH_3COH_3) would show a strong > C = O stretching at ~ 1710 cm^{-1}.

(ii) Both methanol (CH_3OH) and ethanoic acid (CH_3COOH) would show an OH stretching band at 3300 cm^{-1} and 3000 cm^{-1} respectively. But ethanoic acid will also show an intense > C = O stretching band at about 1725 cm^{-1}.

(iii) Both ethyl methyl ketone ($CH_3CH_2COCH_3$) and methyl vinyl ketone ($CH_3-CO-CH=CH_2$) show strong > C = O stretching band. This band appears at 1710 cm^{-1} in the former compound and at 1680 cm^{-1} in the latter compound due to α, β unsaturation.

(iv) Cyclohexane will show a strong > C = O stretching band at ~1780 cm^{-1}.

Example 4: Two isomers *A* and *B* of the molecular formula C_3H_6O give an *IR* absorption band at 1650 cm^{-1} and 1710 cm^{-1} respectively. Assign structural formula to *A* and *B* isomers.

The molecular C_3H_6O has the following isomers:

(i) CH_3COCH_3

(ii) $CH_2=CH-CH_2OH$

(iii) CH_3CH_2CHO

(iv) 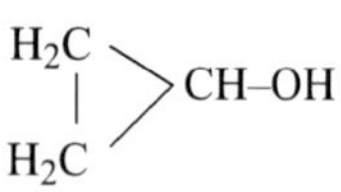

Solution: Since a band is obtained at 1710 cm^{-1} it should correspond to a > C = O group therefore *B* is CH_3COCH_3. An *IR* band at 1650 cm^{-1} corresponds to a compound.

(i) The compound *A* is $CH_2=CH-CH_2OH$

(ii) The compounds will give an *IR* band at ~1740 cm^{-1}.

(iii) The compound will not give an *IR* band at ~1650 cm^{-1} so both these structures are ruled out.

Example 5: Identify a particular acid derivative in each of the following carbonyl stretching frequencies

(a) 1700 – 1815 cm^{-1}

(b) 1720 – 1750 cm^{-1}

(c) 1630 – 1690 cm^{-1}

(d) 1740 – 1790 cm^{-1} and 1800 – 1850 cm^{-1}.

Solution:

(a) Acid chloride

(b) Ester

(c) Amide

(d) Anhydride

Example 6: The carboxylic acid and its derivatives have common C = O group. What are the distinctive features of each of the derivatives which can distinguish it from the other.

Solution:

(i) The carboxylic acid shows a strong and broad O – H structure frequency in the region 2500 – 3000 cm^{-1} which is absent in esters. Further, it shows C – O stretching at 1250 cm^{-1} whereas C – O stretching in esters occurs as two strong bands in the region 1050 – 1300 cm^{-1}.

(ii) Amides show N – H stretching in 3050 – 3540 cm^{-1} and also a strong absorption band due to N – H bending in the region 1645 cm^{-1}. Acids and esters do not show these bands.

Example 7: Arrange the following non-conjugated carbonyl compounds in order of increasing carbonyl stretching frequency: aldehydes, esters, amides acyclic ketones, acid chlorides.

Solution: The carbonyl stretching frequency depends on the inductive and resonance effect. The order is

Amides < Acyclic ketones < Aldehydes < Esters < Acid chlorides.

Example 8: An organic compound (*A*) with molecular formula C_3HgN shows the following peaks in the *IR* spectrum:

1. 3012 cm^{-1} (m), 2.3423 cm^{-1} (s), 3.3236 cm^{-1} (m) and 1615 cm^{-1} (m)

When the compound *A* is treated with nitrous acid, we get a compound *B* which shows a strong peak at 3430 cm^{-1}. Which are *A* and *B* and explain the reaction involved.

Solution: The two bands at 3423 cm^{-1} and 3236 cm^{-1} and due to unsymmetrical and symmetrical N – H stretching. Clearly, the compound contains $-NH_2$ group. The band at 3012 cm^{-1} is due to C – H stretching. The band at 1615 cm^{-1} is due to M – H bonding. Probably structure of (*A*) is consistent with the given molecular formula and data is CH_3 CH_2 CH_2 NH_2.

When it is treated with nitrous acid, $-NH_2$ is converted into OH group which forms a strong peak at 3430 cm^{-1}. The reaction involved is written as

$$\underset{n\text{-Propyl amine}}{CH_3CH_2\,CH_2NH_2} + HONO \longrightarrow \underset{n\text{-Propyl alcohol}}{CH_3CH_2CH_2OH} + N_2 + H_2O$$

Probably structure of *B* $\longrightarrow$ $CH_3CH_2CH_2OH$

Example 9: Calculate the number of vibrational degree of freedom in (a) CO_2 (b) C_6H_6.

Solution:

(a) CO_2 is a linear molecule, the number of vibrational degree of freedom can be calculated as follows

Number of atoms (n) = 3

Total degree of freedom = $3n = 3 \times 3 = 9$

Translational = 3
Rotational = 2
Vibrational degree of freedom = 9 – 3 – 2 = 4

It can be calculated using the following formula for linear molecule.

$$3n - 5 = 3 \times 3 - 5 = 4$$

(b) C_6H_6 is a nonlinear molecule.
Therefore, the number of vibrational degree of freedom is $3n - 6$

$$= 3 \times 12 - 6 = 30$$

Example 10: An carboxylic acid is formed from an aldehyde on replacing hydrogen atom by OH group. OH has strong $-I$ effect and as a result C = O stretching absorption for acids should occur at higher wave number as compared to aldehydes. But actually it occurs at lower wave number. Explain.

Solution: The C = O stretching absorption is lowered in acids because the lone pair on oxygen of OH group forms internal conjugation with C = O group and hence the absorption wave number decreases.

5.23 COMPARISON BETWEEN UV-VISIBLE AND IR SPECTROSCOPY

UV-visible spectroscopy	IR spectroscopy
1. The spectra are usually expressed in terms of wavelength in nm units	1. The spectra are usually expressed in terms of frequencies in wave number (ν) units
2. UV absorption spectrum exhibits broad absorption bands	2. *IR* spectrum exhibits sharp peaks
3. The region of UV-visible spectroscopy extends from 10 – 800 nm	3. The region of IR spectroscopy extends from 800 – 20,000 nm
4. UV-visible radiations possess more energy, they are strong and so can cause electronic excitation	4. *IR* radiations are weak and so cannot produce electronic excitation. They cause bending and stretching vibrations of organic bonds
5. Unsaturated and especially conjugated groups only absorb in this region	5. All organic bonds and groups absorb in this region
6. Gives an idea about the presence of carbonyl group and a conjugated system	6. Gives an idea about the nature of the functional group in the molecule. It helps in identifying the complete structure of the compound

5.24 ROTATIONAL AND VIBRATIONAL SPECTRUM OF A DIATOMIC MOLECULE

Free atoms do not rotate or vibrate. For an oscillatory or a rotational motion of a pendulum, one end has to be tied or fixed to some point. In molecules such a fixed point is the centre of mass. The atoms in a molecule are held together by chemical bonds. The rotational and vibrational energies are usually much smaller than the energies required to break chemical bonds. The rotational energies correspond to the microwave region of electromagnetic radiation (3×10^{10} to 3×10^{12} Hz) energy range around 10 to 100 J/mole) and the vibrational energies are in the infrared region (3×10^{12} to 3×10^{14} Hz; energy range around 10 kJ/mole) of the electromagnetic radiation. For rigid rotors (no vibration during rotation) and harmonic oscillators (wherein there are equal displacements of atoms on either side of the centre of mass) there are simple formula characterizing the molecular energy levels. In real life molecules rotate and vibrate simultaneously and high speed rotations affect vibrations and vice versa.

5.24.1 Rotational Spectra of Diatomic

The two independent rotations of this molecule are with respect to the two axes which pass through *c* and are perpendicular to the bond length, *r*. The rotation with respect to the bond axis is possible only for classical objects with large masses. For quantum objects, a rotation with respect to the molecular axis does not correspond to any change in the molecule as the new configuration is indistinguishable from the old one (Fig. 5.45).

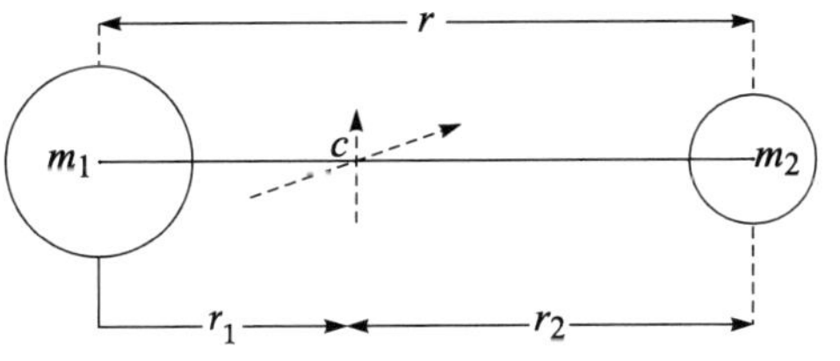

Fig. 5.45 *A rigid diatomic with masses m_1 and m_2 joined by a thin rod of length $r = r_1 + r_2$. The centre of mass is a-c*

The centre of mass is defined by equating the moment on both segments of the molecular axis.

$$m_1 r_1 = m_2 r_2 \qquad \ldots(1)$$

The moment of inertia is defined by

$$I = m_1 r_1^2 + m_2 r_2^2 \qquad \ldots(2)$$

$$= m_1 r_2 r_1 + m_1 r_1 r_2$$

$$r_1 r_2 (m_1 + m_2) \qquad \ldots(3)$$

Since

$$m_1 r_1 = m_2 r_2 = m_2(r - r_1)$$
$$= (m_1 + m_2) r_1 = m_2 r$$

Therefore

$$\left.\begin{aligned} r_1 &= \frac{m_2 r}{m_1 + m_2} \\ r_2 &= \frac{m_1 r}{m_1 + m_2} \end{aligned}\right\} \quad \ldots(4)$$

Substituting the Eq. (4) in Eq. (3), we get

$$\left.\begin{aligned} I &= \frac{m_1 m_2 r^2}{(m_1 + m_2)} = \mu r^2 \\ \therefore \quad \mu &= \frac{m_1 m_2}{m_1 + m_2} \end{aligned}\right| \quad \ldots(5)$$

where μ, the reduced mass is given by

$$\frac{1}{\mu} = \frac{1}{m_1} + \frac{1}{m_2} \quad \ldots(6)$$

The rotation of a diatomic is equivalent to a rotation of a mass at μ distance of r from the origin c. The kinetic energy of this rotation motion,

$$k \cdot \varepsilon = L^2/2I$$

where $L \rightarrow$ Angular momentum
$w \rightarrow$ angular (rotational) velocity in radians/sec

The operator for L^2 is the same as the operator L^2 for the angular momentum of hydrogen atom and the solutions of the operator equations, $L^2 \gamma l_m = l(l+1)\gamma l_m$ where γl_m are the spherical harmonics.

The quantized rotational energy levels for this diatomic are

$$E_j = \frac{h^2}{8\pi^2 I} J(J+1) \quad \ldots(7)$$

The energy differences between two rotational levels is usually expressed in cm^{-1}. The wave number corresponding to a given $\Delta\varepsilon$ is given by

$$v = \Delta E/hc \quad \ldots(8)$$

The energy levels in cm^{-1} are therefore,

$$E_j = BJ(J+1) \quad \text{where} \quad B = \frac{h^2}{8\pi^2 Ic} \quad \ldots(9)$$

The rotational energy levels (Fig. 5.46) of a diatomic molecule are shown in Fig. 5.46.

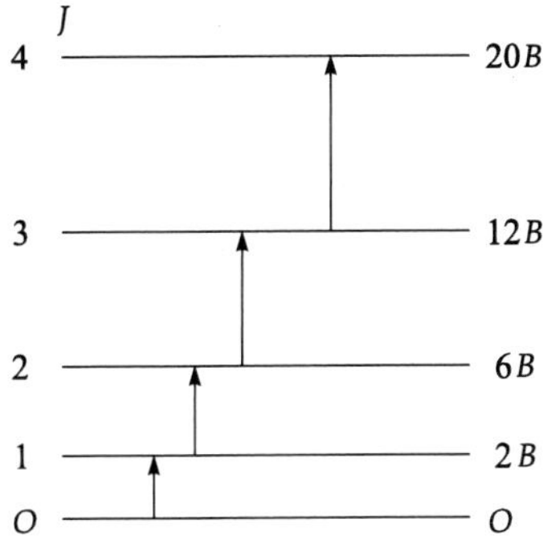

Fig. 5.46 *Rotational energy levels of a rigid diatomic molecule and the allowed transitions*

The selection rule for a rotational transition is

$$\Delta J = \pm 1 \qquad \ldots(10)$$

In addition to this requirement, the molecule has to possess a dipole moment. As a dipolar molecule rotates, the rotating dipole constitutes the transition dipole operator, μ. Molecules such as HCl and CO will show rotational spectra while H_2, Cl_2 and CO_2 will not. The rotational spectrum is shown in Fig. 5.47.

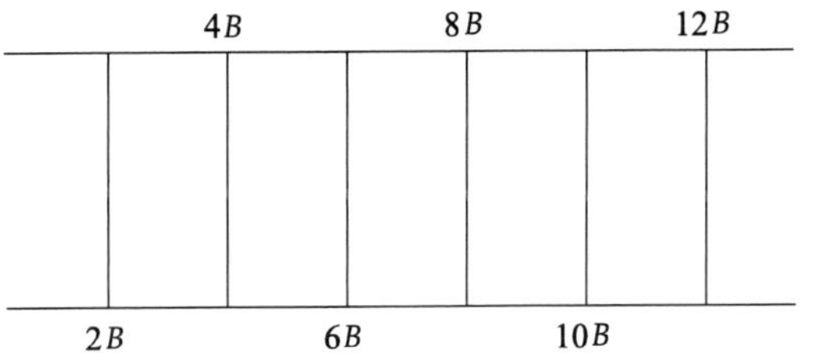

Fig. 5.47 *Rotational spectrum of a rigid diatomic values of B are in cm^{-1}. Typical values of B in cm^{-1} are 1.92118 (CO), 10.593 (HCl), 20.956 (HF), $^1H_2$ (60.864), $^2H_2$ (30.442), 1.9987 (N_2).*

From the values of B obtained from the rotational spectra, moments of inertia of molecules, I, can be calculated. From the value of I, bond length can be deduced.

Example: Calculate the value of I and r of CO

$$B = 1.92118 \text{ cm}^{-1}$$

$$I = h/8\pi^2 BC$$

$$= \frac{6.626 \times 10^{-34}}{8 \times 3.1415^2 \times 1.92118 \times 3 \times 10^{10}}$$

$$= 1.45579 \times 10^{-46} \text{ kg m}^2$$

Since the value of B is in cm^{-1}, the velocity of light C is taken in cm/s. $I = \mu r^2$.

The atomic mass of C ≡ 12.000 amu

$$O = 15.9994 \text{ amu}$$

$$1 \text{ amu} = 1.6604 \times 10^{-27} \text{ kg}$$

The reduced mass of CO can be calculated to be

$$1.13836 \times 10^{-27} \text{ kg}$$

Therefore, $r^2 = I/\mu$

$$= \frac{1.45579 \times 10^{-46}}{1.13826 \times 10^{-27}} \text{ m}^2$$

$$r = 1.131 \text{ Å}$$

The rotational levels are degenerate. Just as there are three p-orbitals for $l = 1$, for $J = 1$, there are 3 degenerate rotational states. The degeneracy for a given value of J is $2J + 1$.

The Boltzmann factor gets modified due to this degeneracy as follows

$$N_j/N_j = [(2J + 1)/(2J + 1)]e^{-\Delta E/kBT} \quad \ldots(11)$$

The implication of this is that the rotational population of the $J = 1$ level is often more than the population of the $J = 0$ levels since their degeneracies are 3 and 1, respectively, when molecules rotate with great speed, they cannot be treated as rigid any more. There are distortions due to centrifugal and other forces. The modification of rotational energies by considering the centrifugal distortion alone is

$$E_j(\text{in cm}^{-1}) = BJ(J + 1) - DJ^2(J + 1)^2 \quad \ldots(12)$$

where the centrifugal distortion constant D is given by

$$D = h^3/32\pi^4 I^2 r^2 kc)\,\text{cm}^{-1} \quad \ldots(13)$$

5.25 ROTATIONAL SPECTRA OF POLYATOMICS

Linear molecules have spectra similar to diatomics. In diatomics as well as linear triatomics $I_A = I_B$; $I_C = 0$. I_A, I_B and I_C are the three moments of inertia of molecules along three independent axes of rotation. Just as any translation can be decomposed into three independent components along three axes such as X, Y and Z any rotation can be decomposed into rotations along three axes A, B and C. The way to choose these axes is to have the simplest value of I_A, I_B and I_C. Since triatomics are heavier than the constituent diatomics, their moments of inertia are larger and the values of rotational constants (B) are smaller in the range of 1 cm^{-1}. The value of I_A or I_B determined from the B value gives the total length of the triatomic. To determine the two bond lengths in the linear triatomic we need to determine the moment of inertia (I_A) of an isotope of the triatomic. From two values of I_A and $I_{A'}$, we can determine the two bond lengths.

The rotational spectra of asymmetric molecules for whom $I_A \neq I_B \neq I_C$ can be quite complicated. For symmetric tops, two moments of inertia are equal

i.e., $$I_A = I_B \neq I_C \quad I_C \neq O \quad \ldots(14)$$

For example, CH_3Cl. The main symmetry axis is the C – Cl axis. We need two quantum numbers to describe the rotational motion with respect to I_A and I_C, respectively. Let J represent the total angular momentum of the molecule and $K \rightarrow$ angular momentum with respect to the C – Cl axis of the symmetric top. J takes on integer values and K cannot be greater than J. The $(2J + 1)$ degeneracy is expressed through the $2J + 1$ values that K can take

$$K = J, J - 1, \ldots 0 - (J - 1), -J \qquad \ldots(15)$$

The rotational energies of a symmetric top are given by

$$E_{JK}/hc = BJ(J + 1) + (A - B)k^2, \text{cm}^{-1} \qquad \ldots(16)$$

The moments of intertia are related to B and A as

$$B = \frac{h}{(8\pi^2 I_c C)} \quad \text{and} \quad A = \frac{h}{(8\pi^2 I_A C)} \qquad \ldots(17)$$

As the energy depends on K^2, energies for states with $+K$ and $-K$ are doubly degenerate. Thus, there will be $J + 1$ levels and $(2J + 1)$ states for each value of J.

The selection rules for symmetric top are

$$\Delta J = \pm 1 \quad \text{and} \quad \Delta K = 0 \qquad \ldots(18)$$

It can be easily shown that

$$E_{J+1K} - E_J, k)/hc = 2BJ(J + 1) \qquad \ldots(19)$$

This implies that the spectrum is independent of the value of K. The physical meaning is as follows: K refers to the rotation about the symmetry axis such as the C – Cl axis. A rotation about this axis does not change the dipole moment. Rotation along the axes A and B changes the dipole moment and thus induces the transition.

By using rotational or microwave spectroscopy, very accurate values of bond lengths can be obtained. For example, in HCN, the C – H length is 0.106317 ± 0.000005 nm and the CN bond length is 0.115535 ± 0.000006 nm. The principle of the microwave oven involves heating the molecules of water through high speed rotations induced by microwaves. The glass container containing water however remains cold since it does not contain rotating dipoles.

5.26 VIBRATIONS AND ROTATIONAL SPECTRA OF A DIATOMIC MOLECULE

The quantized vibration energies (E_v) of a harmonic oscillator are

$$\varepsilon_v = (u + 1/2)h\nu \qquad \ldots(20)$$

$$\nu = 0, 1, 2, \ldots$$

The vibrational frequency (ν) is related to the force constant (k) through

$$V = \frac{1}{2\pi}\sqrt{\frac{k}{\mu}}, \text{Hz} \qquad \ldots(21)$$

The vibrational motion occurs under the action of a binding potential energy. The potential energy $(\rho\varepsilon)$ curve for a harmonic oscillator is shown Fig. 5.48.

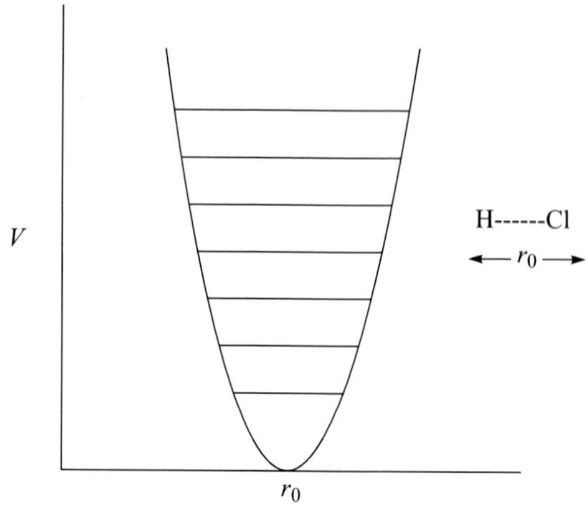

Fig. 5.48 *The potential energy of a harmonic oscillator, $V = K(r - r_0)^2$. The force constant K in N/m for a few molecules are, CO (1902), HF (966), HCl (516), HI (314)*

On either side of the equilibrium bond length r_0, the $\rho\varepsilon$ rises as a symmetric quadratic function (a parabola). The vibrational wave functions can be obtained by solving the Schrödinger's equation. The Hamiltonian operator (for energy) now consists of a kinetic energy term and a potential energy term V as shown in Fig. 5.48 and the solutions for energy, ε_v have already been given in Eq. (20). The selection rules for the anharmonic oscillator are

$$\Delta V = \pm 1 \qquad \text{...(22)}$$

We will see several equally spaced lines (spacing $h\nu$) corresponding to the transitions $0 \rightarrow 1 \rightarrow 2, 2 \rightarrow 3$ and so on. The first transition will be the most intense as the state with $V = 0$ is the most populated. In actual diatomics, the potential is anharmonic. A good description of anharmonic oscillator is given by the Morse function.

$$\text{P.E.} = D_{eq}\,[1 - \exp\{a(r_0 - r)\}]^2 \qquad \text{...(23)}$$

In Eq. (22), D_{eq} is the depth of the PE curve and r_0 is the bond length. A plot of the Morse curve (Fig. 5.49) and the energy levels for the Morse potential are given in Fig. 5.49. The formula for the energy levels of this anharmonic oscillator is

$$\varepsilon_V/hc = e_V = (V + 1/2)\,V - (V + 1/2)^2 V\,X_e,\ \text{cm}^{-1} \qquad \text{...(24)}$$

Here X_e is called the anharmonicity constant whose value is near 0.01. It can be easily deduced from the above formula that the vibrational energy levels for large U start bunching together.

Often, one observes a combined vibrational rotation spectrum. A combined set of vibrational and rotational energy levels of a diatomic is given by

$$\varepsilon_{\text{total}} = BJ(J + 1) + (V + 1/2)\,V - X_e(V + 1/2)^2\,V\ \text{cm}^{-1} \qquad \text{...(25)}$$

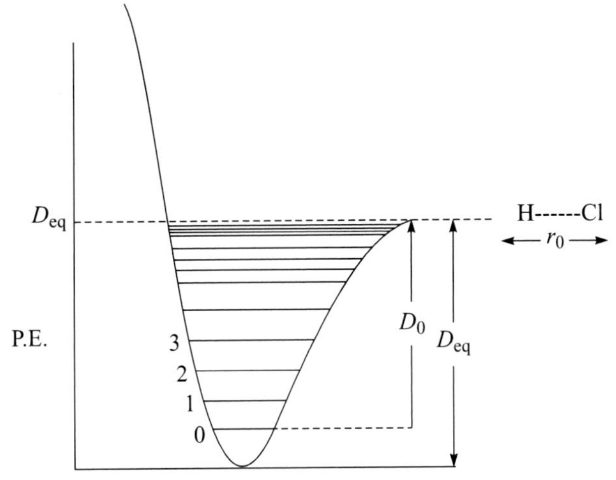

Fig. 5.49 *The Morse potential and the energy levels therein. Note the difference between the dissociation energy D_o and the depth D_{eq}. All molecules have a minimum of the zero point energy of $h_v/2$ corresponding to the $V = 0$ state. This is a consequence of the uncertainity principle.*

The energy level diagram and the spectrum corresponding to the diagram is shown in Fig. 5.50.

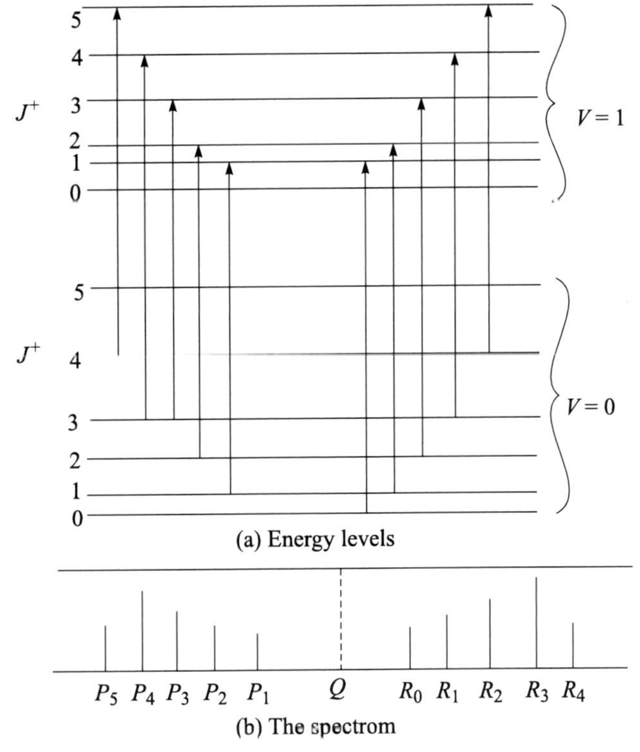

Fig. 5.50 *The vibrational rotational spectrum*

The selection rules are $\Delta V = \pm 1, \pm 2 \ldots$

$\Delta J = \pm 1$. $\Delta J = -1$ corresponds to the R branch on the right at higher frequencies and $\Delta J = J'' - J' = -1$ corresponds to the P branch on the left. The dashed line Q for which $\Delta J = 0$ is not seen. The difference between adjacent R_0 and P_1 is $4B$ and the difference between adjacent R lines and adjacent P lines is $2B$.

5.27 VIBRATIONAL SPECTRA OF POLYATOMICS

An atom moving in three dimensions has three degrees of freedom corresponding to the freedom in movement in say the X, Y and Z directions. A collection of N unbound atoms will have $3N$ degrees of freedom. If the N atoms are bound through the formation of a molecule, the $3N$ degrees of freedom get redistributed into translational, rotational, and vibrational modes. Since the molecule can be translated as a unit, there are three translational modes (degrees of freedom). Similarly, there are three rotational modes with respect to three independent axes of rotation. The remaining $3N - 6$ are the vibrational modes. For a linear molecule, since there are only two rotational modes with respect to the two axes perpendicular to the molecular axis, there are $3N - 5$ vibrational modes.

If the potential energy functions for all the motions can be assumed to be harmonic then the $3N - 6$ modes can be categorized into $3N - 6$ normal modes. For example, H_2O.

There are three atoms and $3N - 6 = 3$ normal modes. In terms of the potential energy functions for vibration, there are three functions: One each corresponding to each O – H bond and one each corresponding to each H – O – H bond bending. In terms of the individual bond vibrations, the vibrational motion can appear quite complex.

The total potential energy P.E.

$$\text{P.E.} = 1/2K(r_1 - r_{10})^2 + 1/2k(r_2 - r_{20})^2 + 1/2K(\theta - \theta)^2 \qquad \ldots(26)$$

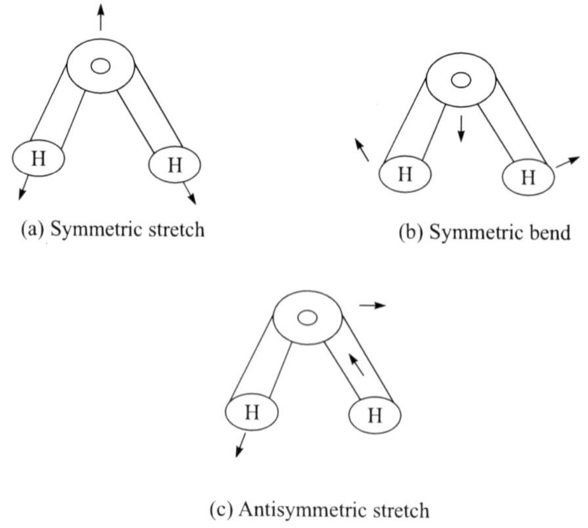

(a) Symmetric stretch

(b) Symmetric bend

(c) Antisymmetric stretch

Fig. 5.51 *Normal modes of vibrations of water*

Here r_{10} and r_{20} are the equilibrium bond lengths of the two O – H bonds and θ_0 is the equilibrium bond angel. A normal mode of vibration is defined as a vibration in which all atoms oscillate with the same frequency and pass through their equilibrium positions at the same time. The centre of mass is unchanged during a normal mode. The three normal modes of vibration of water are shown in Fig. 5.51.

The three normal modes of vibrations of water are the symmetric stretch (V_1 = 3651.7 cm^{-1}) the antisymmetric stretch (V_2 = 3755.8 cm^{-1}) and the symmetric bend (V_3 = 1595.0 cm^{-1}). Bending requires less energy and thus its frequency is lower. The asymmetric stretch requires greater reorganization than the symmetric stretch and hence a larger frequency. Molecular CO_2 is a linear triatomic and has $3N - 5 = 4$ normal modes of vibration. The symmetric stretch (V_1 = 1330 cm^{-1}) asymmetric stretch (V_2 = 2349.3 cm^{-1}) and bending (V_3 = 66.3 cm^{-1}) are shown in Fig. 5.52. The bending mode is doubly degenerate, owing to the two independent bending modes in two perpendicular planes containing the molecular axis.

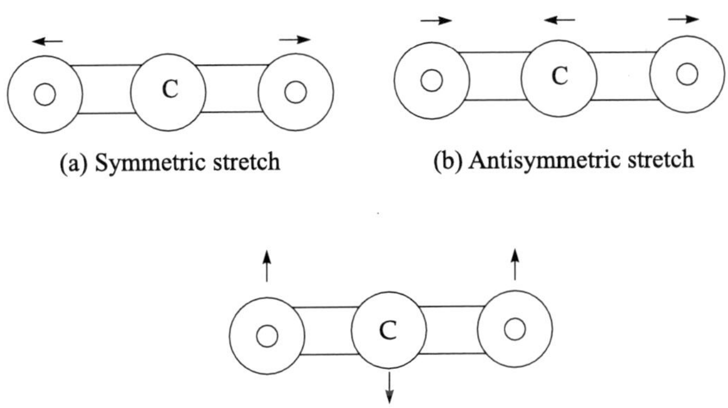

Fig. 5.52 *Normal modes of vibrations of CO_2*

Different molecules can be easily identified by their normal mode frequencies. In addition to these modes, overtones ($2V_1$, $3V_2$, etc.) Combination bonds ($V_1 + V_2$, $2V_1 + V_2$, V_3 ...) and difference bands ($V_1 - V_2$, $V_1 + V_2 - V_3$) can be observed. Since a large number of rotational and vibrational levels are closely spaced they provide a rich base for setting up lasers when the upper levels are populated. As in the case of diatomics, rotational lines are richly dispersed in vibrational spectra of polyatomics. The concept of normal modes can be extended to solids and liquids too. Since in a solid there are a very large number of atoms (of the order of Avogadro number), there are $3N - 6$ normal modes. These are characterized as phonons, which correspond to collective motions of atoms in a solid.

5.28 NMR SPECTROSCOPY

This spectroscopic method deals with the interaction between an oscillating magnetic field of electromagnetic radiation and the magnetic energy of some nuclei having non-zero spin. The transitions occur between magnetic sublevels of the nuclei of a molecule.

The magnetic energy levels are produced by keeping the nuclei in a magnetic field and the energy required for transition is available in radio frequency region (4 to 900 MHz).

5.28.1 Spin of Nuclei and Angular Momentum

Magnetic field is also associated a with moving electrical charges. The nuclear spin may vary from one nucleus to another. The spin of a nucleus is represented by *I* called the spin quantum number. Spin quantum is associated with the angular momentum by the following relation

$$I = \sqrt{I(I+1)}\frac{h}{2\pi}$$

$h \rightarrow$ Planck's constant

Table 5.19 Rules for nuclei showing NMR spectroscopy

Angular momentum (I)	Atomic mass	Atomic number	Examples
Half integer	Odd	Odd or even	${}^{1}_{1}H(1/2)$ ${}^{17}_{8}O(5/2)$
Integer	Even	Odd	${}^{2}_{1}H(1)$ ${}^{14}_{7}N(1)$
Zero	Even	Even	${}^{12}_{6}C(0)$ ${}^{16}_{8}O(0)$

5.28.2 Principle

The nuclei carry a charge, this charge spins on the nuclear axis and this circulation of nuclear charge generates a magnetic dipole along the axis.

The rotation of atomic nuclei along its axis (nuclear spin) behaves as tiny bar magnets and therefore a dipole moment is created and the values of angular momentum, $I = 0, \frac{1}{2}, 1, \frac{3}{2}, 2$ ......

$$I = 0 \text{ denotes no spin}$$

$$I = 1/2 \text{ denotes 1 spin (proton spin)}$$

The intrinsic magnitude of the generated dipole is expressed in terms of nuclear magnetic moment.

The NMR spectra depends upon different types of proton environment. The different type of proton environment is directly proportional to the number of NMR signal.

For example, ^{1}H NMR spectrum of toluene ($C_6H_5CH_3$) gives two NMR signals because toluene has two groups of hydrogen atom.

1. The methyl hydrogens
2. The ring hydrogen

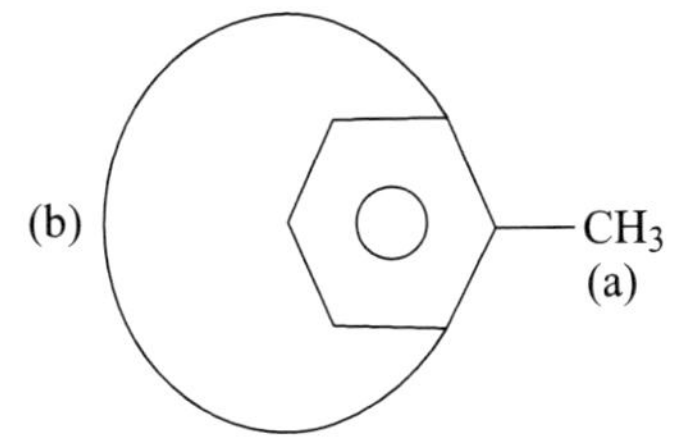

Fig. 5.53

5.28.3 Study of NMR Spectra

NMR spectra may be studied under the following headings.

5.28.3.1 Number of Signals

The number of signals in the NMR spectrum tells the number of different sets of equivalent protons in a molecule. A set of protons with the same chemical environment are called equivalent protons. All equivalent protons form a type or a kind of protons and the NMR spectrum consists of as many signals as there are kinds of protons (*a*, *b*, *c*, *d* protons in the following listed examples).

For example:

(i) (a) CH_3–(b) CH_2–Cl → 2 NMR signal

(ii)

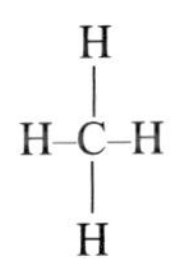

→ 1 NMR signal

(iii) (a) CH_3–(b) OH → 2 NMR signal

(iv)

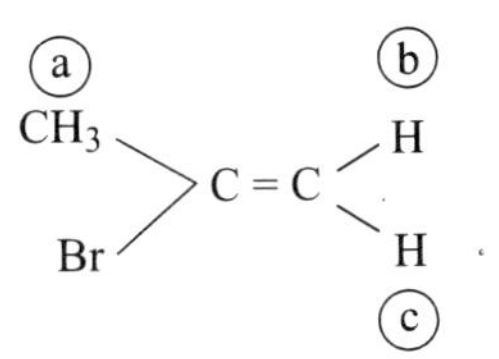

→ 3 NMR signal

(v)

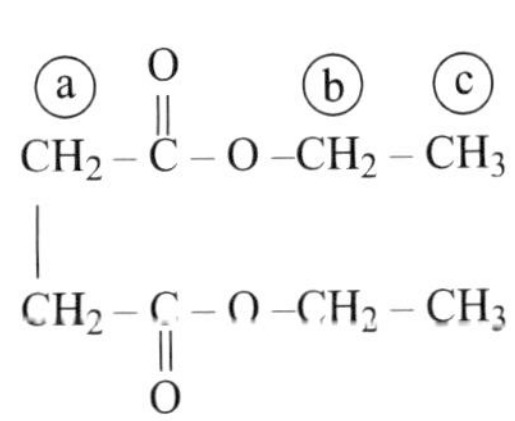

→ 3 NMR signal

(vi) 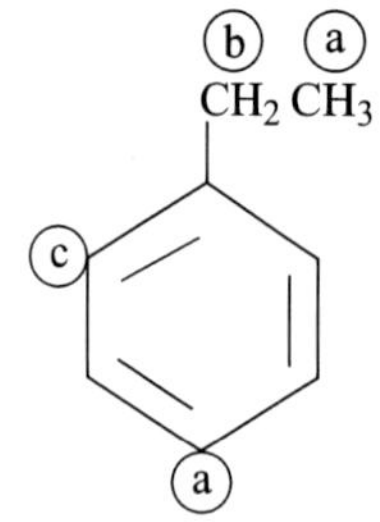 → 3 NMR signal

(vii) 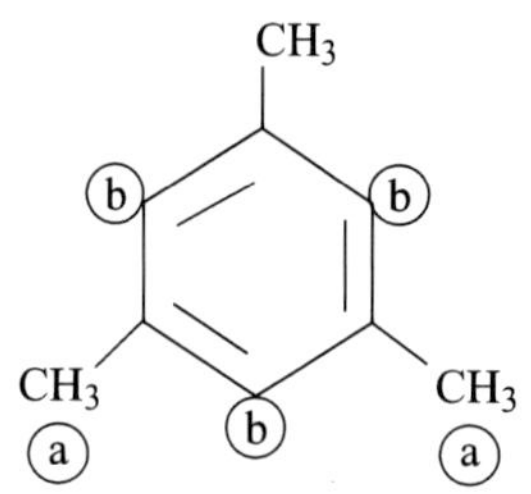 → 2 NMR signal

(viii) 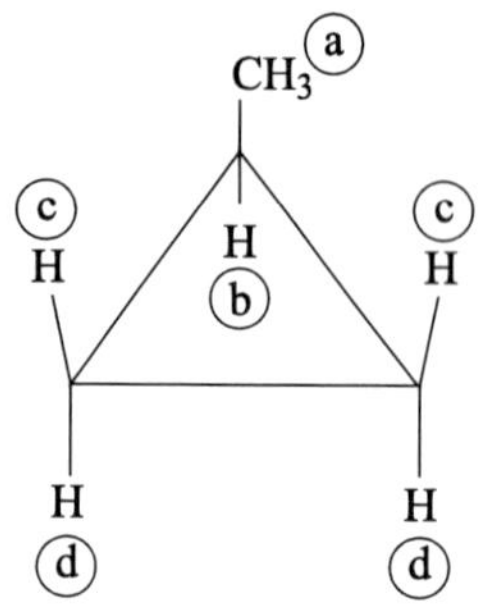 → 4 NMR signal

(ix) $\underset{(d)}{CH_3}-\underset{(c)}{CH_2}-O-\underset{(b)}{CH_2}-\overset{O}{\overset{||}{C}}-\underset{(a)}{OH}$ → 4 NMR signal

(x) 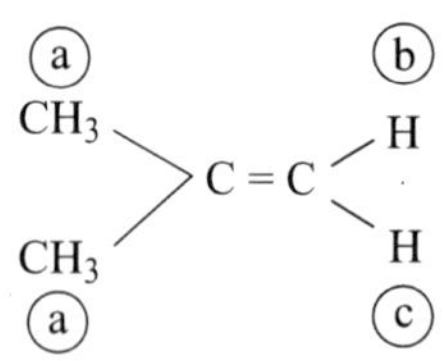 → 2 NMR signal

5.28.4 Position of Signal (Chemical Shift)

The number of signals in an NMR spectrum tells that how many kinds of protons a molecule contains. The position of the signal helps to tell us if they are aromatic, aliphatic, benzylic, acetylenic, etc. These different kinds of protons have different electronic environment. When a molecule is placed in a magnetic field – as it is when one determines an NMR spectrum its electrons are caused to circulate and in circulating, they generate another magnetic field, i.e., induced magnetic fields.

Shielded: Circulation of electrons about the proton itself generates a field aligned in such a way that at the proton, it opposes the applied field. The field felt by the proton is thus diminished and the proton is said to be shielded.

Deshielded: Circulation of electrons – specifically π electrons – about nearby generates a field that can either oppose or reinforce the applied field at the proton, depending on the proton's location. If the induced field opposes the applied field, the proton is shielded as before. If the induced field reinforces the applied field, then the field felt by the proton is augmented and the proton is said to be deshielded.

Chemical shift: For measuring the chemical shift of various protons, the signal for tetramethyl silane (TMS) is used as an internal standard. TMS protons resonate at exactly 60 MHz when the field strength is 1.4092 T, and we can measure the position of other signals in relation to TMS in parts per million (ppm) at higher frequency than TMS down field from the TMS signal. Thus, the difference from the TMS signal and the signal is measured. The use of TMS has the following advantages:

1. It is chemically inert and has a low boiling point.
2. Its resonance signal is very sharp due to twelve equivalent protons.
3. It is soluble in most of the organic solvents and can be added to the sample solution as an internal standard.

With this standard substance two scales are used for measuring chemical shift.

(a) δ scale: This sets the resonance of TMS arbitrarily at zero and counts the resonance at lower fields as positive.

$$\delta_{\text{sample}} = \frac{V_{\text{sample}} - V_{\text{TMS}}}{V_0} 10^6 \text{ ppm}$$

where $V_0 \rightarrow$ oscillator frequency

Illustration: Suppose a proton signal comes at 100 Hz relative to TMS at an oscillating frequency of 60 MHz.

$$\delta_{\text{sample}} = \frac{100}{60 \times 10^6} \times 10^6 \text{ ppm } = 1.67 \text{ ppm}$$

In this δ scale, a positive of δ means down field absorption by the sample, relative to TMS. Absorption at a lower field corresponds to a higher δ value.

(b) τ scale (tau scale): This scale sets the TMS resonance at 10, so that low field resonances are measured at values lower than 10. Absorption at a lower field corresponds to a lower τ value. It can be readily seen that $\tau = 10 - \delta$.

5.28.5 Splitting of Single-Spin Coupling

Each signal in an NMR spectra represents one set of protons in a molecule. However, in certain molecules, a single peak (singlet) may not be observed for a particular set of protons but instead the signal may split into a group of peaks (multipet).

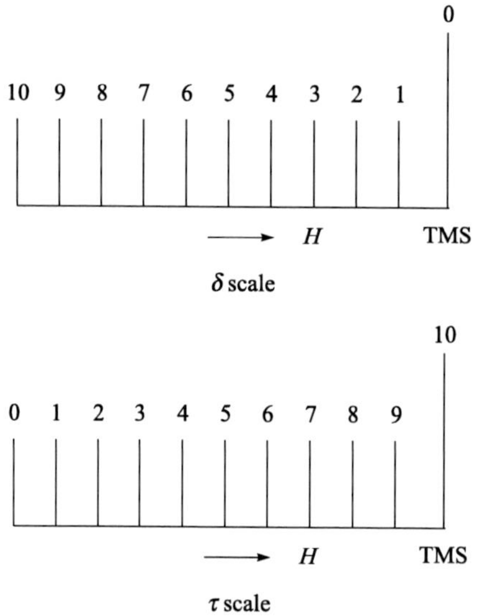

Fig. 5.54

The splitting of a singlet signal into multiplet is due to spin-spin coupling interaction between neighbouring protons. The simplest rule to find the multiplicity of the signal from a group of protons is $(n + 1)$.

where n is the number of neighbouring protons.

For example:

(i) (c) (b) (a)
CH_3-CH_2-OH

It contains three types of protons, namely, three methyl protons, two methylene protons, and one hydroxylic proton. Hence, three NMR signals are obtained under low resolution. However, under high resolution the proton signal from CH_3 splits into a triplet and the signal from CH_2 splits to a quarter (four peaks). This is due to the spin interaction or coupling between the adjacent sets of non-equivalent protons. The signal of the methyl protons splits into a triplet due to three types of orientations of the adjacent methylene protons as shown in Table 5.20. The intensities of the lines in the triplet are in the ratio of 1:2:1 corresponding to the number of like orientations in each case.

Table 5.20

Orientation of protons	Number of like orientations (degeneracy)	Total spin
(i) $\alpha\alpha$ i.e., ↑↑	1	$1/2 + 1/2 = 1$
(ii) $\alpha\beta$, $\beta\alpha$ i.e., ↑↓, ↑↓	2	$1/2 - 1/2 = 1$
(iii) $\beta\beta$ i.e., ↓↓	1	$-1/2 - 1/2 = -1$

α represents clockwise spin

β represents anticlockwise spin

Similarly, we can draw the four possible orientations of the three methyl protons, which are responsible for splitting of the signal into a quarter with the intensity ratio 1:3:3:1.

Orientation of protons	Numbers of like orientations
(i) ↑↑↑	1
(ii) ↑↓↓↑↓↑↑↑↓	3
(iii) ↓↓↑↓↑↓↑↓↓	3
(iv) ↓↓↓	1

Thus, the relative intensities of the individual lines of a multiplet correspond to the numerical coefficient of the lines in the binomial expression where $n \rightarrow$ number of protons.

For example,

$n = 1\ (1 + x)^1 = 1 + x +$ intensity ratio is 1:1 (doublet)

$n = 2\ (1 + x)^2 = 1 + 2x + x^2$ intensity ratio is 1:2:1 (triplet)

$n = 3\ (1 + x)^3 = 1 + 3x^2 + 3x + x^3$ intensity ratio 1:3:3:1

$n = 4\ (1 + x)^4 = 1 + 4x + 6x^2 + 4x^3 + x^4$ intensity ratio 1:4:6:4:1 (quartet)

$n = 5\ (1 + x)^5 = 1 + 5x + 10x^2 + 10x^3 + 5x^4 + x^5$ intensity ratio 1:5:10:10:5:1 (sextet)

$n = 6\ (1 + x)^6 = 1 + 6x + 15x^2 + 20x^3 + 15x^4 + 6x^5 + x^6$ intensity ratio = 1:6:15:20:15:6:1 (septet)

(ii) 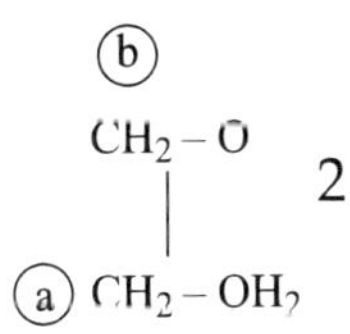

2 NMR signal

Signal (a) corresponds to four protons and will appear as quintet.

Signal (b) splits into a triplet

(iii) 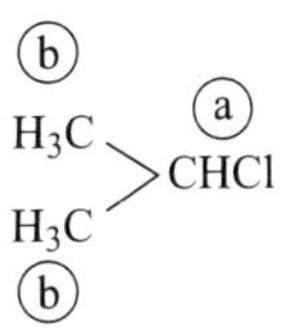

2 NMR signal

The signal given by the proton (a) splits into a septet. The methyl protons (b) given an intense doublet.

5.28.6 Applications of NMR Spectroscopy

Following are the applications of NMR spectroscopy

1. To study the isotopes other than proton
2. Determination of activation energy

3. To study the keto-enol tautomerism
4. NMR spectrum can tell us how many protons are present in a molecule
5. The spectra can easily evaluate the nuclear parameters
6. Identification of organic compound in an environment of protons
7. The knowledge of chemical interest can easily be gained.

5.28.7 Limitations of NMR Spectroscopy

The various limitations of NMR spectroscopy are as follows:

1. In some compounds, two different types of hydrogen atoms resonate at similar resonance frequencies. This results in an overlapping of spectra which makes the interpretation of spectra difficult.
2. While characterising the organic compounds, no information about molecular weight is given. However, the relative number of different protons present is only known.
3. In most of the cases, only liquids can be studied by NMR spectroscopy, though polymers when pre-heated with various solvents frequently become fluids which can be treated as liquids.
4. One of the serious problems with NMR is its lack of sensitivity. The minimum sample size is about 0.1 mL having minimum concentration of about 1%.
   In spite of the above limitations, NMR is one of the most useful analytical tools developed for the elucidation of many organic and inorganic structural problems.

EXAMPLES

Example 1: How many NMR signals are observed in the spectrum of

(i) $CH_3-\overset{\overset{O}{\|}}{C}-CH_3$

(ii) 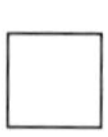

(iii) CH_3OH

(iv) $C_6H_5CH_2OH$

Solution:

(i) $\underset{ⓐ}{CH_3}-\overset{\|}{\underset{O}{C}}-\underset{ⓐ}{CH_3}$ One NMR signal

(ii) 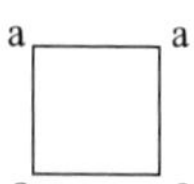 One NMR signal

(iii) $\underset{ⓐ}{CH_3}-\underset{ⓑ}{OH}$ Two NMR signals

(iv) $\underset{a}{C_6H_5}\underset{b}{CH_2}\underset{c}{OH}$ Three NMR signals

Example 2: How many proton signals would you expect in the NMR spectra of the following compounds:

(i) Cyclobutane
(ii) Cyclonhexane
(iii) 1, 2-Dichloroethane
(iv) Methyl ethyl ether
(v) 2-chloropropane

Solution:

(i)
```
(a)    (a)
CH2 – CH2
 |      |
CH2 – CH2
(a)    (a)
```
All the hydrogen atoms are equivalent

(ii) 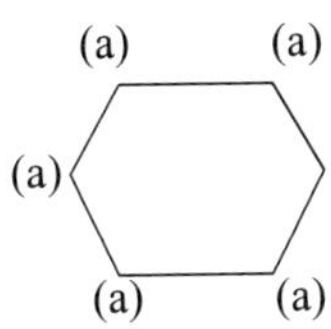 (a) One NMR signal

(iii)
```
H2C – CH2
 |     |
 Cl    Cl
```
One NMR signal

(iv) $\overset{a}{CH_3} – O – \overset{b}{CH_2} – \overset{c}{CH_3}$ Three NMR signals

(v) $\overset{(a)}{CH_3} – \overset{Cl\;(b)}{CH} – \overset{a}{CH_3}$ Two NMR signals

Example 3: A compound has molecular formula $C_{10}H_{14}$. It gives the following NMR data (i) 0.888 (9 H, singlet) (ii) 7.288 (5 H, singlet aromatic protons)

Assign the structural formula to the compound.

Solution:

(i) The nine proton singlet at 0.888 indicates the presence of tertiary butyl group.
(ii) Five-proton singlet at 7.888 confirms the phenyl ring.
Hence, the structural formula of the compound may be

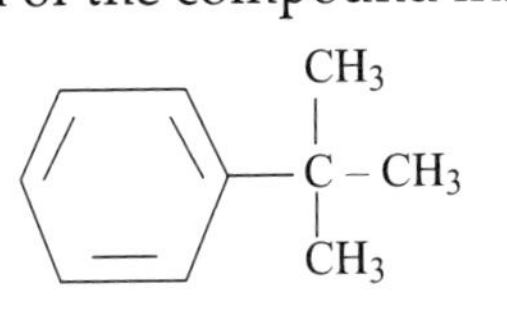

Example 4: The following compounds show only one signal in the NMR spectrum. Write their structural formulas.

(a) C_5H_{12} (b) C_2H_6O (c) $C_2H_4Br_2$ (d) C_4H_6

Solution:

(a) C_5H_{12}: Single signal means all the 12 hydrogens are equivalent. Hence, these can be in four equivalent methyls. The structure is $(CH_3)_4C$.

(b) C_2H_6O: Two equivalent methyl groups account for the single signal. The structure is CH_3-O-CH_3

(c) $C_2H_4Br_2$: Two equivalent methylene groups account for the single signal.

$$Br-CH_2-CH_2-Br$$

(d) C_4H_6: Two equivalent methyl groups, with two carbons having no hydrogen account for the single signal

$$H_3C-C\equiv C-CH_3$$

Example 5: How will you distinguish between the following two compounds on the basis of NMR spectroscopy.

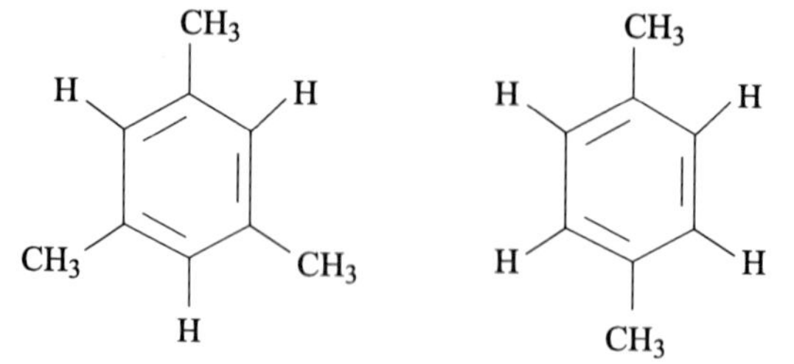

Mesitylene
2 sets of equivalent protons
2 signals

Xylene
2 sets of equivalent protons
2 signals

Solution: Both the compounds would give two signals but the intensities of the peak would be different. In mesitylene the ratio of methyl protons to ring protons is 9:3 whereas as in *P*-xylene it is 6:4.

Example 6: How would you identify ethane, ethene and ethyne by NMR spectroscopy because each of these compounds contains only one type of proton we will get one NMR signal in each case. But these protons are in different environments in different compounds, therefore their relative position is different as given below:

Ethane CH_3-CH_3 $\delta 0.9$ ($\tau 9.1$)
Ethene $CH_2=CH_2$ $\delta 0.5$ ($\tau 7.5$)
Ethyne $CH\equiv CH$ $\delta 2.5$ ($\tau 7.5$)

Example 7: Explain the splitting pattern in the NMR spectrum of the following compounds.

(i) CH_3-CH_3

(ii) CH_3-CH_2Br

(iii) $CH_3-CH_2-CH_2Cl$

Solution:

(i) $CH_3 - CH_3$: Since all six hydrogens are equivalent in ethane, no signal splitting is observed.

(ii) $\overset{a}{CH_3} - \overset{b}{CH_2} - Br$: There are two types of protons, *a* and *b*. The three equivalent *a* protons of the methyl group are under the influence of *b* protons (— CH_2 —). Thus, the NMR band is split in $n + 1 = 2 + 1 = 3$ peaks with relative intensities 1:2:1. The two methylene protons (*b* type) see three neighbouring protons their NMR bond will split into $n + 1 = 2 + 1 = 3 + 1 = 4$ peaks giving rise to a quarter with relative intensity of 1:3:3:1.

(iii) $\overset{a}{CH_3} - \overset{b}{CH_2} - \overset{c}{CH_2} - Cl$: In this molecule *a* type protons ($-CH_3$) will split into $2 + 1 = 3$ peaks giving a triplet with relative intensities of 1:2:1. Similarly, a triplet is obtained from *c* type of proton ($-CH_2-$) with relative intensities of 1:2:1. The *b* type of protons ($-CH_2$) are under the influence of the *a* type and *c* type of protons simultaneously. So the *b* type of protons can interact with $3 + 2 = 5$ protons, so the multiplicity of the peaks is given by $n + 1$ is $5 + 1 = 6$. Thus, the *b* type of proton will exhibit a sextet with relative intensity of 1:5:10:10:5:1.

5.29 NANOMATERIAL

Nanotechnology is an interdisciplinary science involving chemistry, physics, materials science, biology and medicine. In broader sense, nanotechnology is the study of all the phenomena and processes involved in the synthesis, properties and applications of nanostructures and nanomaterial.

5.30 WHAT ARE NANOMATERIALS?

Nano is a Greek word meaning *dwarf* or *something extremely small* and has at least one dimension in the nanometre scale so that it may be defined. Size of a nanomaterial varies from a nanometre to hundreds nanometres. A nanometre (nm) is one millionth of a millimetre (10^{-9} m). One nanometre is approximately the length equivalent to ten hydrogen or five silicon atoms aligned in a line. They are categorised according to their dimensions.

For example, a nanomaterial with three dimensions and less than 100 nm are nano-particles, quantum dots, etc. Nanomaterials with two dimensions and less than 100 nm are nanotubes, fibres and nanowires. Nanomaterials with one dimension and less than 100 nm are thin fibres, layers and surface coatings.

5.31 ADVANCES IN NANOMATERIALS

Nanoparticles occur naturally, for example, volcanic ash, soot from forest fires, are the incidental by-products of combustion processes (e.g., welding, diesel engines) are usually physically and chemically heterogeneous and often termed ultrafine particles. They may be in the form of particles, tubes, rods or fibres. Nanomaterials exist in nature in the form of biological materials or as by-products of human activities. For instance, haemoglobin is 5.5 nanometres in diameter and biological ion channels can be as small as a few tenths of a nanometre. Scientists have already copied the nanostructure of several natural items including the lotus leave to create water repellent surfaces beings used to make stain proof clothing and spider silk which is naturally reinforced with nanoscale crystals. Nanoscaled smoke particles were formed during the use of fire by early humans.

5.32 CLASSIFICATION OF NANOMATERIALS

Nanomaterials have extremely small size which having at least one dimension 100 nm or less. Nanomaterials can be nanoscale in one dimension (for example, surface films), two dimensions (e.g., strands or fibres) or three dimensions (for example, particles). The terms one-dimensional material, two-dimensional material are meaningless. However, these terms are used to express specifically named highly anisotropic properties such as one-dimensional conductivity, one-dimensional connection or one-dimensional disorder. They can exist in single, fused, aggregated forms with spherical tubular and irregular shapes. This classification is based on the restriction imposed by the size of nanomaterials to the flow of charge carriers. Common types of nanomaterials include nanotubes, dedrimers, quantum dots and fullerne. According to Siegel, nanostructural materials are classified as zero-dimensional, one-dimensional, two-dimensional, three-dimensional nanostructures. Nanoparticles, nanoclusters and nanocrystals are called zero-dimensional because the movement of charge carriers is confined in all three directions whereas in case of nanorods, nanowires, nanotubes movement of charge carriers is confined in two directions therefore they are all called one-dimensional nanomaterials. In case of nanofilms, charge carriers are free to move in two directions but confined in only one direction, they are said to be two dimensional.

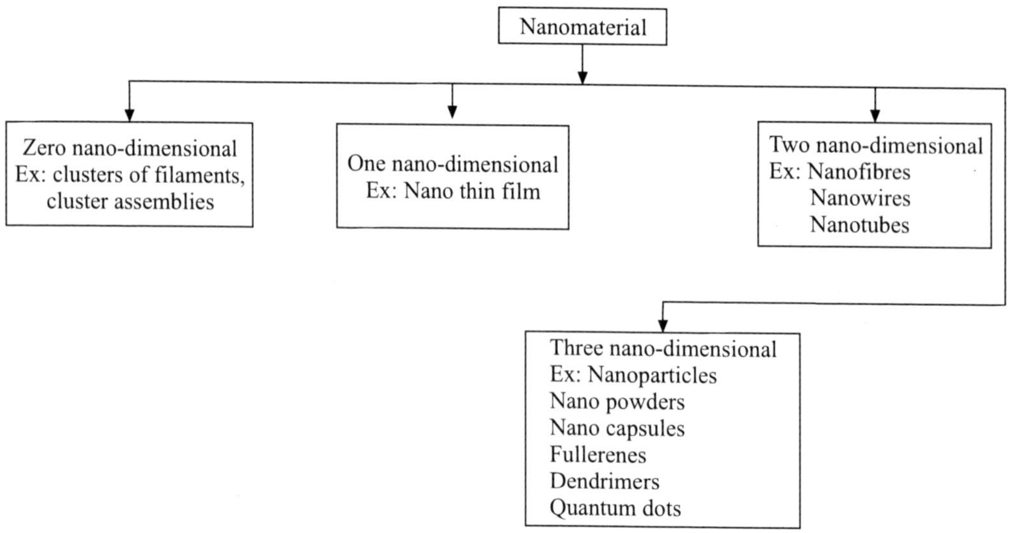

Fig. 5.55 *Schematic representation of the classification of the nanomaterials*

5.33 SIZE DEPENDENT PROPERTIES OF NANOMATERIALS

Materials in the nanometre scale exhibit physical properties distinctively different from the bulk. Properties of any bulk material are independent of its size. It can be altered only by altering their structure and composition.

For example, melting point of gold remains same whether the size of gold is 1 gram or 1 kg. A piece of gold is golden in colour, however, big or small it is. But in nanoscale range, a material with same structure and composition will show different properties when the size of the material is slightly varied. A colloid of gold nanoparticles is no longer "golden" but ruby red in colour. Only the materials in the nano size range exhibit these size dependent properties because a transition from atoms or molecule to bulk form takes place in this size range.

Nanomaterials exhibit several size dependent properties; few of them are given below:

Table 5.21 Characteristic dimensions of some common objects

Object	Size
1. Nanomaterials	1–100 nm
2. Resolution of electron microscope	~0.1 nm
3. Wall thickness of carbon nanotubes	~2–12 nm
4. Water molecule	~0.3 nm
5. Complex organic compounds	2–3 nm
6. Simple gas molecule (N_2, CO, O_2, etc.)	~ 0.3 nm
7. Simple chemical element	0.5–0.8 nm
8. Inorganic compounds	0.5–15 nm
9. Nanowires, nanotubes–diameter length	1–100 nm 102 up to 106 nm
10. Nanoclusters, nanoparticles	1–10 nm
11. Nanocrystals	4–100 nm
12. Diameter of human hair	~ $(5 - 10) \times 10^4$ nm
13. Head of a pin	~10^6 nm
14. Quantum-dot transistor	~300 nm
15. Thickness of a book page	~ 10^5 nm

5.33.1 Optical Properties

The discrete electronic states of nanomaterials allow absorption and emission of light of specific wavelengths. Hence, nanomaterials exhibit unique colours; different from bulk materials.

The optical emission and adsorption occurs when the transition of the electrons occur between these two states. For example, bulk gold appears yellow in colour. Nano sized gold appears red in colour, because the particles are so small that electrons are not free to move about as in bulk gold and this movement is restricted; the particles react differently with light.

Colloidal suspense of gold nanoparticles have a deep red colour which becomes progressively more yellow as the particle size increases.

Example: Gold spheres of 2–5 nm exhibit yellow colour.
Gold spheres of >20 nm exhibit purple colour.

Similarly,

Silver particle of 40 nm exhibits blue colour.
Silver particle of 100 nm exhibits yellow colour.
Prism shaped silver particle exhibits red colour.

5.33.2 Mechanical Properties

Mechanical properties are size dependent as well. Many of the mechanical properties of nanomaterials are different from those of the bulk materials including hardness, elastic modules, fracture toughness, scratch resistance and fatigue strength. The yield strength and hardness of a microstructured (metal and alloys) and the macrostructural metal contain a higher fraction of grain boundary volume. For example, for a grain size of 10 nm, between 14 and 27% of all atoms reside in a region within 0.5–1.0 nm of a grain boundary, therefore, grain boundaries play a significant role in the material properties.

Changes in the grain size result in a high density of incoherent, interfaces or other lattice defects such as dislocations, vacancies, etc.

As the grain size of the solid decreases, the proportion of atoms located at or near grain boundaries relative to those within the interior of a crystalline grain scales as $1/d$. This has important implications for properties in ultra-fine grained materials which will be principally controlled by interfacial properties rather than those of the bulk.

(Crystals contain internal interfacial defects known as grain boundaries where the lattice orientation changes.)

5.33.3 Magnetic Properties

Magnetic nanoparticles are used in a range of applications like imaging bio-processing refrigeration as well as high storage density magnetic memory media. The large surface area to volume ratio results in a substantial proportion of atoms having different magnetic coupling with neighbouring atoms leading to differing magnetic properties. Bulk gold and platinum are non-magnetic but at the nano size they act as magnetic particles. Gold nanoparticles become ferromagnetic when they are capped with the appropriate molecules such as thiol.

Giant magneto resistance is a phenomenon observed in nanoscale multilayers consisting of strong ferromagnetic (Fe, CO, Ni) and a weaker magnetic or non-magnetic buffer (Cr, Cu). It is usually employed in data storage and sensing.

5.33.4 Surface Area

Many physical and chemical properties of a material depend on its surface properties. If a bulk material is subdivided into individual nanomaterials, the total volume remains the same, but the collective surface area is greatly increased. For example, the collective surface area of a cube of 1 m^3 volume (each side of 1 m) is 6 m^2. If this cube is progressively cut into smaller and smaller cubes of 1 m^3, one can get 10^{27} cubes. The collective surface area of 10^{27} cubes is 6000 km^2.

Thus, surface area is enormously increased on moving from bulk to nanoscale. Nanomaterials have a significant proportions of atoms existing at the surface. Properties like catalytic activity, gas adsorption and chemical reactivity depend on the surface area. Therefore, nanomaterials can show specific surface related properties that are not observed in bulk materials.

For example, catalytic activity of gold.

Bulk gold is catalytically inactive. But, gold nanoparticles are catalytically very active for selective redox reactions.

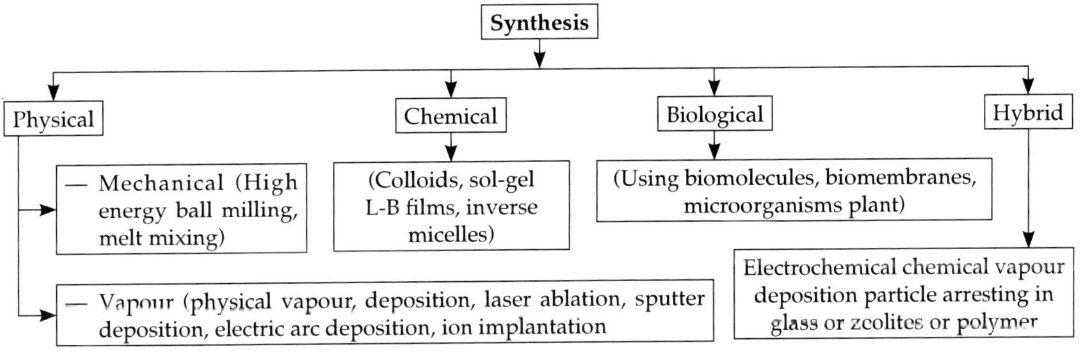

Fig. 5.56 *Different methods for p-synthesis of nanomaterials*

5.33.5 Electrical Properties

The effects of size on electrical conductivity of nanostructures and nanomaterials are complex, since they are based on distinct mechanisms. These mechanisms can be generally grouped into four categories: surface scattering, including grain boundary scattering, quantized conduction including ballistic conduction, coulomb charging and tunnelling and widening and discrete of band gap and change of microstructures. In addition, increased perfection, such as reduced impurity, structural defects and dislocations, would affect the electrical conductivity of nanostructures and nanomaterials.

5.34 SYNTHESIS OF NANOMATERIALS

Many techniques are available to synthesize nanomaterials which can be generally grouped as top-down and bottom-up approaches. These approaches play very important roles in the synthesis of nanomaterials.

Top-down approach: Bigger materials are broken down into small nano-size materials by means of various physical methods. The main physical methods used for this purpose are sputtering pulsed laser deposition, high energy ball milling, and ion-beam implantation.

There are few major drawbacks in this approaches:

1. These methods cause crystallographic damage and introduce surface defects and internal stress in the resulting nanomaterials.
2. These methods require expensive apparatus.
3. Nanomaterials obtained from these methods are generally contaminated.

Bottom-up Approach: Bottom-up approach means small to big; by orderly assembly of growth species like atoms or ions or molecules. These are simple chemical methods and are in use since several years for the synthesis of various chemicals.

For example, synthesis of polymers by connecting individual monomers, the growth of single crystals, deposition of films in electronic industry and production of industrial chemicals all involve bottom-up approaches.

Main advantages of these approaches:

1. These methods provide better control over chemical composition, size, microstructure and aspect ratio of nanomaterials.
2. The methods are simple, versatile and economically viable.
3. Nanostructures with less defects and high purity can be synthesized.

Some of the most widely used bottom-up approaches are given below.

5.34.1 Sol-gel Process

The sol-gel process is a wet chemical solution deposition widely used in the fields of materials science and ceramic engineering. In materials science, the sol-gel process is a method for producing solid materials from small molecules. The method is used for the fabrication of metal oxides especially, the oxides of silicon and titanium. The process involves conversion of monomers into a colloidal solution that acts as the precursor for an integrated network of discrete particles.

The following steps are involved in the synthesis of nanomaterials by sol-gel process.

1. Preparation of stable sol solution.
2. Conversion of sol to gel by a polycondensation or polyesterification.
3. Aging of a gel (syneresis), during which the polycondensation reactions continue until the gel transforms into a solid mass, accompanied by contraction of the gel network and expulsion of solvent from gel can exceed to 7 days and is critical to the prevention of cracks in gels that have been cast.
4. Drying of the gel, when water and other volatile liquids are removed from the gel network. The drying process has itself been broken into four distinct steps:
   (i) The constant rate period
   (ii) The critical point

(iii) The falling rate period
(iv) The second falling rate period.

If isolated by thermal evaporation, the resulting monolith is called an xerogel. If the solvent (such as water) is extracted under supercritical or near supercritical conditions, the product is an aerogel.

5. Dehydration, at temperature as high as 8000°C is used to remove M – OH groups for stabilizing the gel, i.e., to protect it from rehydration.
6. Densification and decomposition of the gels at high temperature ($T > 8000°C$), i.e., to collapse the pores in the gel network and to drive out remaining organic contaminants.

The typical steps that are involved in sol-gel processing are shown in Fig. 5.57.

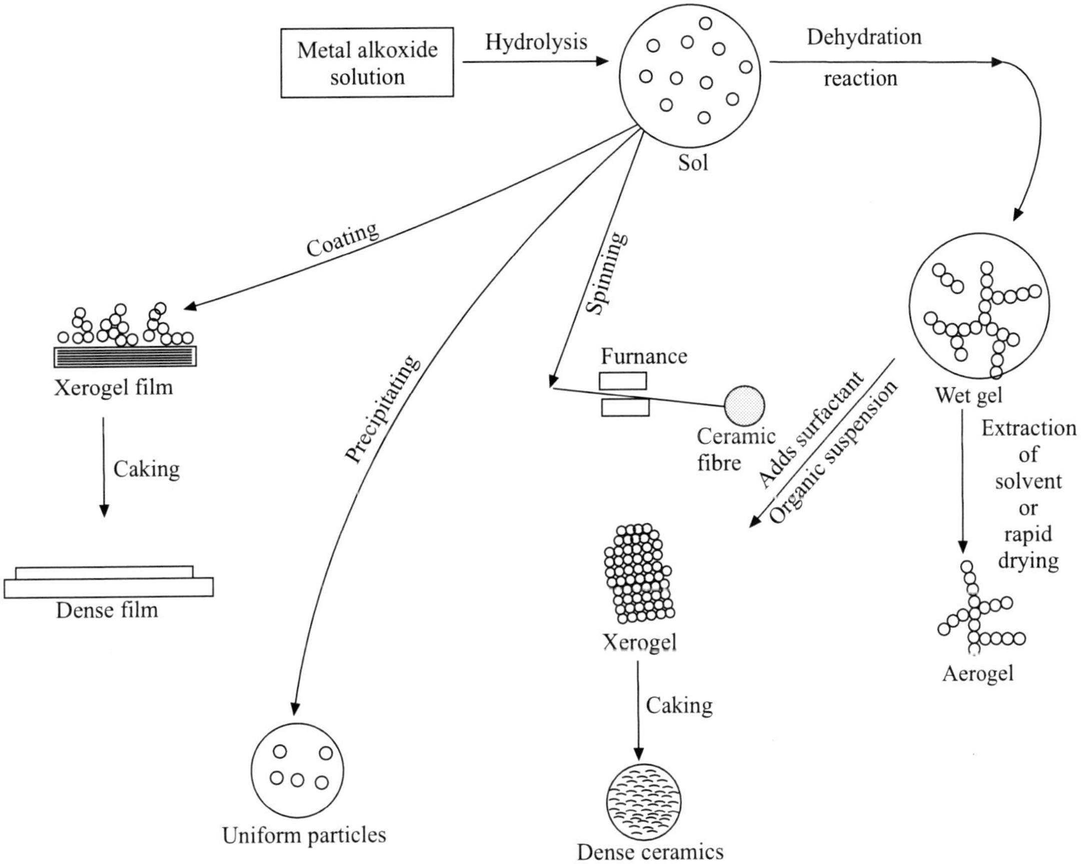

Fig. 5.57 *Sol-gel process of synthesis of nanomaterials*

Working: In sol-gel process, either a metal salt or metal alkoxide is used as percursor or starting reactant to synthesize nanoparticles of a metal oxide. First, a sol is prepared by dispersing precursors in a solvent. (A sol is a colloidal suspension of small solid particles in a solvent.) Sol is further converted into a gel by hydrolysis and condensation of precursors. Hydrolysis and condensation reactions are initiated by addition of an acid or a base catalyst. Hydrolysis and condensation reactions are complex multiple-step processes. The reactions involved in the sol-gel chemistry based on the hydrolysis and condensation of metal alkoxides $M(OR)_z$ can be described as follows:

Hydrolysis condensation: $MOR + H_2O \longrightarrow MOH + ROH$

$$MOH + ROM \longrightarrow M\text{-}O\text{-}M + ROH$$

Particle size, morphology and microstructure of the nanoparticles can be made by controlling the concentration of precursor, rate of hydrolysis and condensation reaction and aging time. Nanoparticles formed by sol-gel process commonly have a size ranging from 1 to 100 nm.

Advantages

1. It is useful for the synthesis of glasses, glass ceramics or ceramic materials at lower temperature.
2. Samples can be prepared at lower temperatures.
3. Nanomaterials of high purity with good homogeneity can be obtained (due to purity of starting materials).

Disadvantages

1. Avoiding the agglomerating of newly formed particles.
2. Controlling the growth of the particles.

5.34.2 Precipitation Method

Precipitation is the creation of a solid in a solution during a chemical reaction when the reaction occurs in a liquid solution, the solid formed is called the precipitant. Precipitation may occur if the concentration of a compound exceeds its solubility (such as when mixing solvents or changing their temperature. This method relies on the precipitation of nano-metre sized particles within a continuous fluid solvent. An inorganic metal salt, such as chloride, nitride is dissolved in water. Metal cations exist in the form of metal hydrate species.

For example: $Al(H_2O)^{3+}$, $Fe(H_2O)_6{}^{3+}$. These hydrates are added with basic solutions such as NaOH or NH_4 OH.

For example: Synthesis of CUO nanoparticles. Copper oxide (CUO) nanoparticles are synthesized by aqueous precipitation method using copper sulphate penta-hydrate as a precursor and NH_4Cl as stabilizing agent. This gives large-scale production of nano-particles.

Method: 55 g of copper sulphate penta-hydrate was mixed with 25 g of hydroxyl ammonium chloride in 125 mL distilled water. The mixture was allowed to cool in a cold water bath with swirling well and solution of 40 g of sodium hydroxide in 750 mL distilled water was added. The precipitate was allowed to settle and the supernatant liquid was poured off. The oxide is transferred to a 250 mL flask and the volume is made up. The content is washed by repeated decantation until the rinsing is chloride-free. The suction filtration is applied and the residue is washed with 95% alcohol and ether. The residue is dried at 200 to 250°C in air oxidized condition to obtain Cu_2O and then the temperature is increased to 300°C which leads to the formation of CuO. For the formation of Cu_2O nano-particles from Cu_2O precipitant, the following reaction mechanism can be formulated and represented as:

$$2CuSO_4 + 2NH_2OH.\ HCl + 6\ NaOH \longrightarrow Cu_2O + N_2 + 2NaCl + 2Na2SO_4 + 7H_2O$$

$$Cu_2O \xrightarrow{\Delta} Cu$$

5.34.3 Inert Gas Condensation Method

The inert gas evaporation – condensation is one of the simplest techniques for the production of metal nanoparticles. This method is commonly used to obtain metal nanomaterials with sizes in the range 2–100 nm.

Method: In this method metal is vaporized by heating at high temperature inside a chamber. Chamber is preciously evacuated to about 10^{-7} torr and backfilled with inert gas to a low pressure. Metal atoms present in the vapours collide with the inert gas molecules and lose their energy due to collisions. Metal atoms, cooled down rapidly, become supersaturated and then nucleate homogeneously producing nanoparticles which are collected at the bottom.

For the formation of nanoparticles, the hot vapours of metal have to be cooled down quickly. Higher the cooling rate, higher will be the supersaturation resulting in the formation of large number of smaller particles. Rate of cooling can be controlled by varying pressure of inert gas. This is only a physical process involving evaporation and condensation and no chemical reaction is involved.

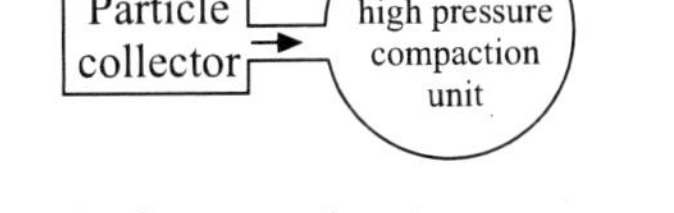

Fig. 5.58 *Schematic diagram of an inert gas condensation method*

5.34.4 Chemical Vapour Condensation

Chemical vapour condensation process was developed in Germany in 1994. It is a well-known process in which a solid is deposited on a heated surface via a chemical reaction from the vapour or gas phase. The reaction is activated by a high temperature above 900°C.

In this method precursor is vaporized and mixed with an inert carrier gas like nitrogen and the mixture is fed into the reactor. Reactor is maintained at a temperature required for the reaction. Precursors undergo reaction, producing a product which is deposited over the substrate. The by-products and leftovers from the reactions are passed on to the gas phase, which are easily removed from the reaction chamber. Reaction is often catalyzed by the catalyst present on the substrate.

In this method proper choice of precursors is very important. Precursor used should be highly pure, and stable at room temperature but with sufficient volatility. Most commonly used precursors are halides, carbonyls, organometallic and hydrides of metals. By this method it is possible to produce almost any metal and non-metallic material as well as compounds such as oxides, carbides, nitrides and many others.

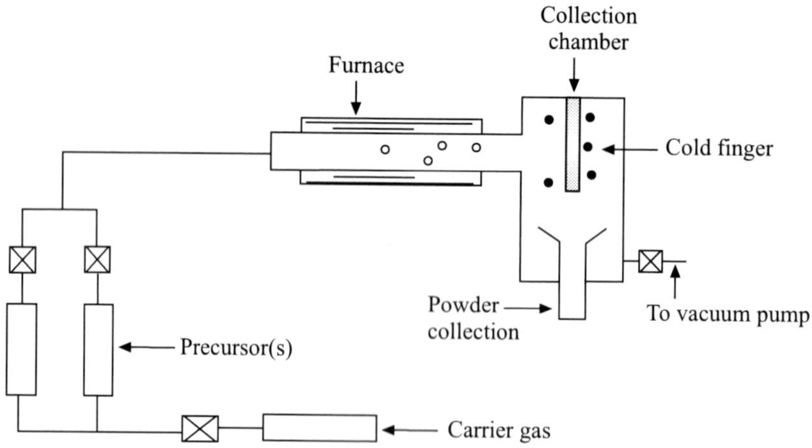

Fig. 5.59 *Schematic diagram of a chemical vapour deposition apparatus*

Advantages

1. Does not require ultrahigh vacuum.
2. With this method, it is possible to deposit nanomaterials of almost any shape and any size.
3. Product is highly pure.

Disadvantage

Precursors used are volatile, which are usually hazardous and extremely toxic. The by-products of these precursors are also toxic.

5.34.5 Hydrothermal Process

Hydrothermal synthesis can be defined as a method of synthesis of single crystals that depends on the solubility of minerals in hot water under high pressure. The advantage of this method over others includes the ability to create crystalline phases which are not stable at the melting point and create materials which have a high vapour pressure near their melting points.

Method: In the process, the decomposition of the precursors in a particular solvent depends on the temperature and pressure which was related to the filling fraction of the solvent within the reaction vessel and was kept constant for all the experiments. The crystal growth is performed in an apparatus consisting of a steel pressure vessel called an autoclave which is a thick-walled steel cylinder hermetically sealed in order to withstand high temperature and pressures for prolonged periods of time, is used for hydrothermal method.

For example: Synthesis of ZnS nanoparticles.

The ultra pure chemical such as $ZnCO_4 \cdot 5H_2O$ powder and $Na_2S \cdot 7H_2O$ powder used as starting materials were dissolved separately in de-ionized water and stirred at room temperature for 30 min. Then they were mixed by dripping the Na_2S solution in drops into $ZnSO_4$ solution while stirring. Following chemical reaction occurs:

$$ZnSO_4 \longleftrightarrow Zn^{2+} + SO_4^{2-}$$

$$Zn^{2+} + Na_2S \longleftrightarrow ZnS + 2Na^+$$

The white gels were charged into the Teflon-lined stainless steel autoclave, which was then filled with the mixed solution up to 50% of its volume. The closed chamber was placed inside a box furnace at a temperature of 220°C for 12 hr and then cooled down to room temperature naturally. The solution containing products are then centrifuged and the supernatant layer poured off, washed with double distilled water. The wash procedure was repeated ten times to remove impurities in the sample. The final white products are dried at 60°C in air for 12 hr.

Advantage

Significant improvement in the chemical activity of the reactant, and in producing materials that cannot be obtained via solid-state reaction.

Disadvantages

1. Not for all materials
2. Expensive autoclaves
3. Difficult to control morphology, size, size distribution
4. Could not monitor and observe the reaction.

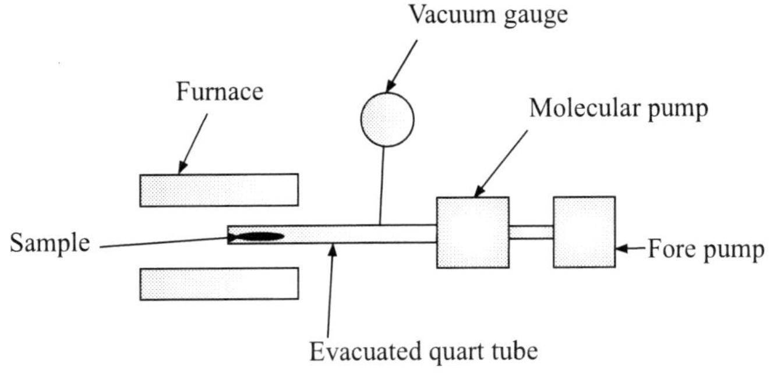

Fig. 5.60 *Schematic diagram of the thermolysis process*

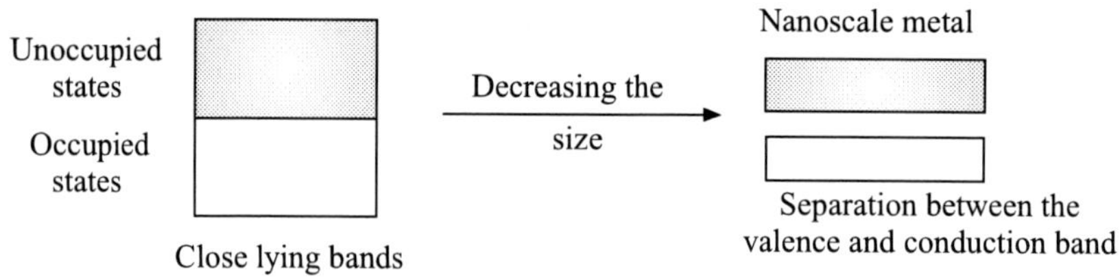

Fig. 5.61 *Origin of the properties*

5.34.6 Thermolysis

The process in which nanoparticles are formed by decomposing the solids into metal cations and molecular anions or metal organic compounds at high temperature is called thermolysis.

For example: SiO_2 nanoparticles can be synthesized from $SiCl_4$.

$SiCl_4$ is vaporized, mixed with dry air and hydrogen and then fed into the flame reactor. The $SiCl_4$ hydrolyses to SiO_2 and HCl under the influence of oxygen-hydrogen reaction. The silica so formed is an aerosol of nano sized particles.

$$2H_2 + O_2 \longrightarrow 2H_2O$$
$$SiCl_4 + 2H_2O \longrightarrow 4HCl$$

5.35 NANOSCALE MATERIAL

Nanoscale materials are a broadly defined set of substances that have at least one critical dimension less than 100 nanometres and possess unique optical, magnetic or electrical properties. Nanoparticle, nanocluster and nanocrystal are three terms used interchangeably.

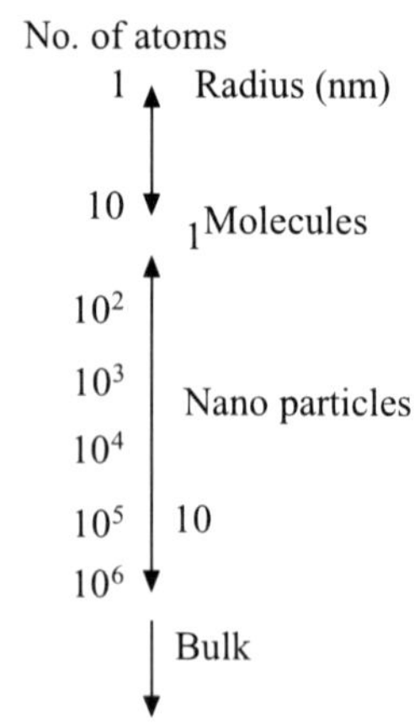

Fig. 5.62 *Distinction between molecules, nanoparticles and bulk according to the number of atoms in cluster*

5.35.1 Nanoparticles

Nanoparticles are microscopic particles with a diameter range between 1 and 100 nm. They are said to be zero-dimensional because the movement of charge carriers is confined in all three spatial dimensions (i.e., charge carriers are not free to move in any direction). Most of the properties of nanoparticles vary with size and can be fine-tuned by variation of size of particle. But in case of bulk materials properties vary with structure and composition and not on the size of material.

5.35.2 Nanoclusters

A nanocluster is a grouping of a number of nanoparticles. They have at least one dimension between 1 and 10 nm and a narrow size distribution. The size of nanoclusters ranges from 0.5–1.5 nm. These clusters are of interest because they can explain the transition from atomic properties to bulk material properties.

For example: A nanocluster of gold formed by 13 atoms, Au_{13} has 1 cosahedral symmetry while the crystal structure of bulk gold is face centred cublic (FCC).

There are different types of nanoclusters as explained below.

1. Metal Nanoclusters: Metal clusters have complicated bonding that varies from metal to metal. Some metals bond primarily with their outer valence SP electrons others bond with the *d* orbitals below the valence orbitals. Because of this variation, the bond strength varies from around 0.5 eV to 3 eV per atom.

Applications:

1. *S-Block metal*: Example: (alkali and alkaline earth metal) bonding is metallic (delocalized and non-directional) involving mainly the valence S-orbitals.
2. *Sp-metal*: (Example: aluminium) bonding has some covalent character.
3. *Transition metal*: Greater degree of covalency and directionality in the bonding. It involves the valence *d*-orbital.

1. Production of non-porous structures and fabrication of thin membrane nanoporous material.
2. Manufacture of thin films for optical and electronic devices.
3. Photo-catalyst reduction of organic substances.

2. Van der Waals Nanoclusters: Inert gas atoms from nanoclusters that are weakly bound by the van der Waals forces.

- Long range atomic attraction is due to the induced dipole force.
- Balance between induced dipole force and quantum closed shell interaction.
- Noble gases form icosahedral van der Waals cluster.
- Short range repulsion is due to quantum closed shell electronic interactions and the binding energy per atom is less than 0.3 eV.

3. Ionic Nanocluster: These are formed by the electrostatic force. The electrostatic bonds in ionic clusters are around 2–4 eV per atom. This is ten times as strong as the bond of a van der Waals nanocluster and the bonding is predominantly ionic (electrostatic). Heteroatomic cluster is composed from atoms with large differences in electronegativity.

Example: Sodium chloride, alkali metal holides $(Na \times Cl_y)^{(x-y)+}$, magnesium oxides $(Mg \times O_y)^{2(x-y)+}$.

4. Nanocrystalline Materials: Materials with microstructural features of nanometric dimensions are referred to in literature as nanocrystalline material. Nanocrystalline materials with a grain size of a few nanometres (1 nm = 10^{-9} m = 10 Å) typically less than 100 nm composed of atoms in either a single or polycrystalline arrangement. These materials can be synthesized by a number of different techniques and the grain size, morphology and composition can be controlled by controlling the process parameters. Consequently, nanocrystalline materials exhibit properties that are significantly different from and often improved over, their conventional coarse-grained polycrystalline materials.

Classification: Nanocrystalline materials can be classified into different categories depending on the number of dimensions in which the material has nanometre modulations. Thus, they can be classified into the following categories:

1. *Layered or Lamellar Structures:* A layered or lamellar structure is a one-dimensional (1D) nanostructure in which the magnitudes of length and width are greater than the thickness, that is, only a few nanometre in size.
2. *Filamentary Structures:* Filamentary structure is a two-dimensional (2D) rod-shaped nanostructure and in this the length is substantially larger than the width or diameter, which are of nanometre dimensions.
3. *Equiaxed Nanostructured Material:* Equiaxed, i.e., all the three dimensions are of nanometre size is termed nanostructured crystallites (three-dimensional (3D) nano-structures. The nanostructured materials may contain, crystalline quasicrystalline or amorphous phases and can be metals, ceramics, polymers or composites. If the grains are made up of crystals, the material is called nanocrystalline. On the other hand, if they are made up of quasicrystal line or amorphous (glassy) phases they are termed nano quasicrystals and nano glasses, respectively.

Synthesis: The synthesis of most important cadmium-selenide (CdSe) semiconductor nanocrystal and gold nanocrystal is given below.

Synthesis of CdSe Semiconductor Nanocrystal

A solution of dimethyl cadmium and a surfactant, trioctyl phosphine oxide is heated to 300°C. Now add the tributyl phosphine selenide to the above mixture with continuous stirring. CdSe nanoparticles rapidly nucleate and then grow in size to become nanocrystals. The product CdSe nanocrystal capped with To Po. After synthesis the nanocrystals are filtered and washed with acetone to remove impurities, size of a nanocrystal depends on the amount of reactant in the solution.

$$Me_2Cd + Bu_3PSe + Oct_3PO \longrightarrow (Oct_3\ PO)\ (CdSe) + C_2H_6 + Bu_3p$$

$$Me \longrightarrow \text{Methyl}$$

$$Bu \longrightarrow \text{Butyl}$$

$$Oct \longrightarrow \text{Octyl}$$

Synthesis of Gold Nanocrystal

Tetrachloroaurate (III) anion is taken in a non-aqueous solution of tetraoctyl ammonium bromide in diethyl ether. In this phase Au (III) is reduced to Au(O) by using sodium borohydride as reducing agent. This reduced gold rapidly nucleates and grows to a size determined by the amount of precursor in solution. The product is Au (SR), alkanethiolate-capped Au nanocrystals that is filtered, washed with acetone. The following reactions are involved:

$$H\,AuCl_4 + Oct_4\,NBr\,(Et_2O) \longrightarrow Oct4\,NAuCl_4\,(Et_2O)$$

$$Oct_4\,NAuCl_4\,(Et_2O) + RSH + NaBH_4 \longrightarrow Au(SR)$$

Properties

The properties of nanocrystalline materials are very often superior to those of conventional polycrystal coarse-grained materials. These are:

1. Increased strength
2. Enhanced diffusivity
3. Improved ductility/toughness
4. Reduced elastic modulus
5. Reduced density
6. Increased specific heat
7. Higher electrical resistivity
8. Superior soft magnetic properties
9. Higher thermal expansion coefficient.

Applications

1. Self-organized smart material
2. Illumination
3. Optical and infrared lasers
4. Drug manufacture
5. Flat-panel displays
6. Protein analysis
7. Bio-tags for gene identification
8. Removal of pollutants and toxins.

5.36 FULLERENE

Fullerenes are a class of molecules made of only carbon atoms in the form of a hollow sphere, ellipsoid, tube and many other shapes. Fullerenes can be of a different type C_{60}, C_{70}, etc., depending on the content of carbon atoms. The most stable is C_{60} containing 60 carbon atoms: because the diameter of the molecules is 1 and having a closed shell with 20 hexagonal faces, 12 pentagonal faces and 60 vertices. C_{60} molecule which is commonly called Buckminster fullerene. The name Buckminster fullerene comes from the name of an architect Richard Buckminster fuller, who had built the geodesic dome with spherical shape. The C_{60} molecule has spherical shape resembling a soccer ball (football). It is also called Buck ball.

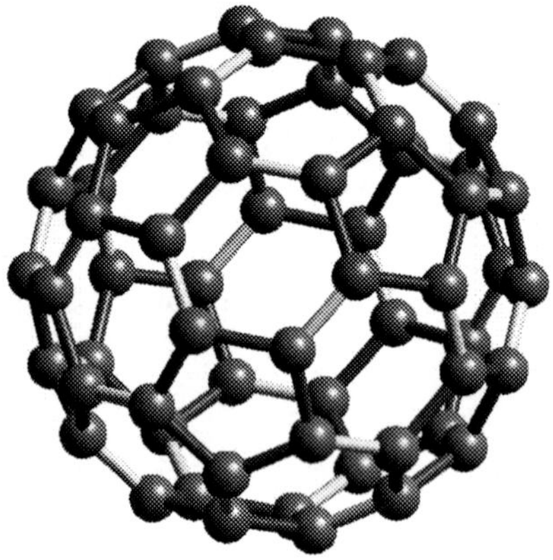

Fig. 5.63 *Structure of the C_{60} molecule (Buckminster Fullerene)*

The average carbon-carbon bond length for C_{60} is 1.44 Å which lies between that of graphite at 1.42 Å and diamond at 1.54 Å, whereas the bonding in graphite is strictly planar and can be described completely by sp^2 hybridization, the bonding between adjacent carbon atoms in fullerenes occurs over a curved surface and leads to some mixture of sp^3 bonding, characteristic of tetrahedrally bonded diamond. The mean diameter of the molecule is measured to be 7.10 Å. At room temperature C_{60} is a molecular solid having a face centred cubic crystal structure and a density of 1.72 g 1 cm^3. The centres of the C_{60} molecule are 10.02 Å apart.

Synthesis: Fullerenes are generally prepared by vaporization of graphite at very high temperature and condensing the vapours in inert atmosphere. Graphite is vaporized by setting up an electric arc between two graphite electrodes in a controlled atmosphere of helium gas. The temperature at the tip of electrode is more than 4000°C and pressure of He gas is 150–200 torrs. A mixture of various fullerenes is obtained by condensing the evaporated carbon. The main product is the fullerene C_{60}. It is extracted and crystallized using benzene as solvent.

Types of Fullerene

1. Buckball clusters:
   (a) Smallest member is C_{20} (unsaturated version of dodecahedrane)
   (b) The most common member is C_{60}.

Nanotubes

1. Hollow tubes of very small dimensions
2. Having single or multiple walls
3. Potential applications in electronics industry.

Megatubes

1. Larger in diameter than nanotubes and prepared with walls of different thickness.
2. Potentially used for transport of a variety of molecules of different size.

Polymers

1. Chain
2. Two-dimensional and three-dimensional polymers are formed under high pressure
3. High-temperature conditions
4. Single-strand polymers are formed using the atom transfer radical. Addition polymerization route.

Applications

1. Used for the conversion of diamond.
2. Used in making carbon films, tunnel diodes, storage devices, etc.
3. Used in secondary batteries, non-aqueous batteries and fuel cell electrodes.
4. Used as gas sensor, temperature sensor, and in the detection of organic vapours.
5. Used as lubricants, adhesives, charge transfer complexes, cosmetics, catalyst, etc.

5.37 NANORODS

In nanotechnology, nanorods are one morphology of nanoscale objects. Each of their dimensions range from 1–100 nm. It is a one-dimensional structure with the aspect ratio less than 10. Aspect ratio is defined as ratio of length of a particle to its width.

$$\text{Aspect ratio} = \frac{\text{Length of a particle}}{\text{Width of a particle}}$$

Length and width of nanorods usually vary from 10 to 100 nm. The aspect ratio of a nanorod varies from 0.1 to 10. Standard aspect ratios (length divided by wdith) are 3–5. It is one dimensional, because the movement of charge carriers is allowed in only one direction and confined in two other directions. It is very useful to study the effect of shape and size on the magnetic, electronic, optical and chemical properties of the material at nanoscale. It is produced by direct chemical synthesis. A combination of ligands acts as shape control agents and bond to different facets of the nanorod with different strengths. This allows different faces of the nanorod to grow at different ratres, producing an elongated object is just behind the fullerence.

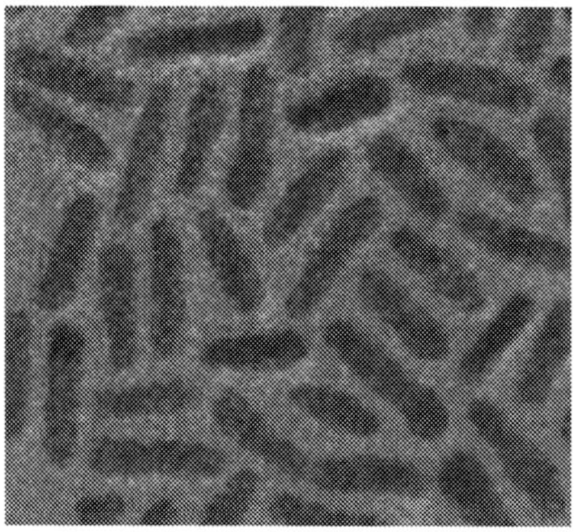

Fig. 5.64 *Typical nanorod structure*

For example: Zinc oxide (ZnO) nanorod also known as nanowire, has a direct band gap energy of 3.37 meV, which is similar to that of GaN, and it has an excitation binding energy of 60 meV. More interestingly, the optical band gap of ZnO nanorod can be toned by changing the morphology, composition, size, etc.

Synthesis of Nanorods

Synthesis of nanorods is carried out inside the pores of the template. A template is an inert material with uniform pore size and can be removed after the synthesis. Dimension of the nanorods depends upon the dimension or geometry of the template, once the synthesis is completed, nanorod is obtained by removing template. Most commonly used templates are carbon nanotubes and porous anodic aluminium oxide (AAO).

Example: TiO_2, nanorods are prepared by sol-gel method using AAO as template. A solution is obtained by dissolved in a mixture of ethanol and water is added to the first solution to obtain a sol. Then AAO membrane is dipped in this sol. The required volume of HCl is added to control hydrolysis and condensation reactions resulting in gel. Then, the sample is dried by evaporation. AAO is removed by washing with NaOH solution Nanorods of TiO_2 are obtained by calcining the sample at 400°C for 24 hours.

Applications

1. It is used in ultra-bright light-emitting diode (LED).
2. It is used to fabricate nano-scale electronic devices.
3. It is used in field effect transistor, ultraviolet photodetector, Schottky diode.

5.38 NANOWIRES

A nanowire is an extremely thin wire with the aspect ratio (length/width) greater than 10. Width of a nanowire is in the range of 1–100 nm but length can be more than nanoscale. It possesses higher aspect ratio than nanorods. It can be considered an elongated nanorod. At this small scale the quantum mechanical effect comes into consideration and hence nanowires are also referred as "quantum wires". These are normally one-dimensional material, because the movement of charge carriers is allowed in only one direction (along the length) and confined in two other directions.

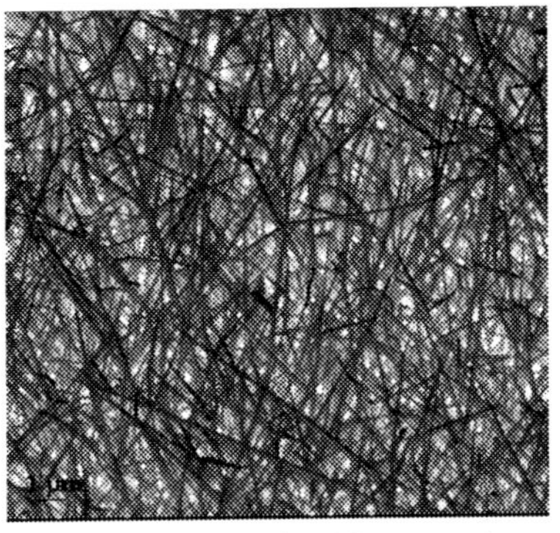

Fig. 5.65 *Metal oxide nanowires*

It shows peculiar shapes. They may show non-crystalline order, i.e., pentagonal symmetry or helicoidal (spiral) shape. The lack of crystalline order is due to one-dimensional periodic arrangement along its axis and can assume any order in plane. Therefore, nanowires exhibit fivefold only for clusters of a few atoms. This fivefold symmetry is equivalent to icosahedral symmetry of atomic clusters. The icosahedrons is an energetically favourable shape for clusters. The icosahedrons is an energetically favourable shape for clusters of a few atoms.

Synthesis of Nanowires

There are two basic approaches for synthesizing nanowires:

1. Top-down
2. Bottom-up

1. Top-down: It reduces a large piece of material to small pieces, by various means such as electrophoresis.

2. Bottom-up: A bottom-up approach synthesizes the nanowire by combining constituent adatoms.

Most synthesis techniques use a bottom-up approach. Initial synthesis via either method may often be followed by a nanowire thermal treatment step.

For Example:

Synthesis of silver nanowire: Silver nanowires with high aspect ratio can be obtained by using polyvinylpyrrolidone (PVP) as a surfactant. This surfactant binds to a particular crystal face and allows the wire to grow in a particular direction, while the growth of the other is inhibited. In a typical synthesis, silver seeds are first generated by heating $AgNO_3$ at 160°C in ethylene glycol, which serves both as reducing agent and the solvent. If a separate solution of $AgNO_3$ and PVP in ethylenes glycol is then added dropwise to this seed solution, highly elongated silver structures result.

Applications

1. **In the environment**
   - Using silver chloride nanowires as a photocatalysis to decompose organic molecules in polluted water.
   - Using an electrified filter composed of silver nanowires, carbon nanotubes and cotton to kill the bacteria in water.
2. Used in biomolecular nanosensors.
3. Used as field emitters.
4. Used as additive for nanocomposites.
5. **In electronics**
   - Using nanowires to build transistors without *p-n* junctions.
   - Using electrodes made from nanowires that would enable flat panel displays to be flexible as well as thinner than current flat panel display.
   - Sensor using zinc oxide nanowire detection elements capable of detecting a range of chemical vapour.

5.38.1 Carbon Nanotubes

Carbon nanotube is a relatively new carbon allotrope discovered by Sumiolizima in 1991. It is a tube-shaped material, made of carbon, having a diameter measuring on the nanometre scale. A nanometre is one-billionth of a metre or about 10,000 times smaller than a human hair. Carbon nanotubes are unique because the bonding between the atoms is very strong and the tubes can have extreme aspect ratios. A carbon nanotube can be as thin as a few nanometres yet be as long as hundreds of microns. It is a one-dimensional material like nanowires but with the aspect ratio (length/width) greater than 1000. Nanotubes are typically longer than a micrometre with diameters ranging from 1 to 20 nm.

Carbon nanotubes are one of the most commonly building blocks of nanotechnology, with one hundred times the tensile strength of steel, thermal conductivity similar to copper, but with the ability to carry much higher currents, and hence are wonder materials. Nanotubes come in a variety of types, viz., long, short, single walled, multi-walled, open, closed, with different types of spiral structures, etc.

5.38.1.1 Types of Carbon Nanotube

There are many different types of carbon nanotubes but they are normally categorized as either single-walled (SWCNT) or multi-walled (MWCNT).

1. Single-walled carbon nanotubes (SWCNTs): A single-walled carbon nanotube is just like a regular straw. It has only one layer or wall. It is formed by rolling up of single graphite layer. Diameter of SWCNT is 1.4 and length can go up to few micrometres.

2. Multi-walled carbon nanotubes (MWCNTs): It is a collection of nested tubes of continuously increasing diameters. They can range from one outer and one inner tube (a double-walled nanotube) to as many as 100 tubes (walls) or more. Each tube is held at a certain distance from either of its neighbouring tubes by interatomic forces. Diameter of MWCNTs varies from 30 to 50 nm and length can go up to a few micrometres.

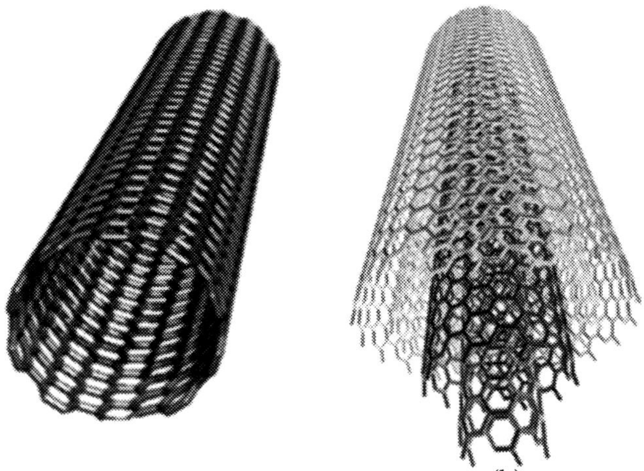

Fig. 5.66 *(a) Single-walled carbon nanotubes (SWCNT) (b) Multi-walled carbon nanotubes (MWCNT)*

3. Synthesis of carbon nanotube: The most popular and simplest way to grow carbon nanotubes in the laboratory is to use chemical vapour deposition (CVD) from hydrocarbons using iron or nickel or cobalt catalysts.

CVD system for CNT growth injects a vaporized hydrocarbon compound (methane or ethane) into a high temperature zone in a furnace. The hot zone contains a substrate on which has been pre-deposited a thin film of iron, nickel or cobalt that has either separated or pre-patterned into nanoscale islands of the metal. These nanoscale islands catalyze the growth of the carbon nanotubes. The catalyst is the key to the whole process and careful attention must be given to its deposition. Both single and multi-walled CNTs can be produced via chemical vapour deposition method.

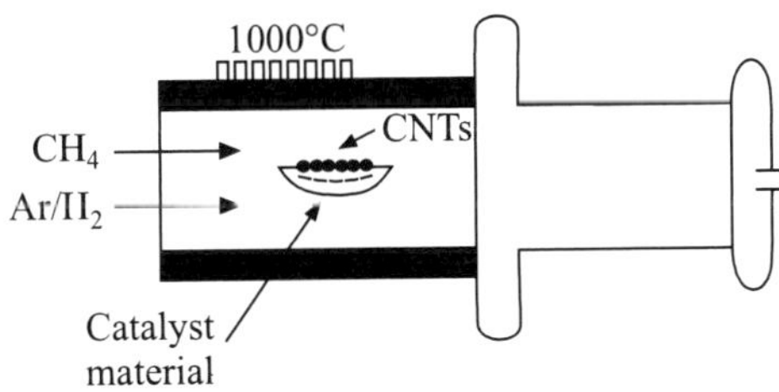

Fig. 5.67 *Chemical vapour deposition method for CNT synthesis*

Properties:

1. It is good for electron filed emitters
2. A low thermal expansion coefficient
3. High thermal conductivity
4. High electrical conductivity

5. Very high tensile strength
6. Very elastic, 18% elongation to failure. Highly flexible. It can be bent considerably without damage.

Applications of carbon nanotube

1. Energy storage
2. Biomedical application
3. Conductive or reinforced plastics
4. Field emitters/emission
5. CMT based transistors
6. CMT based ceramics
7. CMT based fibre and fabrics
8. Electric cables and wires
9. In fabrics
10. Solar cell

5.39 NANOCOMPOSITES

A nanocomposite is a matrix to which nanoparticles have been added to improve a particular property of the material. Such a material will have properties of both nanophase as well as matrix. This allows a combination of properties in one material that would never exist together in nature. Mechanically, nanocomposites differ from conventional composites due to the exceptionally high surface to volume ratio of the reinforcing and/or its exceptionally high aspect ratio.

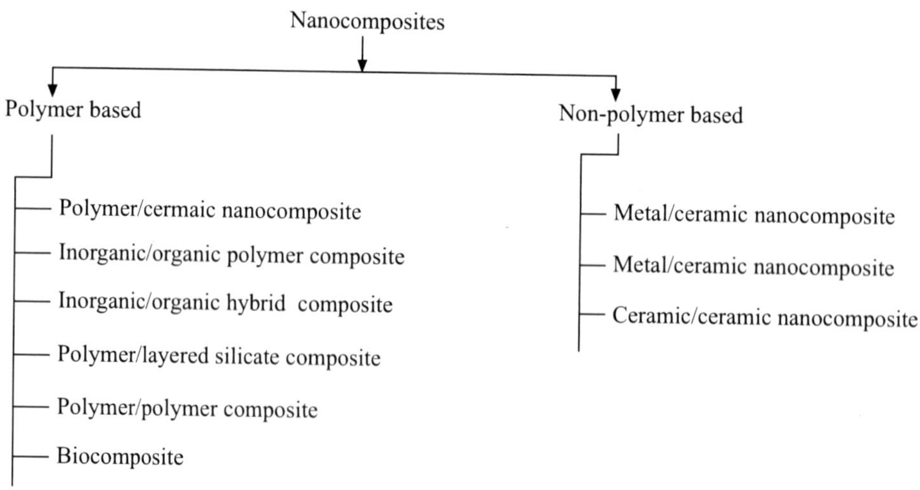

Fig. 5.68 *Schematic representation of the classification of nanocomposites*

Classification of nanocomposites

It can be classified in two categories:

1. Polymer based nanocomposite
2. Nonpolymer based nanocomposite

Polymer based nanocomposite

It is of the following types:

1. **Polymer/ceramic nanocomposite**
   Layer structure
   Example: Barium-titanate with polymers
2. **Inorganic/organic polymer nanocomposite**
   Clusters
   Example: Polymer nanofibre with zero valent nanoparticles.
3. **Inorganic/organic hybrid nanocomposite**
   Nanocrystal
   Example: CdS nanocrystal, poly N-Vinyl carbozole-photorefractivity.
4. **Polymer/layered silicate nanocomposites**
   Example: Nylon-6 (N_6) montmorillonite (MMT) with silicate
5. **Polymer/Polymer nanocomposites**
   Fillers
   Example: Poly (*p*-phinylene oxide)- plastic scrap recyling.
6. **Biocomposites**
   Example: Elastic collagen.

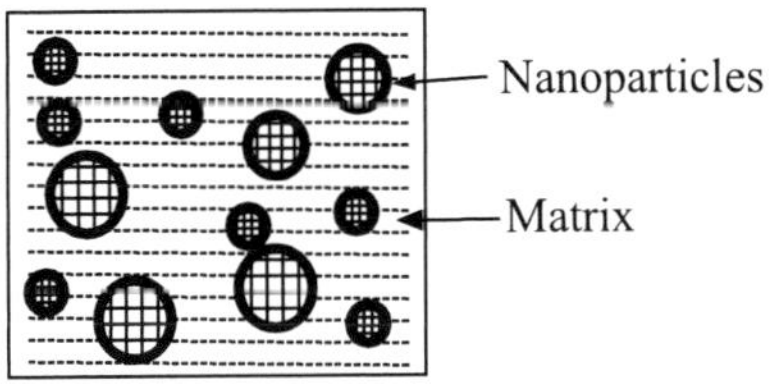

Fig. 5.69 *Isolated nanoparticles of various sizes distributed in a matrix of zero-dimensional nanocomposites*

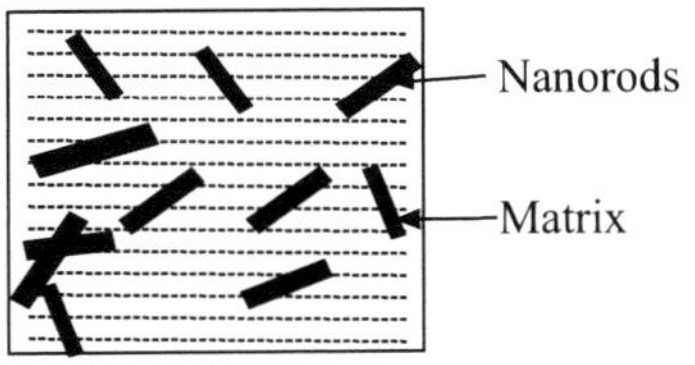

Fig. 5.70 *Nanorods and nanotubes distributed in a matrix of one-dimensional nanocomposite*

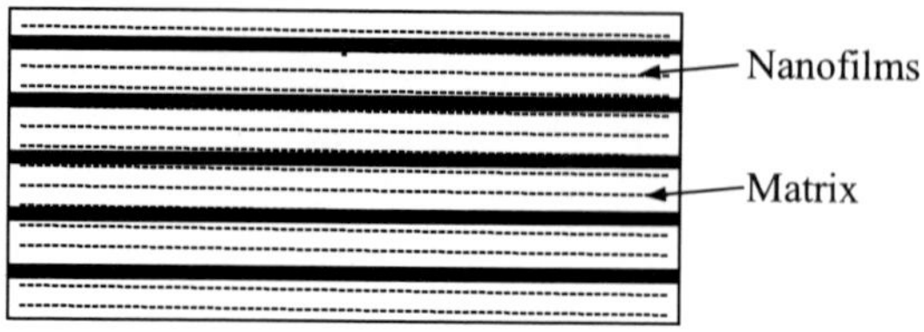

Fig. 5.71 *Nanofilms distributed in a matrix of two-dimensional nanocomposite*

Non-polymer based nanocomposite

It is of the following types:

1. **Metal/Metal nanocomposites:** It is either in the form of alloy or core-shell structure
   Example: Pt – Ru
2. **Metal/ceramic nanocomposites:** It is either in the form of nanotubes or complicated nanostructure.
   Example: Polysilazane/Polysiloxane.
3. **Ceramic/ceramic nanocomposites:** It is in the form of alloy or ceramic.
   Example: Zirconia-toughened alumina.

Polymer nanocomposites are superior to non-polymer nanocomposites

1. Interacting zone is increased such that we can get expected property.
2. Increase in electrical breakdown strength of polymers.
3. Melting temperature, colour, magnetization and charge capacity is more.

Types

Depending on the dimension of nanomaterials distributed on the matrix. These are mainly of three types:

1. *Zero-dimensional composites:* Isolated particles are distributed in a matrix interaction between different particles is minimized by maintaining a sufficient distance between particles.
2. *One-dimensional composites:* In which nanotubes, and nanorods are distributed in a matrix.
3. *Two-dimensional composites:* In which thin nanofilms are placed between two matrix layers.

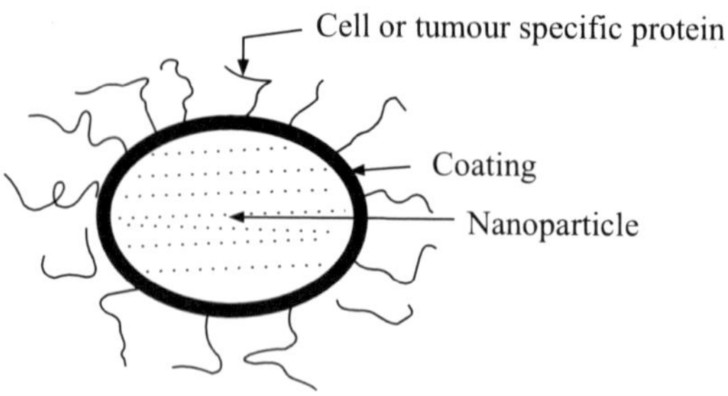

Fig. 5.72 *Coated nanoparticles for medical applications*

Example: Composite materials of various nanomaterials with polymer as a matrix possess excellent mechanical and thermal properties and are used widely in the automotive industry.

Individual nanoparticle can be coated with other material to obtain a coated nanocomposite particle. The resulting nanocomposite particle will have properties of nanoparticle as well as the coated material.

For example: Coating a luminescent ceramic material over ferromagnetic Fe_2O_3, we get a nanocomposite particle with both the properties. By selecting appropriate polymer for outermost coating it is possible to adjust the hydrophilicity or hydrophobicity of nanocomposite particle.

Coated particles are widely used in biology and medicine. Surface of the coated particle may be further attached with biological molecules to obtain biologically functionalized particles which can be used as biological carriers.

Applications

1. Fuel tanks
2. Films
3. Food packing
4. Flameability reaction
5. In artificial joints, economically beneficial
6. Environmental protection
7. Light weight materials, less fuel consumption
8. Erosion and corrosion applications
9. Marine applications
10. Carbon nanotubes, can be made as nanocomposite fibres
11. Electrocatalyst in batteries for energy saving.
12. Abrasion and wear applications.

5.40 DENDRIMERS

D.A. Tomalia coined the word dendrimer from two Greek words *dendros* means tree and *meros* means part. They are highly branched, star-shaped macromolecules with nanometre-scale dimensions. It is defined by three components.

1. A central core.
2. An interior dendritic structure (the branches).
3. An exterior surface with functional surface group. The combination of these components yields products of different shapes and sizes with shielded interior cores that are ideal candidates for applications in both biological and material sciences while the attached surface group affects the solubility and chelation ability, the varied cores impart unique properties to the cavity size, absorption capacity and capture-release characteristics.

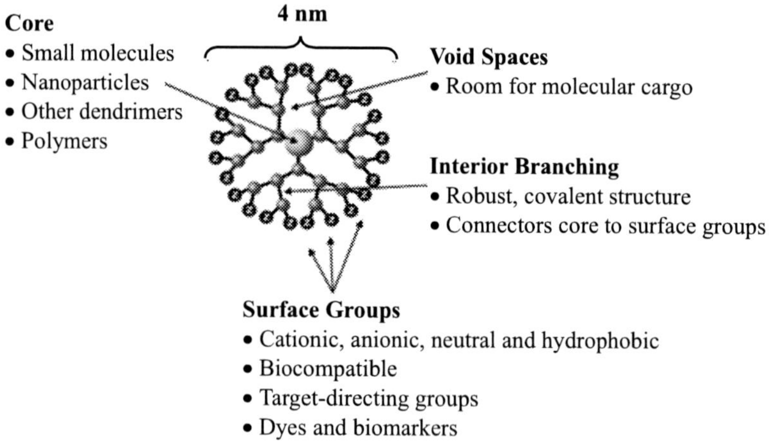

Fig. 5.73 *General structure of dendrimers*

Surface Groups

1. Cationic, anionic neutral and hydrophobic
2. Biocompatible
3. Target-directing groups
4. Dyes and biomarkers

Interior Branching

1. Robust, covalent structure
2. Connects core to surface group.

Each step of addition of branches to a dendrimer is called a generation. A dendrimer can be grown up to 10–15 generations. Low-generation dendrimers have a disc-like structure while higher generations have a globular or spherical shape. As a dendrimer grows from first to tenth genertation its size varies approximately from 1 to 25 nm. In this sense, it is a nanoscale material.

Synthesis of a Dendrimer

There are two fundamentally different concepts for the construction of dendrimers. Both methods are based on the repetition of a sequence of reactions with each sequence creating a new dendrimer generation.

1. **Divergent Process:** Growth of dendrimer molecule proceeds from "inside outward". Starting from the central core, branches grow simultaneously in all directions.

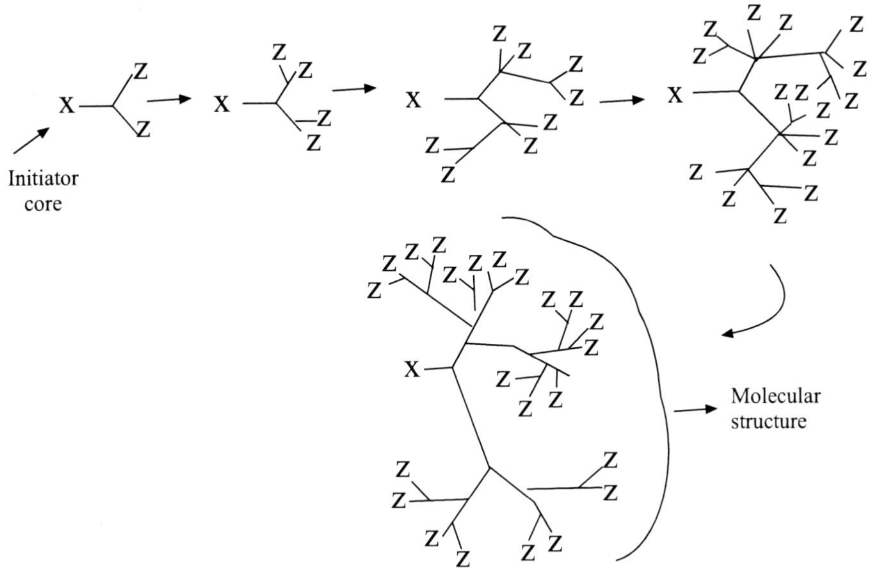

Synthesis from the core to the surface

2. **Convergent Process:** Growth of dendrimer molecule proceeds from the "outside inward". Each branch is grown separately and in the last step of synthesis all branches are joined to the central core.

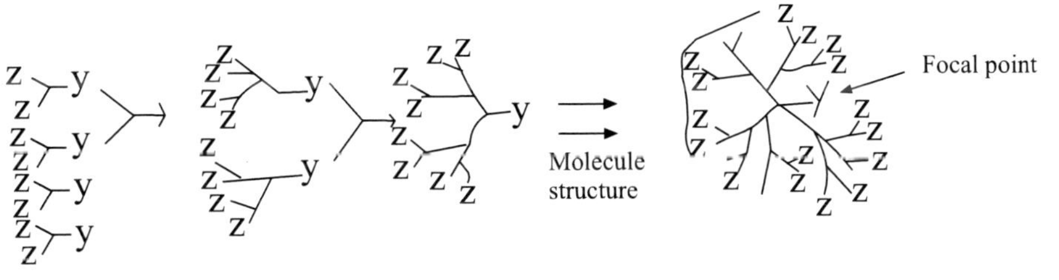

Dendrimers of lower generations have highly asymmetric shape and possess more open structures as compared to higher generation dendrimers. As the chains growing from the core molecule become longer and more branched dendrimers adopt a globular structure. Dendrimers become densely packed as they extend out the periphery which forms a closed membrane-like structrure. When a critical branched state is reached dendrimers cannot grow because of a lack of space.

Example: *Poly (propylenecimine) Dendrimers*: This dendrimer is synthesized using ethylenediamine with four functionalities (each-NH_2 is bifunctional) is used as core molecule. Acrylonitrile with –CN functional group is used as branched monomer (AB_2). In the first step, four acrylonitrile monomers are bonded to two –NH_2 groups of core molecule. In the activation step all four CN groups are reduced to amine groups (–NH_2) by reducing with CO(II) and H_2. At the end of this step, a first generation dendrimer is obtained.

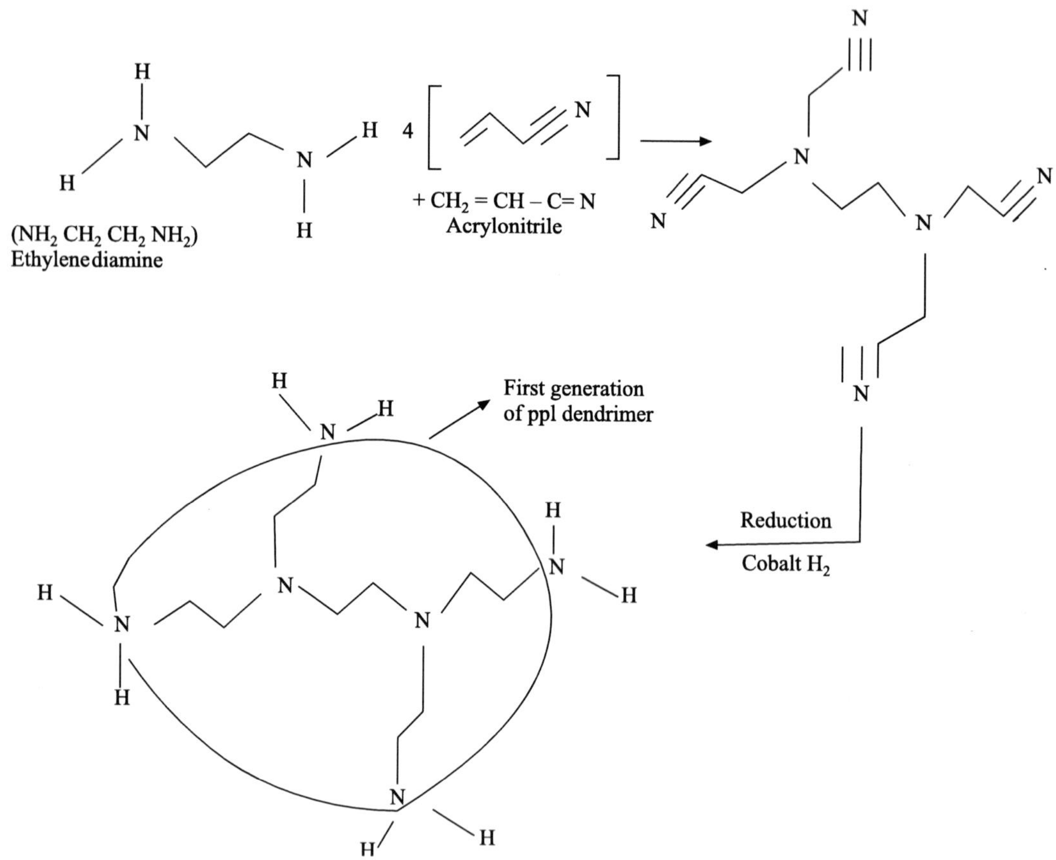

Fig. 5.74 *Poly (Propyleneimine) dendrimer*

In the second step, eight acrylonitrile monomers are bonded to four $-NH_2$ groups of first generation dendrimer and then all eight –CN groups are reduced to amine groups to get second generation dendrimer. Like this, addition of monomer to $-NH_2$ groups and reduction of –CM groups is repeated many times to get higher generation dendrimers.

Properties of Dendrimers

1. Viscosity
2. High aqueous solubility
3. Monodispersity
4. Low compressibility
5. High solubility in non-polar solution
6. Nanoscale size and shape
7. Non-crystalline and have low glass temperature
8. They have some unique properties because of their globular shape and the presence of internal cavities.

Applications of Dendrimers

1. Due to the high surface area and high solubility of dendrimers they are used as nano-scale catalysts in industrial processes.
2. Dendritic sensors
3. Anticancer drugs
4. Used as contrast agents in MRI imaging. MRI imaging is used to produce anatomical images of organs and blood vessels.
5. They can act as carriers called vectors in gene therapy vectors transfer through the cell membrane into nucleus.
6. Dendrimers are used as coating agents to protect drug and deliver it to specific sites in the body.

Review Questions

1. What is colorimetry? What forms the basis for colorimetric determination?
2. What is the basic principle of colorimetry?
3. What is the role of filter in colorimetry?
4. What is potentiometric titration?
5. What is conductance of an electrolyte?
6. What is the principle underlying conductometric titration? Discuss the titration curves obtained in the titration of:
   (a) a strong acid with a strong base, and
   (b) a strong acid with a weak base.
7. Explain the terms: conductance, specific conductance, equivalent conductance and molar conductance as applied to solution of electrolyte. In which units are these quantities expressed. How is specific conductance related to equivalent conductance and molar conductance.
8. When you take a mixture of acids (say HCl and CH_3COOH), which is the acid that neutralizes first)? Give reasons.
9. Discuss the basic principle of atomic absorption spectroscopy.
10. How do you find out the equivalence point of potentiometric titration?
11. State Beer-Lambert's law. Explain how this law can be used to determine the concentration of coloured solutions?
12. What is flame photometry?
13. Explain the components of a flame photometer and their functions.
14. Write a brief note on atomic absorption spectroscopy.
15. How do you estimate Na of a solution adopting flame photometric technique?
16. What are the advantages of using spectroscopy methods?
17. Define electronic spectroscopy. What is its absorption range? Write the relationship between wavelength, frequency and wave number.
18. State Beer's law. If the path length of a beam of light through the sample is doubled and the concentration is made half, what will be the change in the value of absorbance.

19. Name two molecules which undergo $n \rightarrow \pi^*$ electronic transition. Discuss the effect of solvent on these transition.

20. Assign the following figure to the respective isomer on the basis of the following UV spectral information.
α-isomer shows a peak at 228 nm (ε = 1400) while β-isomer has a band at 296 nm (ε = 1100)

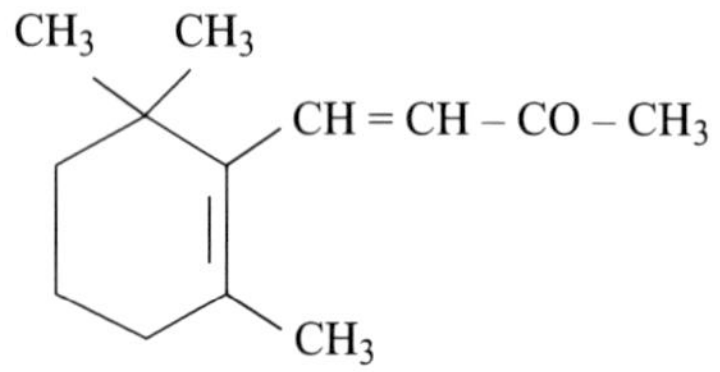

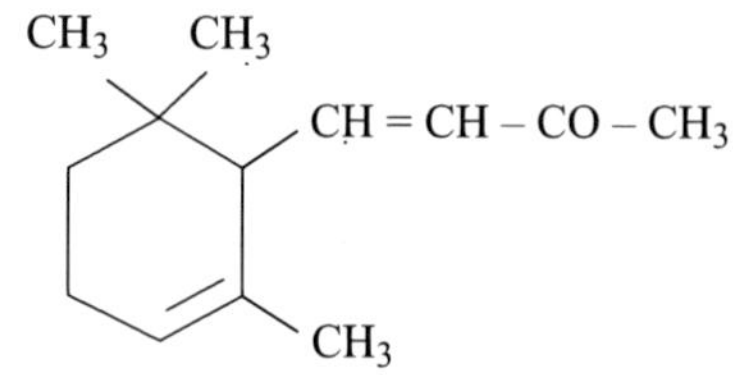

21. The wavelength of maximum absorption for methyl chloride is λ_{max} = 173 nm while for methyl iodide it is λ_{max} = 253 nm. Explain.

22. Criticise the statement, infrared radiation and the ultraviolet radiation have different wavelengths and so they must travel at different speeds.

23. Describe the Woodward-Fieser rules for calculating the absorption maximum in dienes.

24. Increase in polarity of the solvent shifted $\pi \rightarrow \pi^*$ band to longer wavelength but $n \rightarrow \pi$ and $n \rightarrow \pi^*$ bonds to shorter wavelength. Comment on the statement.

25. Explain the various applications of the ultraviolet spectroscopy.

26. What are chromophores? Name some chromophores and explain how you can identify a particular chromophore in a compound.

27. What are auxochromes? Why and how do auxochromes increase the colouring power of a chromopore.

28. The position of absorption of acetone shifted in different solvents are 279 nm in hexane, 272 nm is ethanal and 264.5 in water. Explain.

29. Predict the λ_{max} of the following compounds:

(a) CH_4 (b) CH_3-CH_3 (c) CH_3Cl (d) CH_3I
(e) CH_3OH (f) CH_3NH_2 (g) $(CH_3)_3N$ (h) CH_3-O-CH_3.

30. The UV spectrum of acetone shows two peaks at λ_{max} = 189 nm and λ_{max} = 273 nm. Identify the electronic transition for each peak.

31. List the electronic transition possible when UV light is absorbed by the following:

(a) CH_4 (b) CH_3Cl (c) HCHO

32. Why the λ_{max} for diene I is observed at lower nm than II.

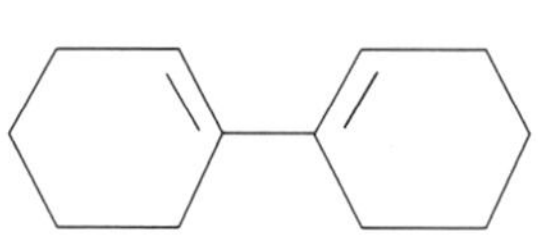

(I)

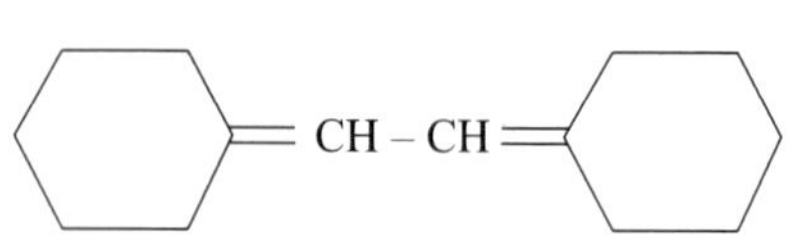

(II)

33. Measurement of UV spectra of four compounds (ketones) gave λ_{max} at 221, 249, 233 and 258 nm. Assign structures to appropriate λ_{max}.

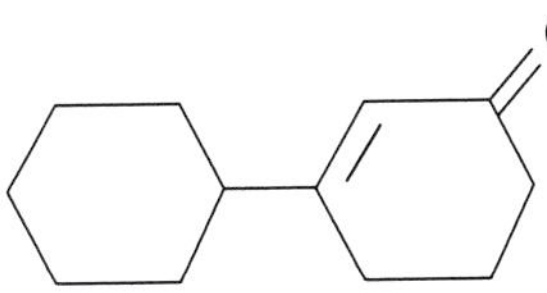

(I)

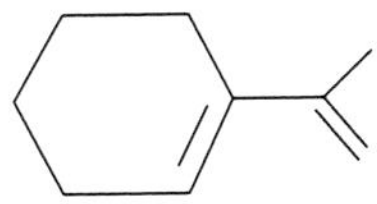

(II)

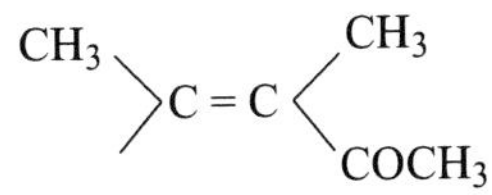

(III)

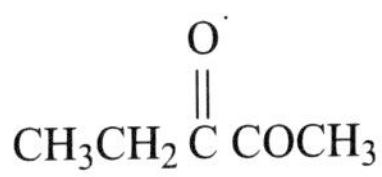

(IV)

34. Biphenyl exhibits intense absorption at λ_{max} = 252 nm, but 2, 2-dimethyl derivative absorbs at λ_{max} = 262, similar to O-xylenes. Explain.

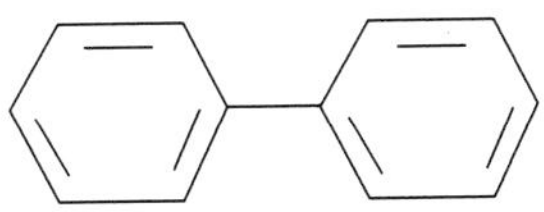

Biphenyl $\lambda_{max} = 252$

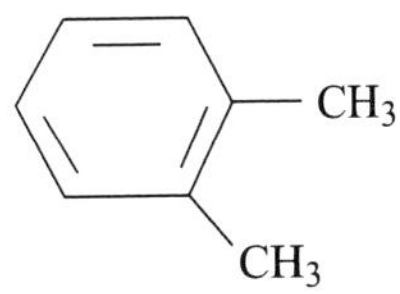

O-xylene $\tau_{max} = 252$

35. How will you distinguish between benzene and anthracene by UV spectroscopy?

36. How will you distinguish between 1, 3-pentadienl and 1, 4-pentadienl by UV spectroscopy?

37. Discuss the principle of *IR* spectroscopy. How will you distinguish between the following pairs of compounds on the basis of *IR* spectroscopy?

(a) Ethyl alcohol and diethyl ether

(b) Acetic acid and ethyl acetate

38. *IR* spectra is often characterized as molecular finger prints, comment on it.

39. Define infrared spectroscopy. Describe the various molecular vibrations in the technique.

40. Which of the following molecular will show *IR* spectrum and why?

$$N_2, O_2, H_2, HCl, CH_4, CO_2, H_2O_2, N_2O_4, C_2H_4$$

41. How will you distinguish the following pairs of compounds with the help of infra-red spectrum?

(a) CH_3CH_2OH and $CH_3 - O - CH_3$

(b) CH_3CH_2CHO and $CH_3 - CO - CH_3$

42. Give the approximate position of the important *IR* absorption frequencies for the following compounds:
    (a) CH_3OH
    (b) $C_6H_5NH_2$
    (c) $C_6H_5CH_3$
    (d) CH_3COOH
    (e) CH_3NH_2
    (f) CH_3CH_2OH
    (g) $CH_3 - CO - CH_3$
43. Why is wave number in cm^{-1} mostly used to measure the positions of a given infrared absorption?
44. How is the *IR* region of electromagnetic spectrum used for qualitative analysis?
45. Write a note on modes of vibrations of polyatomic molecules. Illustrate your answer with suitable examples.
46. What are the different regions of *IR* radiations? Explain various types of stretching and bending vibration with suitable examples.
47. Why is water not a good solvent for *IR* spectroscopy?
48. Why does conjugation increase the wavelength of absorption of a particular group?
49. Discuss the various types of stretching and bending vibrations which arise in aromatic compounds in their *IR* spectra.
50. How will you distinguish between intermolecular and intramolecular hydrogen bonding by *IR* spectroscopy?
51. What do you understand from the number of fundamental vibrations? Calculate in the theoretical number of vibrational degrees of freedom in benzene, methane and toluene. Why are the theoretical number of fundamental vibrations or bands not observed in *IR* spectra?
52. In the *IR* spectrum of a saturated hydrocarbon (molecular formula C_6H_{14}) the significant absorption bands were stretching of saturated C – H at 2960 cm^{-1}, methylene ($-CH_2-$) bending at 1460 cm^{-1} terminal methyl at 1380 cm^{-1} and the rocking ethyl group at 775cm^{-1}. Assign the structure of the compound.
53. Give the characteristic *IR* absorption frequencies of the following
    (a) Aromatic ring
    (b) Alcohols, ethers, carboxylic acid, esters
    (c) Aldehydes, ketones, carboxylic acid, esters
    (d) O – H monomeric alcohols, phenols
    (e) O – H hydrogen bonded alcohols, phenols
54. Distinguish between ethyl alcohol and diethyl ether on the basis of *IR* spectroscopy.
55. Why do ethane and ethyne, unlike propene and propyne have r_o C to C multiple band and stretching bands?
56. What are the various types of molecular vibrations? Write the stretching frequencies of the following functional group.

$$-OH, > C = O, NH_2, \text{ether}, \equiv C - H, -C - H.$$

57. *IR* spectrum of acetone gives two maxima due to C – H vibration at ~1360 cm^{-1} and ~3000 cm^{-1}. Identify the stretching and bending modes. Express the maxima in μ units.

58. What do you understand by molecular spectroscopy? Name the different types of molecular spectra.

59. What are vibrational and rotational quantum numbers?

60. (a) What are the factors that affect the width of spectral line;
(b) What are the factors which affect the intensity of spectral lines.

61. Show vibrational, rotational and electronic energy levels on an energy level diagram. To which region of electromagnetic spectrum do rotational, vibrational and electronic spectra are concerned?

62. What type of molecules give pure rotational spectra?

63. Briefly explain, how a vibrational rotational spectrum is obtained? What do you understand by fundamental bands and overtone bands?

64. What type of molecules exhibit vibrational rotational spectra and why?

65. What do you mean by spin active nuclei? Which of the following nuclei do not show nuclear magnetic resonance

$$^{1}H,\ ^{2}H,\ ^{12}C,\ ^{13}C,\ ^{14}N,\ ^{15}N,\ ^{16}O,\ ^{19}F,\ ^{31}P$$

66. What do you understand by the position of the signals in a NMR spectrum? How many signals are expected in each of the following compounds?
(a) Propane
(b) Isobutane
(c) Ethanol
(d) Cyclobutane
(e) Ethylmethyl ether
(f) Ethyl acetate
(g) Glycol.

67. Explain clearly the chemical shift. How does it help in the elucidation of the structure of organic compounds?

68. Discuss briefly the shielding and deshielding effect involved in NMR spectroscopy with suitable examples.

69. Explain clearly the various factors, which are responsible for effecting the magnitude of chemical shift? Illustrate your answer with suitable examples.

70. Write short notes on:
(a) Chemical shift (b) Spin-spin coupling
(c) Shielding and deshielding (d) Coupling constant

71. Why does a signal for a particular set of protons split into a multiplet? Explain with suitable examples the term spin-spin coupling.

72. Why is TMS used as a standard reference in NMR spectroscopy?

73. Predict the number of NMR peaks with relative areas ratio in the low resolution spectra of (a) Acetaldehyde, (b) Toluene and (c) 1, 3 dichloropropane.

74. Define chemical shift? What is its significance in the determination of the structure of molecules?

75. The low resolution NMR spectrum of the compound $C_4H_8O_2$ exhibits three peaks at 8.1, 5.2 and 1.3 ppm with relative intensity ratio 1 : 1 : 6. Deduce the structure.

76. What do you mean by splitting of NMR signal? How many NMR signals are expected in the NMR spectrum of acetaldehyde? Discuss their splitting.

77. (a) Why is an internal standard is used in obtaining an NMR spectrum? Why TMS is a good standard?

(b) What is the multiplicity of each proton in

(1) Isobutyl bromide

(2) Secondary butyl bromide?

78. Indicate diagrammatically the splitting of signals in NMR spectra of

(i) $CH_2Br - CHBr_2$

(ii) $CH_3 - CHBr_2$

(iii) $CH_3 - CH_2Br$

What is the reason for the multiplicity of peaks?

79. (a) Explain shielding and deshielding effect of protons in benzene and acetylene towards an applied field HO.

(b) What are the different kinds of electronic transitions.

80. What type of nuclei will give NMR spectra? Give two examples of each type.

81. Explain the applications of NMR spectroscopy to:

(a) Structural diagnosis

(b) Hydrogen bonding

(c) Keto-enol tautomerism

82. Give the high resolution HNMR spectrum of ethanol.

83. What types of information is obtained by studying the UV, IR and H-NMR spectra of an organic compound?

84. How will you verify that a particular signal in NMR spectrum arises from –OH, –NH or –SH groups?

85. Write short notes:

(1) spin-spin coupling or splitting

(2) chemical shift

(3) coupling constant

86. What is meant by the term chemical shift? How will you distinguish between *cis* and *trans* isomer with the help of NMR spectroscopy?

87. What are the advantages of taking NMR spectrum at more than one radio frequency.

88. The low resolution NMR spectrum of the compound $C_4H_8O_2$ exhibits three peaks at 8.1, 5.2 and 1.3 ppm with relative intensity ratio 1 : 1 : 6. Deduce the structure.

89. A compound has molecular formula C_2H_3Cl. It cannot show geometrical isomerism. The compound has two NMR signals. The splitting under high resolution of NMR shows one doublet and one triplet. Identify the compound with the help of proper explanation.

90. Explain the NMR spectrum of CH_3CH_2OH molecule. What is spin-spin coupling, explain with the help of split signals of the above molecule.

91. What is nanotechnology? Give two examples of each of material naturally occurring at nanoscale and man-made materials.

92. Discuss the sol-gel method for preparation of nanomaterials with an example.

93. Write a brief note on precipitation method for preparation of nanomaterials.

94. Write a short note on nanowire.

95. Give the synthesis, properties and applications of carbon nanotube.

96. What is a nanocomposite? How different types of nanocomposites are obtained?

97. Give the structural features, properties and applications of Buckminster fullerene.

98. What are dendrimers? How are they constructed?

6

Engineering Materials and Organometallics

A. POLYMER

6.1 INTRODUCTION

The word polymer derives from the classical Greek words *poly* meaning "many" and *meros* meaning "parts." A polymer is a giant long-chain molecule that is composed of a large number of repeating units of identical structure. In other words, a polymer may be defined as a compound whose molecules are made up of a large number of repeating units. It may be expressed as M_n, where M is a monomer and n is the number of repeating units, also known as degree of polymerization. For example, certain polymers, such as proteins, cellulose, and silk, are found in nature, while many others, including polystyrene, polyethylene, and nylon, are produced by synthetic routes. Polymers that are capable of high extension under ambient conditions find important applications as elastomers. In addition to natural rubber, there are several important synthetic elastomers including nitrile and butyl rubber. Other polymers may have characteristics that enable their fabrication into long fibres suitable for textile applications. The synthetic fibres, principally nylon and polyester, are good substitutes for naturally occurring fibres such as cotton, wool, and silk.

In contrast to the usage of the word polymer, materials other than elastomers and fibres that are derived from synthetic polymers are called plastics. A typical commercial plastic resin may contain two or more polymers in addition to various additives and fillers. These are added to improve a particular property such as processability, thermal or environmental stability, or mechanical properties.

6.2 SOME BASIC DEFINITIONS

Polymers

A polymer is a giant or macromolecule formed by the combination of a large number of small molecules (monomers).

The properties of polymers are entirely different from those of its monomers. Average molecular mass of a commercial polymer is normally in the range of $10^3 - 10^7$.

Examples: Polyethylene, Polystyrene, Teflon, etc.

$$\underset{\text{Low molecular mass compound}}{nCH_2{=}CH_2} \longrightarrow \underset{\text{High molecular mass compound}}{\left(CH_2{-}CH_2\right)_n}$$

Monomers

The small molecules which constitute the repeating units in a polymer are called monomer units.

OR

A monomer is a molecule that may bind chemically to other molecules to form a polymer.

$$\underset{\text{Ethylene (Monomer)}}{nCH_2{=}CH_2} \longrightarrow \underset{\text{Polyethylene (Polymer)}}{\left(CH_2{-}CH_2\right)_n}$$

Monomers are often called building blocks of a polymer chain. All simple molecules cannot behave as monomers but only those with two or more binding sites can act as monomers. Thus, molecules like ammonia, water, ethanol, etc., are not monomers. Alkenes, vinyl chloride, adipic acid belong to the family of monomers because they have two active centres in the form of double bonds or functional groups.

Polymerization

The fundamental chemical process in which a monomer (low molecular weight) is converted to the polymer (high molecular weight) is called the polymerization reaction. All polymerization reactions need suitable initiators.

$$\underset{\text{Ethylene (Monomer)}}{nCH_2{=}CH_2} \longrightarrow \underset{\text{Polyethylene (Polymer)}}{\left(CH_2{-}CH_2\right)_n}$$

Example:

	Monomer	Polymer
1.	Vinyl chloride	Polyvinyl chloride
2.	Vinyl acetate	Polyvinyl acetate
3.	Ethylene	Polyethylene
4.	Styrene	Polystyrene

Functionality

Functionality of a monomer is defined as the number of bonding sites or functional group present in a molecule of the monomer.

Depending on the number of bonding sites, the monomers are called bifunctional, trifunctional or polyfunctional.

Example: All double bonded compounds $CH_2{=}CH_2$ (ethene), $CH_3{-}CH{=}CH_2$ (propene), $CH_2{=}CHCl$ (vinyl chloride) have two bonding sites, the double bond. Their functionality is 2 and are bifunctional. For phenol, its 2, 4, 6 positions act as active centres. Therefore, its functionality is 3, i.e., it is trifunctional.

Table 6.1 Functionality of some commercial monomers

Monomer	Functional group Reactive centres	Functionality
Vinyl compounds	Two reactive centres at the double bond	Two
Ethylene glycol	Two alcoholic OH groups	Two
Glycerol	Three alcoholic OH groups	Three
Adipic acid	Two –COOH groups	Two
Phenol	Three active centres at ortho and para positions	Three
Hexamethylene diamine	Two $-NH_2$ groups	Two
Formaldehyde	Two active centres	Two
Bisphenol A	Two phenolic –OH	Two
Di-isocyanate	Two isocyanate groups	Two

Degree of Polymerization (DP)

Degree of polymerization is the number which expresses the total number of repeating units in a polymer chain. It is denoted by DP.

$$\underset{\text{Ethylene}}{n[CH_2{=}CH_2]} \longrightarrow \underset{\text{Polyethylene}}{\left(CH_2{-}CH_2\right)_n}$$

where n is a whole number and it is called the degree of polymerization.

The value of DP indicates the length of a polymer chain. The DP can also be used to compute the molecular weight of a polymer. The molecular weight of a polymer is the product of the number of repeat units (DP) times the molecular weight of repeat unit.

Molecular weight of a polymer = DP × Mol. wt. of repeat unit

For example: If the DP of polystyrene is 1000 its molecular weight is calculated as under:

Mol. wt. of polystyrene = DP × Mol. wt. of styrene

= 1000 × 104 = 1,04,000

6.3 CLASSIFICATION OF POLYMERS

Polymers can be classified in the following ways:

1. Origin or source: There are two types of polymers on the basis of origin or source.

(i) Natural polymers (ii) Synthetic polymers

(i) Natural Polymers: The polymers which are obtained from natural sources such as plant and animal.

Example: Cotton, wool, silk, starch, cellulose, etc.

(ii) Synthetic Polymers: The polymer which is synthesized from simple molecules.

Example: Polyethylene, polystyrene, polyesters.

2. Nature of the chemical reaction: There are two types of polymer on the basis of the nature of the chemical reaction involved in the formation of the polymers.

(i) Addition polymers

(ii) Condensation polymers

(i) Addition Polymer: The polymers which are formed by the addition polymerization reaction in which self addition of several olefinic monomers to each other takes place without elimination of by-products are called addition polymers.

Example: PVC, polystyrene, polyethylene

$$\underset{\text{Polyvinyl chloride}}{n(CH_2{=}CH{-}Cl)} \longrightarrow \left[CH_2{-}\underset{\underset{\displaystyle Cl}{|}}{CH} \right]_n$$

(ii) Condensation Polymer: The polymers which are formed by intermolecular condensation reaction, i.e., reaction between two or more monomer molecules with the elimination of simple molecules like water, ammonia and hydrogen chloride, etc.

Example: Nylon, Polyesters, Phenol — formaldehyde and Urea — formaldéhyde resin, etc.

$$\underset{\text{Hexamethylene diammine}}{nNH_2{-}(CH_2)_6{-}NH_2} + \underset{\text{Adipic acid}}{nHOOC{-}(CH_2)_4{-}COOH}$$

$$\xrightarrow{\text{Initiator}} 2nH_2O + \underset{\text{Nylon 6, 6}}{\left[NH{-}(CH_2)_6{-}NH{-}CO{-}(CH_2)_4{-}CO \right]_n}$$

3. According to structure: Polymers are classified into four types:

(i) Elastomers (ii) Fibres (iii) Resins (iv) Plastics

(i) Elastomers: The polymer which undergoes a very large elongation when pulled but returns to the original length on release of force is called elastomers.

Example: Wool, silk, cotton, jute, etc.

(ii) Fibres: The polymer, which are long thin threads like polymer chains which do not undergo stretching or deformation like elastomers. These long chain molecules are lined up and held together by hydrogen bonding.

Example: Wool, silk, cotton, nylon, polyesters, etc.

(iii) Resins: Low molecular mass polymers either in the liquid or solid form are called resin. Resins are used as adhesive or moulding powers.

Example: Epoxy resin, phenol – formaldehyde and urea – formaldehyde resin.

(iv) Plastics: Plastics are high polymers which can be moulded by the application of heat and pressure. There is no chemical change during the moulding process. They have linear structure.

Example: Polyethylene, polystyrene, polyvinyl chloride, nylons, polyesters, etc.

4. Based on environment: Polymers can be classified as follows:

(i) Ecofriendly: Cellulose, starch

(ii) Non-ecofriendly: Polythene, polyvinyl chloride, Teflon

5. Based on synthesis: Polymers are of two types.

(a) Homopolymer: The polymer made of single repeating units is called homopolymer.

Examples: -A-A-A-A-A-A-A-A-A-A-

Polyvinyl chloride: $$n(CH_2{=}\underset{Cl}{\underset{|}{CH_2}}) \longrightarrow \lbrack CH_2{-}\underset{Cl}{\underset{|}{CH}}\rbrack_n$$

(b) Copolymer: A monomer can also interact with a monomer of different chemical structure to get copolymer and the process is called copolymerization. Copolymers are of four types.

(i) Random Copolymer: The monomers are arranged in a random manner.
-A-A-A-B-B-A-B-A-

(ii) Alternating Copolymer: The monomers A and B are arranged in the alternative form.
-A-B-A-B-A-B-A-B-A-

(iii) Block Copolymer: A sequence of one monomer is followed by a sequence of other in copolymers.
-A-A-A-B-B-B-A-A-A-B-B-B-A-A-A-

(iv) Graft Copolymer: These are branched polymers and contain main chain of exclusively one monomer with the branches of second monomer.
-A-A-A-A-A-A-A-A-A-A-

6.4 TYPES OF POLYMERIZATION

The monomer can undergo polymerization by two types, with each one having a different and distinct mechanism. They are:

1. Addition polymerization
2. Condensation polymerization.

6.4.1 Addition or Chain Polymerization

The polymer which is formed by the addition polymerization reaction in which self addition of several olefinic monomers to each other takes place without elimination of by-products is called addition polymers.

Example: PVC, polystyrene, polyethylene

$$n(CH_2{=}\underset{Cl}{\underset{|}{CH}}) \longrightarrow \lbrack CH_2{-}\underset{Cl}{\underset{|}{CH}}\rbrack_n$$

where, Y = H, ethylene or CH_3, propylene or Cl, vinyl chloride or C_6H_5, styrene or CN, acylonitrile

Characteristics

- Only olefin compounds can undergo addition polymerization.
- The double bond provides required two bonding sites. There is absolutely no elimination of any by-products.
- The self-addition of monomers takes place rapidly.
- The addition polymerization is brought about by free radicals, ionic or coordination mechanism.
- Linear polymers are produced.
- The polymer produced has the same elemental composition as that of the monomer.
- The molecular weight of a polymer is an integral multiple of that of the monomer.

6.4.2 Condensation Polymerization

The polymers which are formed by intermolecular condensation reaction, i.e., reaction between two or more monomer molecules with the elimination of simple molecules like water, ammonia and hydrogen chloride, etc.

Example: Nylon, polyesters, phénol-formaldéhyde and urea = Formaldéhyde resin, etc.

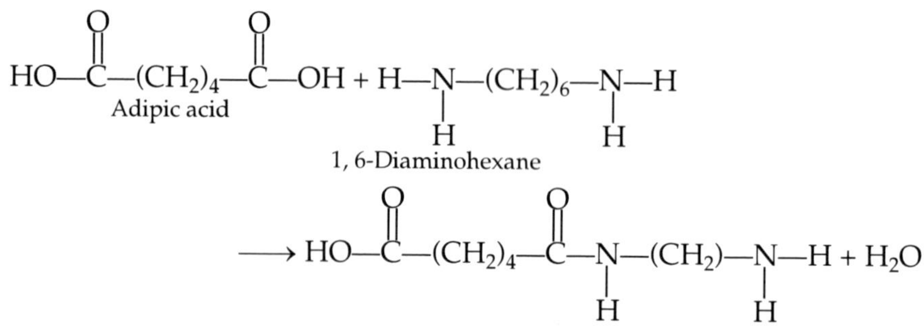

The monomers that are involved in condensation polymerization are not the same as those in adition polymerization. Monomers for condensation polymerization have two main characteristics:

1. Instead of double bonds, these monomers have functional groups (like alcohol, amine or carboxylic acid groups).
2. Each monomer has at least two reactive sites which usually means two functional groups.

B. CEMENT

6.5 INTRODUCTION

Cement is an inorganic binding material containing mixture of calcareous and argillaceous matter possessing adhesive and cohesive properties, capable of binding materials like stones, bricks, building block, etc. It is a fine grey powder and possesses the property of setting and hardening in water by virtue of certain chemical reactions.

6.5.1 Composition of Cement

Cement broadly consists of the following two types of constituents:

1. Calcareous compounds like lime, chalk and marine shells which have a heavy amount of calcium in them.
2. Argillaceous compound like clay, marl and shale having a good amount of silica in them.

So, the elemental composition of cement is lime (obtained from calcareous compounds) and silica (obtained from argillaceous compounds) which covers a significant portion of the cement body.

6.5.2 Types of Cement

There are different types of cement depending on the use and the composition of cement.

1. **Hydrophobic cement:** The cement is prepared by intergrinding cement clinkers with water repellent film forming substance like stearic acid or oleic acid. This film protects the cement from interacting with the moisture present in the atmosphere and the cement remains fresh for use even after storage and long transportation.
2. **Natural cement:** It is obtained by calcining naturally occurring argillaceous lime stone (20-40% clay matter) at high temperature. During calcination, CaO combines with Al_2O_3 and SiO_2 to produce calcium aluminate and calcium silicate.
3. **Slag cement:** It is obtained by just mixing blast furnace slag containing silicates of calcium and aluminium with slaked lime.
4. **Rapid hardening cement:** This type of cement develops early strength as it has a higher rate of gain of strength and should not be confused with quick setting cement as it takes normal time to set but attains its strength quickly. The strength attained by rapid hardening cement in 33 days is equal to the strength obtained in 77 days for normal cement.
5. **Extra rapid hardening cement:** This cement is formed by intergrinding clinkers of the rapid hardening cement with calcium chloride. The process of mixing cement transportation of cement and placing with compaction should be done within twenty minutes. The strength attained by this cement is 10-15%.
6. **Quick setting cement:** The cement is obtained by adding a small amount of alumina in finely ground cement clinkers and reducing the proportion of gypsum. This cement is used for grouting operations and underwater concreting as well.
7. **White cement:** The colour of the cement is mainly due to iron oxide. So while making this type of cement, the materials are completely freed from iron oxide. Various colour cements are developed out of white cement by adding colour pigments.
8. **Waterproof cement:** This is also called hydrophobic cement. During the manufacture of cement in order to keep the cement waterproof, calcium stearate, aluminium stearate, etc., are added to keep it unaffected even in humid conditions.
9. **Portland cement:** An extremely fine ground product obtained by calcining together an intimate and properly proportioned mixture of argillaceous (clay containing) and

calcareous (lime containing) raw materials with the addition of retarder gypsum prior to calcination at 1500°C.

6.5.3 Constituent of Portland Cement

Compound	Chemical formula	Percentage
Lime	CaO	62% – 67%
Silica	SiO_2	17% – 25%
Alumina	Al_2O_3	3% – 8%
Calcium sulphate	$CaSO_4$	3% – 4%
Iron oxide	Fe_2O_3	3% – 4%
Magnesium oxide	Mgo	1% – 3%
Sulphur	S	1% – 3%
Alkalis	Na_2O, K_2O	0.2% – 1%

6.5.4 Function of Constituents

Lime (62%–67%)

It imparts strength and soundness to the cement and if it is present in excess, it makes the cement unsound and causes it to expand and disintegrate. Whereas its deficiency, imparts low strength and causes the cement to set quickly.

Silica (17–25%)

Silica also imparts strength to the cement, similar to that of lime. As the amount of silica is increased in cement, the strength of cement also increases but the setting time of cement decreases. So, when we need cement for an early setting, the percentages of silica and lime also have to be changed.

Alumina (3–8%)

Alumina is used for imparting quick setting property to the cement. Also, it acts like flux in reducing the clinkering temperature during the preparation of cement. If alumina is present in excess, then cement reduces its strength as well.

Calcium sulphate (3–4%)

Calcium sulphate in cement is present in the form of gypsum $CaSO_4 \cdot 2H_2O$. It helps in increasing the initial setting time of the cement, so that cement remains workable for a more extended period.

Iron oxide (3–4%)

It imparts three essential things in cement, which are strength, hardness and colour.

Magnesium oxide (1–3%)

As iron oxide, magnesia also imparts strength, hardness and colour in the cement, but excess magnesia can lead to the unsoundness of cement.

Sulphur (1–3%)

Sulphur is responsible for volume changes in the cement. A very small amount of sulphur makes the cement sound, and an excess of sulphur imparts unsoundness to the cement.

Alkalis (0.3%–1%)

Alkalis are responsible for staining in cement or white coloured efflorescence of the structure which is using cement of higher concentration. This is also considered cancer of cement. So when all of these constituents of cement are mixed and burnt together with each other, it leads to the formation of some complex chemical compounds which are not formed simultaneously and are called bogues compounds. These are tricalcium aluminate, tetra calcium aluminoferrate, tricalcium silicate and dicalcium silicate.

6.5.5 Properties of Portland Cement

1. **Heat of hydration:** When water is mixed with cement, heat is liberated due to hydration and hydrolysis of various constituents.

Constituents	C_3A	C_3S	C_4AF	C_2S
Heat of hydration kJ/kg	880	500	420	250

On an average, the quantity of heat liberated during complete hydration of cement is in the order of 500 kJ/kg. In order to avoid cracking or shrinkage during the prepartion of mortar. It is necessary to dissipate the heat generated.

2. **Setting and hardening:** When cement is mixed with water, various hydration and hydrolysis reactions of the constituent compounds take place. The product compounds thus formed are responsible for setting and gradual hardening. The process of solidification consists of three steps, namely, initial setting, final setting, and hardening.

Initial and final setting: When portland cement is mixed with about half its weight of water, a plastic cement paste results. This paste gradually loses its plasticity and finally becomes stiff and hard.

In this setting process, two stages, namely, initial setting and final setting are recognised based on the ability of a weighed needle which penetrates the mass under a prescribed set of conditions. The initial set takes place when the needle does not penetrate the mass at all. The time durations taken for these two steps are respectively known as initial and final setting time.

For ordinary cements, the initial setting time should not be less than 45 minutes and final setting time should not be more than 10 hours.

Hardening: It is a relatively slow process. After the final set, the solid begins to gain strength and it is this increase in strength which is referred to as hardening.

The constituents of cement are anhydrous, and on mixing with water, a number of hydrolysis and hydration reactions takes place.

(i) $3\,CaO \cdot Al_2O_3 + 6H_2O \rightarrow 3CaO \cdot Al_2O_3 \cdot 6H_2O$ (crystal)

or $C_3A + 6H_2O \rightarrow C_3A \cdot 6H_2O$

These crystals prevent hydration of other constituents. In order to prevent this flash-set gypsum or plaster of Paris is added during pulverisation of cement clinkers. Gypsum retards dissolution of C_3A interacting with it, forming insoluble complex sulphoaluminate. This complex does not possess much hydrating property.

$$3CaO \cdot Al_2O_3 + xH_2O + y\,CaSO_4 \cdot 2H_2O \rightarrow 3CaO \cdot Al_2O_3\,y\,CaSO_4\,zH_2O$$

(where $y = 1$ to 3 and $z = 10$ to 33)

Initial strength of C_3A is good but ultimate strength is quite low.

(ii) Then tricalcium aluminoferrite (C_4AF) reacts with water more slowly compared to C_3A forming both gel and crystalline compound as follows:

$$4CaO \cdot Al_2O_3\,Fe_2O_3 + 7H_2O \rightarrow \underset{\text{gel}}{3CaO \cdot Al_2O_3 \cdot 6H_2O} + \underset{\text{crystal}}{CaOFe_2O_3 \cdot H_2O}$$

C_4AF does not contribute much to the strength of cement because both early strength and ultimate strength are poor and lowest among all the constitutional compounds.

(iii) The gel formed by C_4AF shrinks with passage of time and leaves some capillaries. Water reaches C_3S and C_2S through these capillaries and hydration and hydrolysis reactions start. Rate of hydration of C_3S is medium but that of C_2S is very low.

$$3CaOSiO_2 + xH_2O \rightarrow \underset{\text{(gel)}}{2CaOSiO_2\,(x-1)H_2O} + Ca\,(OH)_2$$

$$2CaOSiO_2 + xH_2O \rightarrow \underset{\text{(gel)}}{2CaOSiO_2\,xH_2O}$$

Strength of cement is increased by gel formation steadily over a length of time. C_3S develops the highest early strength. Its ultimate strength is also the highest. C_2S develops quite low early strength but its ultimate strength is very high, only slightly less than C_3S as shown in the graph.

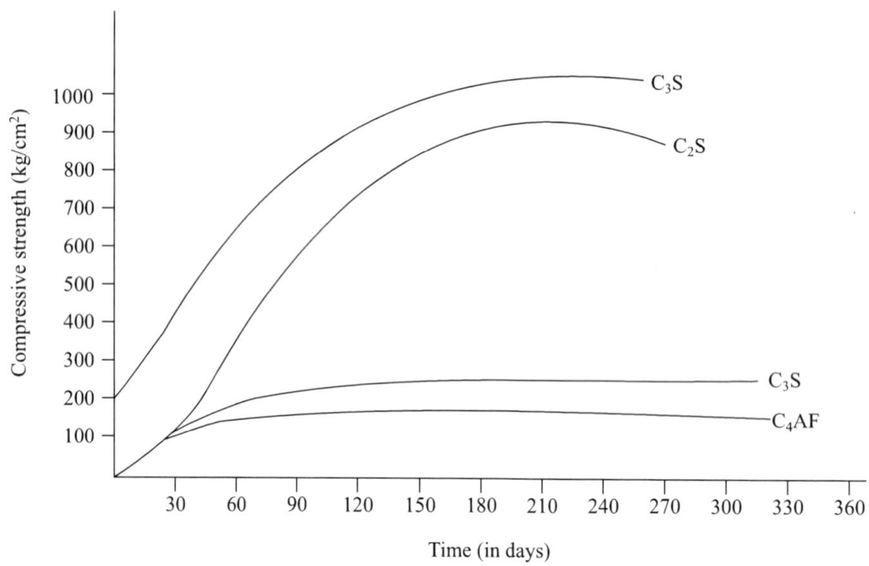

Fig. 6.1 *Growth of cement strength with time*

The setting and hardening (development of strength) of cement may be summarised as follows in the form of a flow diagram.

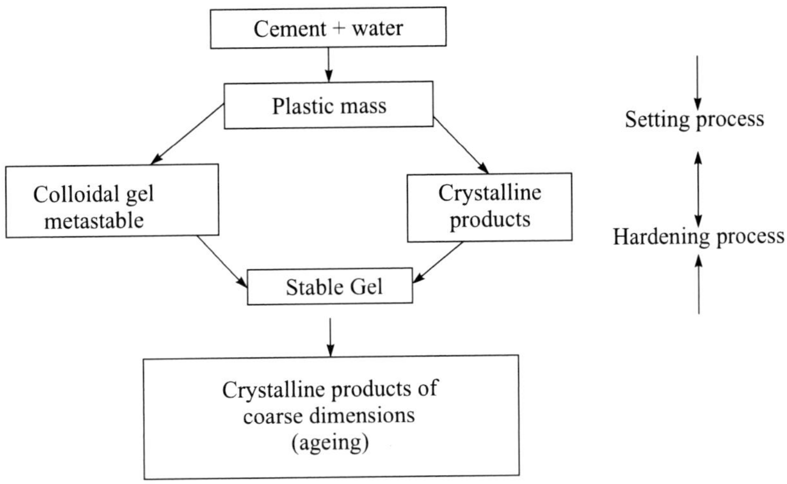

Fig. 6.2 *Flow diagram for strength development in cement*

6.6 DETERIORATION OF CEMENT

Erosion is the deterioration of concrete surface as a result of particles in moving water scrubbing the surface. When concrete surface is exposed to the water-borne sand and gravel, the surface gets deteriorated by particles scrubbing against the surface.

C. ORGANOMETALLIC COMPOUND

6.7 INTRODUCTION

In 1830, Danish pharmacist William C. Zeise isolated a salt $KCl \cdot PtCl_2 \cdot C_2H_4 \cdot H_2O$ or $[K(PtCl_3 \cdot C_2H_4)]\ H_2O$ from the reduction of $PtCl_4$ with alcohol. At that time Zeise had no way of determining the structure of his new compound but today it is known that the same compound can be prepared by the direct action of ethylene or $PtCl_2$ in alcoholic hydrochloric acid.

The attachment of the ethylene carbon atom to the central platinum atom qualifies Ziese's salt as the first organometallic compound.

"The term organometallic generally denotes compounds in which organic groups are linked directly to the metal through at least one carbon atom. It can be viewed as a bridge between organic and inorganic chemistry.

For example:

$(RO)_4 - M$	$R - M - (OR)_3$
This is not organometallic as it does not contain M — C bond (here metal is bonded to O)	This is an organometallic compound, as it contains one M — C bond (here metal M is bonded to C of the R group)
$Ti - (OC_3H_7)_4$	$(OC_3H_7)_3$ 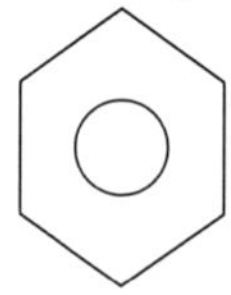
This is not an organometallic compound, as there is no Ti — C bond	This is an organometallic compound as there is a Ti — C bond.

6.7.1 Why are Organometallic Compounds Special

Most of the time, a group of carbons called an alkyl group is bound to an atom that is more electronegative like oxygen or chlorine. This means that the carbon has a partial positive charge.

$\overset{\delta+}{C}-\overset{\delta-}{O}$: The interesting thing about organometallic compounds is that the alkyl group is now bound to an atom that is less electronegative. So that carbon now has a partial negative charge. This means that organometallic compounds have a completely different reactivity.

$\overset{\delta-}{C}-\overset{\delta+}{Mg}$ ***Reactivity:*** The stability and reactivity of organometallic compounds are associated with the nature of the organic ligands and the metal to which they are attached. In group 1, 2 and 13 – 15, the thermal stability of a given type of organometallic compound generally decreases from the lightest to the heaviest element in a group.

Example: $LiCH_3$ is much more stable than KCH_3.

The organometallic compounds form electron rich or anionic carbon atoms, i.e., carbanions during their reactions. This means they will function as bases or nucleophiles. Thus, it is reasonable to write the general formula of these compounds as R^-M^t

$\mathbf{CH_3 \rightarrow Cl}$ Alkyl halide	$\mathbf{CH_3 - Li}$ Organometallic alkyl lithium
Methyl group has lower $_{-1}e^0$ density and hence is an electrophile.	The methyl group has higher $_{-1}e^0$ density and hence it is a nucleophile.

For example: Organolithium and organomagnesium compounds are strong bases since the negative charge is on carbon. It is due to the carbonic nature of organometallics that is responsible for their usefulness as synthetic reagent.

6.8 CLASSIFICATION

The organometallic compounds can be classified on the basis of the nature of metal-carbon bond.

1. **Ionic compounds of electropositive metals:** The essential ionic organometallic compounds are organosodium and organopotassium. In these type of compounds the bonding between the metal and the carbon is ionic. The carbon of the organic group carries a negative charge which is strongly attracted by the positively charged metal ion by electrostatic forces of attraction. The formation of these compounds is favoured if the negative charge on the carbon atom is stabilized, i.e., it is delocalized over several carbon atoms.

   For example: Potassium cyclopentadienyl ($K^+C_5H_5^-$), the negative charge is stabilized due to delocalization over the five carbon atoms of cyclopentadienyl ring.

   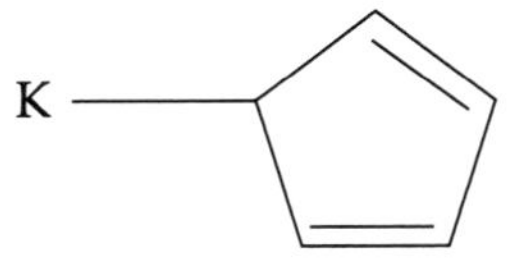

2. **Covalent organometallic compounds:** The covalent organometallic compounds are generally metal-alkyl compounds. These compounds are very common and are generally formed by Zn, Cd, Hg and representative elements of group III, IV and V. The polarity of the covalent bond between metal and alkyl groups depends upon the difference in electronegativity between metal atom and carbon atom.

   Example: $Pb(C_2H_5)_2$, $(C_6H_5)_2Zn$, $(CH_3)_2Cd$, C_4H_9Li, etc.

3. **Electron deficient organometallic compounds:** These type of compounds include the compounds having bridging alkyl groups.

   Example:

   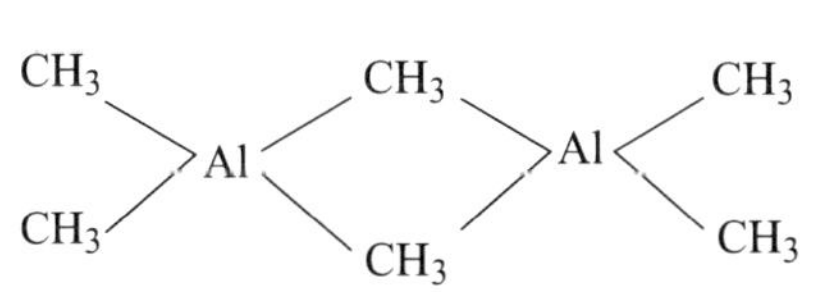

   Dimeric dialkyl aluminium

   (Dimeric trialkyl aluminium which have bridging alkyl groups).

4. **Organometallics involving π bonds:** This type of compounds having metal ligand bond between a metal and the π-orbitals of organic ligands such as alkenes, benzene ring, etc.
   Ferrocene was the first of many π complexes which came to be known as metallocenes. In these complexes, the metal ion is sandwitched between two parallel carbocyclic rings came to be known as sandwich compounds.

For example:

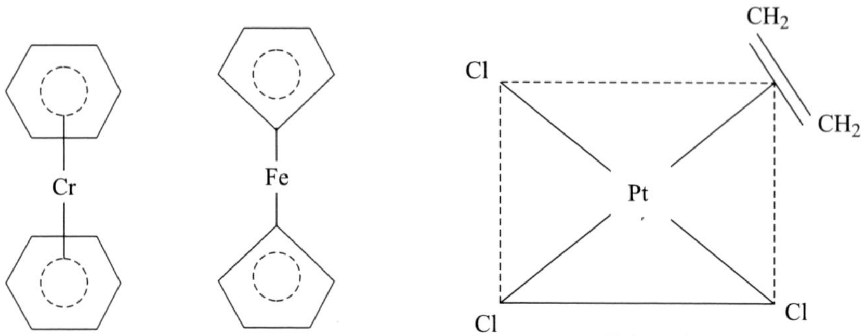

$(C_6H_5)_2Cr$ $(C_5H_5)_2Fe$ Zeise salt (π complex)

5. **Transition metal organometallic compound:** A large number of organometallic compounds are based on transition metals. A transition metal complex consists of a transition metal to which are attached groups called ligands. A ligand can be element (O_2, N_2) or a compound (NO) or an ion (CN^-). It can be inorganic as in the examples just cited or it can be an organic ligand. Ligands differ in the number of electrons that they share with the transition metal to which they are attached.

   For example: Carbon monoxide is a frequently encountered ligand in transition-metal complexes and contributes two electrons, it is best thought of in terms of the Lewis structure $:C^- \equiv O^+:$ in which carbon is the reactive site.

6.8.1 Preparation of Organometallic Compounds

Following methods are used for the preparation of organometallic compound.

1. Replacement Method: In this method Grignard reagent (R – Mg – X) is treated with a compound of transition metal to obtain corresponding transitional organometallic compounds.

$$C_2H_5Br + Mg \xrightarrow{\text{ether}} C_2H_5MgBr$$

$$2C_2H_5MgBr + CdCl_2 \longrightarrow (C_2H_5)_2Cd + MgCl_2 + MgBr_2$$

$$3C_6H_5MgCl + CrCl_3 \longrightarrow (C_6H_5)_3Cr + 3MgCl_2$$

$$2C_6H_5MgBr + ZnBr_2 \longrightarrow (C_6H_5)_2Zn + 2MgBr_2$$

In some cases organometallic compound of one metal is treated with the metal or metal salt.

$$2C_5H_5Na + FeCl_2 \longrightarrow (C_5H_5)_2Fe + 2NaCl$$

2. Addition Method: When unsaturated hydrocarbon reacts with metallic hydrides, organometallic compounds are formed.

$$B_2H_6 + 6H_2C = CH_2 \xrightarrow{100°C} 2(C_2H_5)_3B$$

$$2AlH_3 + 6H_2C = CH_2 \longrightarrow 2(C_2H_5)_3Al$$

$$HSiCl_3 + C_6H_{13}CH = CH_2 \longrightarrow C_8H_{17}SiCl_3$$

$$HSiCl_3 + CH \equiv CH \longrightarrow H_2C = CH \cdot SiCl_2$$

3. Direct Method: In this method, direct reaction of a metal with a suitable reagent takes place.

$$2CH_3Cl + Si \xrightarrow[CU]{300°C} (CH_3)_2SiCl_2$$

$$2C_2H_5I + 2Cu - Zn \longrightarrow (C_2H_5)_2Zn + Zn + 2CuI$$

Some hydrocarbons which are acidic in nature or active enough to behave like halogen atoms and bring out direct reaction with highly electropositive metals.

$$2C_5H_6 + Fe \longrightarrow (C_5H_5)_2Fe + H_2$$

$$2C_5H_6 + FeO \longrightarrow (C_5H_5)_2Fe + H_2O$$

$$2C_5H_6 + CuO \longrightarrow (C_5H_5)_2Cu + H_2O$$

6.8.2 Types of Organometallic Compounds

1. **Simple organometallic compounds** have only alkyl radical or hydrogen atom attached to the metal atom. **Example:** $(C_2H_5)_4Pb$, $(CH_3)_3SnH$, etc. If both or all the hydrocarbon radicals are same then it is called a symmetrical compound. If the attached alkyls are different then it is called an unsymmetrical compound. **Example:** $C_2H_5MgC_4H_9$.
2. **Mixed organometallic compounds** have groups other than alkyl radicals or hydrogen atom attached directly to the metal atom. **Example:** C_2H_5MgBr, $(C_4H_9)_2SnCl_2$, $C_6H_5SbO(OH)_2$.

6.8.3 General Properties

1. **Physical properties and covalent character:** They are typically low-melting solids liquids or gases at ordinary temperatures and are usually soluble in non-polar or weekly polar organic solvents such as alkanes, toluene, ether or even alcohols.
2. **Stability to oxidation:** All organometallic compounds are thermodynamically unstable to oxidation. Many are also kinetically unstable to oxidation at room temperature. **For example:** Me_2Hg and Group IV B derivatives are inert, Group IV A elements are rapidly oxidised.
3. **Stability to hydrolysis:** Organo-derivatives of Group IV B and V B are kinetically stable to attack by water, whereas those of the Group I A and II A elements of Zn, Cd, Al, Ga and In are readily hydrolysed. These hydrolysis reactions usually involve the nucleophilic attack by water and are thus facilitated by the presence of empty low-lying orbitals on the metal (**Example:** Al) which attract the electron pair of the nucleophile.

4. **Reaction with carbon dioxide:** Grignard reagent react with excess CO_2 to yield acids.

$$O=C=O + R\,MgX \longrightarrow R-\overset{\overset{\displaystyle O}{\|}}{C}-O-O\,MgX \xrightarrow{H_2O} R-\overset{\overset{\displaystyle O}{\|}}{C}-OH$$

5. **Reaction with ether:** Alkyl of metal with electronegativities less than 0.9 generally react with ethers, producing a hydrocarbon and an alkoxide and the later is converted to an alcohol by hydrolysis.

$$RNa + R'-O-R'' \longrightarrow R'ONa + R'-R''$$
$$R'ONa \xrightarrow{H_2O} R'OH + NaOH$$

6. **Reaction with non-metallic elements (other than oxygen) and their inorganic compounds:** Halogens decompose all reactive organometallic compounds to form hydrocarbon halides.

$$C_2H_5Li + Br_2 \longrightarrow C_2H_5Br + LiBr$$
$$Cl_2 + H_3C-\overset{|}{\underset{|}{Si}}- \longrightarrow ClH_2C-\overset{|}{\underset{|}{Si}}- + HCl$$

Some important organometallic compounds

Preparation: Trialkyls are prepared by heating aluminium powder, hydrogen and an alkenes under pressure.

$$3(CH_3)_2C=CH_2 + Al + 3/2H_2 \longrightarrow \left[\begin{matrix} CH_3 \\ CH_3 \end{matrix}\!\!>CH\cdot CH_2\right]_3 Al$$

Tr-isobutyl aluminium

Mono, di- and trialkyl aluminium compounds also prepared by reaction between aluminium chloride in the appropriate amount of Grignard reagent.

$$CH_3MgCl + AlCl_3 \longrightarrow CH_3AlCl_2 + MgCl_2$$
$$2CH_3MgCl + AlCl_3 \longrightarrow (CH_3)_2AlCl + 2MgCl_2$$
$$3CH_3MgCl + AlCl_3 \longrightarrow (CH_3)_3Al + 3MgCl_2$$

Some organometallic compounds of this type have a tendeney to exist as dimers or even as polymers

For example:

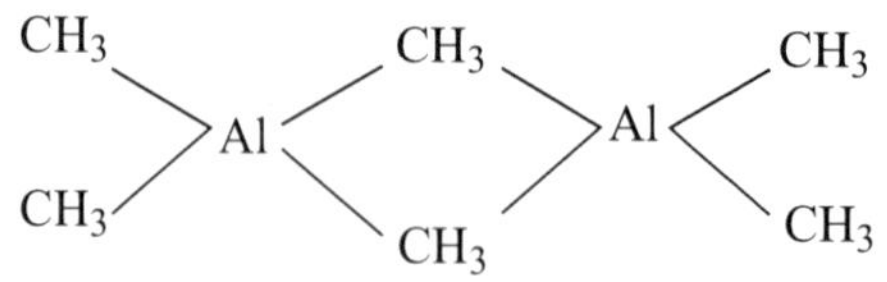

Trimethyl aluminium (dimer)

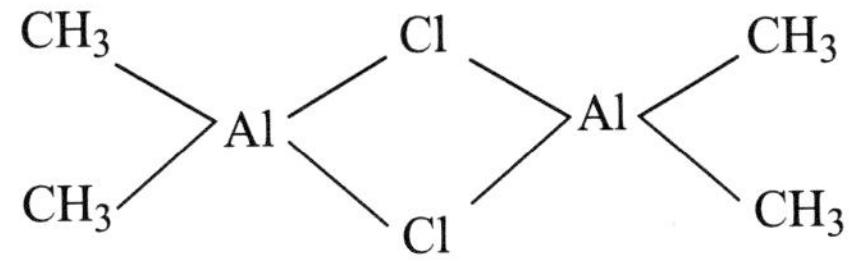

Dimethyl aluminium chloride (dimer)

Function: The function of aluminium alkyl are mainly:

(1) The formation of active metal species $TiCl_3$ (in Ziegler – Natta catalyst)

(2) The replacement of one of the chloride ions in a Ti^{3+} ion at the surface by an alkyl radical derived from it.

Uses:

1. Long straight chain alcohols are manufactured from triethyl aluminium and ethylene.

$$(C_2H_5)_3Al + 3nC_2H_4 \xrightarrow[\text{under pressure}]{\text{Heat at } 100^\circ} Al[(C_2H_4)_nC_2H_5)]_3 \xrightarrow{O_2} Al[O(C_2H_4)_nC_2H_6]_3$$

$$\xrightarrow{H_2O} 3C_2H_5(C_2H_4)_nOH + Al(OH)_3$$

2. Di-isobutyl alminium hydride is a very useful reducing agent for various functional groups, viz., aldhydes ketones, esters and cyanides.

$$R_2CO + [(CH_3)_2CH \cdot CH_2]_2AlH \rightarrow R_2CHO\ Al[CH_2CH(CH_3)_2]_2 \xrightarrow{H_2O} R_2CHOH$$

3. The most important use of trialkyis aluminium is as a catalyst (Ziegler/Natta type catalysts) for the low temperature polymerisation of alkenes.

$$CH_3CH_2Al(C_2H_5)_2 \xrightarrow{CH_2=CH_2} CH_3CH_2CH_2CH_2Al(C_2H_5)_2$$

$$\xrightarrow{CH_2=CH_2} CH_3(CH_2)_5Al(C_2H_5)_2 \xrightarrow{CH_2=CH_2} CH_3(CH_2)_5Al + CH_3(CH_2)_2CH_3$$

6.8.4 Organometallic Compound of Alkali Metal

Alkali metals, especially Li, form many organometallic compounds.

Preparation: It is generally obtained as follows:

$$Li(n-C_3H_7) + C_4H_5I \xrightarrow[\text{ether}]{\text{Solvent}} Li(C_6H_5) + nC_2H_7I$$

$$4Li(C_6H_5) + Sn(CH_2-CH=CH_2)_4 \xrightarrow[\text{ether}]{\text{Solvent}} Sn(C_6H_5)_4 + 4Li(CH_2-CH=CH_2)$$

$$2Li + C_2H_5I \xrightarrow[\text{ether}]{\text{Solvent}} LiI + Li(C_2H_5)$$

Organometallic compounds of other alkali metals such as $Na(C_6H_5)$, $Ph_2C(Na)\ C(Na)Ph_2$, $CSCH_2CH_2CS$ are also known. These are generally prepared from the corresponding mercury compounds.

Example: $2K + HgR_2 \longrightarrow Hg + 2\ KR$

Properties: The monomeric organometallic compounds of alkali metals are not very stable thermodynamically however, polymeric compounds are comparatively stable.

Structure: Alkyl and aryl compounds of lithium are polymeric having considerable covalent character in their Li – C bonds whereas such organometallic compounds of other alkali metals have predominately ionic character so that these can be represented as M^+R^-. The alkyl and aryl sodium, potassium, etc. are much more reactive than the corresponding alkyls and aryls of lithium.

Uses: Lithium alkyls and organometallic compounds of other alkali metals are frequently used in the synthesis of a number of organic compounds. These uses are evident from the following reactions.

$$3\,LiR + BCl_3 \longrightarrow 3\,LiCl + Br_2$$

$$Li(C_6H_5) + (CH_3C_6H_4)_2CO \longrightarrow (CH_3C_6H_4)_2\,C(C_6H_5)OLi \xrightarrow{H_2O} (CH_3C_6H_4)_2\,C(OH)\,C_6H_5 + LiOH$$

Tertiary carbinols

$$Li(C_6H_5) + CO_2 \xrightarrow[\text{solvent}]{\text{Nonpolar}} [C_6H_5CO_2\,Li] \xrightarrow{H_2O} C_6H_5COOH + LiOH$$

$$LiR + R'CON(CH_3)_2 \longrightarrow LiN(CH_3)_2 + R'COR$$

Ketones

6.8.5 Organometallic Compounds of Magnesium

It forms three types of organometallic compound:

(1) Compound of MgR_2 type (R = alkyl or aryl group)

(2) Grignard reagent, RMgX

(3) Organometallic compounds containing one alkyl (R) and one alkoxy group (OR′), R-Mg — OR′

The dialkyl and diaryl magnesium compounds can be prepared in the following manner

$$LiR + R\text{-}Mg\text{-}X \xrightarrow[\text{ether}]{\text{Solvent}} MgR_2 + LiX$$

$$HgR_2 + Mg \xrightarrow[\text{ether}]{\text{Solvent}} MgR_2 + Hg$$

Dimethyl magnesium is polymeric. It contains Mg — CH_3 — Mg bridge in its structure. Diphenyl magnesium dissolves in ether to yield $Ph_2Mg(O\ C_2\ H_5)$ where Ph → phenyl groups. It reacts with phenyl lithium to give $Li(Mg\ Ph_3)$ as shown in the following reaction

$$LiPh + MgPh_2 \rightarrow Li[Mg\ Ph_3]$$

Another class of organometallic compound of magnesium called Grignard reagents can be synthesized by refluxing alkyl or aryl halides with magnesium in dry solvent ether.

$$R\,Br + Mg \xrightarrow[\text{ether (dry)}]{\text{Solvent}} R\,Mg\,Br$$

Organometallic compounds containing one alkyl and the other alkoxy group attached with Mg can be easily synthesised by reacting dialkyl magnesium compounds with alcohols in the following manner

$$MgR_2 + R'OH \longrightarrow R - Mg - OR' + ROH$$

These compounds are polymeric and contain alkyl and alkoxy bridges between Mg atoms. The degree of polymerisiation depends upon the bulk of the alkyl and alkoxy groups.

CH_3— Mg — OR (R is CH_3 or C_2H_5) exist in two polymeric forms:

(a) Chain polymer

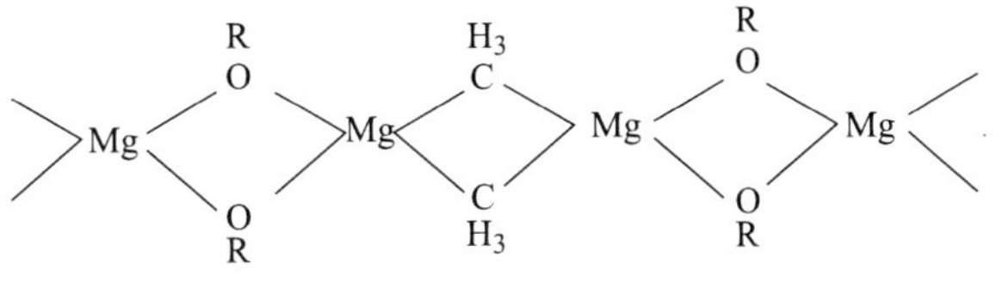

(b) Three-dimensional polymer

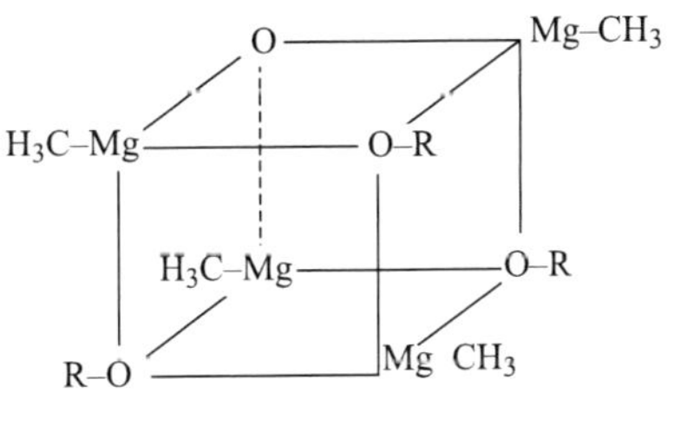

6.9 ORGANOSILICON COMPOUNDS

Oganosilicon compounds are extensively studied due to wide range of commercial applications as water repellents, lubricants and sealants. The first organosilicon compound was prepared by Friedel and Craft (in 1863) by the addition of silicon tetrachloride on diethyl zinc.

$$2\,Zn(C_2H_5)_2 + SiCl_4 \longrightarrow 2\,ZnCl_2 + Si(C_2H_5)_4$$

The major contribution of organosilicon compound was given by the Professor F.S. Kipping (1899 – 1944). All the organosilicon compounds are named with respect to silane SiH_4 (as methane). Following is the simplest member of the group.

$$CH_3SiCl_2(C_6H_5)_2SiCl_2$$

Methyldichlorosilane diphenyldichlorosilane

H_2SiOH	$H_2Si(OH)_2$	$HSi(OH)_3$
Silanol	Silanediol	Silanetriol

$$H_3Si(OSiH_2)n.\ OSiH_3$$

Siloxanes

6.10 ALKYL SILANES

It is derived from the silicon hydride SiH_4 by replacing one or more hydrogen atoms by alkyl groups. They are classified as mono, di, tri and tetra alkyl silanes.

$$SiH_3R,\ SiH_2R_2,\ SiHR_3,\ SiR_4$$

Following methods are used for the preparation of alkyl silanes:

1. By the action of silicon tetrachloride on dimethyl zinc.

$$SiCl_4 + 2ZnMe_2 \longrightarrow Me_4Si + 2ZnCl_2$$

2. By the action of $SiCl_4$ and Grignard reagent

$$SiCl_4 + 4Me\ MgI \longrightarrow Me_4Si + 4Mg\langle^{Cl}_{I}$$

3. By heating silicon tetrachloride and alkyl halide in the presence of sodium.

$$SiCl_4 + 4MeBr + 8Na \xrightarrow{\Delta} Me_4Si + 4NaCl + 4Ma\ Br$$

6.11 CHLOROSILANES

Following methods are used for the preparation of chlorosilane.

1. Grignard method
2. The olefin addition method
3. The direct process
4. Sodium condensation method

(1) Grignard method: Grignard reagent is treated with silicon tetrachloride to give a mixture of chlorosilanes.

$$CH_3MgCl + SiCl_4 \longrightarrow CH_3SiCl_3 + MgCl_2$$

$$CH_3MgCl + CH_3SiCl_3 \longrightarrow (CH_3)_2SiCl_2 + MgCl_2$$

$$CH_3MgCl + (CH_3)_2SiCl_2 \longrightarrow (CH_3)_3SiCl + MgCl_2$$

$$CH_3MgCl + (CH_3)_3SiCl \longrightarrow (CH_3)_4Si + MgCl_2$$

The Grignard method has a great advantage over the direct method since a wide range of organic groups can be attached to silicon by this method.

(2) The olefin addition method: In this method silicon hydrides are added to alkenes under pressure at a temperature below 45° and 400°C in the presence of a catalyst, viz., benzoyl peroxide, platinum, palladium anamine, heat or ultraviolet radiations.

$$RCH = CH_2 + SiHCl_3 \longrightarrow RCH_2CH_2SiCl_3$$

$$CH \equiv CH + SiHCl_3 \longrightarrow CH_2 = CH \cdot SiCl_3$$

$$CH_2 = CH \cdot SiCl_3 + SiHCl_3 \longrightarrow Cl_3SiCH_2CH_2SiCl_3$$

(3) The direct process: In this method the elemental powered silicon is heated with alkyl halide at elevated temperature (250 – 280°) and pressure in the presence of other metal, like copper (as catalyst) to give mixture of alkyl chlorosilane. The main reaction in the process results in the formation of dimethyl chlorosilane.

$$2CH_3Cl + Si \longrightarrow (CH_3)_2SiCl_2$$

However, smaller quantities of trimethyl chlorosilane, methyle trichlorosilane and silicon tetrachloride are also formed. The amount of desired product may be increased to some extent by choice of suitable temperature, rate of flow of methyl chloride reagent and nature and quantity of catalyst.

(4) The Wurtz-Fitting reaction: This method consists in the condensation of silicon tetrachloride and an organic chloride in presence of sodium.

Example: $SiCl_4 + 4RCl + 8Na \longrightarrow SiR_4 + 8\,NaCl$

However, this method is not as important as the other three methods for the preparation of partially substituted chlorosilanes because this method tends to complete substitution to form silicon tetra-alkyls.

Among the organosilicon compounds the chlorosilanes are of great importance because they constitute the intermediates from which the siloxane polymers are prepared. Since they are very reactive and inflammable, they are preserved under dry inert gases such as nitrogen. The most important reaction of chlorosilanes is its hydrolysis to silanols which then undergo self-condensation to form poly-siloxane polymers. Generally, they are known as silicon. For example, a dichlorosilane, an alkane hydrolysis, gives diol which by dehydration and polymersation gives either a linear or cyclic siloxane.

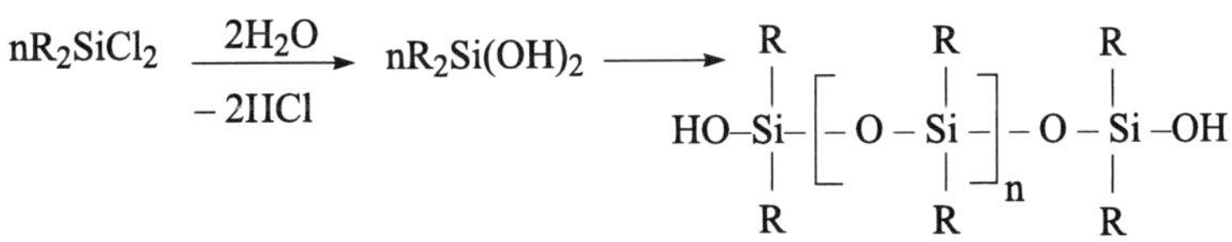

Linear polymer

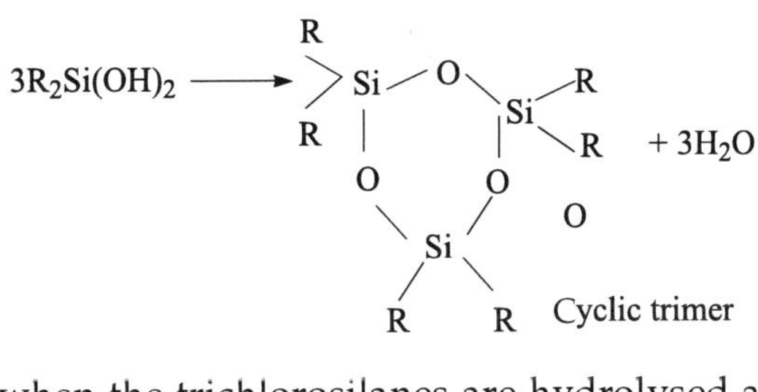

Cyclic trimer

On the other hand, when the trichlorosilanes are hydrolysed and then polymerised a three-dimensional polymer, i.e., a polymer having cross-linkage is produced.

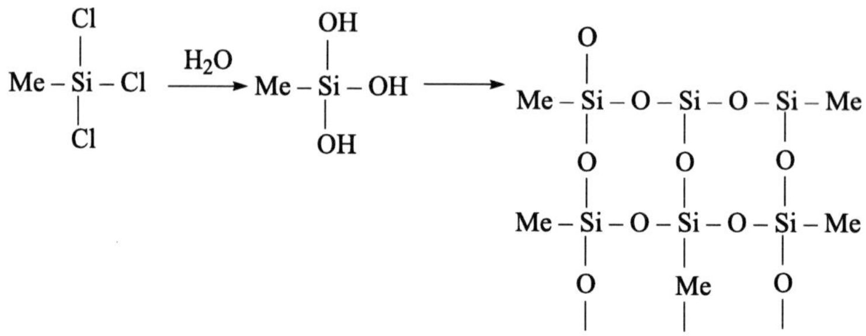

Cross-linked polymer

However, hydrolysis of the trimethyl chlorosilane gives a dimer which is used as a source of chain stopper in the preparation of linear polymer.

$$2(CH_3)_3SiCl \rightarrow 2(CH_3)_3SiOH \rightarrow \underset{\text{Hexamethyldisiloxane}}{(CH_3)_3Si{-}O{-}Si(CH_3)_3}$$

6.12 APPLICATIONS

(1) In biological and agriculture science: To prevent the infection of the young plant, the seeds are treated with organo mercurials such as ethyl mercury chloride or phosphate or with phenyl mercury derivatives. To prevent the cross-infection in hospitals a permanent antiseptic action is desired, is done by treating the objects at the time of manufacture with tri-alkyl tin compounds, such as bis (tri-butyl tin) oxide, $(C_4H_9)_3\ SnOSn(C_4H_9)_3$.

(2) In medicine: Mercurochrome is used as an antiseptic. Mercuhydrin is used in medicine as a diurectic in treatment of oedema. Salvarson or Arsphenamine is effective against spirochetes and trypansomes. NeO solvarson is effective against sleeping sickness.

(3) In physical science: The organometallic compounds of sodium, lithium, magnesium, aluminium, zinc and mercury are used in the synthetics of different compounds. Tetramethyl silane is used as a standard in NMR spectroscopy and a number of organometallic compounds are used as catalyst to speed up reactions. The use of organometallic compounds and catalyst results in the synthesis of dyes, drugs and fine chemicals.

(4) Gasoline additives: For example, tetra ethyl lead is used as an antiknock agent. Ferrocene is used as an additive for promoting smokeless combustion fuels.

(5) Industry and arts: Organolithium and organo aluminium compounds are used as catalysts in industries. Ziegler-Natta catalyst is used in co-ordination polymerization.

(6) In organic polymer: Organometallic compounds are also used in the synthesis of inorganic polymers and elastomers — suitable for use at very high temperature.

(7) Organosilicon compound: Silicones can withstand higher temperature than organic liquids and consequently silicone oils are used in high temperature boilers. These are also used as the suppression of foam in lubricating oil varnishes, etc.

(8) Carbonyl catalysts: The carbonyls of iron, nickel and cobalt are used as catalyst in the oxo process for making alcohols from hydrocarbons. The carbonyls of transition metals are used in the Fischer – Tropsch process for the synthesis of hydrocarbons.

(9) In polymerization:

1. *Alkene dimerization:* Dimerization or polymerization may occur reductively (i.e., with the addition of hydrogen atom)

$$2RCH = CHR' \xrightarrow{+2H} RCH_2\,\underset{\displaystyle R}{\underset{|}{CH}} - \underset{\displaystyle R}{\underset{|}{CH}}\,CH_2R$$

or oxidatively (i.e., with loss of hydrogen atoms)

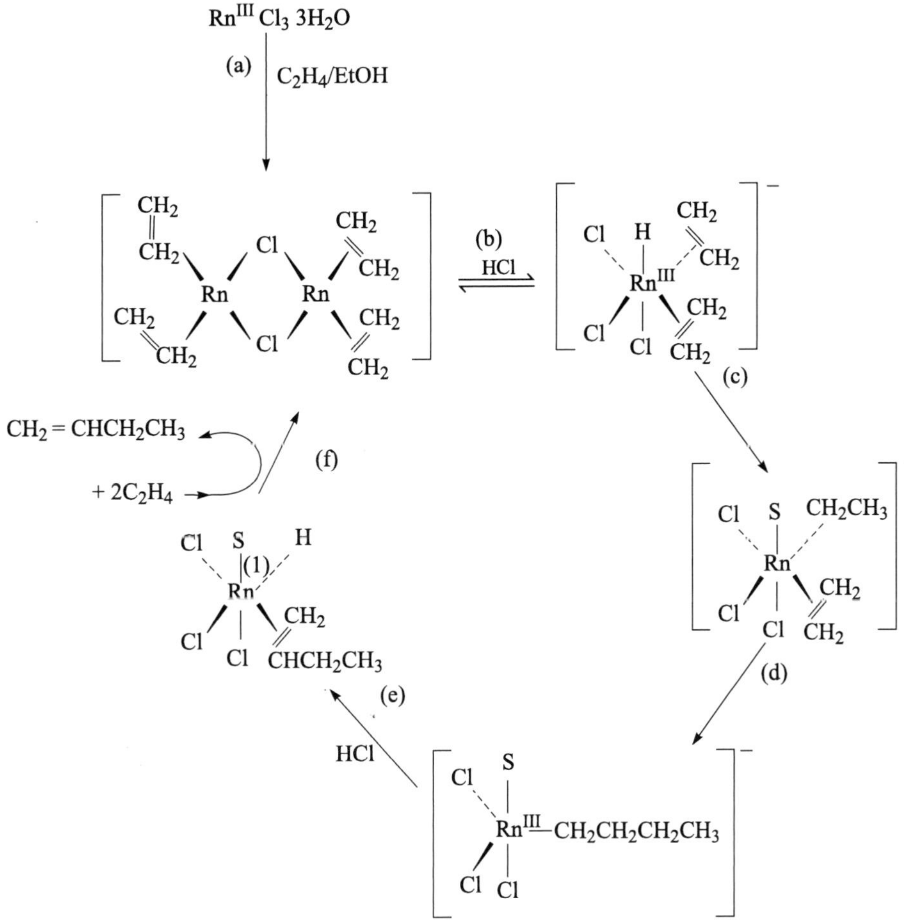

Fig. 6.3 *Mechanism of ethylene dimerization by rhodium trichloride*

$$2RCH = CHR' \xrightarrow{-2H} RCH = \underset{\displaystyle H}{\underset{|}{C}} - \underset{\displaystyle H}{\underset{|}{C}} = CHR'$$

Out of a large number of dimerization processes of simple alkenes which have been catalyzed by a wide variety of dervatives of transition metal (**Example:** CO, Ni, Ru, Rh and Pd). The most studied reaction is probably the dimerization of ethylene by rhodium trichloride for which the mechanism has been proposed on the basis of kinetic and spectroscopy studies. According to this mechanism, the key steps involved in the dimerization of ethylene are:

1. Ethylene coordination and reduction of Rh (III) to Rh (I)
2. Oxidative addition of HCl
3. Insertion of C_2H_4 into Rh — H bond
4. Insertion of C_2H_4 into Rh — C bond
5. Reductive elimination of HCl and reversion to π-butane complex
6. Ethylene coordination and dissociation of the butane ligand

2. *Alkene polymerization:* The polymerization of alkanes to give commercially useful fibres and plastics is a thermodynamically favoured process. However, the alkene must first be activated by a catalyst, so that the polymerization reaction may proceed at a reasonable rate. A variety of organometallic compounds have been frequently used for this activation or catalytic process and the reactions in their presence generally lead to stereoregular polymers.

Ziegler-Natta polymerization of alkenes: It is a heterogeneous catalyst which is formed by the combination of transition metal of group 1 to group 8 with organometallic compound of group 1 to group 3. It consists of a bimetallic coordination complex. These catalysts are made up of a transition metal salt particularly titanium tetrachloride and a metal alkyl triethyl.

Example: triethyl aluminium titanium tetrachloride (Et_3Al, $TiCl_4$). The coordination polymerization has brought about a revolution in the field of polymerization. The essence of the method is the insertion of the alkene into the C – Ti bond of the growing polymer (Fig. 6.2a) which leads to stereoregular isotactic polymers of terminal olefins of high density and high crystallinity chain termination is by hydrogenation. The active catalyst is the titanium complex holding an ethyl group. The long chain of linear polyethylene thus produced lies together in a regular manner in the solid without the defects in regularity imposed by the random branches of the high temperature polymer. In addition to polymerization which proceeds exclusively in head to tail fashion with Ziegler–Natta catalyst, the polymer produced has the methyl groups entirely on side of the zig zag backbone (Fig. 6.2(b)). This isomeric form is termed isotactic where there is significant steric hindrance between methyl appendages, consequently the backbone twists regularly to give a helical shape to the polymer. It is used in the synthesis of polymers of 1-alkenes.

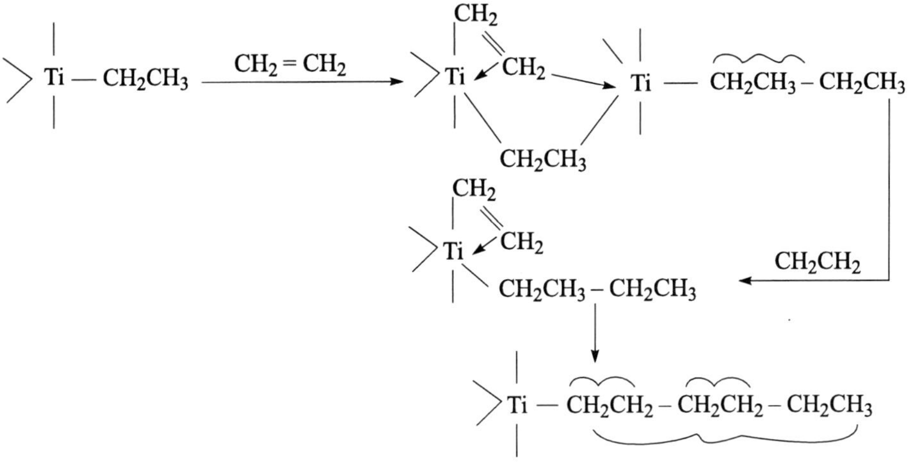

A molecule of polyethylene

(a)

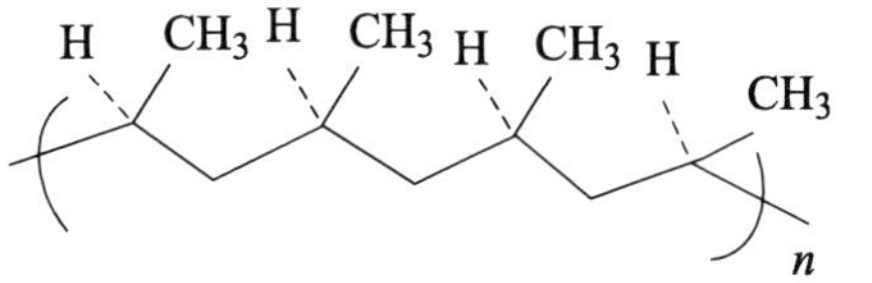

Polypropylene isotactic

(b)

Fig. 6.4 *Mechanism of Ziegler-Natta polymerization of alkene*

Review Questions

1. Define organometallic compounds, their classification and general methods of their preparation.
2. Give the preparation, properties and uses of organometallic compounds of lithium.
3. Discuss the applications of organometallics in homogeneous catalysis.
4. Give mechanism of the following reactions.
   (a) Homogeneous hydrogenation of alkenes
   (b) Waker process
   (c) Hydroformylation
5. What are Ziegler – Natta catalysts? Discuss alkene polymerization using them.
6. Give the mechanism of ethylene dimerization by rhodium trichloride.
7. Discuss various applications of organometallic compounds.
8. Discuss two organometallic compounds used as polymers.

9. How organometallics are used for polymerization reaction?
10. What is the role of $Al(C_2H_5)_3$ in the catalytic system?
11. How Grignard reagent is prepared? Discuss its important use in the formation of silicon polymer.
12. What is the mechanism of Ziegler – Natta polymerization?

7

Organic Reactions and Synthesis of a Drug Molecule

7.1 INTRODUCTION

Organic reactions are chemical reactions involving organic compounds. The basic organic chemistry reaction types are addition reactions, elimination reactions, substitution reactions, pericyclic reactions, rearrangement reactions, photochemical reactions and redox reactions. In organic synthesis, organics are used in the construction of new organic molecules. The production of many man-made chemicals such as drugs, plastics, food additives, fabrics depend on organic reactions. The oldest organic reactions are combustion of organic fuels and saponification of fats to make soap.

7.2 COMMONLY USED TERMS

1. **Nucleophiles:** The name nucleophile means nucleus loving and indicates that it attacks regions of low electron density (positive centres) in the substrate molecule. Nuclophiles are electron rich. They may be negative ions including carbanions or neutral molecules with free electron pair.
   It can be defined as, "a reagent which can donate an electron pair in a reaction is called a nucleophile".
   It can be represented by a general symbol, Nu:
   Example: Cl^-, Br^-, I^-, CN^- OH^-, RCH_2^-, NH_3, RNH_2, H_2O, ROH
   All nucleophiles are also Bronsted bases, but they may be strong bases or weak bases. Only weak bases can exist under acidic conditions. So under acidic conditions, the nucleophiles we tend to see are weak bases only. This means either they are neural or they are anionic but from the IIIrd row of the periodic table or below (usually Cl^-, Br^-, or I^-). Under basic conditions, though any nucleophile can exist.
2. **Electrophiles:** A reagent which can accept an electron pair in a reaction called an electrophile. The name electrophile means electron loving and indicates that it attacks regions of high electron density (negative centres) in the substrate molecule. Electrophiles are electron deficient. They may be positive ions including carbonium

ions or neutural molecules with electron deficient centres. An electrophile can be represented by ε^+.

Example: H^+

Cl^-

Br^+

I^+

NO_2^+, R_3C^+, S^+O_3H, $AlCl_3$, BF_3

7.3 CLASSIFICATION OF THE ORGANIC REACTION

Reaction type	Sub-type	Examples
1. Addition reaction	Electrophilic, nucleophilic	Halogenation Hydrohalogenation hydration
2. Elimination reaction		Dehydration
3. Substitution reaction	Nucleophilic aliphatic substitution Nucleophilic aromatic substitution Nucleophilic acyl substitution Electrophilic substitution Electrophilic aromatic substitution	With SN^1, SN^2 and SN reaction mechanism

7.3.1 Nucleophile Addition Reaction

A nucleophilic addition reaction is a chemical addition reaction in which a nucleophile form a sigma bond with an electron-deficient species.

So there are three fundamental events in a nucleophile addition reaction.

1. Formation of the new sigma bond between the nucleophile (Nu) to the electrophlic C of the C = O group.
2. Breaking of the π bond to the resulting in the formation of an intermediate alkoxide
3. Protonation of the intermediate alkoxide to give an alcohol derivative.
   The carbon oxygen double bond is directly attacked by strong nucleophiles to give rise to the alkoxide. However, when weak nucleophiles are used, the carbonyl group must be activated with the help of an acid catalyst for the nucleophilic addition reaction to proceed.

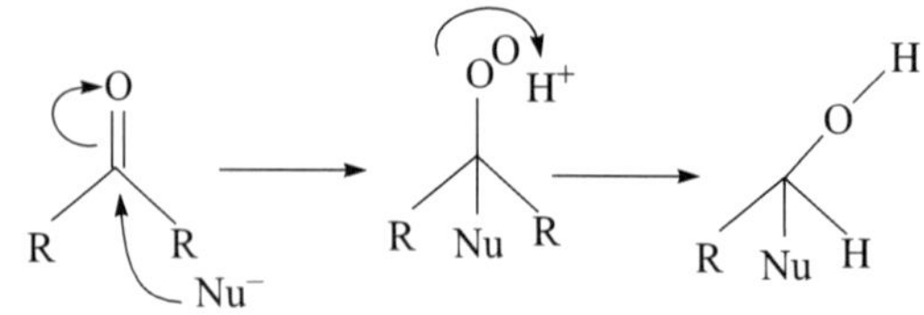

The carbonyl group has a coplanar structure and its carbon is SP^2 hybridized. However, the attack of the nucleophile on the C = O group results in the breakage of the π bond. The carbonyl carbon is now SP^3 hybridized and forms a sigma bond with the nucleophile. The resulting alkoxide intermediate has a tetrahedral geometry. For example, reaction with hydrogen cyanide.

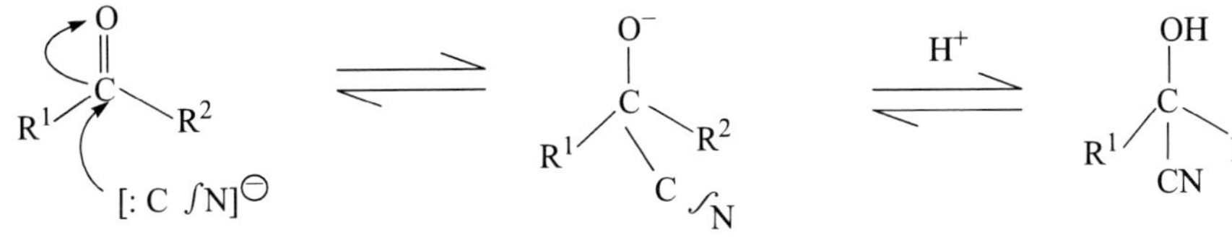

7.4 ADDITION REACTION

An addition reaction is a reaction where two smaller molecules react to form a bigger molecule with no other products.

For example:

ethene + bromine ⟶ 1, 2 dibromoethane

$^{1}C_2H_4 + Br_2 \longrightarrow C_2H_4Br_2$

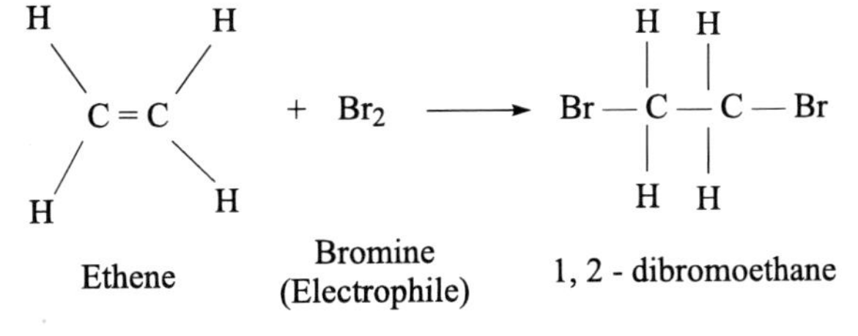

Rules for addition reaction: The following three rules can be used when working with addition reactions:

1. Addition reactions are associated with unsaturated compounds. An unsaturated compound is a compound that contains double or triple bonds.
2. When completed, addition reactions lead to a single compound, which is the final product. Addition reaction occurs when your starting compound or reactant is unsaturated.
3. Atoms can only be added to molecules when forming a product, nothing is eliminated.

7.4.1 Eletrophilic Addition Reaction

An electrophilic addition reaction is an addition reaction where in a chemical compound a π bond is broken and two new σ bonds are formed. The substrate of an electrophilic addition reaction must have a double or triple bond. The driving force for this reaction is the formation of an electrophile X that forms a covalent bond with an electron-rich unsaturated $C = C$ bond. The positive charge on X is transferred to the carbon-carbon bond forming a carbocation.

For example:

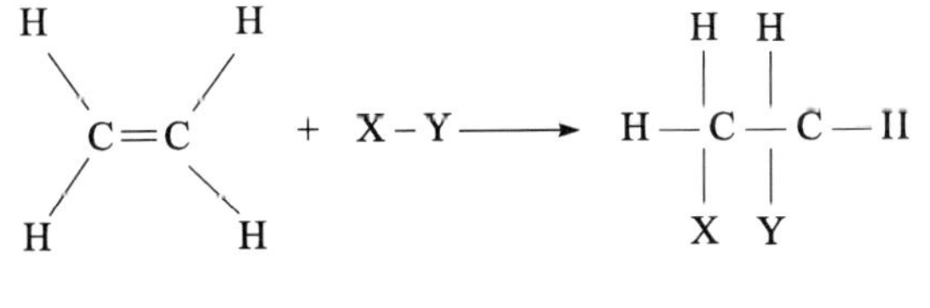

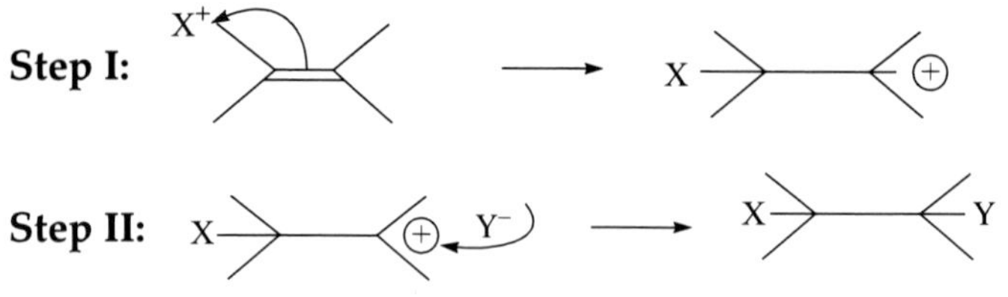

In step I, the positively charged intermediate combines with (X), that is, electron-rich and usually an anion to form the second covalent bond.

In step II, is the same nucleophilic attack process is found in an SN^1 reaction, the exact nature of the electrophile and the nature of the positively charged intermediate are not always clear and depend on reactants and reaction conditions.

Markovnikov's rule: In an addition reaction of a protic acid HX (HCl, HBr or HI) to an alkene or alkyne, the hydrogen atom of HX becomes bonded to the carbon atom that had the greatest number of hydrogen atoms in the starting alkene or alkyne.

For example:

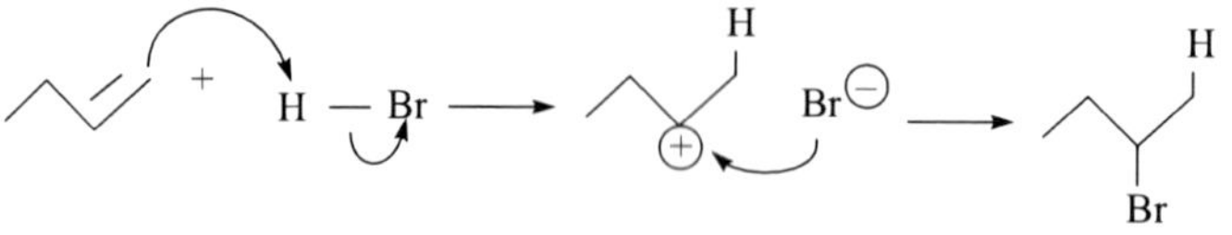

Notice that how the hydrogen from the H — Br is attached at the terminal carbon of the alkene because that carbon originally had 2 hydrogens whereas the carbon on the other side of the double bond has only one.

Anti-Markovnikov's rule: In an addition reaction of a generic electrophile HX to an alkene or alkyne, the hydrogen atom of HX becomes bonded to the carbon atom that has the least number of hydrogen atoms in the starting alkene or alkyne.

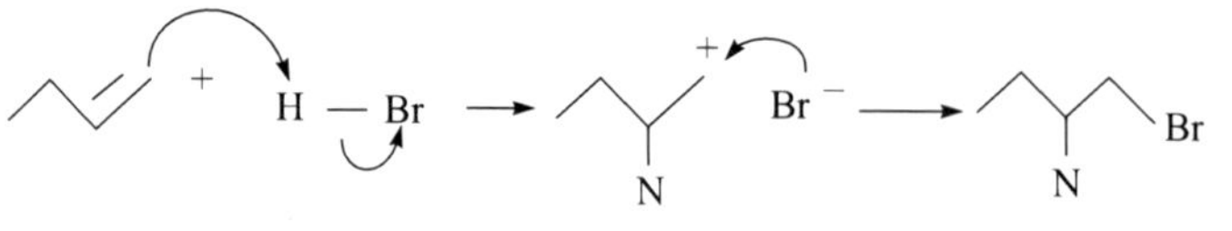

Notice that the hydrogen from the H — Br is now attached at the carbon of the alkene that had only hydrogen thus does not follow Markovnikov's rules and is a minor product.

7.5 SUBSTITUTION REACTION

The reactions in which an atom or group of atoms in a molecule is replaced or substituted by different atoms or group of atoms is called substitution reaction.

Example:

$$CH_4 + Cl_2 \longrightarrow CH_3Cl + HCl$$

(H replaced by Cl)

$$CH_3CH_2I + KOH\,(aq) \longrightarrow CH_3CH_2OH + KI$$

(I replaced by OH)

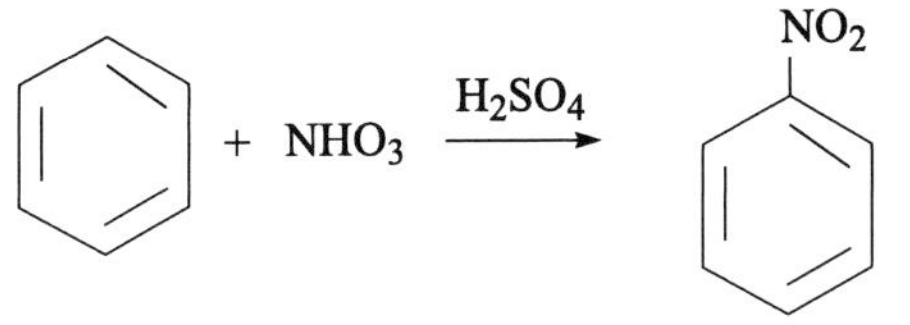

(H of benzene ring replaced by NO_2 group)

7.5.1 Types

There are many types, some of them are described as follows:

(1) Nucleophilic substitution: Nucleophilic substitution is a fundamental class of substitution reactions in which an electron-rich nucleophilic selectively bonds with or attacks the positive or partially positive charge of an atom attached to a group or atom called the leaving group. The positive or partially positive atom is referred to as an electrophile.

The most general form of reaction may be given as the following:

$$NuC{:} + R - LG \rightarrow R - NuC + LG{:}$$

The electron pair (:) from the nucleophile (NuC) attacks the substrate (R — LG) forming a new bond, while the leaving group (LG) departs with an electron pair. The principal product in this case is R — NuC. The nuclophile may be electrically neutral or negatively charged whereas the substrate is typically neutral or positively charged.

Example: $R - Br + OH^- \longrightarrow R - OH + Br^-$

Attacking nucleophile $\longrightarrow OH^-$

Leaving group $\longrightarrow Br^-$

Nucleophilic substitution reaction can be broadly classified as:

1. Nucleophilic substitution at saturated carbon centres.
2. Nucleophilic substitution at unsaturated carbon centres.

2. Nucleophilic substitution at saturated carbon centres: In 1935, Edward D. Hughes and Sir Christopher Ingold studied nucleophilic substitution reaction of alkyl halides and related compounds. They proposed that there were two main mechanisms at work, both of them competing with each other. The two main mechanisms are the SN^1 reaction and SN^2 reaction.

S stands for chemical substitution
N stands for nucleophilic
Number represents the kinetic order of the reaction.

7.5.1.1 Substitution Nucleophilic Unimolecular (SN^1)

The meaning of SN^1 reaction lies in the name itself "substitution, nucleophilic and unimolcular". Therefore, this reaction will follow the first order kinetics. In another term, we can say that the rate determining step in this reaction is unimolucular.

Moreover, the rate of reaction will depend upon the concentration of one species which in this case is holoalkane or alkyl halide.

For example:

$$(CH_3)_3C - Br + OH^- \rightarrow (CH_3)_3C - OH + Br^-$$

$$\text{Nucleophile} \rightarrow OH^-$$

From the reaction, we can conclude that SN^1 reaction is a two-step process and it leads to the formation of carbocation intermediates.

First step: Polarization and cleaving of C – X bond occur to form carbocation intermediates. The first step is a reversible process. It is also the rate determining step of the reaction.

$$(CH_3)_3C - Br \rightleftharpoons (CH_3)_3 C^+ + Br^-$$

Second step: Nucleophile attacks the carbocation to form the respective product.

$$OH^- + (CH_3)_3C^+ \longrightarrow (CH_3)_3 C - OH$$

The planar nature of carbocation results in both inversion and retention configuration.

$$(CH_3)_3 C - Br \rightleftharpoons (CH_3)_3 C^+ + Br^-$$

(Inversion and retention)

Mechanism: The mechanism of an SN^1 reaction occurs in two steps.

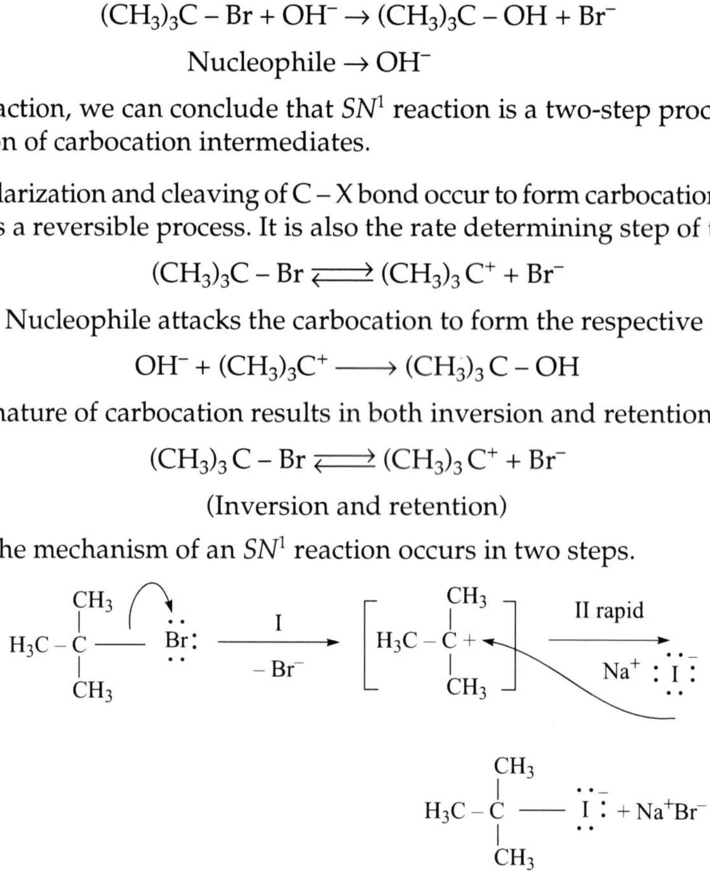

Step I: The slower rate limiting dissociation of the alkyl halide forming a C^+ intermediate.

Step II: A rapid nucleophilic attack on the C^+.

Conditions of SN^1 Reaction: The rate of reaction will depend on the concentration of the alkyl halide and it will not depend upon the nucleophile. This is because the rate always depends on the slowest step which in this case is the breakdown of the C – Br bond in order to form a carbocation. Therefore, the reaction will be a first order reaction. The breakdown energy for breaking the bond is obtained from solvation of the leaving group.

Water and alcohol are types of polar protic solvent. These solvents have the capacity to attract the halogen group thereby facilitating the breakdown of C – X bond in the reaction. This will result in the formation of carbocation intermediate. Additionally, the stabilization of the leaving groups is possible with the help of the protic solvents through hydrogen bonding.

The stability of carbocation will determine the speed of the rate of reaction (more the stability faster the reaction). Therefore, the more easily the nucleophile can attack the compound and leads to substitution process.

Order of Reactivity: Therefore, the reactivity of halides towards SN^1 reaction is R – I > R – Br > R – Cl > R – F. We know that C – I bond of the alkyl iodides can be easily broken and it is easy to release I^- or the leaving group. Therefore, it becomes easy for the nucleophile to attack the alkyl halides and result in the substitution. On the other hand, it is tough to break C – F bond because it is very strong and, in turn, it is difficult for the nucleophile to attack. The reactivity order of alkyl halides in the case of SN^1 reaction is 3° > 2° > 1° > methyl. The same reason is responsible for more reactivity of compounds such as benzylic halide and allylic halides towards SN^1 reaction because it leads to the formation of highly stable resonance structures of carbocation intermediates.

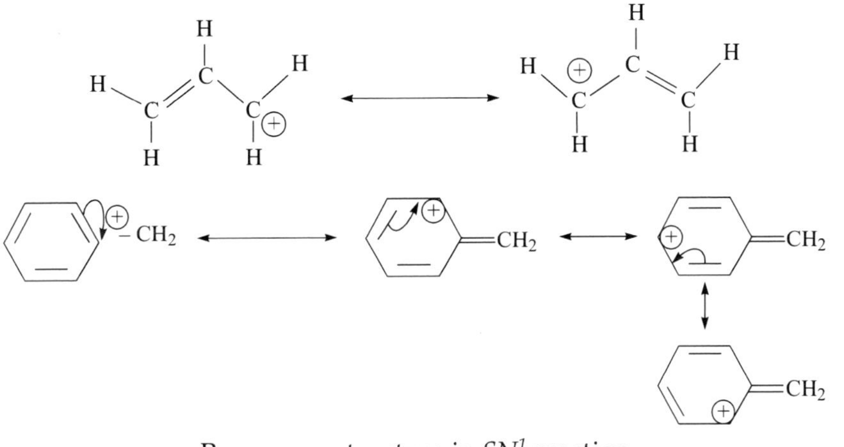

Resonance structure in SN^1 reaction

7.5.1.2 *Substitution Nucleophilic Bimolecular (SN^2)*

It follows second order kinetics. The rate determining step depends on both the concentration of alkyl halides (R – X) and the nucleophile present in the reaction. The SN^2 reaction is a one-step process and there is no formation of intermediates. The basic mechanism of the reaction is as follows

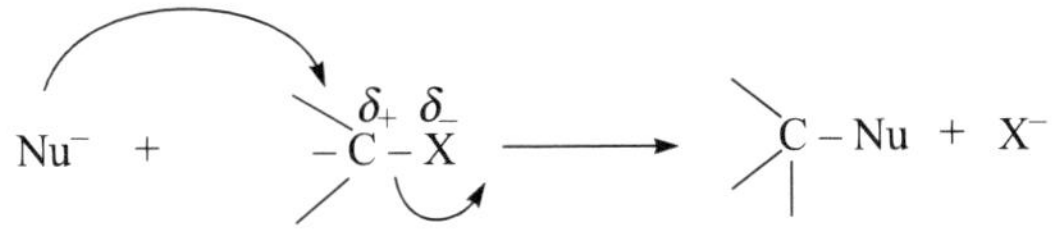

In this reaction the addition of the nucleophile and the elimination of the leaving group take place simultaneously.

Mechanism: For example, CH_3Cl haloalkane reacting with the nucleophile

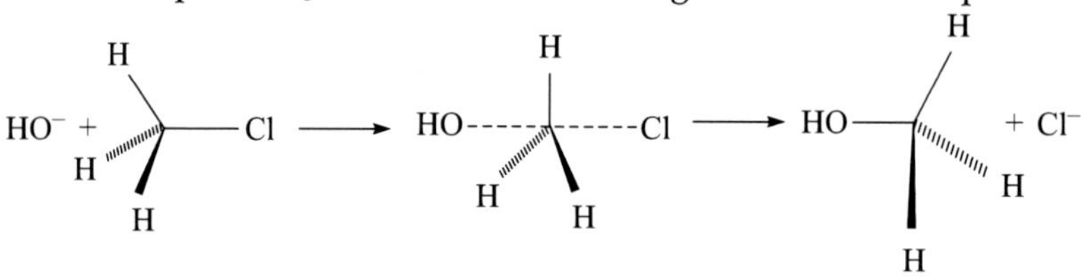

This reaction is a one-step process, unlike SN^1 reaction. In this reaction, there is no formation of intermediates. The reaction undergoes a transition state where the nucleophiles are attached to the alkyl halides. Hence, it is very clear the rate determining step is dependent on the concentration as well as the nucleophile.

Table 7.1 Comparison between SN^1 reaction and SN^2 reaction

SN^1 Reaction	SN^2 Reaction
1. Two-step process	One-step process
2. Formation of carbocation intermediates	Lack of carbocation intermediates formation
3. The strength of nucleophile is not important	Requirement of a strong nucleophile
4. Configuration retention and inversion	Configuration inversion
5. The requirement of polar protic solvent	The requirement of polar aprotic solvent

However, the transition state is highly unstable because in the transition state carbon is bonded to 5 atoms. This is because there is simultaneous breaking of C – Cl bond and the formation of C – OH bond. Moreover, the nucleophile attacks the alkyl halide substrate is a back-side method thereby inverting the configuration of the product.

Order of Reactivity: Since the attack of the substrate occurs from back-side method, it is not favourable to use a bulky nucleophile or a bulky substrate. This is because the SN^2 mechanism is very prone to steric inhibition. Hence, the order of reactivity of alkyl halide towards this type of reaction is $1° > 2° > 3°$.

The order of reactivity follows this particular order because as the crowding around the actual C – X bond increases, the steric inhibition increases. Thus, it decreases the reactivity in the SN^2 reaction. Thus, it is very easy to react if the alkyl halide substrate is methyl halide. It is also quite convenient to undergo SN^2 reaction in ethyl halide but it is difficult to under go SN^2 reaction in case of a secondary and tertiary halide such as isopropyl halide and *t*-butyl halide.

The order of reactivity of halides are:

$$R-I > R-Br > R-Cl > R-F$$

Conditions of the SN^2 chemical reactions: There is a requirement of the strong nucleophile to undergo SN^2 reaction. The reaction takes place in the presence of solvents that are polar aprotic such as DMSO (dimethyle sulfoxide) DMF (dimethyl formamide). The SN^2 reaction does not occur in the presence of polar protic solvent because these types of solvents deactivate the charged nucleophile during the bond formation of polar protic solvent with the strong nucleophile. Thus, it decreases the reactivity of the nucleophile and eases the possibility of an SN^2 reaction.

7.5.1.3 *Nucleophilic Substitution at Unsaturated Carbon Centres*

Nucleophilic substitution via, the SN^1 or SN^2 mechanism does not generally occur with vinyl or aryl halides or related compounds. When the substitution occurs at the carbonyl group, the acyl group may undergo nucleophilic acyl substitution. This is the normal mode of substitution with carboxylic acid derivatives such as acyl chlorides, esters and amides.

7.5.2 Nucleophilic Aromatic Substitution

A nucleophilic aromatic substitution is a substitution reaction in organic chemistry in which the nucleophilic displaces a good leaving group such as a halide on an aromatic ring.

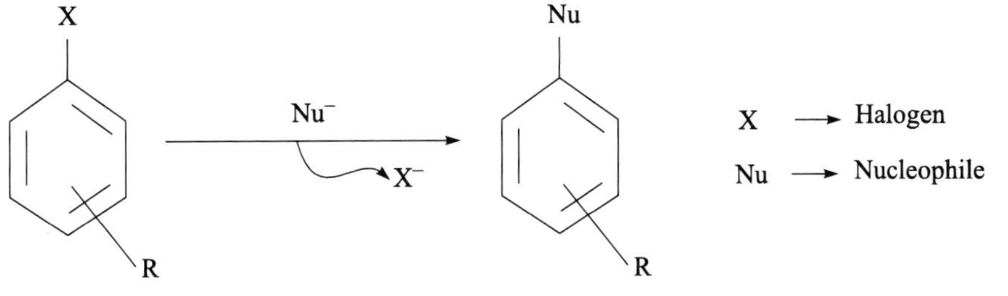

7.5.3 Electrophilic Substitution Reaction

An electrophilic substitution is a reaction in which an electrophile (ε) substitutes for another atom or group (Y) in a compound, R – Y.

$$\varepsilon^+ + R - Y \rightarrow \varepsilon - R + Y^+$$

R – can be either an alkyl or an aryl group.

Following are the electrophilic aliphatic substitutions.

1. Nitrosation
2. Ketone halogenation
3. Keto-enol tautomerism
4. Aliphatic diazonium coupling
5. Carbene insertion into C – H bonds
6. Carbonyl alpha-substitution reaction

7.5.4 Electrophilic Aromatic Substitution

Electrophilic aromatic substitution is an organic reaction in which an atom that is attached to an aromatic system (usually hydrogen) is replaced by an electrophile. Some of the most important electrophilic aromatic substitutions are aromatic, nitration, aromatic halogenations, aromatic sulfonation and acylation.

Example:

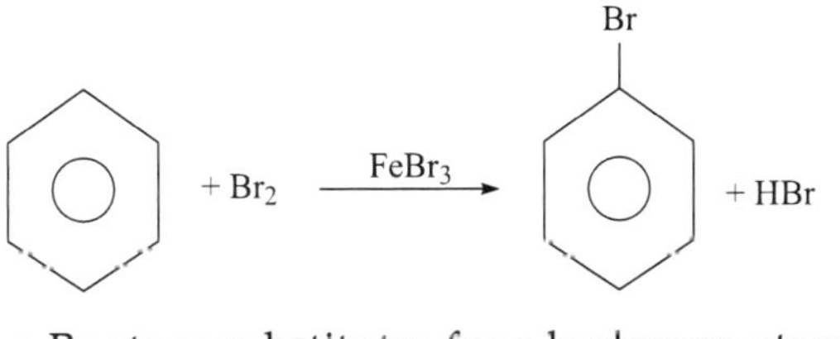

In the above reaction, a Br atom substitutes for a hydrogen atom.

7.6 ELIMINATION REACTION

Elimination reaction is a class of organic chemical reaction in which a pair of atoms or groups of atoms are removed from a molecule usually through the actions of acids, bases or metals and in some cases by heating to a high temperature. It is the main process by which organic compounds containing only single C – C bonds (saturated compound) are transformed to compounds containing double or triple C – C bonds (unsaturated compounds). It is commonly known by the kind of atoms or group of atoms leaving the molecule.

For example:

(1) The removal of a hydrogen atom and a halogen atom is known as dehydrohalogenation.

(2) When both the leaving atoms are halogens the reaction is known as dehalogenation.

(3) The elimination of a water molecule usually from an alcohol is known as dehydration.

(4) When both the leaving atoms are hydrogen atoms, the reaction is known as dehydrogenation.

There are three fundamental events in these elimination reactions.

1. Removal of a proton
2. Formation of the C – C π bond
3. Breaking of the bond to the leaving group

Example:

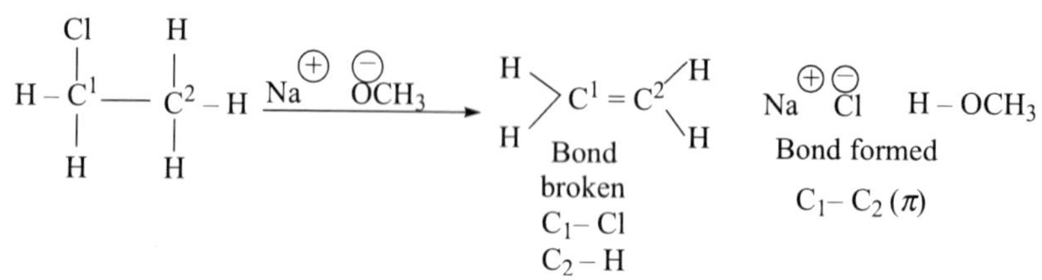

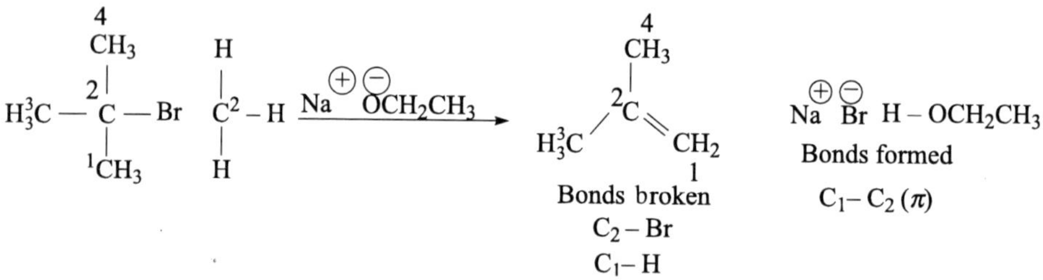

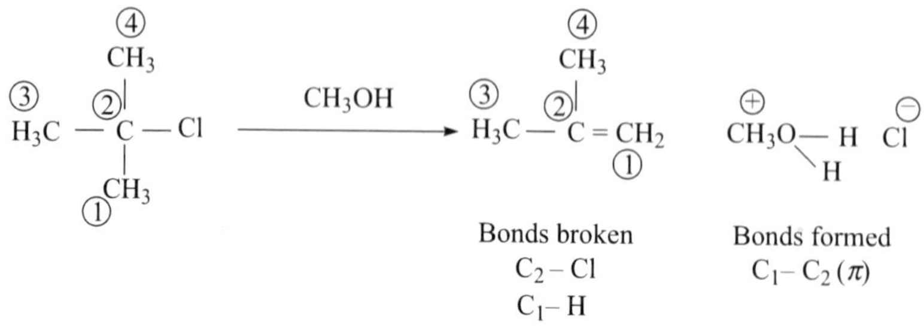

Elimination reactions are also classified as ε_1 or ε_2 depending on the reaction kinetics. In an ε_1 reaction, the reaction rate is proportional to the concentration of the substance to be transformed. In an ε_2 reaction the reaction rate is proportional to the concentration of both the substrate and the eliminating agent.

Concerted (ε_2) Elimination

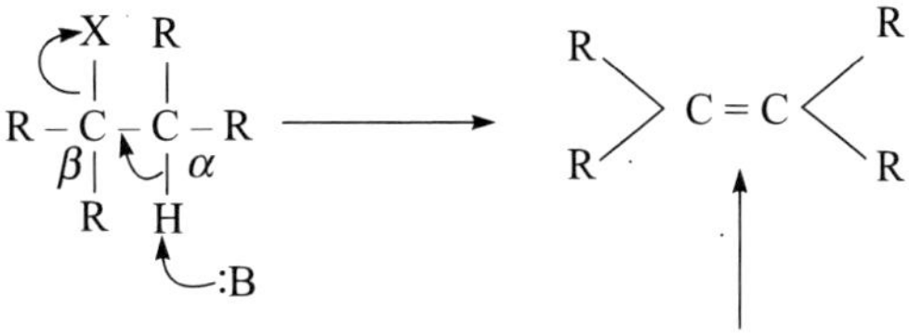

Carbocation (ε_1) Elimination

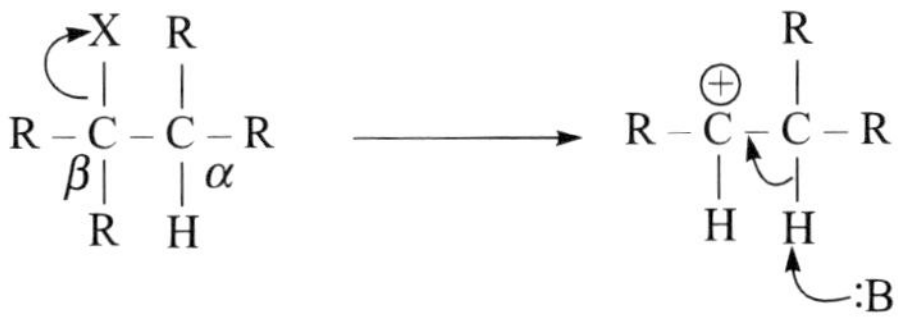

7.7 OXIDATION REACTION

The reaction in which a substance gains oxygen or loses hydrogen is called an oxidation reaction.

For example: The oxidation of primary alcohols to carboxylic acids

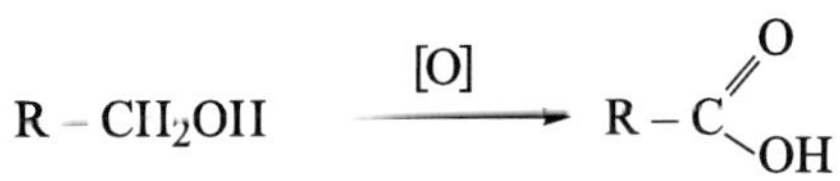

Primary alcohol　　　　Carboxylic acid

$[O] = KMnO_4$

When a primary alcohol is converted to a carboxylic acid, the terminal carbon atom increases its oxidation state by four. Oxidants able to perform this operation in complex organic molecules featuring other oxidation sensitive functional group, must possess substantial selectivity. Potassium permanganate is a very strong oxidant able to react with many functional groups such as secondary alcohols, 1-2 diols, aldehydes, alkenes, oximes and thiols. Under controlled conditions it oxidizes primary alcohols to carboxylic acids very efficiently.

This reaction which was first described in detail by Fournier is typically carried out by adding $KMnO_4$ to a solution or suspension of the alcohol in an alkaline aqueous solution. For the reaction to proceed efficiently the alcohol must be at least partially dissolved in the aqueous solution. This can be facilitated by the addition of an organic co-solvent such as dioxane, pyridine, acetone or *t*-BuOH. It will readily react with a carbon-carbon double bond before oxidizing a primary alcohol. Normally these oxidants are performed under strong alkaline conditions using 1 N NaOH or KOH solution because this promotes a greater oxidation speed and selectivity. In substrate sensitive to strong base, the reaction can be carried out at lower pH or even under acidic conditions at the cost of a greatly decreased reaction velocity. $KMnO_4$ is decomposed in water resulting in formation of manganese dioxide (MnO_2) and gaseous oxygen. This decomposition is catalyzed by acid, base and MnO_2. As the extent of this decomposition is difficult to estimate during the oxidation of primary alcohols, the quantity of $KMnO_4$ must be adjusted during the oxidation by adding it sequentially until the oxidation is complete.

Example: Conversion of alkyl benzenes to benzoic acids using $KMnO_4$.

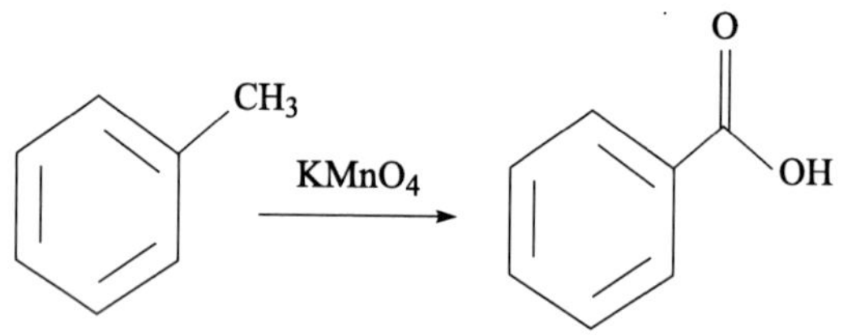

Alkyl benzene Benzoic acid

Chromic acid is also a fairly straightforward reagent. It oxidizes primary alcohols to carboxylic acids and secondary alcohols to ketones. It also oxidizes aldehydes to carboxylic acid.

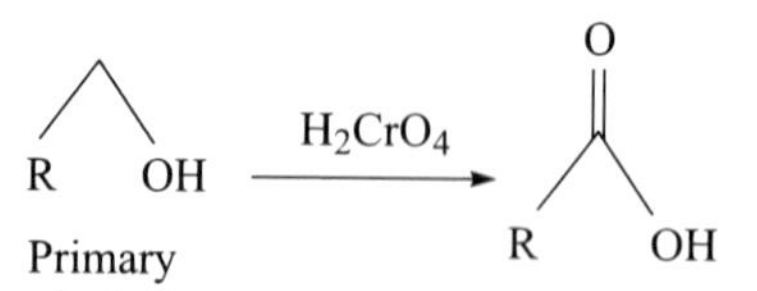

Primary alcohol Carboxylic acid

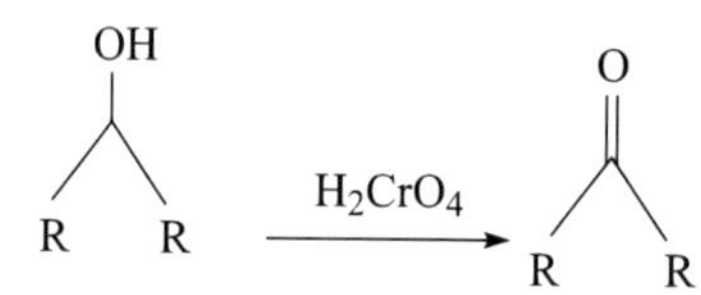

Secondary alcohol Ketone

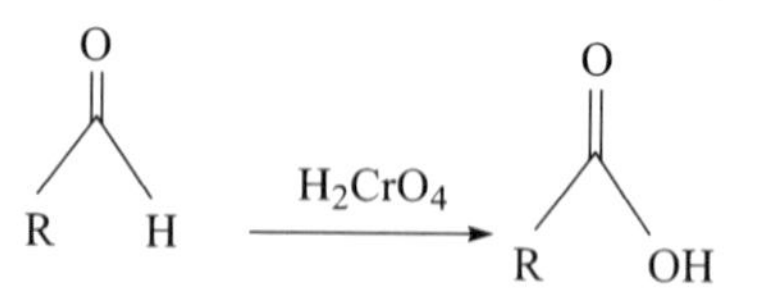

Aldehyde Carboxylic acid

7.8 REDUCTION REACTION

Reduction reactions can be defined by two concepts:

1. Classical or old concept
2. Electronic concept or new concept

7.8.1 Classical or Old Concept

According to the classical concept, reduction reaction is defined as the addition of hydrogen or removal of oxygen.

Example: $Cl_2 + H_2 \rightarrow 2HCl$

$H_2S + Cl_2 \rightarrow 2HCl + S$

7.8.2 Electronic Concept or New Concept

The electronic concept explains reduction reaction more clearly. According to this concept, reduction is a process in which an atom or group of atoms taking part in chemical reaction gains one or more number of electrons. The eventual gain of electrons results in the decrease in positive charge or increase in negative charge of the species.

$$Fe^3 + \overline{e} \rightarrow Fe^{2+}$$
$$Sn^{4+} + 2\overline{e} \rightarrow Sn^{2+}$$

Example: $LiAlH_4$

Preparation: $LiAlH_4$ was first prepared from the reaction between lithium hydride (LiH) and aluminium chloride.

$$4LiH + AlCl_3 \rightarrow LiAlH_4 + 3LiCl$$

Industrial synthesis: The initial preparation of sodium aluminium hydride from the elements under high pressure and temperature.

$$Na + Al + 2H_2 \rightarrow NaAlH_4$$

$LiALH_4$ is then prepared by a salt

$$NaAlH_4 + LiCl \rightarrow LiAlH_4 + NaCl$$

Lithium aluminium hydride ($LiAlH_4$), also abbreviated as LAH, is a reducing agent commonly employed in modern organic synthesis.

1. It is a nucleophilic reducing agent – least used to reduce polar multiple bonds like $C = O$.
2. $LiAlH_4$ can reduce aldehydes to primary alcohols, ketones to secondary alcohols, carboxylic acids and esters to primary alcohols, amides and nitriles to amines, epoxides to alcohols and lactones to diols.
3. $LiAlH_4$ cannot reduce an isolated non-polar multiple bonds like $C = C$. However, the double or triple bonds in conjugation with the polar multiple bonds can be reduced.
4. It is a powerful reducing agent compared to sodium borohydride, ($NaBH_4$), since the Al – H bond is weaker and thus less stable than B – H bond. There is a tetrachedral arrangement of hydrogen around aluminium in aluminium hydride AlH_4^- ion. It is formed by coordination of hydride H^- ions to aluminium Al^{3+} ion. The hybridization in central aluminium is SP^3.
5. The aldehydes or ketones are reduced by $LiAlH_4$ to the corresponding primary or secondary alcohols respectively.

$$H_3C-CHO \xrightarrow[\text{(ii) } H_3O^+]{\text{(i) LiAlH}_4} H_3C-CH_2OH$$

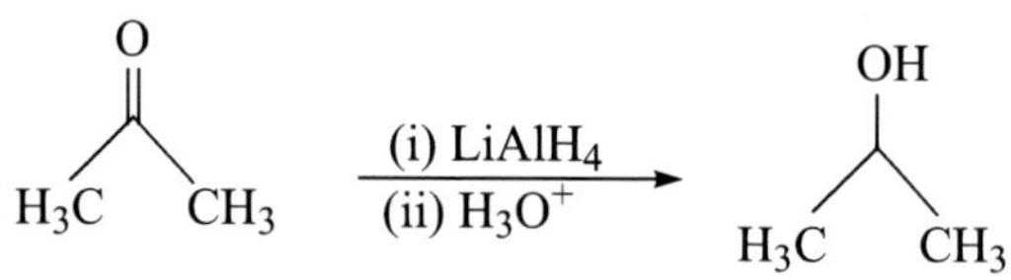

7.8.3 Sodium Borohydride

It is a good reducing agent. Although not as powerful as lithium aluminium hydride ($LiAlH_4$). It is very effective for the reduction of aldehydes and ketones to alcohols. By itself it will generally not reduce esters, carboxylic acids or amides (although it will reduce acyl chlorides to alcohols). It is also used in the second step of the oxymercuration reaction to replace mercury (Hg) with H.

Example:

(1) Reduction of ketones

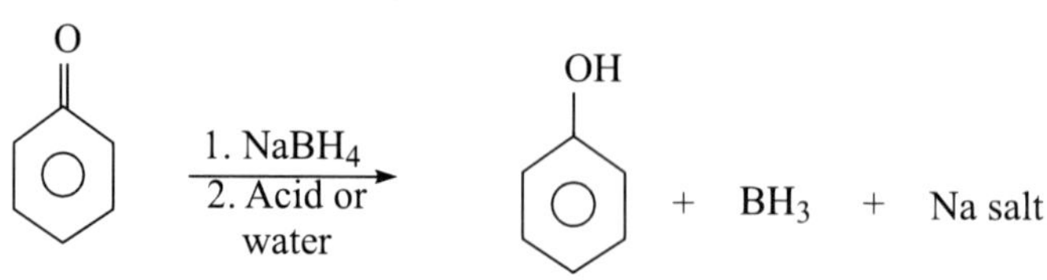

(2) Reduction of aldehydes

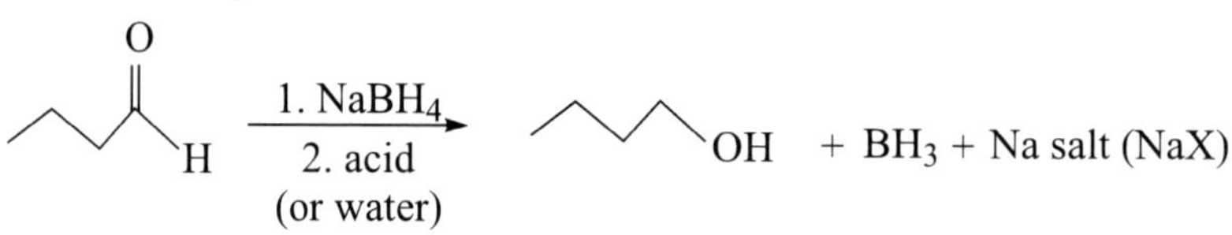

Bond form	Bond Break
C – H	C – O
O – H	B – H
$Na^{\oplus} X^{\ominus}$	$Na^{\oplus} BH_4^{\ominus}$

7.9 SYNTHESIS OF A COMMONLY USED DRUG MOLECULE

7.9.1 Paracetamol

It is also known as acetaminophen or APAP.

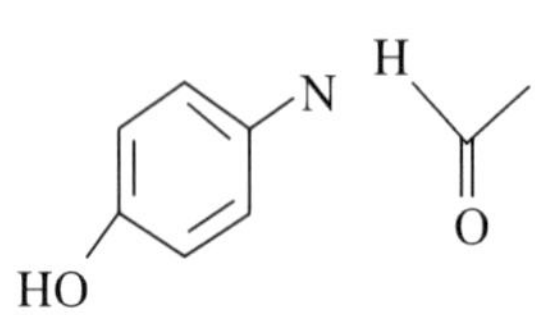

The original method for production involves nitration of phenol with sodium nitrate gives a mixture of two isomers from which the wanted 4-nitrophenol (bP 279°C) can easily be separated by steam distillation. In this electrophilic aromatic substitution reaction, phenol's oxygen is strongly activating, thus the reaction requires only mild conditions as compared to nitration of benzene itself. The nitro group is then reduced to an amine, giving 4-aminophenol. Finally, the amine is acetylated with acetic an hydride.

Industrially direct hydrogenation is used, but in the laboratory scale sodium borohydride serves.

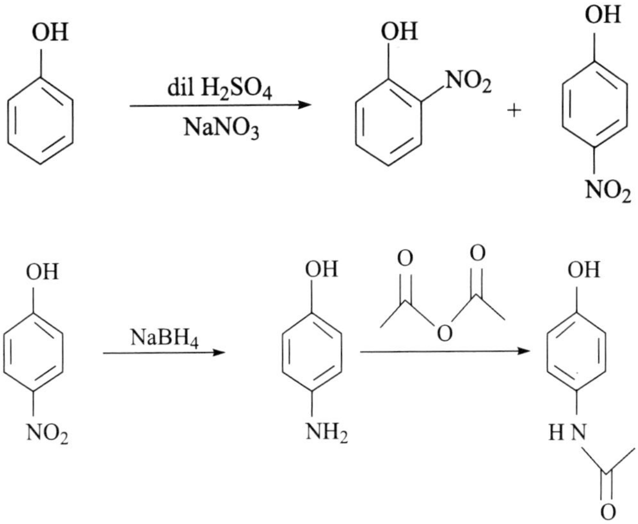

Pharmaceutical Application of Paracetomol

1. It is a medicine used to treat pain and fever.
2. It is typically used for mild to moderate pain relief.
3. Evidence for its use to relieve fever in children is mixed. It is often sold in combination with other medications such as in many cold medications.

7.9.2 Aspirin

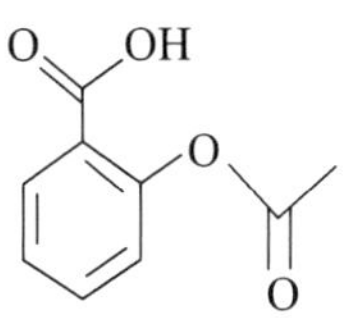

The synthesis of aspirin is classified as an esterification reaction. Salicyclic acid is treated with acetic anhydride an acid derivative causing a chemical reactions that turn salicyclic acids hydroxyl group into an ester group (R – OH → $ROCOCH_3$). This process yields aspirin and acetic acid, which is considered a byproduct of this reaction. Small amounts of sulphuric acid (and occasionally phosphoric acid) are almost always used us a catalyst.

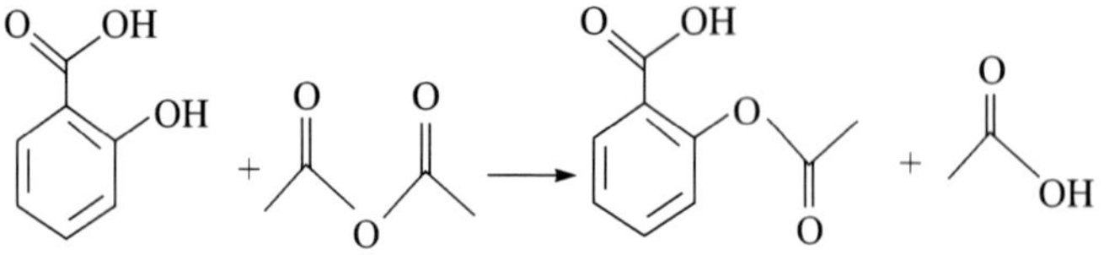

7.9.3 Pharmaceutical Applications of Aspirin

Aspirin is used in the treatment of a number of condititions including fever, pain, rheumatic fever and inflammatory diseases such as rheumatoid arthritis, pericarditis and Kawasaki disease.

Lower doses of aspirin have also been shown to reduce the risk of death from a heart attack or the risk of stroke in some circumstances.

REVIEW QUESTIONS

1. State the kinetic and stereochemical differences between SN^1 and SN^2 reactions. Why is benzyl chloride more reactive than ethyl chloride in both reactions?
2. With the help of example show that the alkylation of acid-catalyzed reaction of ambient temperature.
3. Explain how each of the following factors affects ε_1 and ε_2.
   (a) The structure of alkyl halide
   (b) The concentration of the base
   (c) The strength of the base
   (d) The solvent
4. The addition of bromine to 2-cyclohexyl benzaoate in 1, 2-dichloroethane produces four dibromo derivatives. Propose a mechanism to account for the formation of each compound.

8

Periodic Properties

8.1 EFFECTIVE NUCLEAR CHARGE

The effective nuclear charge is the net charge an electron experiences in an atom with multiple electrons, (or) the attractive positive charge of nuclear protons acting on valence electrons.

It is always less than the total number of protons present in a nucleus due to shielding effect. It is symbolized as zeff or Z^*.

The term effective is used because the shielding effect of negatively charged electron prevents higher orbital electrons from experiencing the full nuclear charge of the nucleus due to the repelling effect of inner-layer electrons (Fig. 8.1). The shielding effect is the name given to the balance between the attraction, between valence electrons and protons and the repulsion between valence and inner electrons. The shielding effect explains the trend in atomic size on the Periodic Table and also why valence electrons are readily removed from an atom. The more shielding that occurs, further the valence shell can spread out. As a result atoms will be larger. It can be approximated by the equation

$$Z_{eff} = Z - S$$

where $Z \rightarrow$ Atomic number

$S \rightarrow$ Number of shielding electrons

Example: Consider a neutral neon atom (Ne), a sodium cation (Na^+) and a fluorine anion (F^-). What is the effective nuclear charge for each.

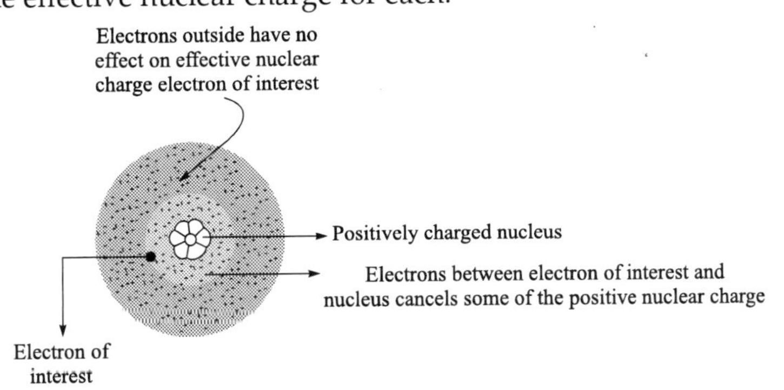

Fig. 8.1 *Effective nuclear charge based on electron shielding*

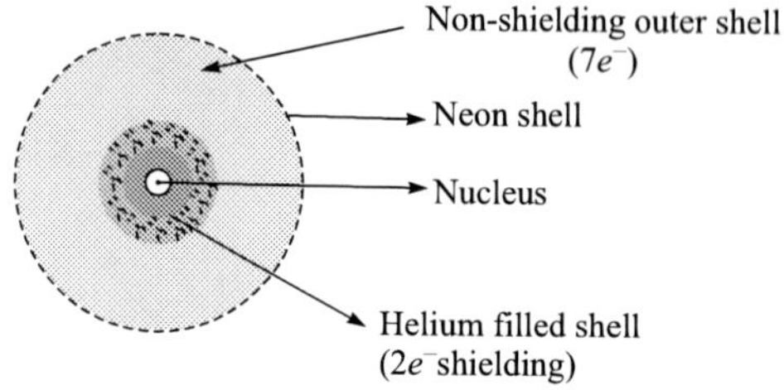

Fig. 8.2 *A fluorine atom showing the extent of effective nuclear charge*

Table 8.1 Effective nuclear charges for selected atoms

Atom	**Sub level**	**Atomic number**	Z_{eff}
H	1*s*	1	1
H_2	1*s*	2	1.69
Li	1*s*, 2*s*	3	2.69, 1.28
Be	1*s*, 2*s*	4	–3.68, 1.91
B	1*s*, 2*s*, 2*p*	5	4.68, 2.58, 2.42
F	1*s*, 2*s*, 2*p*	9	8.65, 5.13, 5.10
Na	1*s*, 2*s*, 2*p*, 3*s*	11	10.63, 6.57, 6.80, 2.51

Ne has 10 electrons
Electronic configuration = $1s^2, 2s^2, 2p^6$
Valence shell = 2
Valence electron = 8

$$Z_{eff}\,(Ne) = 10 - 2 = 8^+$$

Sodium has 11 electrons but the Na^+ ion has lost an electron and thus has 10 electrons.

Once again, the electronic configuration is the same as in previous example and the number of non-valence electron is 2 (by losing one electron, the valence shell becomes the $n = 2$ shell). The atomic number for Na^+ is 11 therefore $Z_{eff}\,(Na^+) = 11 - 2 = 9^+$. Fluorine (Fig. 8.2) has 9 electrons but F^- has gained an electron and thus has 10. The electronic configuration is the same as for neon and the number of non-valence electron is 2. The atomic number for F^- is 9, therefore

$$Z_{eff}\,(F^-) = 9 - 2 = 7^+$$

In each of the above examples (Ne, F^-, Na^+) an atom has 10 electrons but the effective nuclear charge varies because each has a different atomic number. The sodium cation has the largest effective nuclear charge, which results in electrons being held the tightest and therefore Na^+ has the smallest atomic radius (Table 8.1).

8.2 BASIS OF THE EFFECTIVE NUCLEAR CHARGE

An electron might not be able to get that close to the nucleus is because of the other electron. Electrons repel each other and because of the exclusion principle there can only be two electrons in each orbital. For example, lithium electronic configuration is 3. Its first two electrons are in 1*s*. The third electron is in 2*s*. Because the principal quantum number is bigger the 2*s* electron is usually farther from the nucleus than the 1*s* electron. That means that often the 1*s* electrons are between the 2*s* electron and the nucleus. This means that the total coulomb attraction felt by the 2*s* electron is smaller than you would calculate using just nuclear charge and distance because there is also the electron-electron repulsion. This is **the basis of the effective nuclear charge** because of the electron-electron repulsion the outer electrons in a many-electron atom feel less attractive towards the nucleus, so they feel like the nuclear charge is smaller than it actually is. So effective nuclear charge is always smaller than actual nuclear charge. It depends on the type of electrons. Electrons in *s* orbitals, even 4*s* or 5*s* still spend some time right at the nucleus and when they are three, they feel the full nuclear charge, so on average the *s* electrons feel a nuclear charge closer to the actual nuclear charge. Electrons in *d* or *f* orbitals readly do not get very close to the nucleus, so they really get blocked or shielded by inner electrons. They feel a smaller effective nuclear charge than *s* electrons and electrons in *p*-orbitals are in between.

Usually if you do any calculation of orbitals for many-electron atoms you will use effective nuclear charge instead of actual nuclear charge. The size and energy of the orbitals will depend on effective nuclear charge, not on actual nuclear charge. Size and energy of the orbitals determine some very important chemical properties including the size of the element (as an atom ion or in a molecule) and how easily it loses or gains electrons. Electrons take up most of the space in an atom, so orbital size tells you the size. Losing electrons, requires the energy by which they are bound, which is roughly same as the orbital energy. Adding electrons only works if the orbital where they will go is of lower energy than where they came from

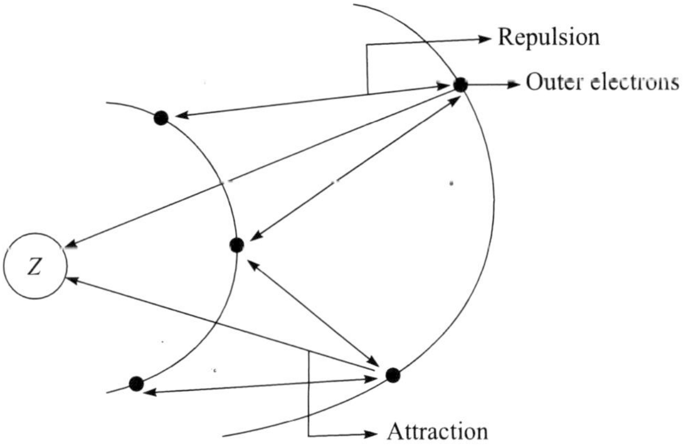

Fig. 8.3 *The effects of electron shielding on outer electron*

8.3 PENETRATION OF ORBITAL

Penetration power is the attracting capacity of any orbital to the electron. The phenomenon or process is related to releasing energy. Orbital penetration refers to how close an electron is able to get to the nucleus while the electron is orbiting around the nucleus (electrons that are orbiting closer to the nucleus are said to be penetrating the nucleus more than the electrons that are orbiting farther away from the nucleus). For example: *s* orbital has maximum penetration effect because when an additional electron enters in the atom then it enters in *s*-orbital penetrating atom. Since *s*-orbital because of being nearest to the nucleus of atom so, nucleus has maximum attraction on that electron and it releases maximum energy whereby value of electron affinity will be maximum for that electron. If additional electron goes in *f*-sub is farthest than *s, p, d* sub-shells so due to far from nucleus *f*-sub shells feels less attraction from the nucleus and releases less energy.

Hence, the order of the penetration effect on the basis of releasing energy will be as following

$$s > p > d > f$$

Pentetration depends on the shell (n) and sub-shell (l) and on their shapes. For instance, if we see the following radial distribution function diagram (about electron density distribution within orbital) among 3*s*, 3*p* and 3*d* orbitals for third shell. (Penetration effect for the same value of shell (n) and their different value of sub-shell.)

If we watch solid line that is on the first peak (electron density) of 3*s* orbital and this peak is near the nucleus so the penetration power of 3*s* orbital for nucleus is more than 3*p* and 3*d* orbital (Fig. 8.5).

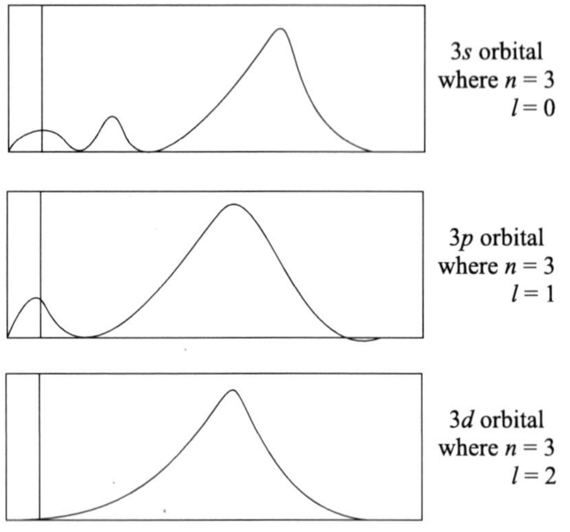

Fig. 8.4 *Penetration of different orbitals*

Similarly, if we watch dotted line, the first peak of electron density of 3*p* orbital is far from solid line and also from nucleus so, the penetration power of 3*p* orbital for nucleus is less than 3*s* and more than 3*d* orbital.

If we watch dashed line, the first peak of electron density of 3*d* orbital is more far from dotted line and also from nucleus (Fig. 8.5) so the penetration power of 3*d* orbital for nucleus is less than 3*s* and 3*p* orbitals.

We come to know from diagram (Fig. 8.6) that penetration effect also depends upon the shape of the orbitals.

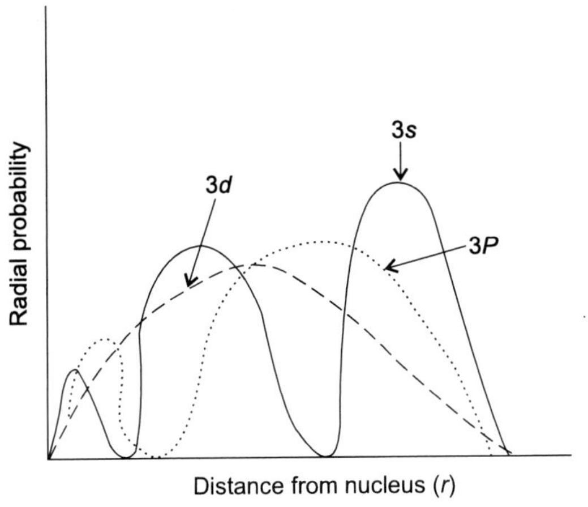

Fig. 8.5 *Penetration of the 1s shielding by the 2s and 2p electrons in lithium significantly lowers those levels*

In a multi-electron system the penetration of the nucleus by an electron is measured by the relative electron density near the nucleus of an atom for each shell and sub shell of an electron.

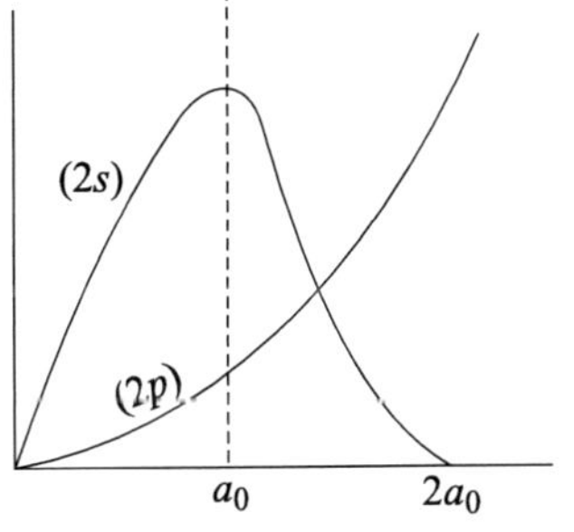

Fig. 8.6 *Penetration of 2s and 2p states inside the first Bohr radius*

1. In multielectron atom due to electron-electron repulsion inner shell electrons repel the coming electrons, so they cannot penetrate deeply.

2. In multielectron atom inner electrons satisfy the nuclear charge so, resultant effective nuclear charge decreases (shielding effect) due to this coming electrons feel less attraction of the nucleus and releases less energy whereby electron affinity gets decreased on going far from the nucleus.

For different values of shell (n) and sub-shell (l) penetrating power of an electron follows this trend.

$$1s > 2s > 2p > 3s > 3p > 4s > 3d > 4p > 5s > 4d > 5p > 6s > 4f > 5d > 6p > 7s$$

8.4 ORBITALS

Orbitals are regions of space where electrons are most likely to be found in an atom. These orbitals are designated as *s*, *p*, *d*, *f* among others. The energy increases as you move up the orbital in such a way that *s*-orbital has the lowest energy and *f*-orbital has the highest energy.

***s*-orbital:** Orbital is symmetrically spherical about the axis and as the energy level increases the electrons are located further away from the nucleus.

***p*-orbital:** These orbitals are oriented in a particular direction, that is, in a given energy level we have three *p*-orbitals, i.e., *Px*, *Py* and *Pz*.

***d*-orbital:** These orbitals are for atoms which exhibit higher energy levels and are composed of five orbitals. That is dxy, dyz, dxz, dx^2y^2 and dz^2.

***f*-orbital:** Principal shells with $n = 4$ can have sub-shell with $l = 3$ and mf values of –3, –2, –1, 0 +1, +2 and +3. These sub-shells consist of seven *f*-orbitals. Each *f*-orbital has three modal surfaces, so their shapes are complex.

8.4.1 Orbital Energies

The energy of atomic orbitals increases as the principal quantum number increases. In any atom with two or more electrons the repulsion between the electrons makes energy of sub-shells. Different values of *l* differ so that the energy of orbital increases within a shell in the order $s < p < d < f$. The 1*s* orbital (at the bottom of the diagram Fig. 8.7) is the orbital with electrons of lowest energy. The energy increases as we move up to the 2*s* and then 2*p*, 3*s* and 3*p* orbitals showing that the increasing values have more influence on energy than the increasing *l* value for small atoms. However, this pattern does not hold for larger atoms. The 3*d* orbital is higher in energy than the 4*s* orbital. Electrons in successive atom on the Periodic Table tend to fill low-energy orbital first (Fig. 8.7). For example: The 5*p* orbital fills immediately after the 4*d* and immediately before the 6*s*. The filling order is based on observed experimental results and has been confirmed by theoretical calculations. As the principal quantum number (*n*) increases the size of the orbital increases and the electron spends more time farther from the nucleus. Thus, the attraction to the nucleus is weaker and the energy associated with the orbital is higher (less stabilized). But this is not the only effect we have to take into account within each shell as the value of *l* increases the electron are less penetrating (meaning there is less electron density found close to the nucleus) in the order $s > p > d > f$ electrons that are closer to the nucleus slightly repel electrons that are farther out offsetting the more dominant electron nucleus attractions slightly. Recall that all electrons have –9 charges but nuclei have +*e* charged electrons in orbitals that experience more shielding are less stabilized and thus higher in energy. For small orbitals (1*s* through 3*p*), the increase in energy due to principal quantum number is more significant than the increase due to *l* however for larger orbitals the two trends are comparable and cannot be simply predicted. The order of increase in energy of various orbitals is given as

$$1s < 2s = 2p < 3s = 3p = 3d < 4s = 4p = 4d = 4f$$

8.4.2 Factors Affecting the Energy of Orbitals

1. For a given value of the principal quantum number the *s* orbital electron will be more tightly bound to the nucleus than *p*-orbital electron, which is more tightly bound in comparison to a *d*-orbital electron.
2. *s*-orbital electrons will have a lesser amount of energy (more negative) than that of *p*-orbital electrons which will have lesser energy than that of *d*-orbital electrons.

3. The extent of shielding from the nucleus is different for electrons in different orbitals. It leads to the splitting of energy levels having same principal quantum number. Thus, the energy of orbitals depends on the values of both principal quantum number (n) and azimuthual quantum number (l). Hence, the lower the value of $(n + l)$ for an orbital the lower is its energy. For the cases where we have two orbitals having the same value of $(n + l)$, the orbital with a lower value of n (principal quantum number) will have lower energy.
4. The energy of the orbitals in the same sub-shell decreases with increase in the atomic number (Z_{eff}).

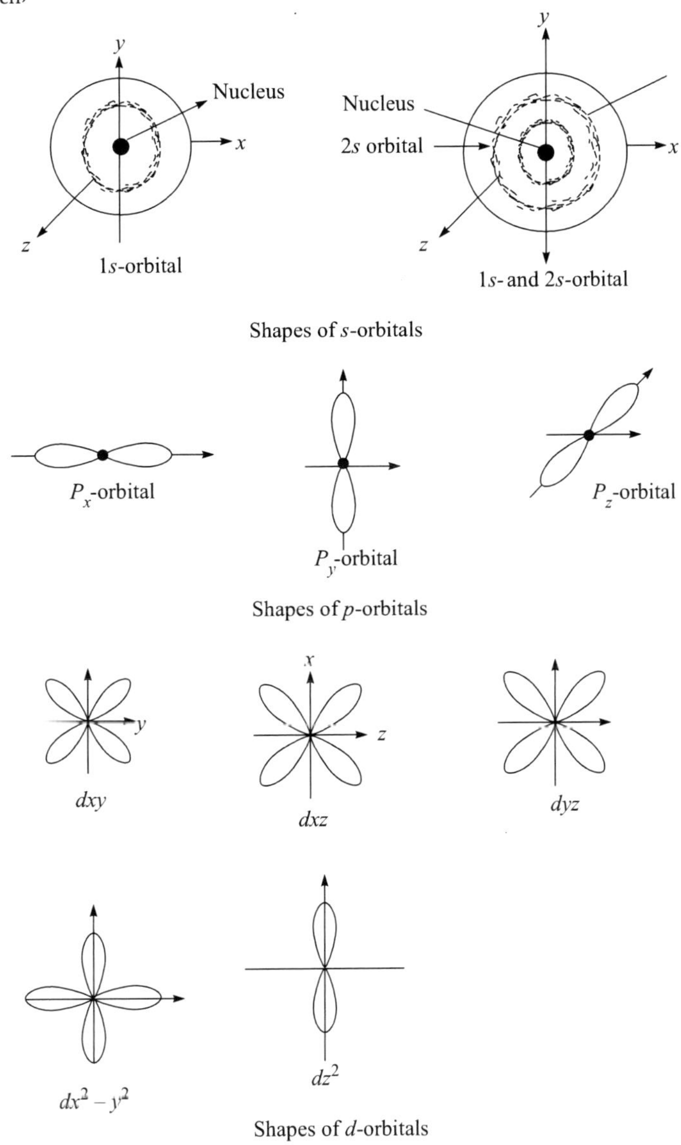

Fig. 8.7 *Shapes of different orbitals*

8.5 ELECTRONIC CONFIGURATION

The electronic configuration of an element describes how electrons are distributed in its atomic orbitals. Electronic configurations of elements describe the types of chemical bonds formed by the element. Elements with similar electron configurations generally exhibit similar properties.

The valency of an element is often determined via its electronic configuration and with the help of a simplified form of the molecular orbital theory.

8.5.1 How to Write Electronic Configuration Shells

The maximum number of electrons that can be accommodated in a shell is based on the principal quantum number (n). It is represented by the formula $2n^2$, where n is the shell number. The shells, values of n, and the total number of electrons that can be accommodated are shown in Table 8.2.

Table 8.2

Shell *n* value	Max. electrons in electronic configuration
K shell, $n = 1$	$2 \times 1^2 = 2$
L shell $n = 2$	$2 \times 2^2 = 8$
M shell $n = 3$	$2 \times 3^2 = 18$
N shell $n = 4$	$2 \times 4^2 = 32$

Sub-shells: The sub-shells into which electrons are distributed are based on the azimuthal quantum number. It is denoted by l.

- This quantum number depends on the value of the principal quantum number (n). Therefore, when n has a value of 4, four different sub-shells are possible.
- When $n = 4$, the sub-shells correspond to $l = 0$, $l = 1$, $l = 2$ and $l = 3$ and are named the *s*, *p*, *d* and *f* sub-shells, respectively.
- The maximum number of electrons that can be accommodated by a sub-shell is given by the formula $2n^2$.
- Therefore, the *s*, *p*, *d* and *f* sub-shells can accommodate a maximum of 2, 6, 10 and 14 electrons, respectively.

All the possible sub-shells for values of n up to 4 are given in Table 8.3.

Table 8.3

Value of principal quantum number	Value of azimuthal quantum number	Resulting sub-shell in the electronic configuration
$n = 1$	$l = 0$	1*s*
$n = 2$	$l = 0$	2*s*
	$l = 1$	2*p*
$n = 3$	$l = 0$	3*s*

	$l = 1$	$3p$
	$l = 2$	$3d$
$n = 4$	$l = 0$	$4s$
	$l = 1$	$4p$
	$l = 2$	$4d$
	$l = 3$	$4f$

Thus, it can be understood that the $1p$, $2d$ and $3f$ orbitals do not exist because the value of the azimuthal quantum number is always less than that of the principal quantum number.

Noted Point

1. With the help of these sub-shell labels, the electronic configuration of magnesium (atomic number 12) can be written as $1s^2, 2s^2, 2p^6, 3s^2$.
2. The electronic configuration of an atom is written with the help of sub-shell labels.
3. These labels contain the shell number (given by the principal quantum number), the sub-shell name (given by the azimuthal quantum number) and the total number of electrons in the sub-shell is in superscript.
4. For example, if two electrons are filled in the s sub-shell of the first shell, the resulting notation is $1s^2$.

8.6 ATOMIC RADIUS

Atomic radius is the distance from the atomic nucleus to the outermost stable electron of a neutral atom.

The atomic radius is a term used to describe the size of the atom. It is impossible to isolate an individual atom and determine its radius because its radius is affected by the association with other atoms. Thus, a number of radii have been defined for an atom depending upon the different types of bonding between the atoms. These are:

1. Covalent radius
2. Van der Walls radius or collision radius
3. Crystal radius (metallic or atomic radius)

8.7 COVALENT RADIUS

Covalent radius is half of the distance between the nuclei of two like atoms joined by a single covalent bond".

For example: We know from spectroscopic data that the distance between the nuclei of two chlorine atoms in a chlorine molecule atom is 1.98 Å. The covalent radius of chlorine atom is therefore,

$$\frac{1}{2} \times 1.98 = 0.99 \text{ Å}$$

In a heteronuclear diatomic molecule *AB*, if both the atoms are linked by a single covalent bond and have nearly the same electronegativity, the bond length (d_{A-B}) is equal to sum of covalent radii of the two atoms.

$$d_{A-B} = r_A + r_B$$

Thus, knowing the bond length and covalent radius of one of the atoms, the covalent radius of other atom can be determined.

$$\begin{aligned} d_{\text{H-Cl}} &= r_{\text{H}} + r_{\text{Cl}} \\ 1.36\ \text{Å} &= 0.37\ \text{Å} + r_{\text{Cl}} \\ r_{\text{Cl}} &= 1.36\ \text{Å} - 0.37\ \text{Å} \\ &= 0.99\ \text{Å} \end{aligned}$$

The bond lengths in free elements and in other covalent molecules can be measured by *X*-ray diffraction, electron diffraction, infrared spectroscopy, nuclear magnetic resonance spectroscopy, etc. This method is used for non-metals.

In some cases, this simple formula is not correct due to the difference in the electronegativities of two different atoms in a molecule. Schomaker and Stevenson (1941) proposed the following relation to calculate the covalent radius.

$$d_{A-B} = r_A + r_B - 0.09\,(X_A - X_B)$$

X_A and X_B are the electronegativities of two atoms *A* and *B*. Later on Pauling gave a general formula for the same as

$$d_{A-B} = r_A + r_B - C\,(X_A - X_B)$$

where *C* is a constant which depends upon the atom involved in the molecule. The value of *C* in *A* is 0.08, 0.06, 0.04 and 0.02 for the *P*-block elements of second, third, fourth and fifth periods, respectively.

8.8 VAN DER WAALS RADIUS (COLLISION RADIUS)

It may be defined as "one half of the distance between the two unbounded atoms when the electrostatic forces between them are balanced". Thus, it is the distance at which the two atoms are in contact without bond formation.

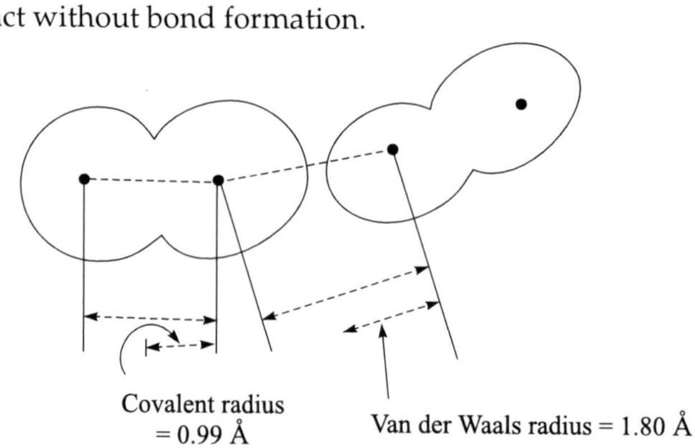

Fig. 8.8 *Covalent and van der Waals radii of chlorine atoms*

For example, (Fig. 8.8) the distance between the nuclei of two chlorine atoms of two neighbouring chlorine molecules is 3.60 Å, the van der Waals radius of chlorine is

Therefore, $$\frac{1}{2} \times 3.60 = 1.80 \text{ Å}$$

The values of van der Waals radii can be obtained by *X*-ray diffraction studies of various elements in the solid state.

8.9 CRYSTAL RADIUS (METALLIC OR ATOMIC RADIUS)

It is defined as "one half of the distance between the nuclei of two adjacent metal atoms in the metallic close packed crystal lattice.

For example: The internuclear distance between two adjacent sodium atoms in a crystal of the metal is 3.72 Å. The crystal radius of sodium $\frac{1}{2} \times 3.72 = 1.86$ Å.

The crystal radii are about 10-15% higher than the single bond covalent radii.

The metallic radii can be calculated by knowing the density of the metal and the nature of the unit cell, i.e., the arrangement of other atoms around the central metal atom.

8.9.1 Variation of Atomic Radius in a Group

In general, atomic radius depends on the following factors:

(1) Number of shells

(2) Nuclear charge

In general, atomic radius increases in a group as the atomic number increases.

The number of shells increases in a group and, therefore the atomic radius increases. On the other hand, due to the increase in the nuclear charge there occurs a contraction of the shells. Both these factors are contrary to each other, but the effect of adding a shell is so large that it overcomes the decrease due to the increased nuclear charge. Thus, as a whole the atomic radius increases in a group from top to bottom. For example, the covalent radii of the element of IA group are given below.

Li = 1.34 Å, Na = 1.57 Å, K = 1.96 Å

Rb = 2.11 Å

Cs = 2.25 Å

8.9.2 Variation of Atomic Radius in a Period

In a period, the atomic radius decreases as the atomic number increases. The number of shells remains the same in a given period. Therefore, there should be no effect on atomic radius due to this factor. On the other hand, there is an increase in the nuclear charge. Therefore the shells are attracted more and more strongly towards the nucleus. Thus, the atomic radius decreases in a period as the atomic number increases.

For example: The covalent radii of the elements of the second period are given as follows:

Element	$_3Li$	$_4Be$	$_5B$	$_6C$	$_7N$	$_8O$	$_9F$	$_{10}Ne$
Electronic configuration	2, 1	2, 2	2, 3	2, 4	2, 5	2, 6	2, 7	2, 8
Covalent radii	134 Å	0.90 Å	0.82 Å	0.77 Å	0.75 Å	0.73 Å	0.72 Å	–

EXAMPLES

Example 1: Van der Waals radius of chlorine is more than covalent radius of chlorine. Explain.

Solution: Van der Waals radius of chlorine is one half of the distance between the nuclei of two nearest atoms of two adjacent molecules of chlorine in the solid state which are non-bonded together whereas covalent radius of chlorine is one half of the distance between the nuclei of two atoms which are bonded together in a single chlorine molecule by a covalent bond.

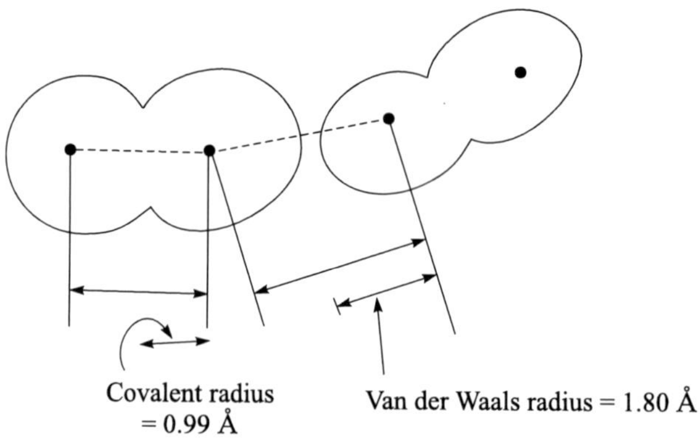

Example 2: Why noble gases have comparatively large atomic radii.

Solution: The noble gases have stable electronic configurations. Therefore, they have no tendency to lose, gain or share electrons and thus they ordinarily do not form any bonds. Being bondless they exist as single atoms. The only forces which keep the atoms together are weak van der Waals forces. Hence, the atomic radii of noble gas atoms are only the van der Waals radii which are comparatively larger as compared to covalent radii.

Example 3: Out of the following which has largest atomic size and which has the smallest: Na, Be, Mg, Li.

Solution: The elements Na, Be, Mg, Li may be arranged in the groups and periods as

Li Be

Na Mg

The size of Na atom will be maximum and that of Be atom will be minimum because in the group, atomic size increases, whereas in the period it decreases with the increase of atomic number.

8.10 IONIC RADIUS

It is defined as "the distance between the centre of the nucleus of an ion and the point up to which the nucleus has influence on its electron cloud" (Figure 8.9). The limit of electron cloud around nucleus cannot be determined with certainty. Therefore, the radius of an ion is determined by determining the internuclear distance in an ionic compound by *X*-ray measurements. The distance is then taken as the sum of the radii of the two ions involved. If we know the radius of one ion we can calculate the radius of the other.

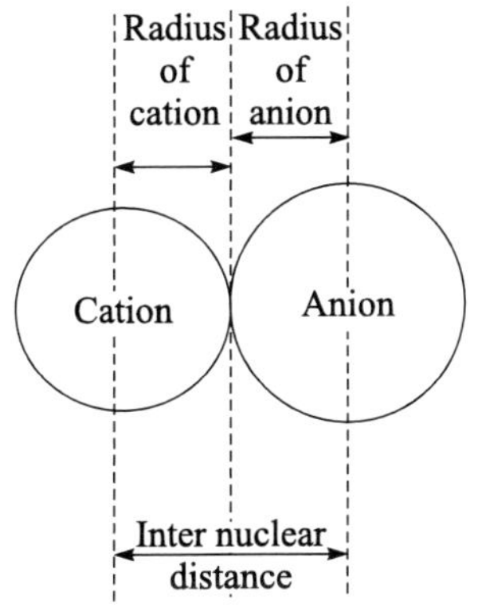

Fig. 8.9

The atom changes into ions either by the loss or gain of electrons. A cation is always smaller than the corresponding atom. Whereas anion is always bigger than the corresponding atom.

8.10.1 Variation of Ionic Radii in a Group

Since the atomic size increases as we move down the group, therefore ionic radius also increases in moving from top to bottom in a group. For example, ionic radius of alkali metals may be represented as

Alkali metal ion	Li^+	Na^+	K^+	Rb^+	CS^+
Ionic radii	0.68 Å	0.95 Å	1.33 Å	1.48 Å	1.69 Å

8.10.2 Variation of Ionic Radii in a Period

Since the atomic size decreases in a period, therefore ionic radius also decreases. In a period, the ionic radii of cations of elements of group I – III and also anions of the elements of group V – VII decrease with the increase in atomic number.

Groups	I	II	III	V	VI	VII
Cations	Na^+	Mg^{2+}	Al^{3+}	N^{3+}	O^{2-}	F^-
Ionic radii	0.95 Å	0.65 Å	0.50 Å	1.71 Å	1.40 Å	1.36 Å

8.10.3 Determination of Ionic Radius

Following methods have been used to determine the ionic radius.

(1) Land's method
(2) Bragg's method
(3) Pauling method

Out of these Pauling method is the most widely used.

Pauling Method: The method is used to determine ionic radii of isoelectronic ions. We know that radius of an ion is inversely proportional to the effective nuclear charge, i.e.,

$$r_{\text{ion}} \propto \frac{1}{Z_{\text{eff}}} \tag{1}$$

The effective nuclear charge is given by the relation

$$Z_{\text{eff}} = z - \sigma \tag{2}$$

where $z \rightarrow$ atomic number

$\sigma \rightarrow$ screening or shielding constant of the other electrons present in the ion.

Therefore, Eq. (1) becomes

$$r_{\text{ion}} \propto \frac{1}{z - \sigma}$$

$$r_{\text{ion}} = \frac{C}{z - \sigma} \tag{3}$$

where, $C \rightarrow$ proportionality constant.

It depends upon the principal quantum number of the outermost electrons.

Applying Eq. (3) to cation and anion, we have

$$r_c = \frac{C}{z_c - \sigma_c} = \frac{C}{(Z_{\text{eff}})_c} \tag{4}$$

$$r_a = \frac{C}{z_a - \sigma_a} = \frac{C}{(Z_{\text{eff}})_a} \tag{5}$$

where r_c and r_a are the ionic radii of cation and anion z_c and $z_a \rightarrow$ atomic numbers.

We know that cation and anion in a crystal are in close contact with each other. Therefore, internuclear distance (d) is the sum of ionic radii

i.e., $$d = r_c + r_a$$

By knowing the value of d we can calculate r_a. For example, the internuclear distance of NaF is 2.31 Å. Both Na^+ and F^- ions have neon type configuration ($1s^2$, $2s^2$, $2p^6$). The screening constant (σ) for both ions is calculated by using Slater's rules and is found to be 4.5. Thus, from Eq. (4) and (5), we have

$$r_{Na^+} = \frac{C}{11 - 4.5} = \frac{C}{6.5} \tag{6}$$

$$r_{F^-} = \frac{C}{9-4.5} = \frac{C}{4.5} \quad (7)$$

$$r_{Na^+} + r_{F^-} = \frac{C}{6.5} + \frac{C}{4.5}$$

$$d_{Na-F} = \frac{C}{6.5} + \frac{C}{4.5}$$

$$2.31 = C\left[\frac{C}{6.5} + \frac{C}{4.5}\right]$$

$$= 2.31 = C\left[\frac{4.5+6.5}{6.5\times 4.5}\right]$$

$$2.31 = \frac{C\times 11}{6.5\times 4.5}$$

$$C = \frac{2.31\times 6.5\times 4.5}{11} = 6.14$$

Thus, from Eqs. (6) and (7)

$$r_{Na^+} = \frac{C}{6.5} = \frac{6.14}{6.5} = 0.99\ \text{Å}$$

$$r_{F^-} = \frac{C}{4.5} = \frac{6.14}{4.5} = 1.36\ \text{Å}$$

Cation is smaller than its atom: A cation is formed by the loss of one or more electrons from the outermost shell of the atom. This generally results in the removal of outer shell of electrons completely. Thus, the number of shells of cations are less than that of its atom. Moreover with the removal of one or more electrons from the outermost shell, the effective nuclear charge is increased. Therefore, the electrons are attracted more strongly towards the nucleus than the electrons of its atom. Hence, because of both the factors, the size of cation is smaller than that of its atom. The size of cation also decreases with the increase in the oxidation number. The decrease in the size when the atom changes into cation is indicated below in some of the cases.

$$Na = 1.57\ \text{Å}$$

$$Na^+ = 0.95\ \text{Å}$$

$$Fe = 1.26\ \text{Å}$$

$$Fe^+ = 0.76\ \text{Å}$$

$$Fe^{3+} = 0.64\ \text{Å}$$

Anion is bigger than its atom: Anion is formed by the addition of one or more electrons to the outermost shell of the atom. The number of the shells of anions and its atom are generally the same. However, with the addition of one or more electrons to the outermost shell, the electrons repel each other and thus, effective nuclear charged is decreased. Therefore, the electrons are attracted less strongly towards the nucleus than its atom. Hence, the size of anion is bigger than that of the corresponding atom. The increase in the size when the atom changes into anion is indicated below in some of the cases.

$$Cl = 0.99 \text{ Å} \quad Br = 1.14 \text{ Å}$$

$$Cl^- = 1.81 \text{ Å} \quad Br^- = 1.95 \text{ Å}$$

8.10.4 Applications in Explaining the Chemical Behaviour

1. The radius ratio of cations and anions gives a possible idea about the crystal structure.
2. The radii of cations and anions predicts the nature of the bond formed between the ions. In general, larger cation and smaller anion prefer to form an ionic bond whereas smaller cation and larger anion prefer to form a covalent bond.
3. The acidity of solutions is due to hydration of cations which is directly related to their radii. In general, small cations are much more hydrated than large cations. Hence, they form more acidic solutions.

EXAMPLES

Example 1: Na^+, Mg^{2+} and Al^{3+} are iso-electronic, yet the radius of Na^+ is maximum and that of Al^{3+} minimum. Explain.

Solution: Electronic configuration of

$$Na = 2, 8, 1 \qquad Na^+ = 2, 8$$

$$Mg = 2, 8, 2 \qquad Mg^{2+} = 2, 8$$

$$Al = 2, 8, 3 \qquad Al^{3+} = 2, 8$$

The number of shells in each case is 2 but Na has 11 protons, Mg^{2+} has 12 protons, whereas Al^{3+} has 13 protons. Thus, the effective nuclear charge in the case of Na^+ is least, whereas in the case of Al^{3+} it is maximum. On account of this, the electrons will be attracted less strongly towards the nucleus in the case of Na^+ ion. Hence, the size of Na^+ ion is maximum and that of Al^{3+} ion it is minimum.

Example 2:

(a) Which has larger size in each pair?

(i) Na^+ or Mg^{2+}

(ii) O^{2-} or F^-

(iii) CU^{2+} or CU^+

(b) Arrange I, I^+ and I^- in order of increasing radius?

Solution: (a)

(i) $_{11}Na = 2, 8, 1$ $_{11}Na^+ = 2, 8$

$_{12}Mg = 2, 8, 2$ $_{12}Mg^{2+} = 2, 8$

The number of shells in each case is 2 but Na has 11 protons, Mg^{2+} has 12 protons.

(ii) The electronic configurations of O^{2-} and F^- are

$$_8O = 2, 6 \quad _8O^{2-} = 2, 8$$

$$_9F = 2, 7 \quad _9F^- = 2, 8$$

Both the ions have the same number of shells but the effective nuclear charge in case of O^{2-} is less than that of F^- ion. Thus, the electrons will be attracted less strongly towards the nucleus in case of O^{2-} ion than F^- ion. Hence, the size of O^{2-} is larger than that of F^- ion.

(iii) The electronic configurations of Cu^+ and Cu^{2+} ions are

$$_{29}Cu = 2, 8, 18, 1$$

$$_{29}Cu^+ = 2, 8, 18$$

$$_{29}Cu^{2+} = 2, 8, 17$$

Both the ions have the same number of shells, i.e., three but the effective nuclear charge in the case of Cu^+ ion (29 protons and 28 electrons) will be less than that of Cu^{2+} ion (29 protons and 27 electrons). Thus, the electrons will be attracted less strongly towards the nucleus in the case of Cu^+ than Cu^{2+} ion. Hence, the size of Cu^+ ion is larger than of Cu^{2+} ion.

(b) $I^+ < I < I^-$

This is because the cation is smaller than the corresponding atom, whereas anion is bigger than the corresponding atom.

8.11 IONIZATION POTENTIAL OR IONIZATION ENERGY

It can be defined as "The minimum energy required to remove most loosely bound electron from a gaseous atom or ion is known as ionization potential or ionization energy". The most loosely bound electron means the electron present in the outermost shell.

This energy is generally measured in electron volt per atom or kilocalories per mole or kilojoules per mole. The electrons are removed in stages one by one. Consequently, there may be first, second, third or *n*th ionization potential, depending upon the removal of first, second, third or *n*th electron from the atom. Thus,

First ionization potential is the energy required to convert

$$M(g) \rightarrow M^+(g)$$

Second ionization potential is the energy required to convert

$$M^+(g) \rightarrow M^{2+}(g)$$

It has been observed that the value of ionization potential increases in the order

First < Second < Third < ...... so on.

This is because the removal of second, third, or *n*th electron is done against greater and greater nuclear attraction.

Measurement of Ionization Potential: For the measurement of ionization potential, the vapours of the element are taken in the discharge tube. The tube is then connected with a source of current. The voltage applied is then slowly increased. When ionization occurs, a current flows at a certain applied voltage. This gives the value of first ionization potential. The voltage is again increased. After some time a stage comes when the sudden rise of current takes place. This gives the value of second ionization potential.

8.11.1 Factors Affecting the Value of Ionization Potential

Ionization potential depends on the following factors.

(1) Size of the atom: Larger the distance of the electron from the nucleus, weaker will be the forces of attraction and thus, lesser will be the amount of energy required to remove the electron. In other words, larger the size of the atom, smaller will be the ionization potential.

(2) The charge on the nucleus: Higher the charge on the nucleus, greater will be the force of attraction and thus, it will be more difficult to remove an electron. Hence, higher will be the value of ionization potential.

(3) Electrons in the inner shells: The electrons in the inner shells produce a screening effect on the nucleus. The electrons of the outermost shells are therefore, attracted with lesser force and hence it is easier to remove the valency electrons. Thus, larger the number of electrons in the inner shells, lesser will be the value of ionization potential.

(4) Completed shells, completed or half completed sub-shells: The atoms having such type of electronic structures are relatively more stable. Therefore, more energy is required to remove electrons in such case. Hence, ionization potential will be high.

(5) Shape of orbital: The shape of the orbitals also affects ionization potential. *s*-orbital is more close to the nucleus than the *p*-orbital of the same orbit. Thus, it is easier to remove electrons from a *p*-orbital as compared to *s*-orbital. The following order for the same orbit is as follows:

$$s > p > d > f$$

8.11.2 Variation of Ionization Potential in a Group

In general, the ionization potential decreases in a group as the atomic number increases. This is because in a group atomic size increases. On account of this, the outer electrons are attracted less and less strongly by the nucleus. For example, the first ionization potential of alkali metals are

Li = 5.4 eV   Na = 5.1 eV   K = 4.3 eV

Rb = 4.2 eV   CS = 3.9 eV

8.11.3 Variation of Ionization Potential in a Period

In general, the ionization potential increases in a period as the atomic number increases. This is because in a period the nuclear charge increases and the atomic size decreases. On account of this, the electrons are attracted more and more strongly by the nucleus.

The first ionization potentials of the elements of the second period are given as

II Period	$_3Li$	$_4Be$	$_5B$	$_6C$	$_7N$	$_8O$	$_9F$	$_{10}Ne$
First ionization potential in (eV)	5.4	9.3	8.3	11.2	14.5	13.6	17.4	21.6
		$1s^2, 2s^2$			$1s^2, 2s^2, 2p^3$			
III Period	$_{11}Na$	$_{12}Mg$	$_{13}Al$	$_{14}Si$	$_{15}P$	$_{16}S$	$_{17}Cl$	$_{18}Ar$
First I.P. in (eV)	5.1	7.6	6.0	8.1	11.0	10.4	13.0	15.8
		$(1s^2, 2s^2, 2p^6, 3s^2)$			$(1s^2, 2s^2, 2p^6, 3s^2, 3p^3)$			

The value of first ionization potential of Be and N in the second period and Mg and P in the third period has higher value than expected. This is because Be and Mg have completed *S*-sub-shells whereas N and P have half completed *p*-sub-shells. These electronic configurations are relatively more stable.

The ionization potential of the noble gases is expected to be very high because in these cases, the electron is to be removed from the stable system. Helium has the maximum value of ionization potential. If a graph is plotted between atomic number and ionization potential of elements, noble gases occupy the peaks whereas alkali metals occupy the lowest positions.

8.11.4 Applications

(1) The elements having lower value of ionization potential. For example, alkali and alkaline earth metals are more reactive, strong reducing agents and are basic in nature.

(2) The value of ionization potential gives an evidence about the number of valency electrons in the atom. For example, in case of sodium the second ionization potential (47.3 eV) is too much higher than the first ionization potential (5.1 eV). This indicates the presence of one electron in valency shell of sodium.

(3) The elements having higher value of ionization potential are less reactive, strong oxidizing agents and are acidic in nature (except inert elements).

(4) The relative value of ionization potential and electron affinity of two elements gives an idea about the nature of bond formed between two elements. If the difference between both the values is greater, the bond will be more ionic and if the difference is less, the bond will be more covalent.

EXAMPLES

Example 1 Explain giving reasons why?

(a) The second ionization potential of lithium is very high as compared to its first ionization potential.

(b) The third ionization potential of magnesium is very high.

Solution:

(a) Electronic configuration of lithiums, $_3Li = 1s^2, 2s^1$

First ionization potential is the energy required to convert $Li \rightarrow Li^+$, i.e., $1s^2, 2s^1$ to $1s^2$
Second ionization potential is the energy required to convert

$$Li^+ \rightarrow Li^{2+}, \text{ i.e., } 1s^2 \text{ to } 1s^1$$

The second ionization potential of lithium is very high as compared to its first ionization potential, because after the removal of outer electron the effective nuclear charge is increased and the removal of the second electron will have to be done against greater nuclear attraction. Secondly, the second electron is to be removed from the completed shell, i.e., from the stable system. Therefore, the amount of energy required for the said removal will be very high.

(b) The electronic configuration of $_{12}Mg = 2, (2, 6)$

$$2 = 1s^2, 2s^2, 2p^6, 3s^2$$

First ionization potential is the energy required to convert $Mg \rightarrow Mg^+$

$$1s^2, 2s^2, 2p^6, 3s^2 \text{ to } 1s^2, 2s^2, 2p^6, 3s^1$$

Second ionization potential is the energy required to convert $Mg^+ \rightarrow Mg^{2+}$

$$1s^2, 2s^2, 2p^6, 3s^1 \text{ to } 1s^2, 2s^2, 2p^6$$

Third ionization potential is the energy required to convert $Mg^{2+} \rightarrow Mg^{3+}$

$$1s^2, 2s^2, 2p^6 \text{ to } 1s^2, 2s^2, 2p^5$$

The third ionization potential of magnesium is very high, because after the removal of two outer electrons, the effective nuclear charge is much increased and the removal of the third electron will have to be done against greater nuclear attraction. Secondly, the third electron is to be removed from the completed shell, i.e., from the stable system. Therefore, the amount of energy required for the said removal will be very high.

Example 2 On moving along a period, the ionization potential of elements increases, then why the first ionization potential of magnesium (atomic number 12) is higher than that of aluminium (atomic number 13).

Solution: The electronic configurations of magnesium and aluminium

$$_{12}Mg = 2, (2, 6)\ 2$$
$$= 1s^2, 2s^2, 2p^6, 3s^2$$

1s	2s	2p	3s
↑↓	↑↓	↑↓ ↑↓ ↑↓	↑↓

$$_{13}Al = 2, (2, 6)\ (2, 1)$$
$$= 1s^2, 2s^2, 2p^6, 3s^2, 3p^1$$

1s	2s	2p	3s	3p
↑↓	↑↓	↑↓ ↑↓ ↑↓	↑↓	↑

The first ionization potential of magnesium is higher than that of aluminium because in the case of magnesium, 3*s* sub-shell is completely filled and hence is more stable. Therefore, the amount of energy required to remove electron from the stable system will be high. Moreover *s*-orbital is more close to the nucleus than the *p*-orbital of the same orbit. Thus, it is difficult to remove electron from *s*-orbital in the case of magnesium as compared to *p*-orbital in the case of aluminium.

Example 3 On moving along a period the ionization potential of elements increases, then why is the first ionization potential of nitrogen (atomic number 7) higher than that of oxygen (atomic number 8).

Solution: The electronic configurations of nitrogen and oxygen are

$$_7N = 2, (2, 3) \qquad _8O = 2, (2, 4)$$

$$= 1s^2, 2s^2, 2p^3 \qquad = 1s^2, 2s^2, 2p^4$$

$$= 1s^2, 2s^2, 2px^2, 2py^1, 2pz^1 \qquad = 1s^2, 2s^2, 2px^2, 2py^1, 2pz^1$$

1*s*	2*s*	2*p*	1*s*	2*s*	2*p*
↑↓	↑↓	↑ ↑ ↑	↑↓	↑↓	↑↓ ↑ ↑

The first ionization potential of nitrogen is higher than that of oxygen; because in the case of nitrogen, 2*p* sub-shell is exactly half completed and hence is relatively more stable. Therefore, the amount of energy required to remove electron from the stable system will be high.

Example 4 The ionization energy of Na^+ is more than Ne although both have the same electronic configuration. Explain?

Solution: The ionization energy of Na^+ is more than Ne although both have the same electronic configuration. This is because in Na^+, the charge on the nucleus (11 protons) is more than that in Ne (10 protons). Therefore, the electrons are attracted more strongly in Na^+ than in Ne. Hence, in Na^+ greater amount of energy will be required to remove the electron than in Ne. As a result, ionization energy of Na^+ is more than that of Ne.

8.12 ELECTRON AFFINITY

It is "The energy released when a gaseous atom accepts an electron forming a free gaseous ion."

$$X(g) + e \rightarrow X^-(g) + \text{Energy released}$$

The energy released is measured in electron volt per atom or kilojoule per mole. Thus, larger is the energy released, the greater is the electron affinity. In other words "It is measure of the force with which the added electron is bound by the nucleus of the atom. There may be first, second or higher electron affinities just like ionization potential.

$$O(g) + e \rightarrow O^-(g) + 1.47 \text{ eV}$$

$$O^-(g) + e \rightarrow O^{2-}(g) - 8.75 \text{ eV}$$

The second and higher order electron affinities are always negative because there is repulsion between negatively charged ion and electron being added and energy has to be given for this addition of electrons.

The electron affinities of some of the elements have been determined by absorption spectroscopy and are given in Table 8.4.

Table 8.4

							He
							O
Li	Be	B	C	N	O	F	Ne
0.58	0	0.33	1.13	0.20	1.47	3.62	O
Na	Mg	Al	Si	P	S	Cl	Ar
0.78	0	0.52	1.39	0.80	2.07	3.82	O
						Br	Kr
						3.54	O
						I	Xe
						3.24	O

8.12.1 Measurement of Electron Affinity

The electron affinity of an element can be determined by Born Haber cycle. With the help of this cycle, the following equation has been deduced for the determination of lattice energy of ionic crystal MX.

$$\Delta H = S + I + 1/2\,D + E + U$$

where $S \rightarrow$ Sublimation energy of metal M
$I \rightarrow$ Ionisation energy of metal M
$D \rightarrow$ Dissociation energy of X_2 molecule
$E \rightarrow$ Electron affinity of an element X
$U \rightarrow$ Lattice energy of ionic crystal MX
$\Delta H \rightarrow$ Heat of formation of ionic crystal MX

The electron affinity of an element X can be calculated, if the values of different thermochemical quantities are known.

8.12.2 Factors Affecting the Value of Electron Affinity

Its depends upon the following factors:

(1) **Size of the atom:** Larger is the size of the atom, lesser will be the forces of attraction between the nucleus and additional electron. Consequently, electron affinity will be low. Thus, bigger is the size of the atom lesser will be the electron affinity.

(2) **Charge on the nuclear:** Higher is the charge on the nucleus, greater will be the forces of attraction between the nucleus and additional electron, consequently electron affinity will be high.

(3) **Electronic configuration:** The atoms having stable electronic configuration, for example, completed shells, completed or half completed sub-shells do not have a tendency to accept additional electron. Such atoms have either zero or very low electron affinity.

8.12.3 Variation of Electron Affinity in a Group

In general, the electron affinity decreases in a group as the atomic number increases. This is because the atomic size increases and therefore, effective nuclear attraction for the electron decreases.

It is observed from Table 8.5, that the elements of the second period, i.e., from lithium to fluorine have lower electron affinity than the next lower members in their respective groups. This abnormal behaviour is due to the fact that in these elements, the added electron will go to the second shell which is comparatively small. Thus, there is a large repulsion between the electrons already present and the electron being added to the atom. In such cases, a small amount of energy released is utilized in overcoming these forces of repulsion.

8.12.4 Variation of Electron Affinity in a Period

In general, electron affinity increases in a period as the atomic number increases. In a period, the nuclear charge increases and the atomic size decreases. Therefore, the tendency of the atom to attract additional electron increases and hence electron affinity increases in a period.

8.12.5 Applications

1. The elements having higher value of electron affinity can easily accept additional electron. These elements are electronegative. They can easily forms anion and ionic compounds.
2. The elements having higher value of electron affinity act as strong oxidizing agents. Example: Halogens.
3. The relative values of ionization potential and electron affinity give an idea about the nature of bond formed between two elements.

EXAMPLES

Examples 1 The electron affinities of Be, Mg and noble gases are zero and those of nitrogen (0.20 eV) and phosphorus (0.80 eV) are very low.

Solution: This unexpected behaviour is due to the completed and half completed sub-shells of these elements which are relatively more stable. In the case of Be and Mg there is a completed *s*-sub-shell in the outermost shell and the added electron has to go to the *p*-sub-shell of higher energy. The noble gases have completed shells and the added electron will have to go to the next shell of considerably higher energy. This accounts for their electron affinities being zero.

$$\left.\begin{array}{l} {}_{4}Be = 1s^2\ 2s^2 \\ {}_{12}Mg = 1s^2, 2s^2, 2p^6, 3s^2 \end{array}\right\} \begin{array}{l}\text{Fully completed} \\ \text{s-sub-shell configuration}\end{array}$$

In the case of nitrogen and phosphorus, there is a half completed *p*-sub shell which is relatively more stable, therefore, the electron affinities of these elements corresponding to the addition of one electron will be less than expected.

$$\left.\begin{array}{l} {}_{7}N = 1s^2, 2s^2, 2p^3 = 1s^2, 2s^2, 2p'x, 2p'y, 2p'z \\ {}_{15}P = 1s^2, 2s^2, 2p^6, 3s^2, 3p^3 \\ \quad = 1s^2, 2s^2, 2p^6, 3s^2, 3p'x, 3p'y, 3p'z \end{array}\right\} \text{Half completed } p\text{-sub-shell configuration}$$

Example 2 Explain with proper reasoning why the formation of $F^-(g)$ from F(g) is exothermic whereas that of $O^{2-}(g)$ from O(g) is endothermic

Solution:

$$F(g) + e \rightarrow F^-(g) \text{ Exothermic} \quad (1)$$

$$O(g) + e \rightarrow O^-(g) \text{ Exothermic} \quad (2)$$

$$O^-(g) + e \rightarrow O^{2-}(g) \text{ Endothermic} \quad (3)$$

The formation of $O^-(g)$ from O(g) is exothermic just like formation of $F^-(g)$ from F(g) but in the formation of $O^{2-}(g)$ from $O^-(g)$, there occurs a repulsion between $O^-(g)$ and electron being added and energy has to be given for this addition of electron. The third step is so endothermic that it overweighs the heat liberated in the second step. Hence, the overall reaction $O(g) + 2e \rightarrow O^{2-}(g)$ is endothermic.

8.13 ELECTRONEGATIVITY

It is "The tendency of an atom in a compound to attract shared electrons towards itself".

1. **Pauling's Scale:** Pauling gave an arbitrary scale to measure the electronegativity of different elements. This scale is based upon the calculation of bond energies. For example, fluorine has the maximum value of electronegativity, i.e., 4.0 whereas noble gases have the minimum value, i.e., 0.0. It is based upon the calculation of bond energy. He considered the formation of *AB* covalent molecule as

$$1/2A_2 + 1/2B_2 \rightarrow AB$$

$$1/2(A - A) + 1/2(B - B) \rightarrow A - B$$

According to Pauling, the bond dissociation energy ε_{AB} of covalent molecule *AB* is always greater than the geometric mean of the bond dissociation energies ε_{A-A} and ε_{B-B} of A_2 and B_2 molecules.

i.e., $$\varepsilon_{A-B} > [\varepsilon_{A-A} \times \varepsilon_{B-B}]^{1/2}$$

The difference between the two values is known as ionic resonance energy of $A - B$ bond and is represented by $\Delta\varepsilon$.

$$\Delta\varepsilon = \varepsilon_{A-B} - [\varepsilon_{A-A} \times \varepsilon_{B-B}]^{1/2}$$

$\Delta\varepsilon$ – measure of polarity of bond $A - B$. It depends upon the electronegative X_A and X_B of atoms A and B as

$$X_A - X_B = 0.18\sqrt{\Delta\varepsilon}$$

In the above relation bond energies are taken in kJ/mole. If bond energies are expressed in kcal/mole the relation becomes

$$X_A - X_B = 0.18\sqrt{\Delta\varepsilon}$$

2. **Mulliken Scale:** The average of ionization potential and electron affinity is known as electronegativity. On this scale, the average of ionization potential and electron affinity is known as electronegativity. Thus,

$$\text{Electronegativity} = \frac{\text{Ionization potential} + \text{Electron affinity}}{2}$$

Mulliken gave the following relation for calculating electronegativity on Pauling's scale

$$\text{Electronegativity} = \frac{0.374[\text{Ionization potential} + \text{Electron affinity}]}{2}$$

$$= 0.187\ [\text{Ionization potential} + \text{Electron affinity}] + 0.17$$

In the above relation, ionization potential and electron affinity both are measured in electron volts. The method is not useful as electron affinities of all the elements are not known.

8.13.1 Factors Affecting the Values of Electronegativity

Its depends upon the following factors:

(1) **Size of the Atom:** Smaller is the size of the atom, greater will be its tendency to attract shared pairs of electrons towards itself and hence has greater electronegativity. Thus, electronegativity decreases with the increase in the size of the atom.

(2) **Oxidation State:** Higher is the oxidation state of an atom, greater is its electronegativity. For example, electronegativity of Fe^{3+} (1.96) is greater than that of Fe^{2+} (1.83).

(3) **Effective Nuclear Charge:** Higher is the effective nuclear charge, greater will be its tendency to attract shared pairs of electrons towards itself and hence has greater electronegativity.

(4) **State of Hybridization:** The electronegativity of an atom is also affected by the state of hybridization. Greater is the s-character in a hybrid orbital, more will be its electronegativity.

For example:

Hybridization state	S-character	Electronegativity
SP	50%	3.29
SP^2	33.3%	2.75
SP^3	25%	2.48

(5) **Nature of other atoms with which the atom is bonded:** Electronegativity is not the property of the isolated atom. It depends on the number and nature of atoms to which it is bonded. Thus, electronegativity is not constant. For example, electronegativity of phosphorus in PF_3 is 2.44 and in PCl_3 it is 2.24, whereas its accepted value is 2.15.

8.13.2 Variation of Electronegativity in a Group

In general, the electronegativity decreases in a group as the atomic number increases. This is because in a group the atomic size increases, consequently the added electron is at a larger distance from the nucleus and is therefore, held less tightly. For example, the electronegativity values on the Pauling's scale of the alkali metals are given below:

Alkali metal:	Li	Na	K	Rb	CS	Fr
Electronegativity:	1.0	0.9	0.8	0.8	0.7	0.7

8.13.3 Variation of Electronegativity in a Period

In general, the electronegativity increases in a period with the increase in atomic number. This is because in a period the nuclear charge increases and the atomic size decreases. The atom, therefore, attracts the electrons more strongly.

For example: The electronegativity values of the elements of the second period are given below:

IInd Period:	Li	Be	B	C	N	O	F
Electronegativity:	1.0	1.5	2.0	2.5	3.0	3.5	4.0

8.13.4 Applications

1. **Metallic and non-metallic character of elements:** The element having lesser value of electronegativity will be a metal whereas an element having higher value of electronegativity will be a non-metal.
2. **Diagonal relationship:** The diagonally placed elements of second and third period of the Periodic Table show many similar properties due to almost same electronegativity of the elements.
3. **Nature of bond:** With the help of electronegativity values, it is possible to predict the nature of bond formed between two atoms. If there is a large difference between the electronegativities of the two atoms, the bond formed will be ionic or polar in character. On the other hand, if the difference is small, the bond formed will be covalent or non-polar in character.

Pauling gave a rule by which the nature of the bond formed between two atoms can be predicted. According to this rule, if the electronegativity difference between two atoms is 1.7 the bond is 50% ionic in character and when the difference is more than 1.7, the bond is correspondingly more ionic.

4. **Stability of bond:** The greater is the difference in the electronegativity of two atoms, more stable will be the bond.

HX molecule:	H–F	H–Cl	H–Br	H–I
Electronegativity difference:	1.9	0.9	0.7	0.4

$\xrightarrow{\hspace{6cm}}$

Stability of H – X bond decreases

5. **Bond length:** Schomaker and Stevenson (1941) gave the following relation to calculate the covalent radius

$$d_{A-B} = r_A + r_B - 0.09\,(X_A - X_B)$$

where r_A and r_B are the covalent radii of atoms A and B and X_A and X_B are their electronegativities.

6. **Acid base character of the oxides of the elements of the same period:** The electronegativity difference between an element and oxygen predicts the nature of oxide formed by the element. If the difference of electronegativity between the two is large, the oxide is basic and if the said difference is small, the oxide will be acidic.

EXAMPLES

Example 1: The ionization potential and electron affinity of fluorine are 17.42 eV and 3.62 eV, respectively. Calculate the electronegativity of fluorine on Pauling's scale.

Solution:

$$\begin{aligned} X &= 0.187\,[\text{ionization potential} + \text{Electron affinity}] + 0.17 \\ &= 0.187\,[17.42 + 3.62] + 0.17 \\ &= 0.187 \times 21.04 + 0.17 \\ &= 3.93 + 0.17 \\ &= 4.10 \end{aligned}$$

Example 2: The covalent radii of carbon and oxygen are 0.77 Å and 0.73 Å and their electronegativities are 2.5 and 3.5, respectively. Calculate the C–O bond length.

Solution:

$$\begin{aligned} d_{\text{C-O}} &= r_\text{C} + r_\text{O} - 0.09\,(X_\text{O} - X_\text{C}) \\ &= 0.77 + 0.73 - 0.09\,(3.5 - 2.5) \\ &= 1.50 - 0.09 \\ &= 1.41 \text{ Å} \end{aligned}$$

Example 3: The bond energies of hydrogen, chlorine and hydrogen chloride are 103.4 kcals/mole, 57.8 kcals/mole and 102.7 kcals/mole, respectively. If the electronegativity of hydrogen is 2.1. Calculate the electronegativity of chlorine.

Solution:

$$\begin{aligned}\Delta\varepsilon &= \varepsilon_{H-Cl} - [\varepsilon_{H-H} \times \varepsilon_{Cl-Cl}]^{1/2} \\ &= 102.7 - [103.4 \times 57.8]^{1/2} \\ &= 102.7 - 77.3 \\ &= 25.4\end{aligned}$$

Also we know that

$$\begin{aligned}X_{Cl} - X_H &= 0.18\sqrt{\Delta E} \\ X_{Cl} &= 2.1 = 0.18\sqrt{25.4} \\ &= 0.18 \times 5.04 \\ X_{Cl} &= 2.1 + 0.9072 \\ &= 3.0072\end{aligned}$$

8.14 POLARIZING POWER AND POLARIZABILITY OF IONS: FAJAN'S RULE

The ability of a cation to distort an anion is known as its polarization power and the tendency of the anion to become polarized by the cation is known as polarizability.

Fajan in 1924 gave certain rules with the help of which it is possible to know the conditions under which electrovalency passes into covalency.

According to this rule "When two oppositely charged ions approach each other closely the cation attraction attracts the electron atmosphere of the anion (electrostatic attraction) and simultaneously repels the anion nucleus". This results in the deformation, distortion or polarization of the anion. Similarly, the anion also polarizes the cation but the effect will be negligible or less pronounced because of smaller size of cation. The net result of this ion polarization is that there occurs some sharing of electrons between the two ions, i.e., the bond becomes partly covalent in character.

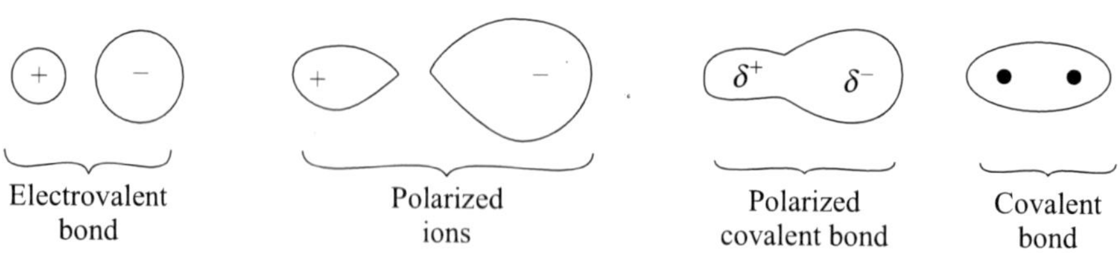

Fig. 8.10 *Transition between ionic and covalent bonding*

The polarizing power and polarizability that enhances the formation of covalent bonds is favoured by the following factors.

(1) Large charge on the ions: The greater the charge on cation or anion or both, greater will be the forces of attraction between the oppositely charged ions and thus, there will be more chances of electrovalency to pass into covalency.

For example: Al^{3+} ion (cationic charge +3) will distort the chloride ions more than Na^+ ions (cationic charge +1). Thus, $AlCl_3$ is covalent whereas NaCl, is ionic.

(2) Small size of the cation: A small cation polarizes the anion more than a large cation. This is because of its greater attractive force for the electron of the anions.

For example: II A group, the Be^{2+} ion has the least size and has the greatest polarizing effect. Thus, $BeCl_2$ is the most covalent among the chlorides of II A group elements.

(3) Large size of the anion: A large anion is more easily polarized by the cation than a smaller anion. This is because the outer electrons of the large anion are less strongly held by its nucleus and therefore, more strongly attracted by the cation.

For example: $AlCl_3$ is covalent whereas AlF_3 is ionic.

(4) A cation with non-noble gas configuration: The cations having non-noble gas type of electronic configuration ($s^2p^6d^{1-10}$) of their outermost shells polarize the anions more strongly than the cations having noble gas type of electronic configuration (s^2p^6). This is because *d*-electrons in such cations screen the nuclear charge less effectively than *s* and *p* electrons of cations having noble gas type of electronic configurations. As a result, the cations having non-noble gas type of electronic configurations behave as if they had greater charge.

For example: The halides of copper (I) and silver (I) are more covalent than the corresponding halides of sodium and potassium although the charge on the ions is the same (+1) and their ionic sizes are similar.

(5) Nature of the solvent: The solvents having high dielectric constant. Example, water weakens the electrostatic forces between the ions and thus decreases the tendency of electrovalency to pass into covalency.

EXAMPLES

Example 1 With the help of Fazan's rules which compound of each of the following pairs is more covalent and why?

Solution: (i) CUO or CUS (ii) LiCl or NaCl

(iii) NaCl or AgCl (iv) $AlCl_3$ or AgCl

(iv) $AlCl_3$ or AlF_3 (v) $SnCl_3$ or $SnCl_4$

(1) CuO or CuS: CuS will be more covalent than CuO. This is because in these compounds cation (Cu^{2+}) is the same but S^{2-} ion is bigger in size than O^{2-} ion. Thus, Cu^{2+} ion will polarize S^{2-} ion more than O^{2-} ion.

(2) LiCl or NaCl: LiCl will be more covalent than NaCl. This is because in these compounds anion (Cl^-) is the same but Li^+ ion has smaller size than Na^+ ion. Thus, Li^+ ion has smaller size than Na^+ ion. Thus, Li^+ ion will polarize Cl^- ion more than Na^+ ion.

(3) NaCl or AgCl: AgCl will be more covalent than NaCl. In these compounds, Ag^+ ion has non-noble gas type of electronic configuration whereas Na^+ ion has noble gas type of electronic configuration

$$_{47}Ag^+ = 2, 8, 18, 18 \quad _{11}Na^+ = 2, 8$$

The *d*-electrons of Ag^+ ion screen the nuclear charge less effectively than *S* and *P* electrons of Na^+ ion. As a result, Ag^+ ion behaves as if it has a greater charge. On account of this, Ag^+ ions will polarize Cl^- ions more strongly than Na^+ ions.

(4) $AlCl_3$ or AlF_3: $AlCl_3$ will be more covalent than AlF_3. In these compounds the cation (Al_3) is the same but Cl^- ion is bigger than F^- ion. Thus, Al^{3+} ion will polarize Cl^- ions, the F^- ion is very small and cannot be polarized by Al^{3+} ion.

(5) $SnCl_2$ or $AnCl_4$: $SnCl_4$ will be more covalent than $SnCl_2$. This is because in $SnCl_4$, the Sn^{4+} ion has greater charge than Sn^{2+} ion in $SnCl_2$. As a result, Sn^{4+} ion will polarize the Cl^- ions more strongly than Sn^{2+} ion.

Example 2 Cu_2Cl_2 is more covalent than NaCl

or

Polarizing power of Cu^+ is greater than Na^+?

Solution: Cu_2Cl_2 is more covalent than NaCl. In these compounds Cu^+ ion has non-noble gas type of electronic configuration, whereas Na^+ ion has noble gas type of electronic configuration

$$Cu^+ = 2, 8, 18$$

$$Na^+ = 2, 8$$

The *d*-electrons of Cu^+ ion screen the nuclear charge less effectively than *S* and *P* electrons of Na^+ ion. As a result, Cu^+ ion behaves as if it has a greater charge. On account of this, Cu^+ ion will polarize Cl^- ions more strongly than Na^+ ions.

Example 3 Covalent character decreases from $BeCl_2$ to $RaCl_2$?

Solution: The covalent character decreases from $BeCl_2$ to $RaCl_2$. This is because in all these compounds anion (Cl^-) is the same and cationic size increases from Be^{2+} to Ra^{2+}. Thus, Be^{2+} ion will polarize the Cl^- more than other cations of the II A group. As a result the covalent character decreases from $BeCl_2$ to $RaCl_2$.

8.15 OXIDATION STATE

The term oxidation was first used by Antoine Lavoisier to signify reaction of a substance with oxygen. Much later it was realized that the substance upon being oxidized loses electrons and the meaning was extended to include other reactions in which electrons are lost, regardless of whether oxygen was involved.

Oxidation state, sometimes referred to as oxidation number, describes the degree of oxidation (loss of electrons) of an atom in a chemical compound. Conceptually, the oxidation state which may be positive, negative or zero is the hypothetical charge that an atom would have if all bonds of atom of different elements were 100% ionic with no covalent component. This is never exactly true for real bonds.

Oxidation numbers are usually written with the sign (+) or (–) first then the magnitude which is the opposite of charges on ion.

Following are the guidelines for determining the oxidation number.

1. Atoms in their elemental state have an oxidation number of zero.
2. Atoms in monoatomic (i.e., single atom) ions have oxidation equal to their charge.
3. In compounds: Fluorine is assigned a (–) 1 oxidation number, oxygen is usually assigned a –2 oxidation number (except in peroxide compounds where it is (–1) and in binary compounds with fluorine where it is positive) and hydrogen is usually assigned a +1 oxidation number except when it exists as the hydride ion H^- in which case rule 2 wins.
4. In compounds all other atoms are assigned an oxidation number, so that the sum of the oxidation number on all the atoms in the species equals the charge on the species.

Example 4 What is the oxidation state of the hydrogen atom in H_2 and H_2O?

Solution: Hydrogen (H_2): We can use rule 1 here. Since H_2 is in its elemental state, it has an oxidation number of zero.

Water (H_2O): Rules 1 and 2 do not apply to this situation, so we can skip them. Rule 3 tells us that hydrogen usually has the oxidation number of +1 except in hydrides. We do not have a hydride here, but even if we are not sure if our hydrogen has a +1 or –1 oxidation state so we can use rule 4 to check.

$$(-2.1 \times 1) + (+1 \times 2) = 0$$

which is what we should except for a neutral molecule. Thus, using rules 3 and 4 we can assign the hydrogen in H_2O an oxidation state of +1.

8.16 COORDINATION NUMBER

It is the number of ligands attached to the central ion (more specially the number of donor atoms). Coordination numbers are normally between two and nine. The number of bonds depends on the size, charge and electron configuration of the metal ion and the ligands. In simple terms, the coordination number of a complex is influenced by the relative sizes of the metal ion and the ligands and by electronic factors such as charge which is dependent on the electronic configuration of the metal ion.

Typically, the chemistry of complexes is dominated by interaction between *S* and *P* molecular orbitals of the ligands and the *d*-orbitals of the metal ion. The *S*, *P* and *d* orbitals of the metal can accommodate electrons. The maximum coordination number for a certain metal is thus releated to the electronic configuration of the metal ion (specifically, the number of empty orbitals) and to the ratio of the size of the ligands and the metal ion. Large metals and small ligands lead to high coordination numbers (Example: $MO(CN)_8^{4-}$) small metals with large ligands lead to low coordination number (Example: $Pt[P(CMe_3)]_2$). Due to their large size, lanthanides, actinides and early transition metals tend to have high coordination number.

8.16.1 Ligands

In coordination chemistry, a ligand is an ion or molecule (functional group) that binds to a central metal atom to form a coordination complex. Virtually every molecule and every ion can serve as a ligand for (or coordinate to) metals. Many ligands are capable of binding metal ions through multiple sites, usually because the ligands have lone pairs on more than one atom.

Classification of ligands: Depending on the number of sites at which one molecule of a ligand is coordinated to the central metallic atom, the ligands have been classified as monodentate (or unidentate) and polydentate (or multidentate) ligands).

1. **Monodentate or unidentate ligands:** The ligands which have only one donor atom or are coordinated through one electron pair are called monodentate or unidentate ligands. Such ligands are coordinated to the central metal ion at one site or by one metal-ligand bond only. These ligands may be all anions and all simple Lewis bases. Thus, the halides and pseudohalides are important anionic ligands. Ammonia, carbon monoxide and water are particularly common charge neutral ligands. Simple organic species are also very common. All unsaturated molecules are also ligands, utilizing their π-electrons in forming the coordinate bond. Also, metals can bind to the σ bonds in for example silanes, hydrocarbons and dihydrogen.

2. **Polydentate or multidentate ligands:** Ligands that bind via more than one atom are often termed polydentate or chelating. A ligand that binds through two sites is classified as bidentate and three sites as tridentate. Chelating ligands are commonly formed by linking donor group via organic linkers. **Example:** A classical bidentate ligand is ethylenediammine, which is derived by the linking of two ammonia groups with an ethylene ($-CH_2CH_2-$) linker. A classic example of polydentate ligand is the hexadentate chelating agent EDTA, which is able to bond through six sites, completely surrounding some metals.

   There are several types of polydentate ligands which can be characterized based on how they interact with the central ion. For example, trans spanning ligands are bidentate ligands that can span coordination positions on opposite sides of a coordination complex. Ambidentate ligands can attach to the central atom in two places but not both. A bridging ligand links two or more metal centres changing the size and electronic properties of ligands can be used to control catalysis of the central ion and stabilize unusual coordination sites.

8.16.2 Geometries

Different ligand structural arrangements result from the coordination number. Most structures follow the pattern as if the central atom were in the middle and the corners of that shape are the locations of the ligands. These shapes are defined by orbitals overlap between ligand and metal orbitals and ligand-ligand repulsion, which tend to lead to certain regular geometries. However, there are many cases that deviate from regular geometry.

For example: Ligands of different sizes and with different electronic effects often result in irregular bond lengths.

Coordination number: This arrangement is not very common for the first row transition metal ion complexes and some of the best known examples are for silver (I). In this case we have a low charge and an ion at the right bond side of the *d*-block indicating smaller size.

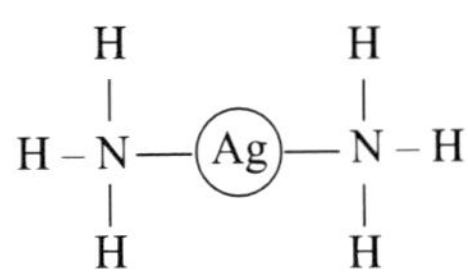

The linear $[Ag(NH_3)_2]^+$ ion

Coordination number 3

1. Trigonal planar geometry: Well known for main group species like CO_3^{2}, etc. This geometry has four atoms in a plane with the bond angles between the ligands at 120 degrees.

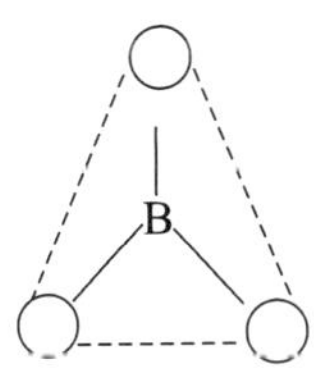

BO_3^{3-}

2. Trigonal pyramidal: More common with main group ions.

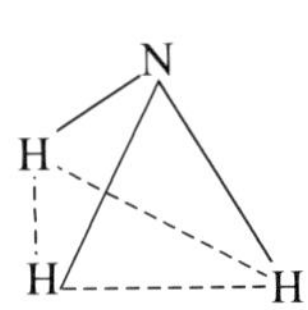

NH_3

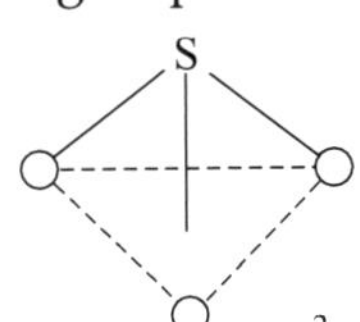

SO_3^{2-}

3. T-shaped: The first example of a rare *T*-shaped molecule was found in 1977.

Coordination number 4: Two different geometries are possible. The tetrahedron is the more common while the square planar is found almost exclusively with metal ions having a d^8 electronic configuration.

Tetrahedral geometry:

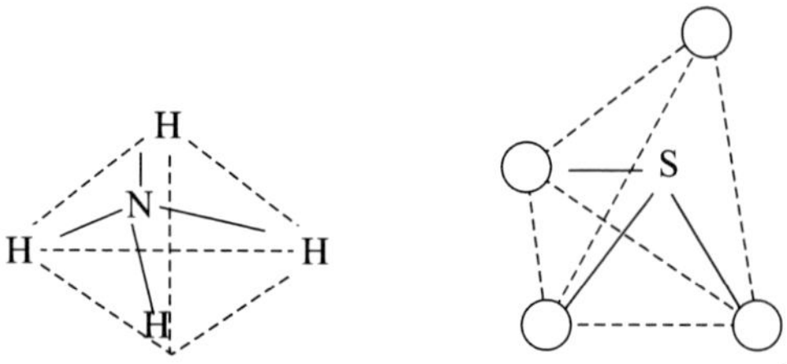

NH_4^+ SO_4^{2-}

Square planer: This is fairly rare and is included only because some extremely important molecules exist with this shape.

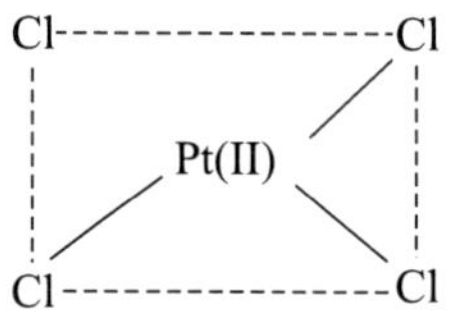

$PtCl_4^{2-}$

Coordination number 5

Square pyramid geometry

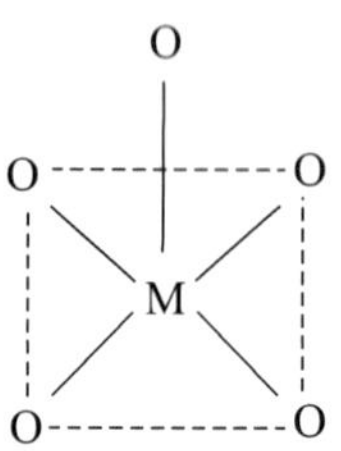

Trigonal bipyramid geometry

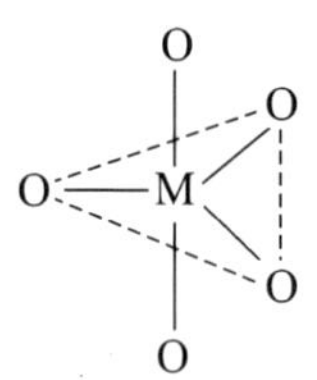

Coordination number 6

Hexagonal planer geometry: Unknown for first row transition metal ions, although the arrangement of six groups in a plane is found in some higher coordination number geometries.

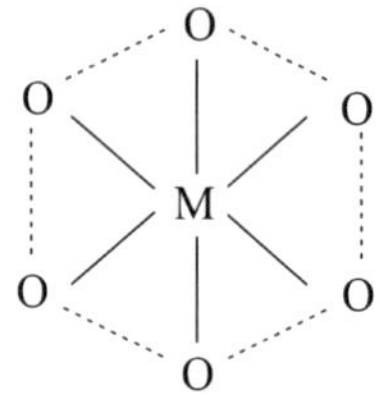

Trigonal prism geometry: Most trigonal prismatic compounds have three bidentate ligands such as dithiolates or oxalates and a few are known for first row transition metal ions.

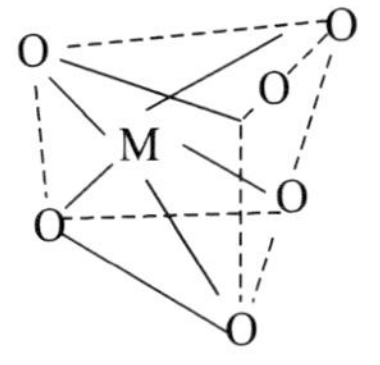

Octahedral: The most common geometry found for first row transition metal ions, including all equal ions. In some cases distortions are observed and these can sometimes be explained in terms of the Jahn-Teller Theorem.

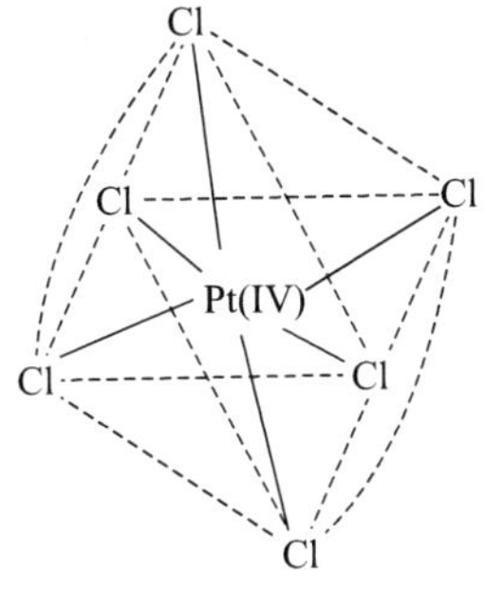

REVIEW QUESTIONS

1. What do you understand by orbital? Discuss the shape and orientations of various orbitals.
2. What do you understand by atomic radius? How is it determined? How does it change within a group and a period of the Periodic Table.
3. Why noble gases have comparatively large atomic radii?
4. Van der Waals radius of chlorine is more than covalent radius of chlorine. Explain.
5. Differentiate between atomic radius and covalent radius of the same element.
6. Out of the following which has largest atomic size and which has the smallest: Na, Ba, Mg, Li.
7. What do you understand by ionic radius? How is it determined, why is cation smaller and anion is larger than its atom? What are its applications in explaining the chemical behaviour?
8. Explain why the size of Na^+ ion is smaller than Na atom and that of Cl^- ion also bigger than Cl atom.

9. Na^+, Mg^{2+} and Al^{3+} are iso-electronic. Yet the radius of Na^+ is maximum and that of Al^{3+} minimum. Explain.

10. Which has larger size in each pair
    (i) Na^+ or Mg^{2+}
    (ii) O^{2-} or F^-
    (iii) Cu^{2+} or Cu^+

11. What do you mean by ionization potential? How is it measured? On what factors does it depend and how does it change in a group and in a period. What are its applications in explaining the chemical behaviour?

12. Explain giving reasons why–
    (a) the second ionization potential of lithium is very high as compared to its first ionization potential?
    (b) the third ionization potential of magnesium is very high?

13. On moving along a period the ionization potential of elements increases, then why is the first ionization potential of nitrogen (atomic number 7) higher than that of oxygen (atomic number 8).

14. What do you mean by electron affinity? On what factors does it depend and how does it change in a group and in a period? What are its applications in explaining the chemical behaviour?

15. What do you mean by electronegativity? How is it determined? On what factors does it depend and how does it change in a group and in a period? Give its application in explaining the chemical behaviour.

16. The ionization potential and electron affinity of fluorine are 17.42 eV and 3.62 eV, respectively. Calculate the electronegativity of fluorine on Pauling's scale.

9

Stereochemistry

9.1 INTRODUCTION

The space relationship of atoms within organic molecules has assumed such an importance that a new branch of chemistry, stereochemistry, has been established.

It involves the study of the different spatial arrangements of atoms in a molecule.

It is commonly referred to as three-dimensional structure of the molecules since it focuses on stereoisomers (chemical compounds) with the same chemical formula but a different spatial arrangement in three dimensions.

One of the most distinctive features of organic compounds is their ability to show isomerism. The term isomer (Greek iso – equal, mers – parts) was coined by Berzelius to different compounds having the same molecular formula. The isomers are defined as: "Two or more compounds having the same molecular formula but different chemical and physical properties and the phenomenon is termed isomerism."

It is of two types:

1. Structural isomerism
2. Stereoisomerism

9.2 STRUCTURAL ISOMERISM

Different arrangements of atoms within the molecule are called structural isomers and the phenomenon is called structural isomerism.

For example:

$$CH_3 - CH_2 - OH \qquad\qquad CH_3 - O - CH_3$$

$$(C_2H_6O) \qquad\qquad (C_2H_6O)$$

9.3 STEREOISOMERISM

"The isomers which differ in relative arrangement of atoms or groups in space around the carbon atom are known as stereoisomers and the phenomenon is known as stereoisomerism". That is, the isomers are different from each other only in the way the atoms are oriented in space.

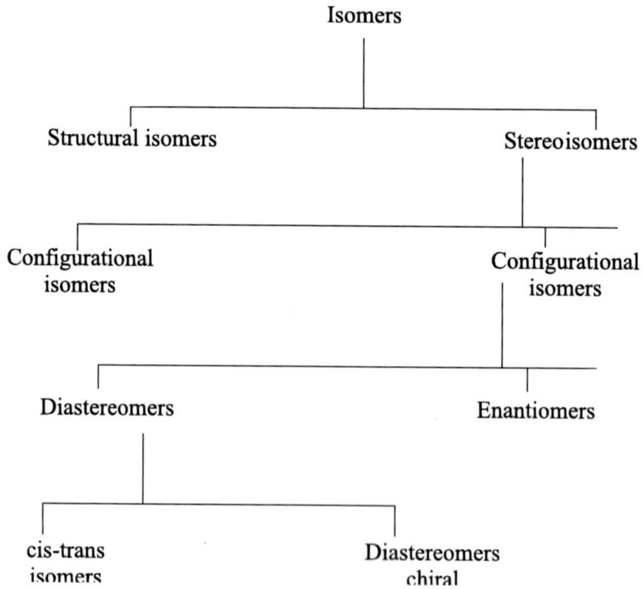

Fig. 9.1

They have two major classes (Fig. 9.1)

1. Configurational stereoisomers
2. Conformational stereoisomers

9.3.1 Configurational Stereoisomers

Configurational stereoisomers are those that cannot be converted into one another by rotation around a single bond. Interconversion in these stereoisomers usually involves a bond-breaking process. This type of isomerism exists either due to rigidity in a molecule (geometrical isomers) or due to right or left handed arrangement of atoms around carbon atom (optical isomers).

9.3.2 Conformational Stereoisomers

The different arrangements of atoms in space that results from free rotation of groups about carbon-carbon bond axis are called conformations or conformational isomers or rotational isomers.

9.4 BASIC CONCEPTS IN STEREOCHEMISTRY

Before we discuss the details of the stereoisomers it is worthwhile to be familiar with certain basic concepts involved.

9.4.1 Chirality (Stereogenicity)

A chiral molecule is defined as that centre of carbon atom which is attached with four dissimilar groups or atoms. Such a carbon atom is called an asymmetric carbon atom (or a chiral).

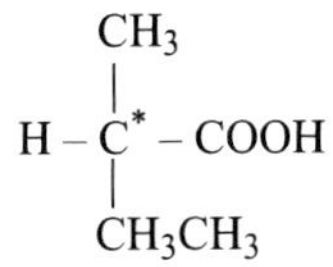

A chiral molecule is one that is not super impossible on its mirror image. A chiral compound can be superimposed on their mirror image. For example, our right and left hands are very similar, yet they are not identical. They are related to each other as mirror images and as such they cannot be superimposed on top of each other. Hence, they are said to be chiral.

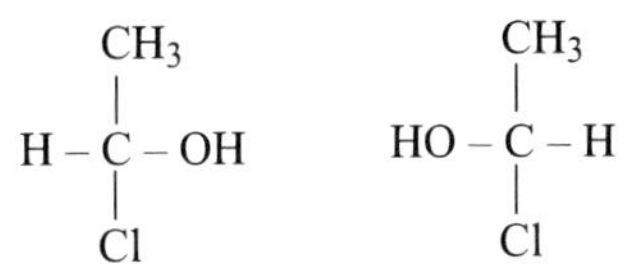

Chiral carbons

For the carbon to be chiral it follows that

- the carbon atom is SP^3 hybridized, and
- there are four different groups attached to the carbon atoms.

9.4.2 Fischer Projection

Fischer projection was introduced by Emil Fischer in (1891). It represents the spatial arrangement of groups around chiral carbon atoms. This convention represents molecular structure in a two-dimensional framework of vertical and horizontal bonds (Figs. 9.2 and 9.3).

In Fischer projections, the vertical carbons are always viewed as being positioned behind the plane of the paper (away from the viewer) while the horizontal carbons are imagined as coming out of the plane (towards the viewer).

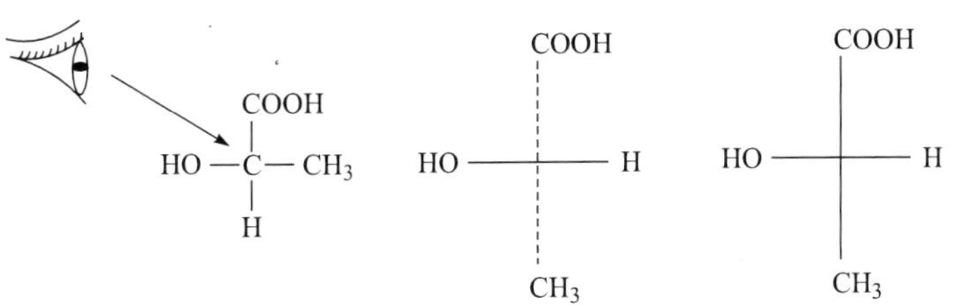

Fig. 9.2 *Fischer projection of (S) – Lactic acid*

Similarly, we represent the isomers of glyceraldehyde as shown in Fig. 9.3.

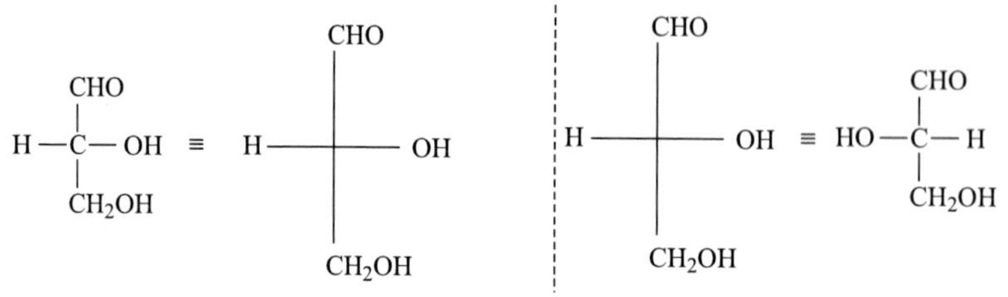

Fig. 9.3

9.4.3 Plane Polarized Light

Plane polarized light is the light in which all wave vibrations have been filtered out except for those in one plane.

Polarized light (Ordinary (non-polarized) light consists of many beams vibrating in different planes)

Plane polarized light (It consist of only those beams that vibrate in the same plane)

Plane polarized light is created by passing ordinary light through a polarizing device. Such devices transmit selectively only that component of a light beam having electrical and magnetic field vectors oscillating in a single plane. The plane of polarization can be determined by an instrument called polarimeter, as shown in Fig. 9.4.

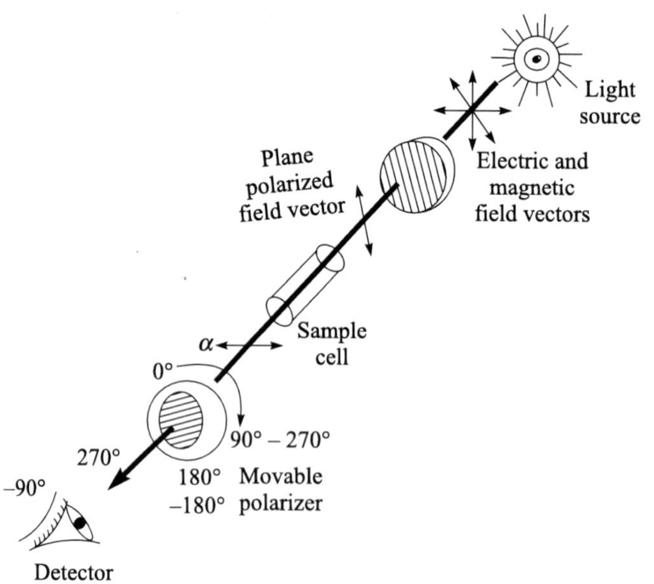

Fig. 9.4 *Polarimeter*

Monochromatic (single wavelength) light is polarized by a fixed polarizer next to the light source. A sample cell holder is located in line with the light beam, followed by a movable polarizer (the analyzer) and an eyepiece through which the light intensity can be observed. In modern instruments an electronic light detector takes place of the human eye. In the absence of a sample, the light intensity at the detector is maximum when the second (movable) polarizer is set parallel to the first polarizer ($\alpha = 0°$). If the analyzer is turned 90° to the plane of initial polarization, all the light will be blocked from reaching the detector.

9.4.4 Optical Activity

It is the ability of a chemical substance to rotate the plane of vibration of polarized light to the right or left. Optically active compounds are chiral compounds which possess the ability to rotate plane-polarized light.

- Polarimeter is used to the measure the optical activity of compounds.
- If a solution containing an optically active substance is placed in the sample tube, it rotates the plane of the polarized light, and the degree of rotation is measured.
- If it rotates the light to the right (clockwise) it is called dextrorotatory and is designated with a + sign. (+) dextrorotatory (Latin *dexter* = right).
- If it rotates light to the left (counterclockwise), it is called leavorotatory and it is designated with – sign (–) leavorotatory (Latin *laevus* = left).

9.4.5 Specific Rotation

It is an intrinsic property of the molecule. It depends on the following factors:

- Concentration of the sample compound
- Length of the sample tube
- Temperature

Specific rotation is the degree of rotation by 1.00 gm of compound in 1.00 mL of solution measured in a tube with a 1 decimetre path length. The amount of rotation is also depends on the wavelength (typically the *D* line of sodium, 589.3 nm) and solvent.

$$[\alpha]_D^{20} = \alpha / lC$$

where α = Observed rotation at 20°C

l = Cell path length in decimetres

C = Concentration in gm/mL

D = Wavelength (typically the D line of sodium, 589.3 nm)

So, it can be expressed as

$$[\alpha]_\lambda^T = \alpha / lC$$

Example

A sample of pure (S)-2-butanol was placed in a 10.0 cm polarimeter tube. Using the *D* line of a sodium lamp, the observed rotation at 20°C was +104°. The density of this compound is 0.805 gmL^{-1}. What is the specific rotation of (S)-2 butanol?

Solution:

$$[\alpha]_\lambda^T = \alpha / lC$$

$$[\alpha]_\lambda^T = \frac{+104°}{1.00\,d_m \times 0.0805\ gml^{-1}}$$

$$[\alpha]_\lambda^T = +129°.$$

9.5 OPTICAL ISOMERISM

Substances which have a power of rotating the plane polarized light are said to be optically active. This property of rotating the plane polarized light is called optical activity. Compounds having similar chemical and physical properties but different behaviour towards polarized light are called optical isomer and this phenomena is called optical isomerism. On the basis of optical activity, the organic compounds are divided into three types.

9.5.1 Dextrorotatory

The optical isomer which rotates the plane of polarized light to the right direction (clockwise) is known as dextrorotatory isomer and desigend (+) or (*d*) Fig. 9.5.

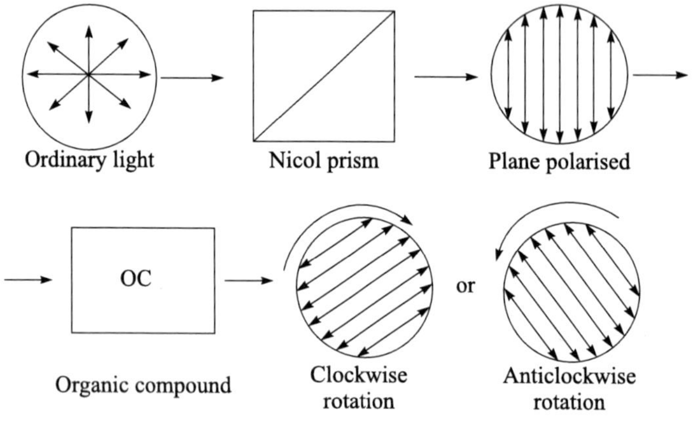

Fig. 9.5

9.5.2 Laevorotatory

The optical isomer which rotates the plane polarized light to the left (anticlockwise) is known as laevorotatory isomer and is designated (–) or (*l*).

9.5.3 Racemic Mixture

If the optical powers of *d* and *l* isomers are equal in magnitude and opposite in sign then an equimolar mixture of the two forms will be optically inactive, the mixture is termed racemic mixtures or *dl* form of (+, –).

For example: Polarimeter (Fig. 9.4) is used to determine and measure the optical activity of a substance in terms of angle of rotation of the planes in degrees.

The angle of rotation depends upon the following factors:

- Nature of substance
- Wavelength of light
- Concentration of solution
- Nature of solvent
- Length of path through which polarized light passes

  Thus, optical activity is expressed in terms of specific rotation (α_{SP}).

$$\alpha_{SP} = \frac{\alpha_{\text{obs}}}{l \times C}$$

where $\alpha_{\text{obs}} \rightarrow$ Observed rotation

$l \rightarrow$ Length of solution in decimetre

$C \rightarrow$ No. of gm per mL in the solution

For an organic compound to show optical activity it should have an asymmetric or chiral carbon atom and not have a plane of symmetry.

9.6 CHIRALITY

The property of non-superimposability of an object on its mirror image is called chirality. Chirality in organic molecule is due to the tetrahedral nature of the carbon atom. The carbon atom attached to four different atoms or groups is called an asymmetric chiral carbon atom indicated by C*. Thus, it can be represented as

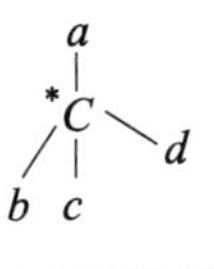

All organic compounds containing one asymmetric or chiral carbon atom are optically active. A chiral object does not have any element of symmetry such as a plane of symmetry, a centre of symmetry or an axis of symmetry. On the other hand, chiral objects or molecules have at least one or more elements of symmetry.

9.6.1 Plane of Symmetry

It is defined as a "plane which divides the molecule into two exactly identical halves". When a molecule has a plane of symmetry it is not optically active even though it has chiral carbon.

Example: Mesotartaric acid having plane of symmetry is optically inactive (Fig. 9.6).

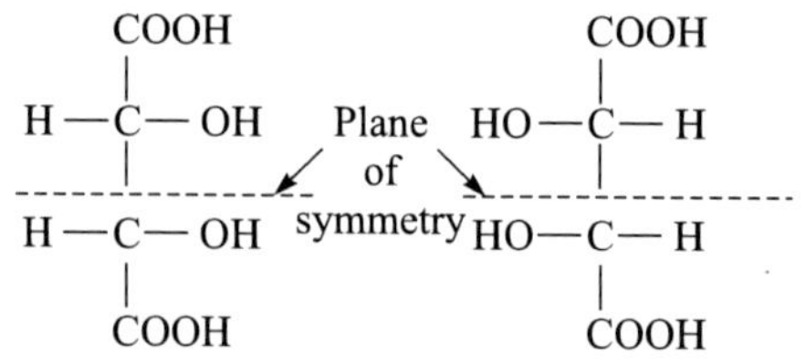

Fig. 9.6 *Molecules containing plane of symmetry are optically inactive*

9.6.2 Centre of Symmetry

Centre of symmetry is a point at which all the straight lines joining identical points in the molecule cross each other.

Example: The molecule 2, 4 dimethyl cyclobutane -1, 3-dicarboxylic acid has a centre of symmetry (at the center of cyclobutane ring) and is therefore optically inactive (Fig. 9.7).

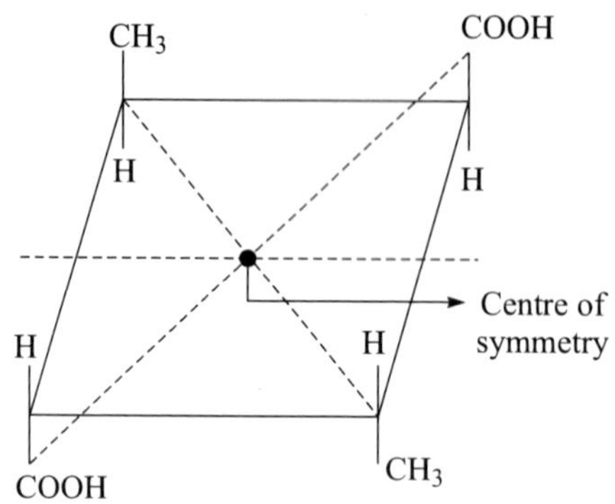

Fig. 9.7 *2, 4-dimethyl cyclobutane-1, 3-dicarboxylic acid molecules containing centre of symmetry are optically inactive*

9.6.3 Alternating Axis of Symmetry

A molecule has alternating axis of symmetry if an arrangement identical to the original is obtained when the molecule is rotated through an angel $360°/n$ about an axis followed by reflection in a plane perpendicular to this axis. A molecule having n-fold axis of symmetry is optically inactive.

Example: If the molecule *A* (Fig. 9.8) is rotated through 180° about the axis, passing through the centre of the molecule, and then reflected in the mirror perpendicular to this axis, then an arrangement superimposable on *A* is obtained.

Thus, the molecule (*A*) is said to posses $360°/180° = 2$-fold axis of symmetry and will be optically inactive.

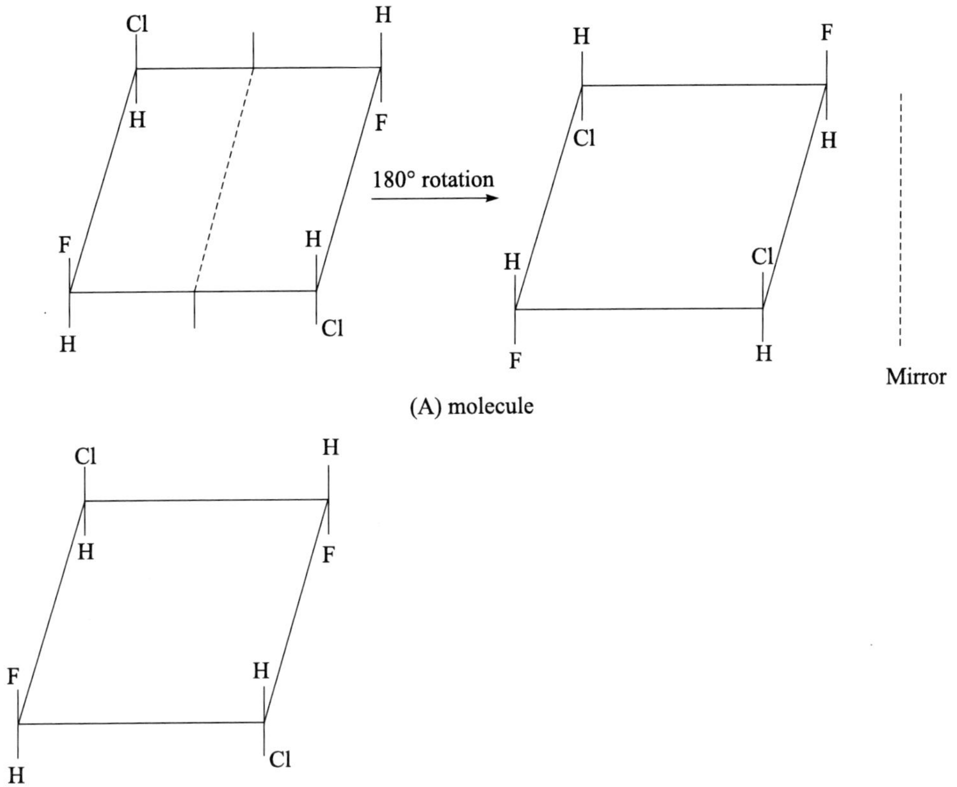

Fig. 9.8 *Molecule A showing alternating axis of symmetry*

9.7 CAUSE OF OPTICAL ACTIVITY

The necessary and sufficient conditions for a molecule to show optical activity are as described below.

9.7.1 Molecular Dissymmetry

It is the fundamental condition for optical activity. A molecule that does not possess any element of symmetry like plane, centre or alternating axis of symmetry is called dissymmetric molecule and is optically active.

For example: A molecule can exist in enantiomeric form only when it has a molecular dissymmetry.

9.7.2 Non-Superimposability

The molecule should be non-superimposable on its mirror image.

So, the concept of molecular dissymmetry and non-superimposability on mirror image are important as these are not related to chiral carbon atoms only. Even when a chiral carbon atom is present, a compound need not necessarily be optically active as in the case of mesotartaric acid, on the other hand, there are optically active compounds having no chiral carbon atoms, such as allenes, biphenyls, spiranes, etc. (structure of mesotartaric acid, 2, 3 pentadiene).

9.8 DIASTEROMERS

The other type of optical isomer are pairs of molecules that are not mirror images but each molecule rotates plane polarized light. These are optically active. Such molecules are called diastereomers. When there are two or more chiral carbons in a molecule it can have more optical isomers.

For example:

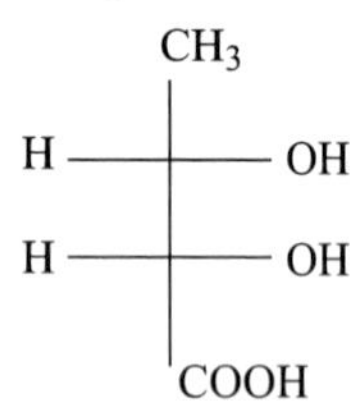

(I)

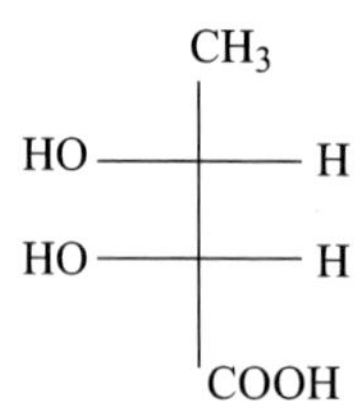

(II)

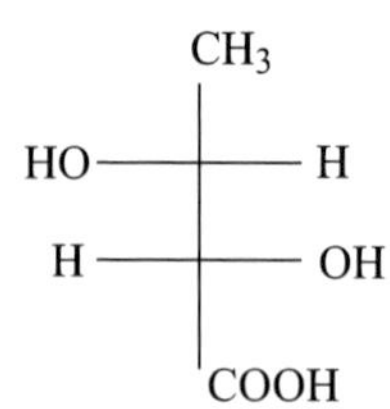

(III)

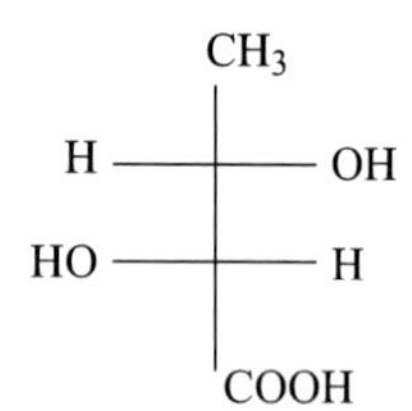

(IV)

This compound has 4 optical isomers and 2 pairs of enantiomers. The pair of enantiomers are stereoisomers of one another. Stereoisomers which are not mirror-image isomers are called diastereomers. The structure I and III are not mirror images but are stereoisomers because they rotate the plane polarized light. Hence, I and III are a pair of diastereoisomers, similarly II and IV are diastereomers.

9.9 MESO COMPOUND

The compound which contains chiral carbons but can be superimposed on its mirror image is known as meso compound. These compounds have two chiral carbons within the molecule but the compound is optically inactive.

Actually, these compounds have two chiral carbons and there is a plane of symmetry that will have half the molecule on one side of the plane of symmetry. While the other half which is the mirror image of the first half on the other side of the plane. Some compounds with two chiral carbons have less than four stereoisomers.

Example:

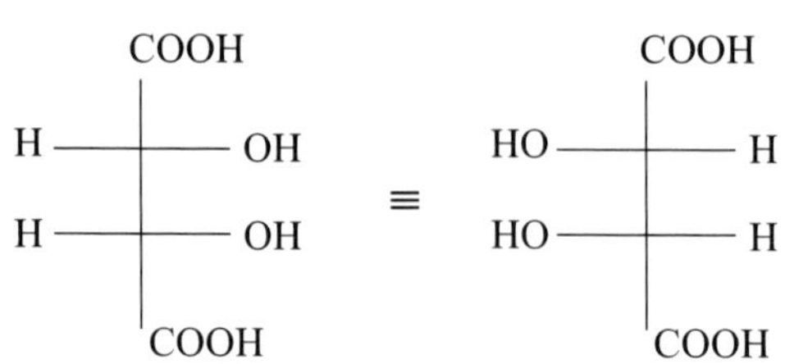

Achiral diasteromers (meso compounds)

This molecule has an internal plane of symmetry. The top half is the mirror image of the bottom half of the molecule. Although there are 2 chiral carbons in this molecule, this particular stereoisomer is a chiral or optically inactive.

Meso compound is optically inactive because as the plane polarized light passes through the molecule, half the molecule will rotate the light to the right and the other half of the molecule will rotate the light back in the other direction by the same number of degrees. This means that the light exists in the sample with no apparent rotation.

9.10 RACEMIC MIXTURE

A mixture consisting of equal parts of any pair of enantiomers is called a racemic mixture (or racemic modification) and is designated by (+/–). A racemic mixture does not rotate plane polarized light because (+) rotation caused by one enantiomer is cancelled by rotation in the opposite direction by the (–) enantiomer. A solution of a racemic mixture of enantiomer is optically inactive. Clearly unequal mixtures of two enantiomers have a lower optical rotation that depends on the enantiomeric excess of the mixture. Mixtures of unequal amounts of the enantiomers are referred to as scalemic.

$$\text{Enantiomeric excess} = (\%\ \text{enantiomer } A - \%\ \text{enantiomer } B)\%$$

Example: Optical isomerism.

9.10.1 Lactic Acid

($CH_3CHOHCOOH$) contains an asymmetric central carbon atom and it exists in two optically active forms, that are related to each other as mirror images and the two are non-superimposable.

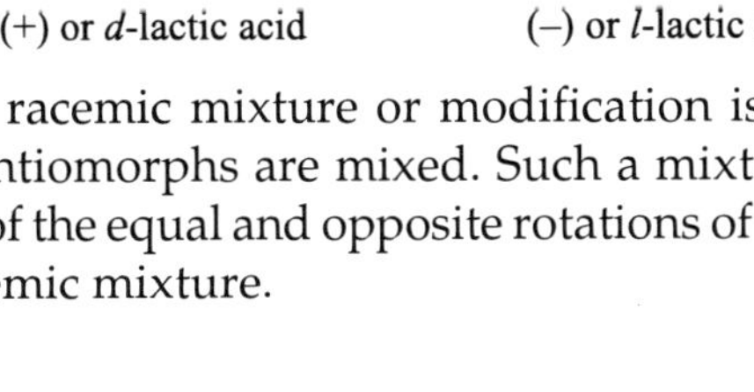

(+) or *d*-lactic acid (–) or *l*-lactic acid

A third form called a racemic mixture or modification is obtained, when equimolar quantities of *d* and *l* enantiomorphs are mixed. Such a mixture is optically inactive as a result of the cancellation of the equal and opposite rotations of the two isomers. The symbol ± is used to denote a racemic mixture.

9.10.2 Tartaric Acid

It has two asymmetric carbon atoms each linked to the same four different groups, that is, –OH, –H, –COOH and –CH(OH) COOH. The spatial arrangement of various groups in tartaric acid can be represented in three ways. So tartaric acid exists in three different forms *d*-tartaric acid, *l*-tartaric acid and racemic tartaric acid. Mesotartaric acid is the inactive variety as the rotation of the upper half is compensated by the rotation of the lower half (being in opposite direction). It cannot be resolved into active constituents so it is inactive by internal compensation. Unlike dextro- and laveo-varieties, it has a plane of symmetry. The molecule of a meso compounds a chiral.

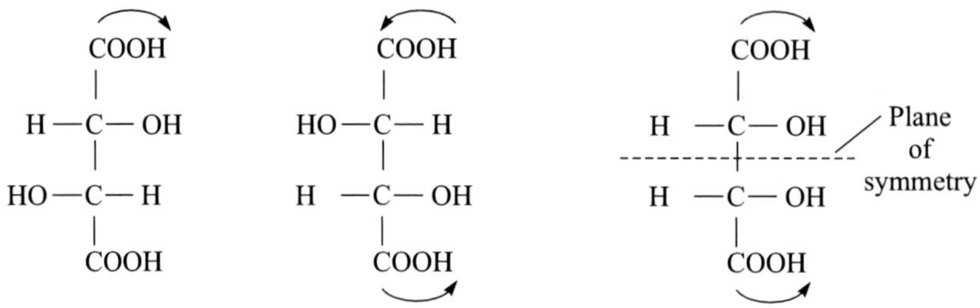

Dextro or (+) (d) — (a) Laevo or (–) (l) — (b) Meso — (c)

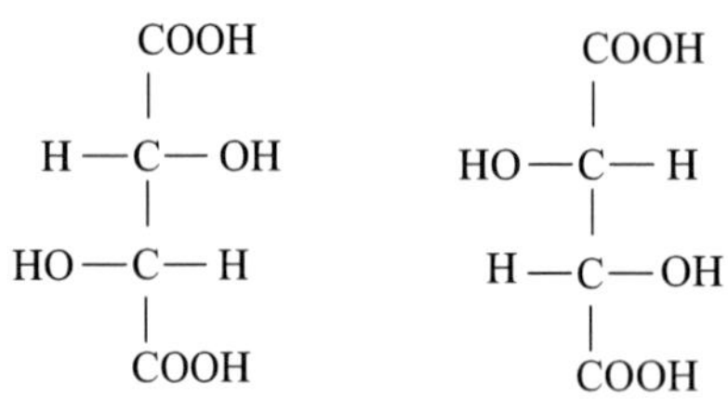

(d) (e)

Racemic mixture

Enantimers 1 and 2 3 and 4
Diasteromers 1 and 4 2 and 3

9.11 CONFIGURATION

The arrangement of atom that characterizes a particular stereoisomer is called its configuration.

or

The arrangement in space of the atoms or constituent group is called its configuration.

9.11.1 Relative Configuration (*D* and *L* Configuration)

D prefix is assigned to enantiomers which are chemically related to a right handed reference compound (glycerldehyde) and *L* prefix to a left handed reference compound.

For example: Glyceraldehyde

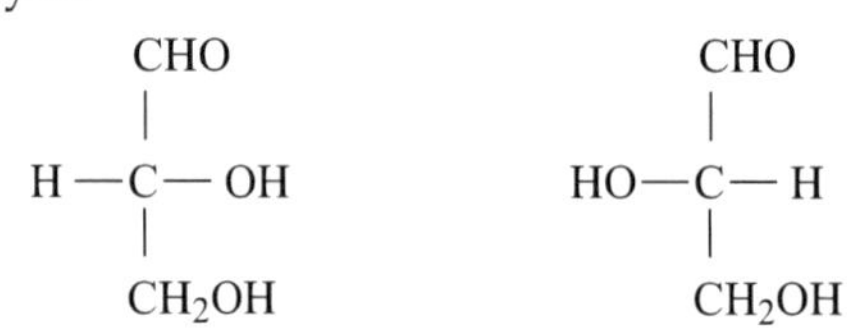

D (+)-Glyceraldehyde (I) L (–)-Glyceraldehyde (II)

In I structure OH group on the right, the hydrogen atom on the left, the –CHO at the top and the $-CH_2OH$ group at the bottom was arbitrarily given the configurational symbol *D*. The mirror image compounds (–)-glyceraldehyde in which –OH group is on the left and hydrogen is on the right was given the configuration *L*.

(+) and (–) sign indicate that the two forms are dextro- and laevorotatory respectively, but it must be noted that it is not necessary that all the compounds belonging to *D* and *L*-series will be dextro- and laevorotatory, respectively.

9.11.2 *R* and *S* (Absolute) Configuration

This system was developed by Cahn, Ingold and Prelog and is based on the actual three-dimensional formula. This system involves the following rules:

1. The four different groups/atoms attached to chiral carbon atoms are arranged in order of priority which is determined by their atomic numbers. The atom of highest atomic number gets highest priority.

   For example: I BrCHCl I (z = 53) gets highest priority and H(z = 1) gets least priority. In case chiral carbon atom has two isotopic atoms, the isotope of higher mass number gets priority **Example:** among deuterium (D) and hydrogen (H), the former gets priority.

2. If two same atoms are attached with the chiral carbon atom then priority is decided by the atomic number of the next atom.

   For example: $(CH_3)_3C— > (CH_3)_2CH— > CH_3CH_2— > CH_3—$
   t.butyl, Isopropyl, Ethyl, Methyl

3. A doubly or triply bonded atom × is equivalent to two or three such atoms.
   For example:

   = C is equivalent to C – C

   ≡ C is equivalent to C(C)(C)

   In the case of –CHO and $-CH_2O$ groups the —C(H) = O group has higher priority than $-CH_2OH$, groups. This is because in the –CHO groups, the C atom is attached with O, O and H atoms while in the $-CH_2OH$ group. C atom is attached with O, H, H atoms.

4. Some common substituents are given the following priority sequence in decreasing order.

   I, Br, Cl, SO_3H, SH, F, OR, OH, NO_2, NR_2, NHCOR, MH_2, CCl_3, COCl, COOR, COOH, $CONH_2$, CHO, CH_2OH, CN, CR_3, C_6H_5, CH_2R, CH_3, D and H.

Example: Bromochloroiodo methane

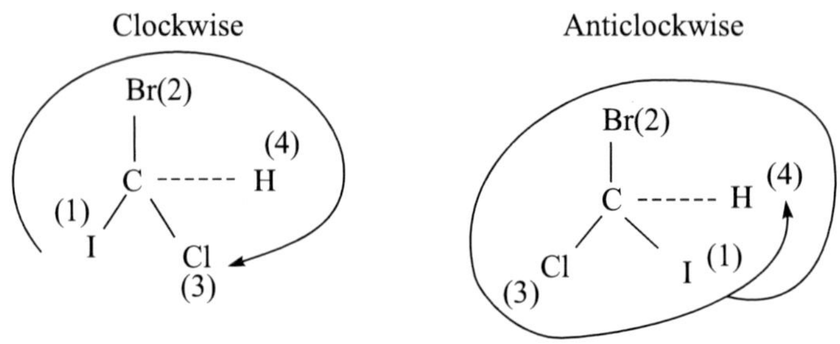

R-configuration (I) S-configuration (II)

The atomic number of I = 53 that of Br = 35, Cl = 17 and H = 1. Clearly the order of priority of these atoms has the order I, Br, Cl, H. In the configuration I, the eye moves in a clockwise direction while tracing a sequence I → Br → Cl and thus it is given R-configuration. In the same way, configuration II is S.

Example: Secondary butyl chloride

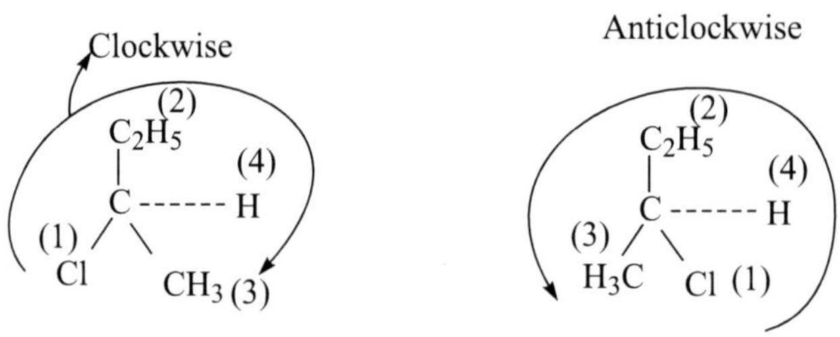

R-configuration S-configuration

In this case the order of priority is Cl, C_2H_5, CH_3, H.

Example: Lactic acid

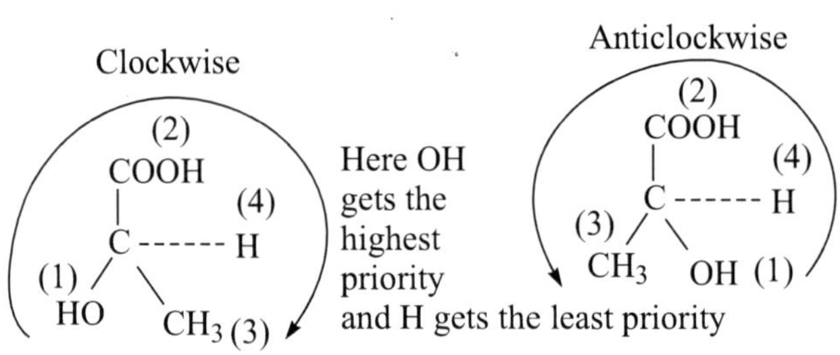

R-configuration S-configuration

Example: Glyceraldehyde

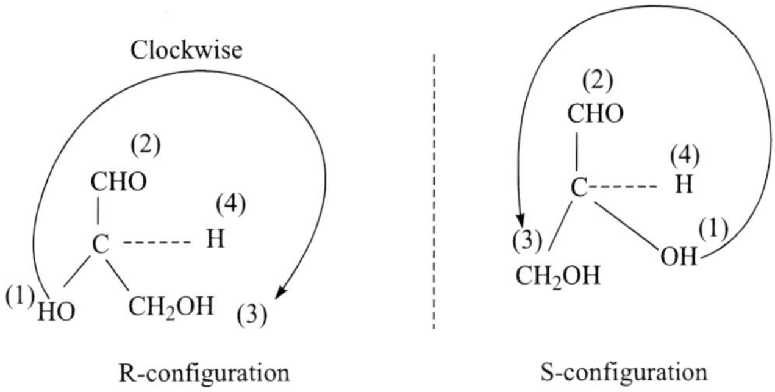

9.12 FISCHER PROJECTIONS

It is used to represent the three-dimensional molecules with an asymmetric carbon in two dimensions. To construct a Fischer projection, the molecule is so oriented that the top and bottom are included equally behind the plane of the paper. The groups on the left and right are similarly included in front of the plane of the paper. The molecule is projected in the shape of a cross as illustrated below for one lactic acid (stereoisomers).

The Fischer projection represents a molecule in such a way that vertical groups are pointing back and the horizontal groups are pointing outwards of the plane of the paper.

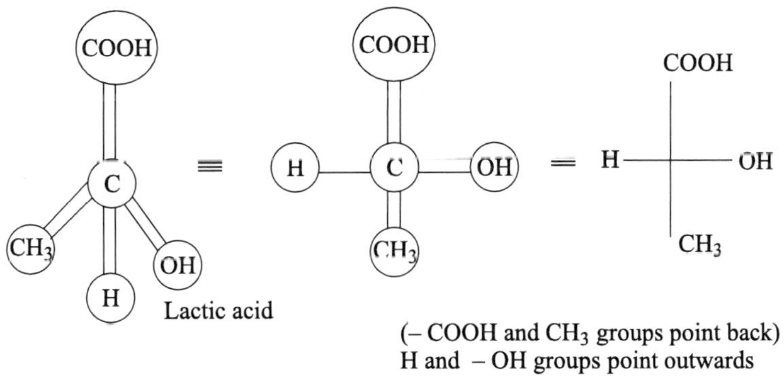

9.12.1 *R* and *S* Configuration from Planar Formula (Fischer Projection)

In case of a planar structure, give priority numbers to all the four atoms or groups attached to chiral carbon and then decide the configuration as follows.

Case I: When the atom or group of lowest priority is at the bottom.

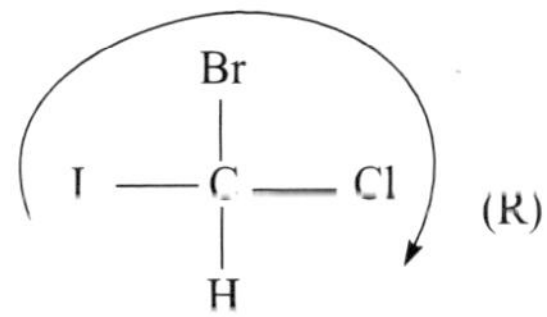

Simply rotate the eye in the order of decreasing priority. Clockwise rotation represents R-configuration and anticlockwise rotation represents S-configuration.

Case II: When the atom or group of lowest priority is at the top. In such a case, rotate the molecule by 180°, so as to bring the atom or group of lowest priority at the bottom. Now decide the configuration by rotating the eye in the order of decreasing priority.

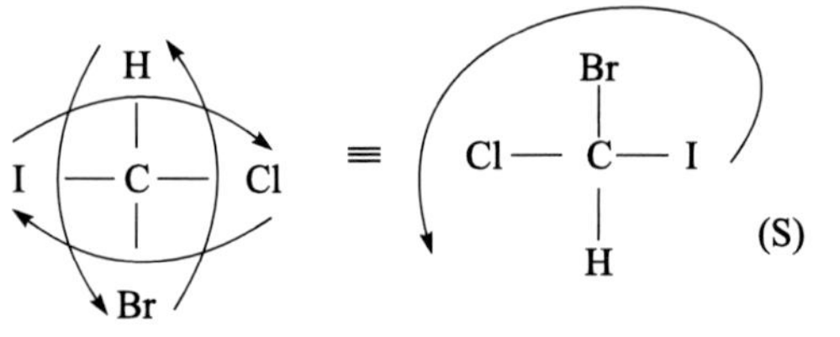

Case III: When the atom or group of lowest priority is at the right-hand side as a horizontal line. In such a case, keep the top position fixed and change the position of other atoms or groups in clockwise direction so that the atom or group of the lowest priority comes to the bottom.

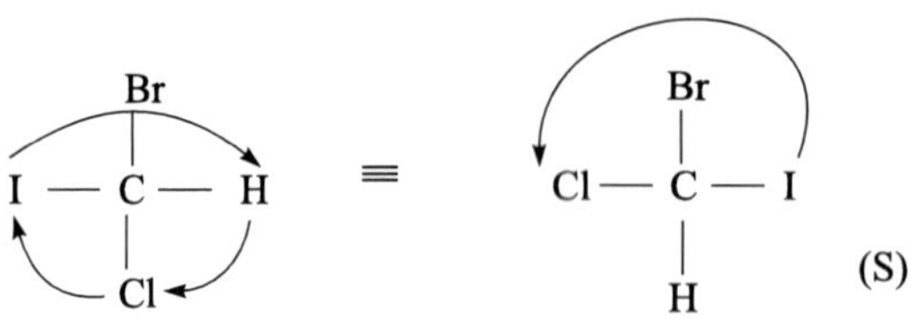

Case IV: When the atom or group of lowest priority is at the left-hand side as a horizontal line. In such a case, keep the top position fixed and change the position of the remaining atoms or groups in the clockwise direction so that the atom or group of lowest priority comes to the bottom.

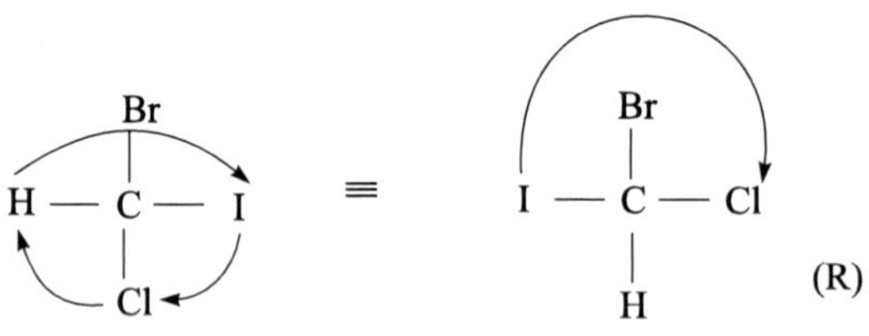

9.13 DIFFERENCE BETWEEN CONFORMATION AND CONFIGURATION

The term conformation should not be confused with configuration which relates to those spatial arrangements of the atoms of a molecule that can be changed only by the breaking and making of bonds whereas the spatial arrangements in conformation are changed simply by rotation about a single bond.

In other words we can say that the various stereoisomers that differ in configuration can be interconverted only by breaking and making of bonds whereas the conformational arrangement may be interconverted by rotation of one part of the molecule with respect to the rest of the molecule about a single bond joining these two parts.

9.14 OPTICAL ISOMERISM DUE TO RESTRICTED ROTATION ABOUT A SINGLE BOND

(1) Substituted biphenyl: In case of substituted biphenyls the benzene rings lie in different planes which are vertical to each other. Their mirror images are non-superimposable on each other. Hence, such molecules exhibit optical activity.

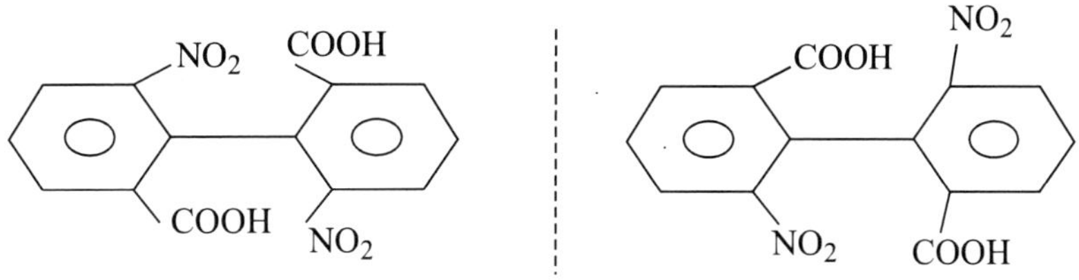

Mirror

(2) Substituted allenes: Compounds containing a carbon atom with adjacent double bonds are *r* known as allenes. They are represented by the formula RCH = C = CH R.

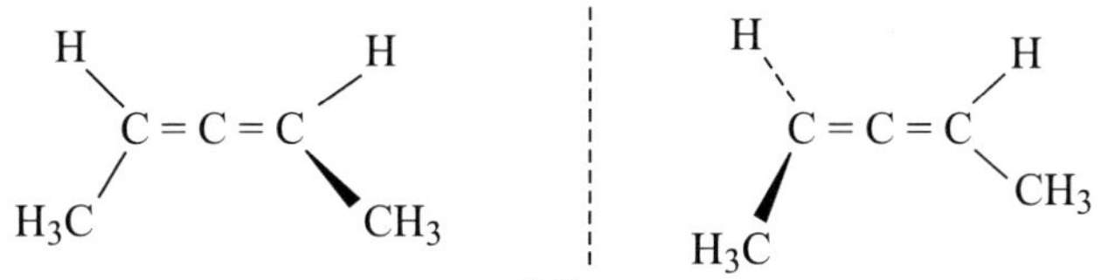

Mirror

Non-superimposable mirror image (optically active)

9.15 D, L-NOMENCLATURE OF RELATIVE CONFIGURATION

Before 1951, there was no method for determining the absolute configuration (actual arrangement of atom in space) of a compound and hence the configuration of all the compounds was studied with respect to glyceraldehyde (relative configuration) (+) glyceraldehyde having the –OH group on the right and the hydrogen atom on the left, the –CHO and $-CH_2OH$ groups being at the top and bottom respectively was arbitrarily given the configurational symbol D. The mirror image compound (–) glyceraldehydes in which the OH group is on the left and hydrogen on right was given the configuration L.

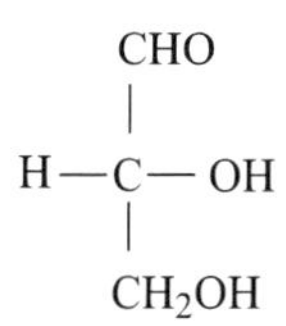

D (+) Glyceraldehyde
(having – OH to the right)

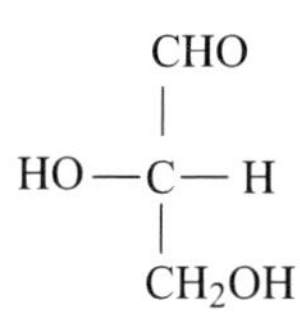

L (–) Glyceraldehyde
(having – OH to the right)

9.16 ABSOLUTE CONFIGURATION

It is actual arrangement in space of atoms or groups constituting a particular stereoisomer. The first compound whose absolute configuration was determined is sodium rubidium salt of (+) tartaric acid. It was not possible to decide whether configuration I is *d*-lactic acid or *l*-lactic acid. Determination of absolute configuration is a cumbersome process and can be accomplished only if the compound forms well defined crystals.
For example:

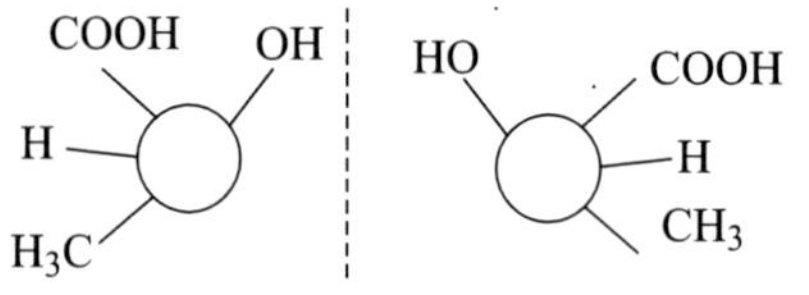

(I) (II)

Vogel determined directly the absolute configuration of sodium rubidium salt of (+) tartaric acid by *X*-ray diffraction method after obtaining its well-defined crystals.

9.16.1 Optical Isomerism due to Restricted Rotation about Single Bond

(1) **Substituted biphenyls**

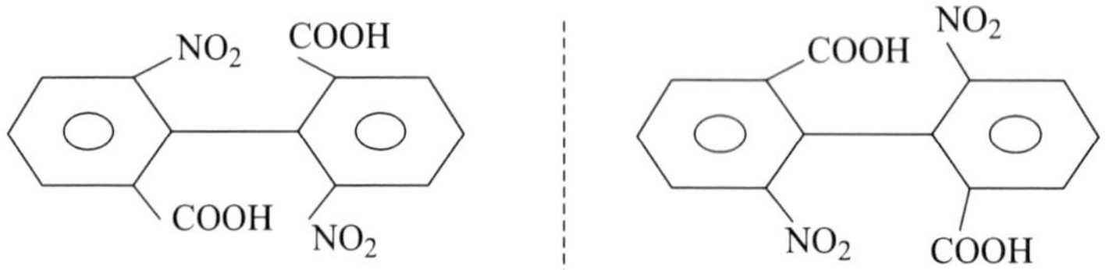

Mirror

In case of substituted biphenyls, the benzene rings lie in different planes which are vertical to each other. Their mirror images are non-superimposable on each other. Hence, such molecules exhibit optical activity.

(2) **Substituted allenes:** Compounds containing a carbon atom with adjacent double bonds are known as allenes or cumulated dienes. They are represented by the formula RCH = C = CHR.

For example:

Non-superimposable mirror image (optically active)

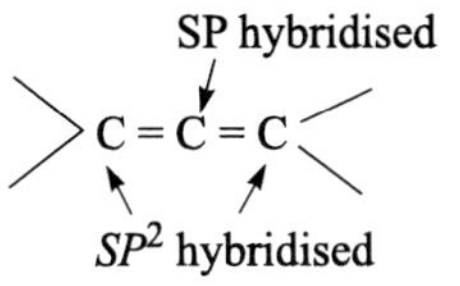

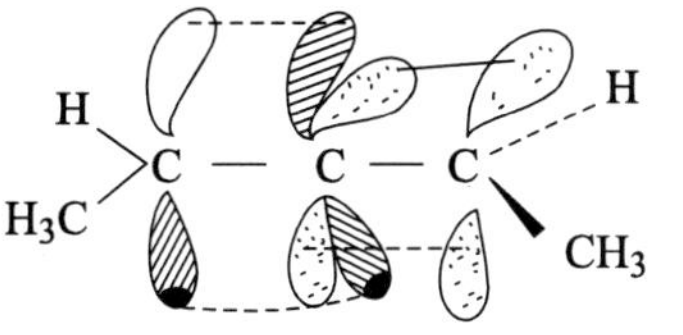

P-molecular orbital picture of cumulated diene part

Such compounds are non-superimposable mirror images of each other. Hence, they are disymmetric and optically active.

9.17 GEOMETRICAL ISOMERISM (RESTRICTED ROTATION)

When the same molecular formula represents two compounds which differ in the spatial arrangement of atoms or groups around carbon-carbon double bond then such isomers are called geometrical isomers. The phenomenon is called geometrical isomerism.

It is due to the restricted or hindered rotation around the double bond.

In case of organic compound, the rotation around (C – C) single bonds is free but the rotation around (C ═ C) double bond is restricted. The two carbons of the C – C double bond are SP^2 hybridized. The three SP^2 hybrid orbitals lie in one plane. Thus, the three σ-bonds formed by each of these two carbon atoms are co-planar. Each carbon atom is left with one unhybridized *P*-orbital at right angels to the SP^2 hybrid orbitals. These two *P*-orbitals overlap to form the π-molecular orbital. The π-cloud spreads over both carbon atoms above and below the plane of sigma bond (C – C). Thus, a C = C double bond consists of one σ-bond and one π-bond.

The overlap of two unhybridized *P*-orbitals is maximum when they are parallel to each other. This is possible only if all the four groups (attached to the two carbon atoms) along with the two carbon atoms are co-planar as shown in Fig. 9.9.

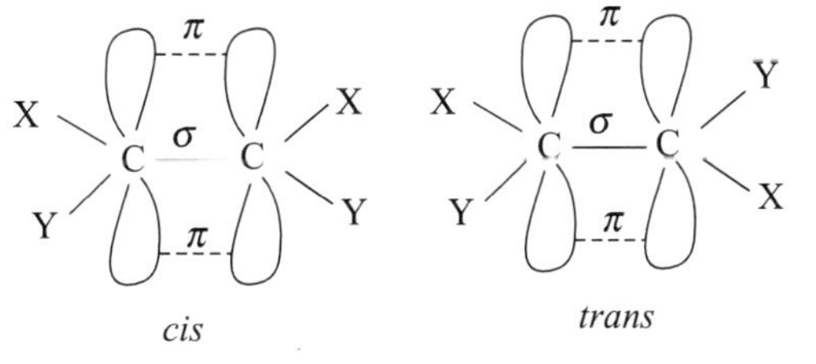

Fig. 9.9 *cis* and *trans* isomer

Hence, any change in this co-planar geometry by twisting the (C = C) bond would make the molecule unstable. Therefore rotation around C – C = bond would be restricted.

On account of the restricted or hindered rotation about the double bond, two different spatial arrangements are possible. These cannot be interconverted easily.

9.17.1 Conditions for Geometrical Isomerism

1. The molecule must have a double bond.

2. There must be two different atoms or groups attached to each carbon atom of the double bond. No geometrical isomerism is possible, if one or both the carbons of the double bond carry two similar atoms or groups.

   Example: Propylene and isobutylene do not show geometrical isomerism.

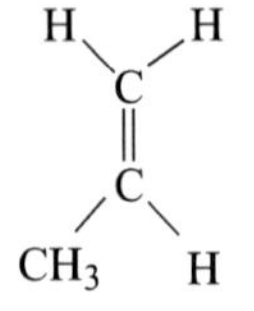

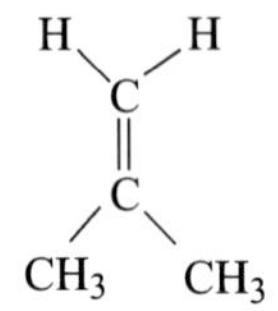

Isobutylene

Propylene

(No geometric isomers possible)

Examples of Geometrical Isomerism

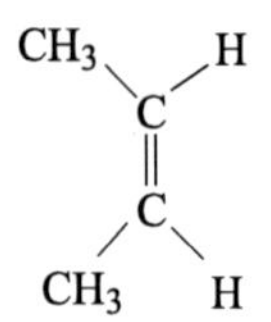

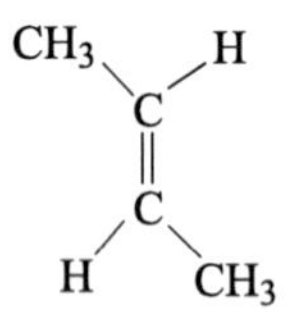

cis-but 2-ene

trans-but 2-ene

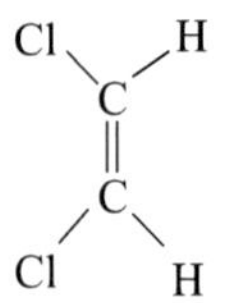

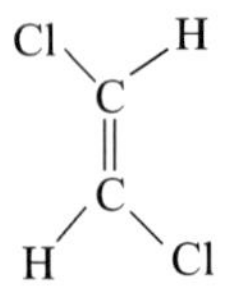

cis
1, 2-dichloroethene

trans
1, 2-dichloroethene

9.17.2 Characteristics of Geometrical Isomers

1. Geometric isomers have similar chemical properties.
2. They have different physical properties such as melting points, dipole moments, refractive index, and adsorption capacity on materials used in chromatographic analysis.
3. The *cis* isomers show a large dipole moment due to their unsymmetrical structure while *trans* isomers have zero or negligible dipole moment due to their symmetrical structure. The individual bond moments in case of *trans*-isomers cancel each other because they act in opposite directions. The higher values of the dipole moment in case of *cis* isomers means higher polarity which is responsible for the higher boiling points of *cis* isomers. The *trans* isomers due to their symmetrical structure fit well into the crystal lattice. Consequently, *trans* isomers have generally higher melting points than the *cis*-isomers (due to their poor symmetry).

9.17.3 Oximes Syn-anti-Isomerism

Aromatic aldoximes and ketoximes having carbon-nitrogen double bond also show geometrical isomerism due to restricted rotation around C = N bond. In aldoximes, the isomer in which H and OH group are on the same side is called syn isomers (analogous to *cis*) and the isomer in which H and OH groups are on the opposite side is called anti-isomer (analogous to *trans*).

For example:

(i) *Consider benzaldoximes.* Its two geometrical isomers are written as

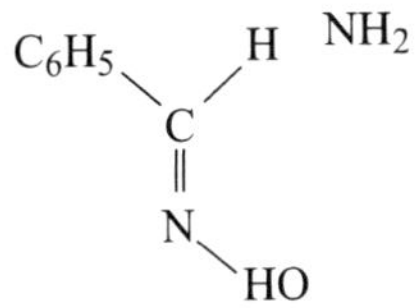

(*cis* or syn)

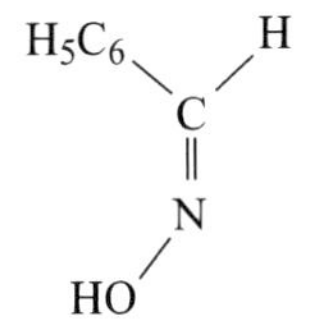

(*trans* or anti)

(ii) Consider the isomers of phenyl-*p*-tolyl ketoxime

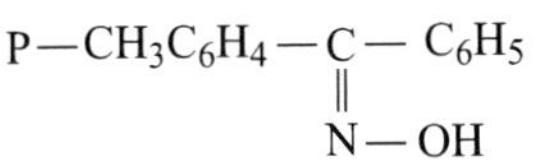

syn-Phenyl-P-tolyl ketoxime
(anti-p-tolyl-phenyl ketoxime)

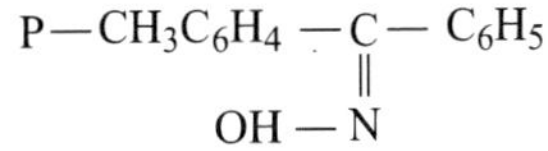

syn-P-tolyl-phenyl ketoxime
(anti-phenyl p-tolyl ketoxime)

It is important to note that ketoximes like benzophenone oxime, $(C_6H_5)_2C$ = NaOH having two similar aryl groups do not show geometrical isomerism.

9.17.4 *E* and *Z* Nomenclature of Geometrical Isomers

A new system of nomenclature known as *E* and *Z* nomenclature has been introduced to name such double and stereoisomers unambiguously. The symbol *E* stands for the German word *entgegen* which means *opposite* while the symbol *Z* stands for Zusammen which means *together*.

The proposed system of nomenclature involves the following two steps:

Step 1: Assign the priority order to the two groups attached to each of the doubly bonded carbon in accordance with the following sequence rules:

(1) *Sequence rule:* If the two atoms attached to the doubly bonded carbons are different, priority depends on atomic number. The atom having higher atomic number gets higher priority. The one with lower atomic number gets the lower priority.

Example: In the compound bromine (at number 35) gets higher priority over chlorine (at number 17) and methyl group (C-atomic number 6) gets higher priority over hydrogen (atomic number 1).

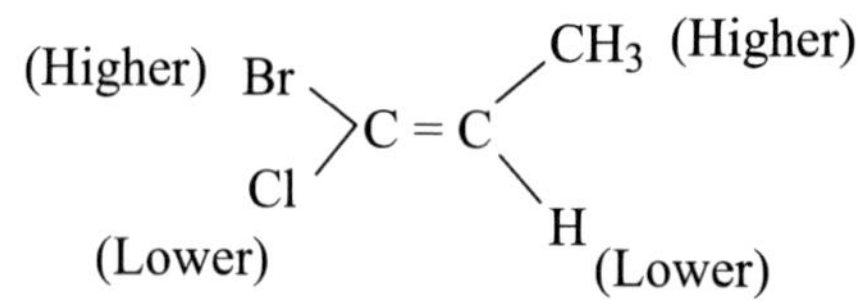

(2) *Sequence rule 2:* If the two atoms directly linked to a doubly bonded carbon are isotopes of the same elements, isotope with higher mass number gets higher priority.

Example: Out of H and D, D gets higher priority.

(3) *Sequence rule 3:* If two atoms directly attached to the doubly bonded carbon are the same, the order of priority is decided by the application of rules (i) and (ii) to the next atoms in the group. If the second atoms from the group provide no choice, third or fourth atoms must be compared until a choice is made. For example, out of C_2H_5 and CH_3, $CH_3 - CH_2$ has higher priority over CH_3 because in CH_3, the second atoms are H, H, H and in $CH_3 - CH_2$ they are C, H, H. Since carbon has higher atomic number than hydrogen, C_2H_5 has higher priority as compared to CH_3. Out of $-CH_2OH$[–C(O, H, H)] and $-C,(CH_3)_3$[–C, C, C)] – CH_2OH has higher priority over $-C(CH_3)_3$.

(4) *Sequence rule 4:*

(a) A doubly or triply bonded atom *X* is considered to be equivalent to two or three *X*'s respectively. Thus, = *X* equals.

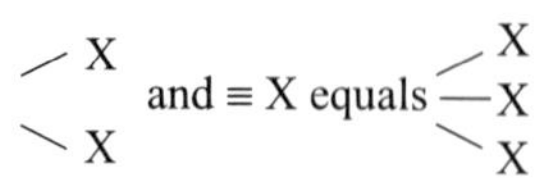

For example: —C(H)=O group equals —C(H)(O)—O and takes priority over $-CH_2OH$

which equals —C(H)(H)—OH.

(b) It may be noted, however, that if in a group two or three atoms of *X* are separately linked to same atom through single bonds, such a group gets higher priority over the group having doubly or triply bonded *X*.

For example: In the groups —C(H)=CH_2 the group —C(CH_3)(H)—CH_3 gets priority over the group —C(CH_3)(H)—CH_3. This is because in —C(H)=CH_2 the two carbons are actually attached to C through single bonds while in —C(H)—CH_3 there is only one carbon attached through double bond.

Step 2: After deciding the sequence of priority of two groups attached to each carbon of carbon-carbon double bond, the compound is assigned *Z* configuration if the atoms/groups of higher priority are on the same side of the double bond. On the other hand, if the atoms/groups of higher priority on each carbon are on the opposite sides of the double bond, the isomer is assigned the configuration *E*.

Note: When no atom is present in a position as in 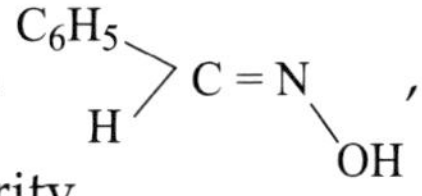 , the position is said to have an atomic number zero and thus lower priority.

For example:

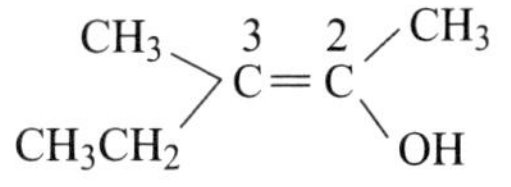

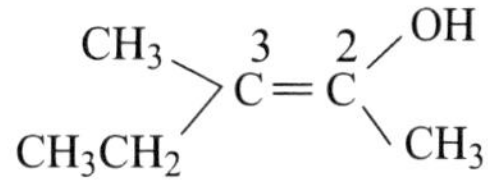

On carbon atom NO_2; OH gets higher priority than CH_3 group. On carbon atom number 3, CH_3CH_2 gets higher priority than CH_3.

Thus, assigning the priority order in terms of *a* and *b*, the configuration can be written as

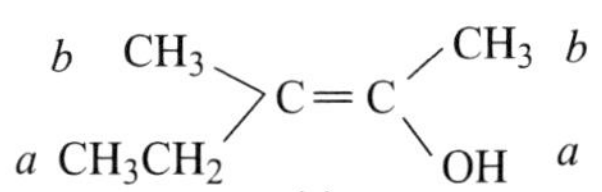

(z)

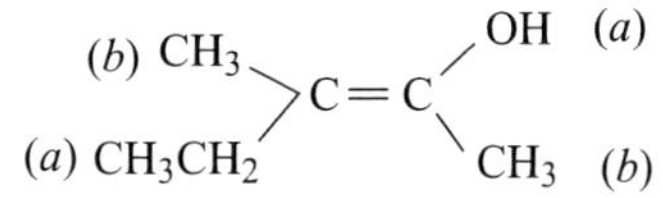

(Two groups are on the same side)

(Two groups on the opposite side)

(2)

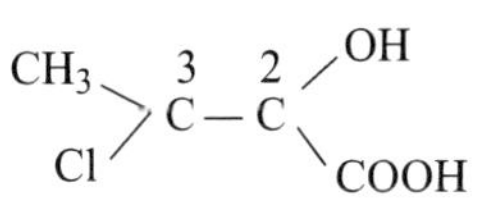

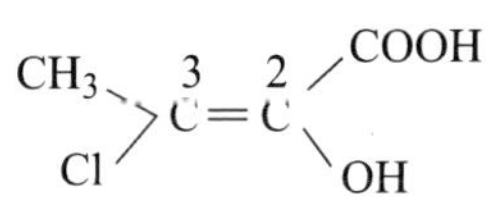

Following the sequence rules, we say that on carbon atom NO_2, OH gets higher priority compared to COOH. On carbon atom number 3, Cl atom gets higher priority than CH_3 group. Thus, assigning the priority order on groups attached to each carbon atom separately. The configuration can be written as

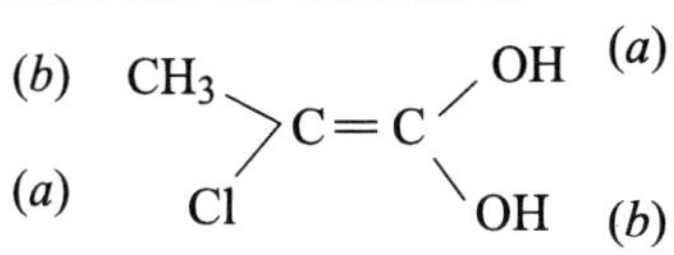

(ε)

(Two groups are on the opposite side)

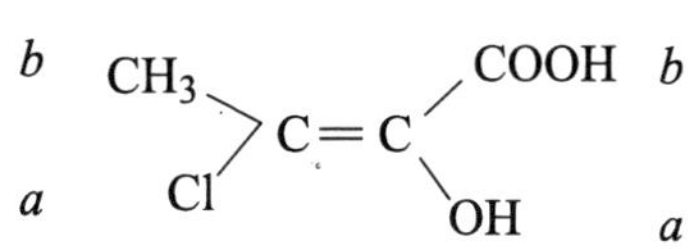

(Z)

(Two groups are on the same side)

9.17.5 Cis and trans Isomerism in Cyclic Compounds

Just like π-molecular orbital of a double bond of alkenes, a closed chain or ring also results in the hindered rotation around C–C single bond and hence many substituted cycloalkanes exhibit *cis-trans* isomerism.

Example: L2-dimethyl cyclopropane can exist in two stereoisomeric forms as shown below:

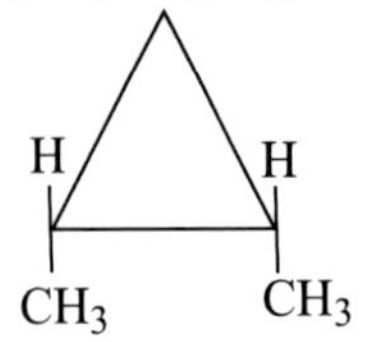

(I)

Cis 1, 2 dimethyl cyclopropane

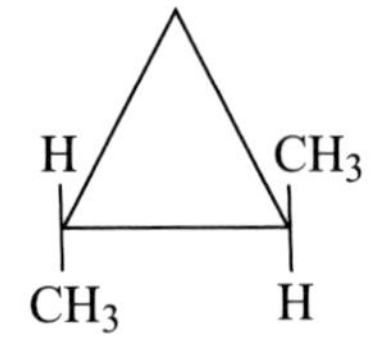

(II)

Trans 1, 2-dimethyl-cyclopropane

Structure in which both the substituents are situated above or below the plane of the ring represent the *cis*-isomer and the structure in which one substituent lies above and the other below the plane of the ring represent the *trans*-isomer.

Some other examples are given below.

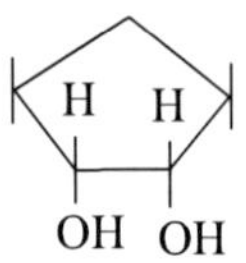

cis

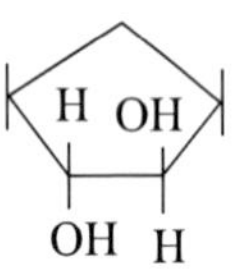

trans

cyclopentane 1, 2-diol

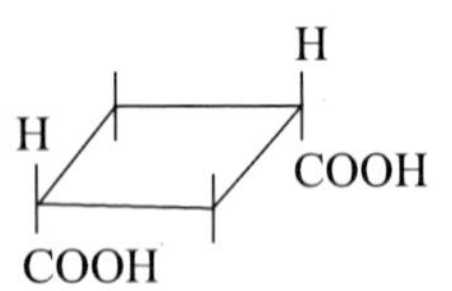

cis

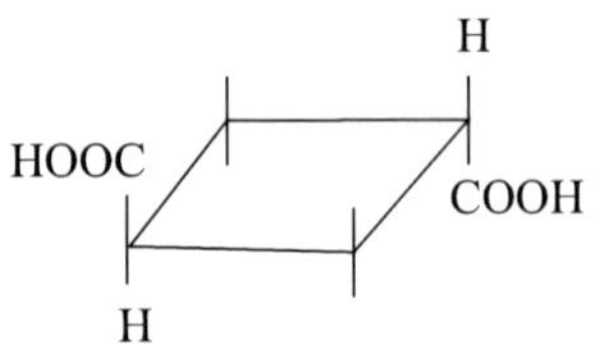

trans

Cyclobutane 1, 3-dicarboxylic acid

9.18 *E* AND *Z* NOMENCLATURE OF DIENES

For polyenes with more than one double bond, the stereochemistry around each double bond is determined on the basis of same rules as has been explained for one double bond. Dienes are capable of existing in several geometric isomers.

For example: 2, 4-Hexadienes have four possible geometric isomers as shown below.

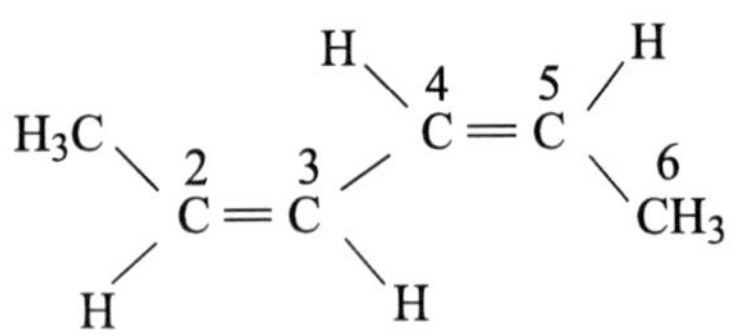

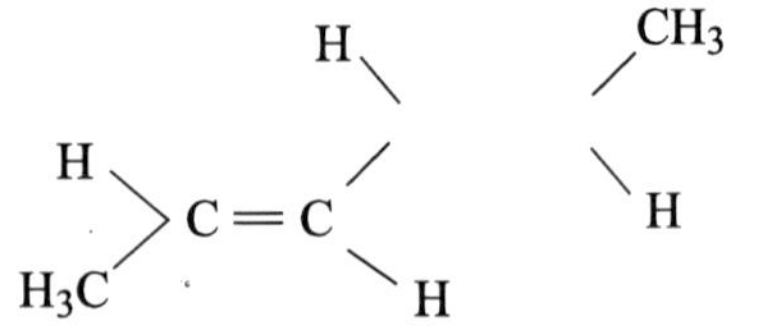

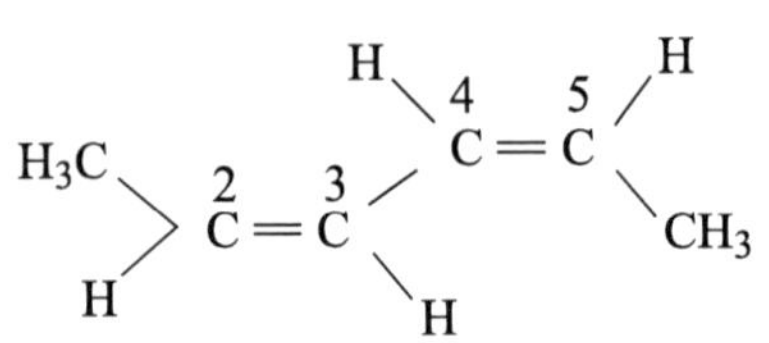

(2Z, 4Z) -2, 4-Hexadiene

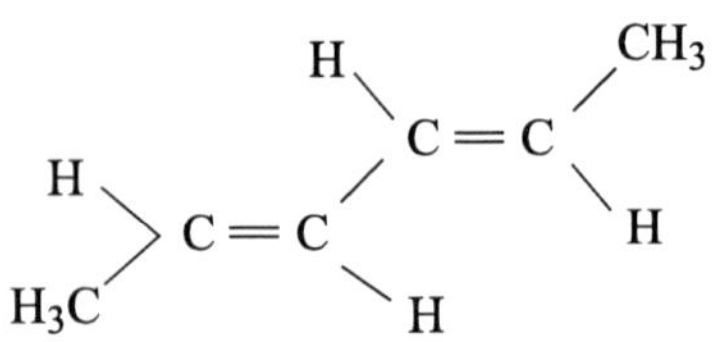

(2E, 4E) -2, 4-Hexadiene

9.19 CONFORMATION OF SMALL RINGS

(a) **Cyclopropane:** Three carbon atoms of cyclopropane are geometrically coplanar and rotation about its C — C bonds is impossible. So conformational analysis in cyclopropane is simpler than in any other cycloalkane. As has already been discussed in Baeyer's strain theory cyclopropane has maximum angle strain. In addition to angle strain cyclopropane is destabilised by torsional strain. Each C — H bond of cyclopropane is eclipsed with two others.

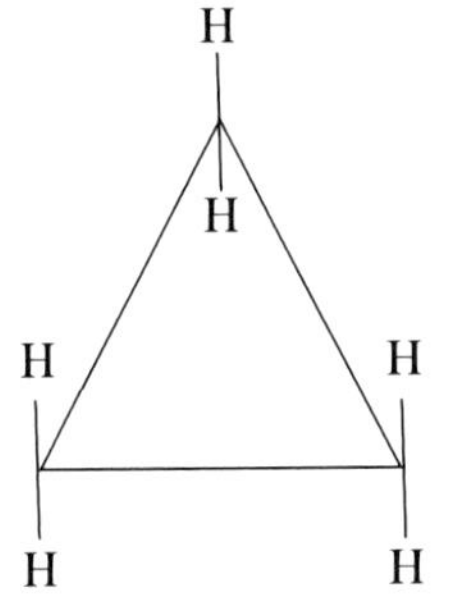

Fig. 9.10 *All adjacent H pairs of bonds*

(b) **Cyclobutane:** Cyclobutane has less angle strain than cyclopropane and can reduce the torsional strain by adopting non-planar puckered conformation rather than planar geometry. It exist in two conformations.

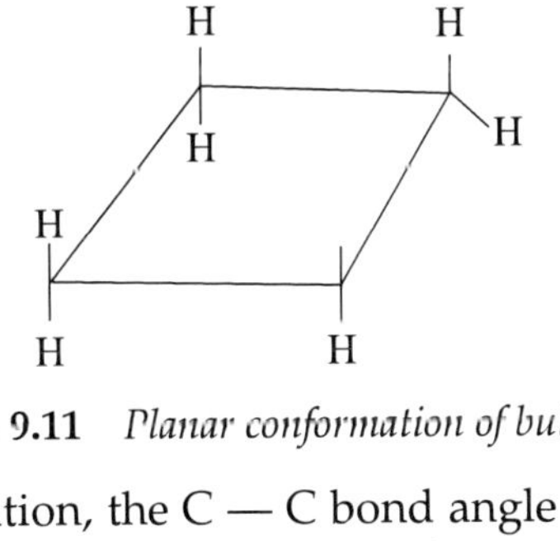

Fig. 9.11 *Planar conformation of butane*

(i) *Planar:* In this conformation, the C — C bond angle is 90°, the carbon atoms lie in the same plane and the H-atoms of the ring eclipse with each other which gives rise to torsional strain along with angle strain.

(ii) *Non-planer or puckered ring conformation*

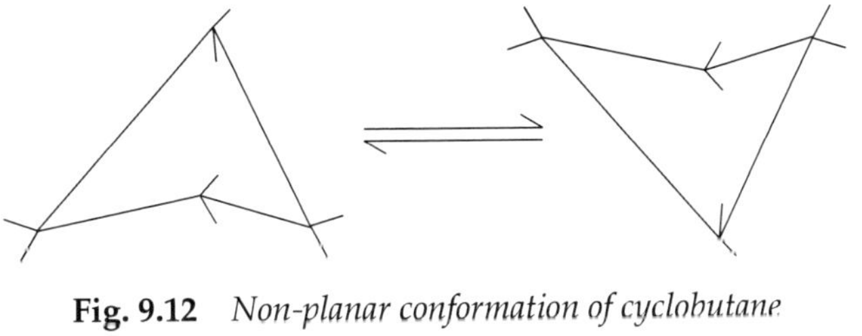

Fig. 9.12 *Non-planar conformation of cyclobutane*

Cyclobutane can reduce angle strain and torsional strain by orienting carbon atoms in different planes.

(c) **Cyclopentane:** Angle strain in planar conformation of cyclopentane is relatively less (< 108°). The torsional strain, however is also there since C — H bond present above and below the planes is eclipsed. Some of this strain is relieved in non-planar conformations, i.e., envelop and half-chair which are of comparable energy.

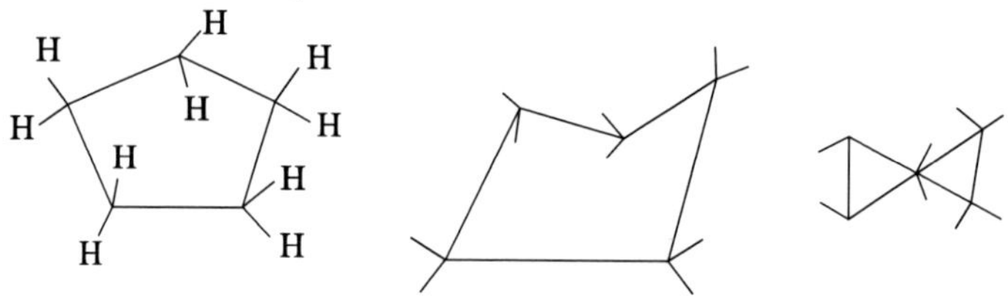

(a) 5C – in one plane (b) 4C – in one plane (c) 3C in one plane

9.20 CONFORMATIONAL ANALYSIS OF DISUBSTITUTED CYCLOHEXANES

Three position isomers of disubstitued cyclohexanes exist depending on the relative position of two substituents.

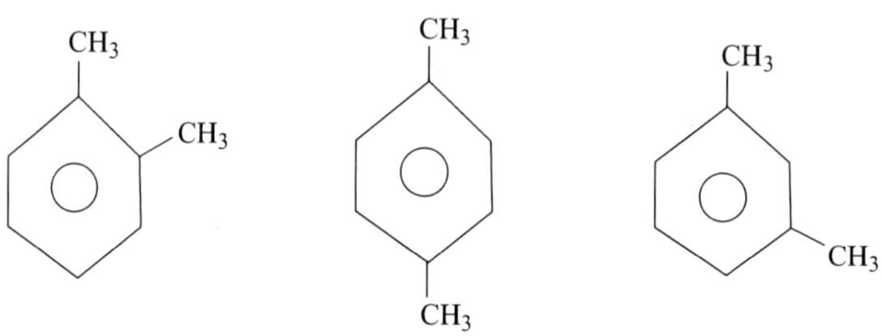

(a) 1, 2 dimethyl cyclohexane (b) 1, 3 dimethyl cyclohexane (c) 1, 4 dimethyl cyclohexane

Considering the chair conformation of cyclohexane (more stable form) as the basic conformation these two methyl groups in each case can either take up both axial or both equatorial or one equatorial and other axial positions. Any $-CH_3$ group or substituent at axial positions will destabilise the corresponding conformation due to 1, 3-diaxial interaction.

1. **Conformations of 1, 2-dimethyl cyclohexane**

These can be classified into two isomers.

(a) Diequatorial or Diaxial 1, 2-dimethyl cyclohexane (*trans*-isomers).

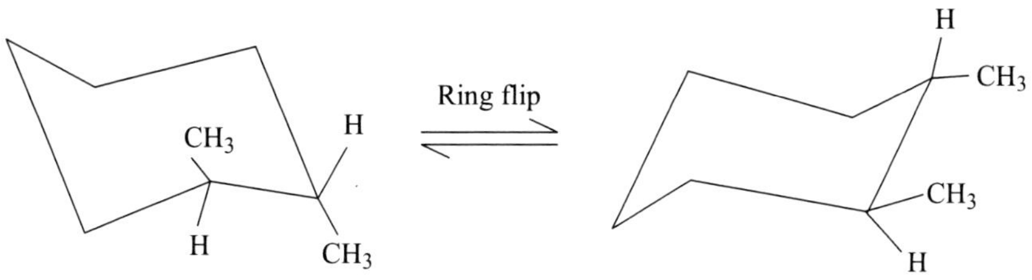

Diaxial | Diequatorial

trans-1, 2-Dimethyl cyclohexane (Less stable) | *trans*-1, 2-Dimethyl cyclohexane (More stable)

In this case there are two equivalent structures having either both $-CH_3$ groups in axial position or both in equatorial position (methyl groups are not oriented in the same plane. Thus, known as *trans*-isomers).

(b) Axial – equatorial or Equatorial – axial –1, 2-dimethyl cyclohexane (*cis*-isomer)

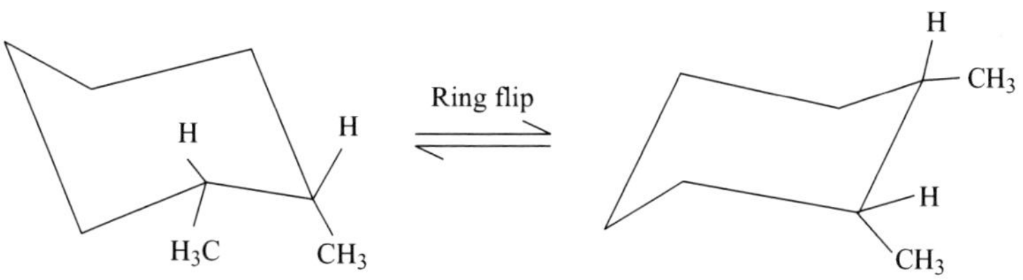

(Axial-equatorial) | (Equatorial-axial)

Cis 1, 2-Dimethyl cyclohexane

Here both $-CH_3$ groups are in the same plane so termed *cis*-isomer.

Because of absence of 1, 3-diaxial interaction is *trans*-isomer, it is more stable than its corresponding *cis*-isomer.

2. Conformations of 1, 3-dimethyl cyclohexane

(a) Diequatorial or diaxial –1, 3-dimethyl – cyclohexane (*cis*-isomer).

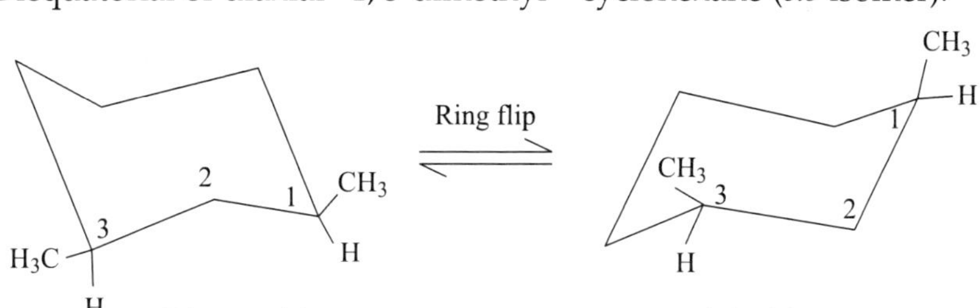

Diequatorial | (Diaxial)

cis-1, 3-dimethyl cyclohexane (More stable) | *cis*-1, 3-dimethyl cyclohexane (Less stable)

(b) Equatorial – axial or Axial equatorial –1, 3-dimethyl cyclohexane (*trans*-isomer)

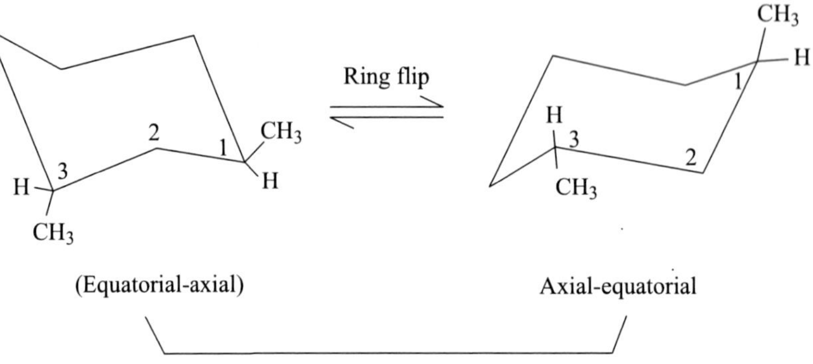

trans-1, 3-dimethyl cyclohexane

The most stable conformation of *cis* 1, 3-dimethyl cyclohexane has both methyl groups equatorial.

3. Conformations of 1, 4-dimethyl cyclohexane

(a) Diequatorial or diaxial –1, 4-dimethyl cyclohexane (*trans*-isomer)

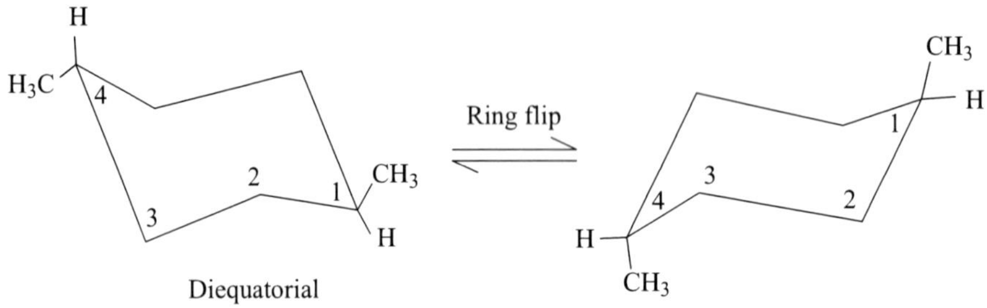

trans 1, 4-dimethyl cyclohexane (More stable)

Both $-CH_3$ groups are oriented in different planes

(b) Equatorial – axial or Axial –equatorial 1, 4-dimethyl cyclohexane (*cis*-isomer)

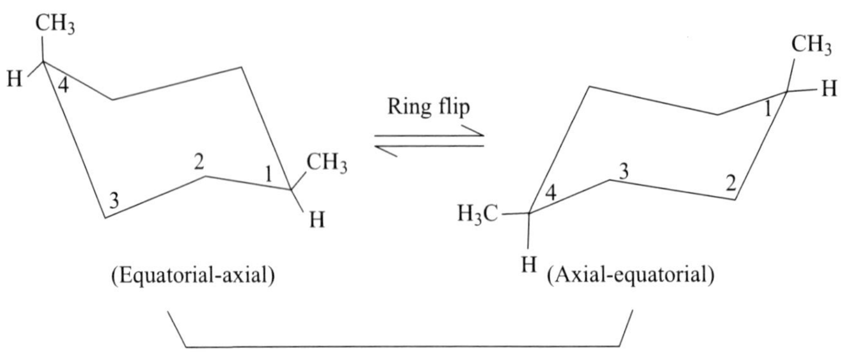

cis 1, 4-Dimethyl cyclohexane

Trans 1, 4-dimethyl cyclohexane is more stable than *cis* 1, 4-dimethyl cyclohexane because both $-CH_3$ groups are equatorial in the most stable conformation of *trans* 1, 4-dimethyl cyclohexane while one $-CH_3$ group must be axial in the *cis*-stereoisomer.

9.21 CONFORMATIONS OF MEDIUM AND LARGE RINGS

The same fundamental principles apply to medium and large rings as are applied to smaller ones. For them, there are more conformational possibilities. Also in case of cycloalkanes there is no angle of strain because the rings have puckered ring structure maintaining C – C – C bond angles equal to 109°28.

(a) *Conformations of cyclooctane:* Because of more angle strain (–12.46°), cyclooctane cannot show planar structure. So two possible stable conformations are there.

(a) Tub-shaped (b) Saddle-shaped

Conformations of cyclooctane

(b) *Conformations of cyclodecane:* Again cyclodecane exists in nonplanar puckered ring structures with no angle strain.

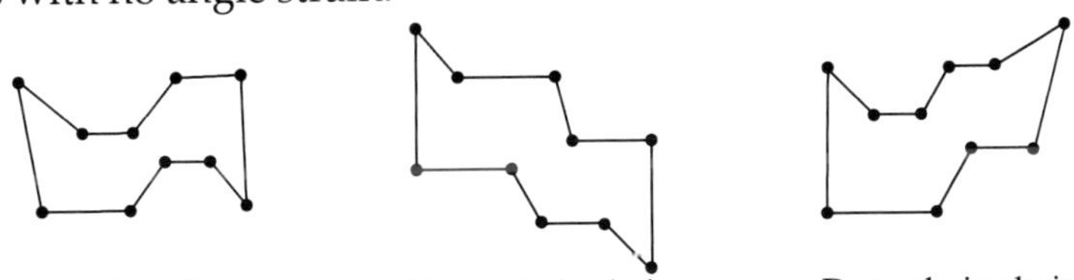

Boat-chair-boat Chair-chair-chair Boat-chair-chair

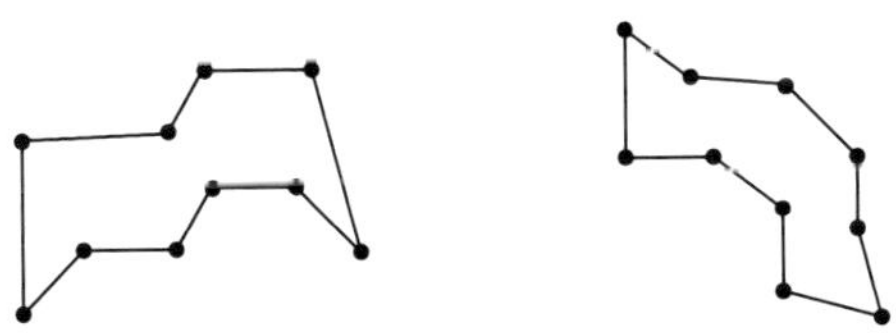

Chair-chair-boat Chair-boat-chair

9.22 CONFORMATION OF POLYCYCLIC-RING COMPOUNDS

The hydrogen atoms at the ring junctions (fused C-atoms) can orient themselves on the same side or opposite side with respect to each other. Therefore, decalin can acquire two conformations.

(a) *Trans-decalin:* In this conformation, the H-atoms are on the opposite side and this is possible only if the fusion of two cyclohexane rings occurs through equatorial positions and axial H-atoms remain as such.

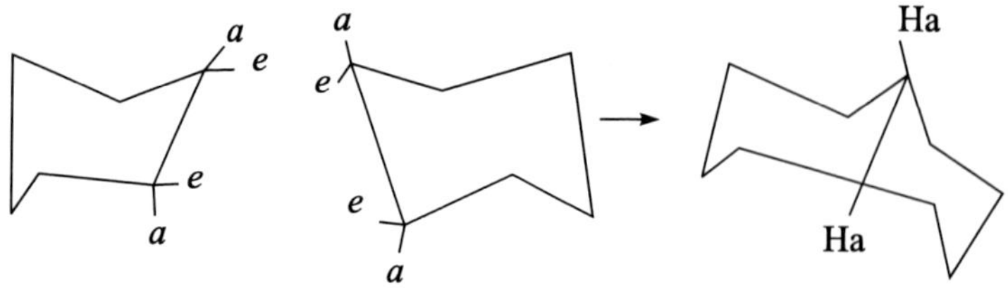

(b) *cis-decalin:* In this system, the two rings fuse in such a way that one axial and equatorial positions fuse together to form bicyclic compound. The H-atoms of fused C-atoms are *cis* – with respect to each other. It has been observed that *trans* configuration is more stable than *cis* because in such a case, 1, 3-diaxial strain is minimum.

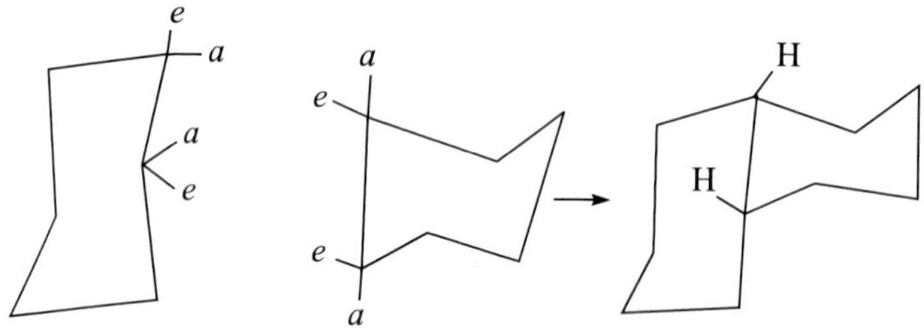

REVIEW QUESTIONS

1. What do you understand by enantiomers?
2. What are *R* and *S* configurations? State and explain the sequence rules with suitable examples.
3. Explain the difference between diastereoisomers and the enantiomers.
4. How will you resolve racemic lactic acid into optically active form? Give the methods for this purpose.
5. Write a brief note on racemisation.
6. What do you mean by resolution? Explain giving a chemical method?
7. What is meant by specification of configuration?
8. What is *E*, *Z* notation? Assign *E* and *Z* configuration to the following compounds ClCH = CHF?
9. How will you justify the statement that disymmetry is the essential and the sufficient condition for optical activity?
10. What are *R* and *S* configurations? State sequence rules.
11. Disymmetry is the only condition for a molecule to show optical activity. Explain?
12. Write down the configuration of $\begin{matrix} \text{CHCOOH} \\ \| \\ \text{CHCOOH} \end{matrix}$ according to *E* and *Z* sequence rules.

13. (a) Give *R* and *S* configurations to the following compounds

(i) 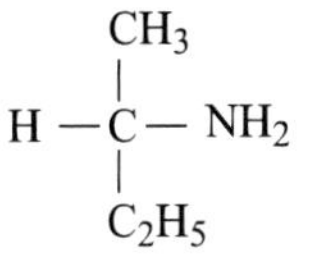   (ii) 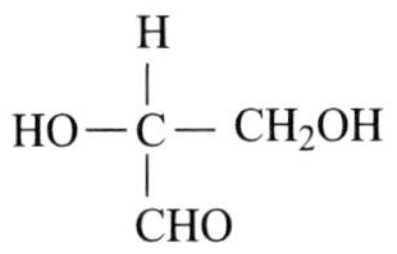

(b) What is restricted rotation about the double bond? What type of isomerism does it lead to?

14. Assign E and Z configuration in the following:

(i) 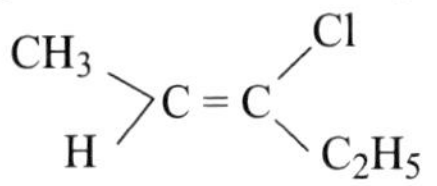   (ii) 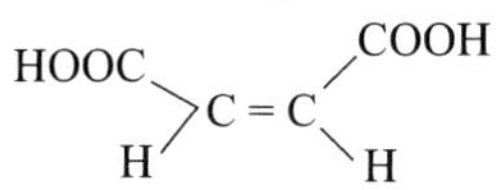   (iii) 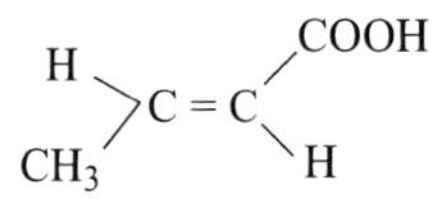

15. (a) Define

(1) Stereoisomerism

(2) Optical isomerism

(b) What is plane polarized light? Explain the concept of chirality.

16. (a) Write the difference between diastereoisomers and enantiomers.

(b) Explain asymmetric synthesis.

17. Explain the following:

(1) Enantiomers and diastereoisomers

(2) Delocalization

(3) Mesoisomer and racemic mixture

18. Draw the possible chair conformational structures for the *cis* and *trans* 1, 3 dimethyl cyclohexane and compare the stability of the more stable conformer for each pair of geometrical isomers.

19. (1) What do the symbols *E* and *Z* stand for? Illustrate briefly the *E* and *Z* system of naming a pair of geometrical isomers. What are the advantages of E, Z system over the conventional *cis-trans* system.

(2) Using *E*, *Z* notation write the IUPAC names for the following compounds.

(i) 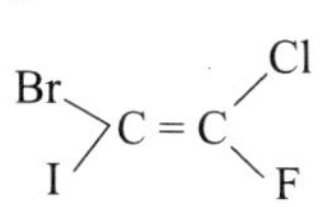   (ii) 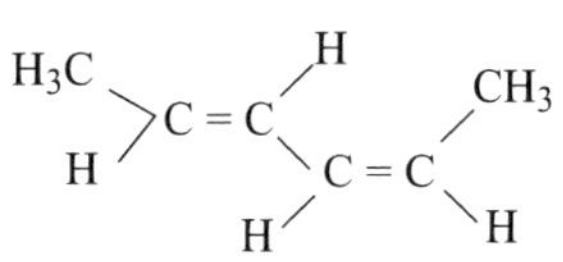

20. Draw the possible chair conformational structure for the *cis* and *trans* 1, 3-dimethyl cyclohexane and compare the stability of the more stable conformer for each pair of geometrical isomers.

21. Discuss the conformation of cyclohexane and their relative stability.

22. Write short notes on:

(1) Sequence rules

(2) Absolute configuration in terms of *D* and *L* notations

23. (a) Taking suitable examples, write different rules followed to assign *R* and *S* configurations.

(b) What do you understand by chair and boat conformations? Why is chair conformation of cyclohexane more stable than the boat conformation?

10

Atomic and Molecular Structure

The concept of wave nature of matter came from the dual character of radiation, which sometimes behaves as a wave and sometimes as a particle. De Broglie pointed out in 1924 that radiation including visible light, infrared, ultraviolet and X-rays behave as waves in propagation experiments based on interference and diffraction. These experiments conclusively proved the wave nature of these radiaitions. However, radiation behaves as a particle in interaction experiments which include black body radiation photoelectric effect and Compton effect. Here radiation interacts with matter in the form of photons or quanta. Radiation cannot exhibit its particle and wave properties simultaneously-

A wave is specified by its

(i) frequency,
(ii) wavelength (λ),
(iii) phase,
(iv) amplitude, and
(v) intensity.

A particle is specified by its

(i) mass (m),
(ii) velocity (v),
(iii) momentum (ρ), and
(iv) energy (ε).

De Broglie's Equation (de Broglie's Matter Waves): Einstein (1905) suggested that light has a dual character and behaves both as a particle and as a wave. L. de Broglie (1924) suggested that all material particles in motion such as electrons, protons, neutrons, atoms, molecules, etc., have dual character. According to L. de Broglie any moving electron in the atom is a material particle associated with wave properties and he compared electrons with photons having negligible masses. According to L. de Broglie the wavelength (λ) of a particle of mass (m) and velocity (v) is given by the relation.

$$\lambda = \frac{h}{m\mu}$$

where $h \rightarrow$ Planck's constant

$m\mu \rightarrow$ Momentum of the particle.

Derivation of de Broglie's Equation: According to the wave mechanical model of the atom an electron behaves as a standing wave which goes around the nucleus in a circular orbit. The only necessary condition for the establishment of such a stationary wave is that the length of the orbit should be a whole number multiple of the wavelength of the electron as shown in Fig. 10.1.

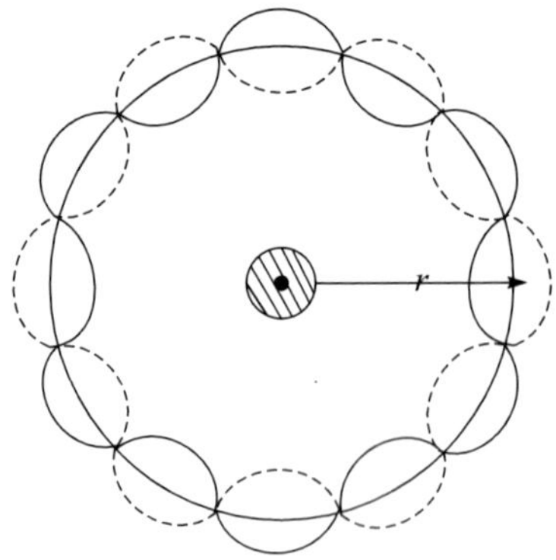

Fig. 10.1

If r is the radius of the circular orbit, then

$$2\pi r = n\lambda$$

$$\lambda = h/mv$$

$$2\pi r = \frac{nh}{mv}$$

$$mvr = nh/2\pi$$

where $n = 1, 2, 3$, etc.

Since $mvr \rightarrow$ Angular momentum of the electron as a particle.

Wave mechanical picture leads naturally to Bohr's postulate that the angular momentum is an integral multiple of $h/2\pi$. Other quantum conditions can also be derived similarly, i.e., the angular momentum is quantized.

Heisenberg's Uncertainty Principle: It state: "It is impossible to determine simultaneously both the position and the momentum of a small moving particle like an electron".

$$\Delta x \times \Delta P \geq \frac{h}{4\pi}$$

where $h \rightarrow$ Planck's constant.

If Δx is very small, i.e., the position of the electron is known more or less exactly, and ΔP would be large, i.e., uncertainty with regard to momentum will be large or vice versa. The relation implies that a simultaneous and precise measurement of both position and momentum (velocity) of a dynamic particle like electron is impossible and the extent of inherent uncertainty in any such measurement is of the order of h (Plank's constant).

Concept of Probability: On the basis of the concept of probability, it is possible to state or predict the probability or relative chance of finding an electron of a particular energy in a given region of space at a given time. The volume in space around the nucleus of the atom in which there is maximum probability of finding an electron is called an orbital.

10.1 SCHRÖDINGER WAVE EQUATION

Erwin Schrödinger in 1926, gave a wave equation to describe the behaviour of electron waves in atoms and molecules. He obtained a mathematical equation describing the state of an electron in terms of mass (m), total energy (E), potential energy (V) relative to atomic nucleus, Planck's constant (h) and a quantity Ψ called the wave function of the electron (or the amplitude of the wave). The equation is called Schrödinger's wave equation. It can be developed from the general differential equation of wave motion in one dimension, i.e., along X-axis.

$$\frac{\partial^2\Psi}{\partial x^2} = -\frac{4\pi^2}{\lambda^2}\Psi \qquad (1)$$

where $\lambda \rightarrow$ wavelength of light

From de Broglie's equation, $\lambda = h/mu$

Thus, from Eq. (1), we get

$$\frac{\partial^2\Psi}{\partial x^2} = -\frac{4\pi^2\Psi}{(h/mu)^2} = -\frac{4\pi^2 m^2 u^2}{h^2}\cdot\Psi$$

$$\frac{\partial^2\Psi}{\partial x^2} + \frac{4\pi^2 m^2 u^2}{h^2}\Psi = 0 \qquad (2)$$

The total energy (E) of a particle (or electron) is given by

$$E = \text{Kinetic energy} + \text{Potential energy}$$

$$E = 1/2\,mu^2 + V$$

$$E - V = 1/2\,mu^2$$

$$u^2 = \frac{2(E-V)}{m}$$

Putting the value of u^2 in Eq. (2), we get

$$\frac{\partial^2\Psi}{\partial x^2} + \frac{4\pi^2 m^2}{h^2} \times \frac{2(E-V)}{m}\Psi = 0$$

$$\frac{\partial^2\Psi}{\partial x^2}+\frac{8\pi^2 m}{h^2}(E-V)\Psi = 0 \qquad (3)$$

Extending Eq. (3) to the three dimensions x, y and z, we get

$$\boxed{\frac{\partial^2\Psi}{\partial x^2}+\frac{\partial^2\Psi}{\partial y^2}+\frac{\partial^2\Psi}{\partial z^2}+\frac{8\pi^2 m}{h^2}(E-V)\Psi = 0} \qquad (4)$$

Equation (4) is known as Schrödinger's wave equation.

So, Schrödinger's equation can be employed for determining the total energy of an electron. The potential energy (V) of an electron is specified in terms of space coordinate not on time, i.e., Schrödinger's wave equation is time independent. Like other differential equations, the Schrödinger wave equation is also governed by boundary conditions. Generally, its solutions are only attained for certain energy values called characteristic or eigenvalues. The corresponding wave function (Ψ) which is generally complex is called characteristic or eigenfunction, so we can write $\Psi = a + ib$ and its conjugate $\Psi^* = a - ib$ and their product, $\Psi\Psi^* = (a - ib)(a + ib) = a^2 + b^2$ which is real.

For a one-dimension system $|\Psi^2|\,dx$ represents the probability of finding an electron within a range of x and $x + dx$. But in three-dimension system $(\Psi^2|dv)$ represents the probability of an electron within the volume range of v to $(v + dv)$.

10.1.1 Significance of Wave Function (Ψ)

The wave function (Ψ) denotes the amplitude of a three-dimensional stationary electron wave.

It has no physical significance except that it shows the amplitude of the electron wave. The value of Ψ^2 gives the probability of finding an electron of a given energy ε, from place to place in a given region around the nucleus.

It is possible to identify the region of space around the nucleus where there is a high probability of finding an electron associated with a certain energy. The space is known as an atomic orbital.

It may be noted that in Bohr's theory the electron associated with a certain energy is supposed to be situated at a certain distance from the nucleus.

In wave mechanics, the equation for calculating the exact energy of the electron has been derived, but there is great uncertainty in finding its exact position. This is in conformity with Heisenberg's uncertainty principle.

10.1.2 Application of Schrödinger's Equation

Particle in a box showing energy quantization (like electrons in metal). Let us consider a particular of mass m which is free to move in a one-dimensional box of length l as shown in Fig. 10.2.

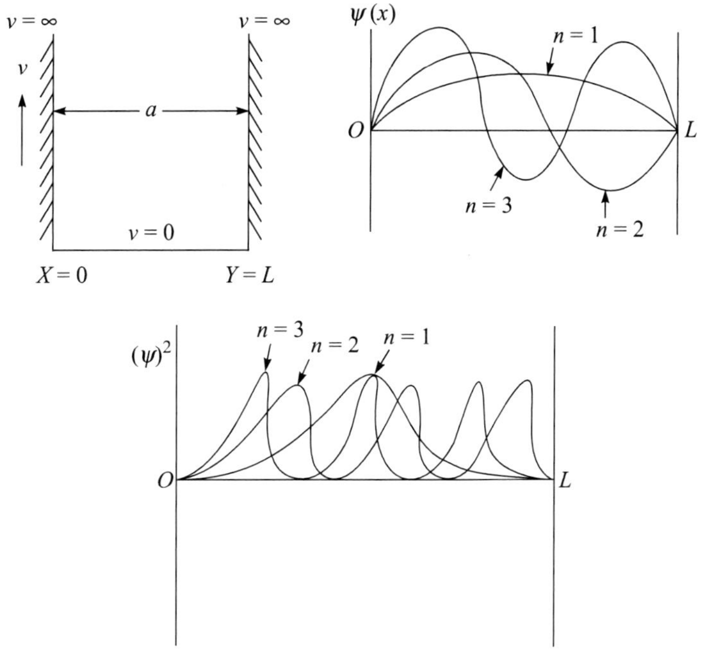

Fig. 10.2

The potential energy (v) of the electron at the bottom of the box is constant and may be taken as zero. Hence, inside the box $v = 0$.

Let the potential energy be infinite for $x < 0$ and $x > L$. So Ψ function has to be zero at $x = 0$ and for all negative values of x, since the particle is not allowed over the walls of the box. Similarly, Ψ function must be zero for all values $x > L$.

A general solution of Schrödinger's equation is

$$\Psi(x) = a \sin kx + b \cos kx$$

For boundary conditions, we have

$$\begin{aligned} \Psi(x) &= 0 \text{ at } x = 0 \text{ or } \Psi(0) = 0 \\ 0 &= a \sin 0 + b \cos 0 \text{ or } b = 0 \\ \Psi(x) &= 0 \text{ at } x = l \\ a \sin kl + b \cos kl &= 0 \end{aligned}$$

It can only be possible when

$$kl = n\pi \quad \text{or} \quad k = \frac{n\pi}{l}$$

where x is called quantum number and is equal to 1, 2, 3, … ∞

Substituting this value in Eq. (4) and also p.

$$\frac{n^2\pi^2}{l^2} = \frac{8\pi^2 m}{h} E$$

E – (total energy of the electron) = K.E. of the electron

$$E_n = \frac{n^2h^2}{8\,ml^2} \quad \text{when } n = 1, 2, 3 \ldots \infty$$

The above equation means that the particle in a box does not possess any arbitrary amount of energy, rather it possesses discrete set of energy values, i.e., its energy is quantized.

A few energy levels are given below:

$$\varepsilon_1 = \frac{h^2}{8\,ml^2}$$

$$\varepsilon_2 = \frac{4h^2}{8\,ml^2}$$

$$\varepsilon_3 = \frac{9h^2}{8\,ml^2}$$

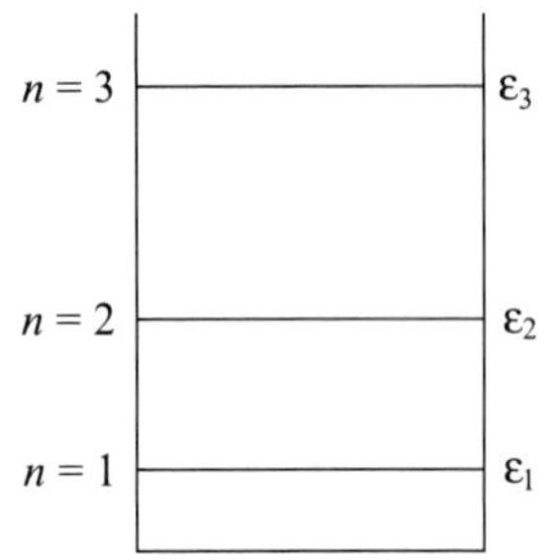

The reason why a particle in a box, i.e., bound particle possesses a quantized energy whereas a free particle cannot easily be deduced from the above equation.

Example: Calculation of minimum energy of a particle in a box from Heisenberg's uncertainty principle.

Here $\Delta x = l$ and the particle bounces back in the box. So, $\Delta P = 2P$ because the momentum changes from $+P$ to $-P$ continuously, so we can write

$$\Delta x \times \Delta P = l \cdot (2P) = h \quad \text{or} \quad P = \frac{h}{2l}$$

$$\varepsilon = 1/2\,mu^2 = \frac{(mu)^2}{2m} = \frac{P^2}{2m}$$

$$\varepsilon = \frac{h^2}{8\,ml^2}$$

10.1.3 Application

(i) In conjugated system: The simplest case is the 1, 3 butadiene molecule.

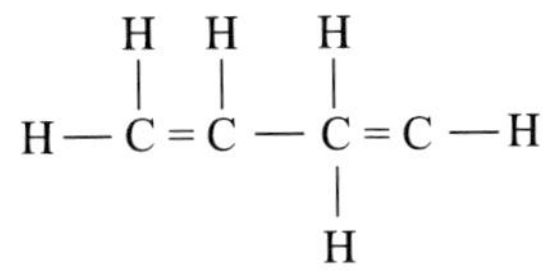

The four π-electrons are assumed to move freely over the four carbon framework of single bonds. We neglect the zig-zagging of the C – C bonds and assume a one-dimensional box. There are totally 6π electrons since each carbon is SP^2 hybridized and thus contributes a P orbital and one π electron to the system. The lowest energy configuration, called the electronic ground state, corresponds to the 6 electrons being in the lowest three orbitals. Higher energy configurations are constructed by promoting a electron from the highest filled orbital (HFO) to a higher vacant orbital. The higher energy arrangement is called an electronically excited state, the lowest or first electronically excited state. The lowest or first electronically excited state, of course, is achieved by promoting an electron from the highest filled orbital to the lowest unifilled orbital (LUO).

The energy difference between these two states is given by the following expression.

$$\Delta\varepsilon = \varepsilon_{\text{LUO}} - \varepsilon_{\text{HFO}} = \frac{h^2}{8\,mL^2}(n_{\text{LUO}}^2 - n_{\text{HFO}}^2)$$

The energy required for this electronic transition can be supplied by a photon of the appropriate frequency given by the Planck relation

$$\varepsilon = h\nu = hC/\lambda$$

where h → Planck constant

ν → Frequency

C → Speed of light

λ → Wavelength

Now in the lowest energy state of butadiene, the 4 delocalized electrons will fill the two lowest unoccupied molecular orbitals. The total π-electron density will be given by

$$P = 2\Psi_1^2 + 2\Psi_2^2$$

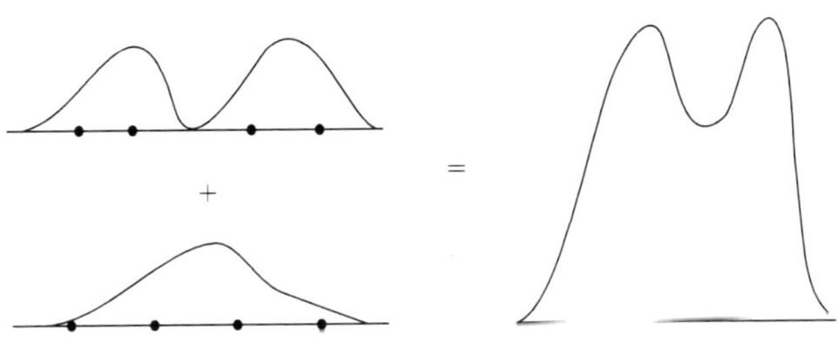

Fig. 10.3 *π-electron density in butadiene*

A chemical interpretation of this picture might be that, since the π-electron density is concentrated between carbon atom 1 and 2, and between 3 and 4, the predominant structure of butadiene has double bonds between these two pairs of atoms. Each double bond consists of a π-bond, in addition to the underlying σ-bond. However, this is not the complete story, because we must also take account of the residual π-electron density between carbons 2 and 3. In the terminology of valence-bond theory, butadiene would be described as a resonance hybrid with the contributing structures $CH_2 = CH - CH_2 = CH_2$ (the predominant structure) and $CH_2 - CH = CH - CH_2$ (a secondary contribution). The reality of the latter structure is suggested by the ability of butadiene to undergo 1, 4-addition reaction.

For example: β-carotene ($C_{40}H_{56}$) is a conjugated polyene with an orange colour and a molecular length of approximately 3.8 nm. Due to β-carotene's high level of conjugation, electrons are dispersed throughout the length of the molecule, allowing one to model it as a one-dimensional particle in a box. β-carotene has 11 C – C double bonds in conjugation, each of those double bonds contains two π-electrons therefore β-carotene has 22 π-electrons, with two electrons per energy level. It can be treated as a particle in a box at energy level $n = 11$, therefore the minimum energy needed to excite an electron to the next energy level can be calculated $n = 12$ as follows:

$$\Delta\varepsilon = \frac{(r^2 f - n^2 i)h^2}{8\,mL^2}$$

$$= \frac{(12^2 - 11^2)h^2}{8\,mL^2}$$

$$= 2.3658 \times 10^{-19}\,\text{J}$$

Using the previous relation of wavelength to energy, recalling both Planck's constant *(h)* and the speed of light (C).

$$\lambda = hC/\Delta\varepsilon$$

$$= 0.00000084 \text{ m} = 840 \text{ nm}$$

This indicates that β-carotene primarily absorbs light in the infrared spectrum therefore it would appear white to a human eye. However, the observed wavelength is 450 nm indicating that the particle in a box is not a perfect model for this system.

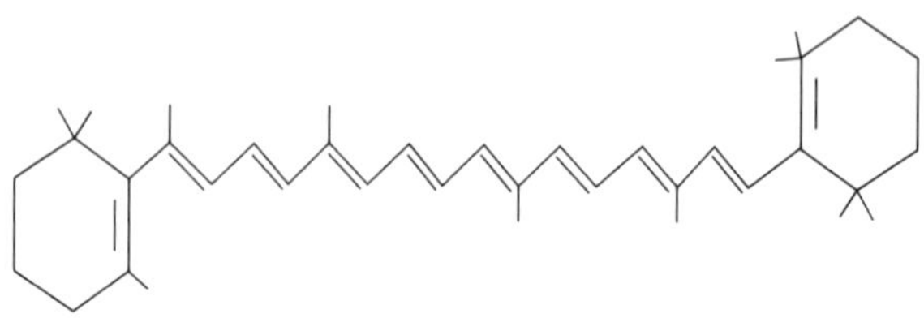

Fig. 10.4 *β-carotene is a conjugated polyene*

(ii) Quantum Dots as a Particle in a Box

Quantum dots are extremely small semiconductors (on the scale of nanometres). Inside the semiconductor used in microelectronics there are small particles of semiconductors. These

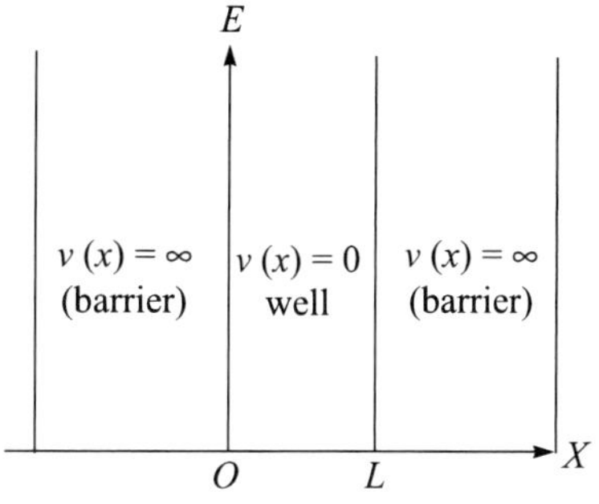

Fig. 10.5

may contain an electron and a hole (absence of an electron), particle may only move backward and forward along a straight line with impenetrable barriers at either end. The walls of a one-dimensional box may be visualized as the region of space with an infinitely large potential energy. Conversely, the interior of the box has a constant zero potential energy. This means that no forces act upon the particle inside the box and it can move freely in that region. However, infinitely large forces repel the particle if it touches the walls of the box, preventing it from escaping.

So, the potential energy,

$$V(x) = \begin{cases} 0 & 0 < x < L \\ \infty \end{cases}$$

where $L \rightarrow$ Length

$x \rightarrow$ Position within the box

The wave function can be found by solving the Schrödinger's equation for the system. Inside the box, no forces act upon the particle which means that the part of the wave function inside the box oscillates through space and time with the same form as a free particle.

$$\Psi(x, t) = [A \sin (kx) + B \cos (kx)]e^{-i\omega t}$$

The energies which correspond with each of the permitted wave numbers may be written as

$$E_n = \frac{n^2 h^2 \pi^2}{2\, mL^2} = \frac{n^2 h^2}{8\, mL^2}$$

The radius of the quantum dot influences the wavelength of the emitted light due to quantum confinement. This equation describes the effect to a change of the radius of the quantum dot on the wavelength (λ) of the emitted light hence,

$$\varepsilon = hC/\lambda$$

where $C \rightarrow$ speed of light

and the general equation is

$$\Delta E(r) = E_{gap} + \frac{h^2}{8r^2}\left(\frac{1}{me} + \frac{1}{mh}\right)$$

$me \rightarrow$ Effective mass of the electron

$mh \rightarrow$ Effective mass of the hole

$E_{gap} \rightarrow$ Energy of the band gap of the semiconductor

$R \rightarrow$ Radius of the quantum dot

For example, when the electrons of two nanometer CdSe quantum dots relax after excitation, blue light is emitted. Similarly, red light is emitted in four nanometre CdSe quantum dots.

Probability Distribution: These are namely:

(i) An angular probability distribution, i.e., how the angular position θ varies from nucleus with respect to a fixed axis.

(ii) A radial probability distribution, i.e., how (r) the distance from the nucleus does vary.

Angular probability distribution of orbitals, i.e., shapes or boundary surfaces of orbitals gives the probability of finding an electron in a particular direction.

S-orbital: An *S*-electron has no preferred direction in space, i.e., there is an equal chance of finding it in any direction around the nucleus this is graphically shown in Fig. 10.6. In this diagram, the nucleus of the atom is at the origin and the surface of the sphere represents the probability of finding the *S*-electron which is therefore same in all directions.

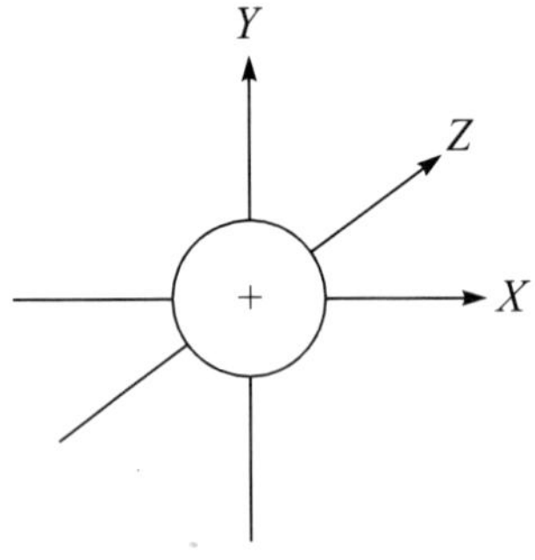

Fig. 10.6

***P*-orbital:** A level has an accommodation for six electrons distributed over three *P*-orbitals to each value of the principal quantum number.

These three orbitals are at right angles to each other and the three angular probability distributions are shaped like dumb-bells along the three coordinate axes and are named as P_x, P_y and P_z orbitals. In the absence of magnetic field these orbitals are equivalent in energy content and are said to be triply degenerate. In the presence of external magnetic field these vary in their energy content depending on the magnetic quantum number. The different shapes are shown in Fig. 10.7.

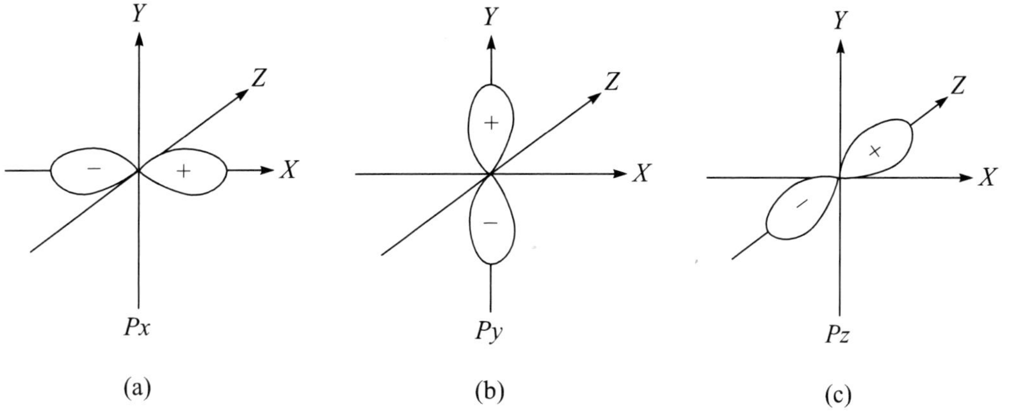

Fig. 10.7

It is most likely that in the P_x orbital the electron will be found in the direction of x-axis and there is no chance of it to be found in any of the directions perpendicular to x-axis, i.e., $y - z$ plane in this case is nodal plane with zero electron density. Similarly, in the case of P_y and P_z orbitals the electron will be found along y and z axes.

***D*-orbital:** The angular probability distribution of five d-orbitals that can accommodate ten electrons is quite complicated. These five orbitals are named dxy, dyz, dxz, $dx^2 - y^2$, dz^2. These are shown in Fig. 10.8.

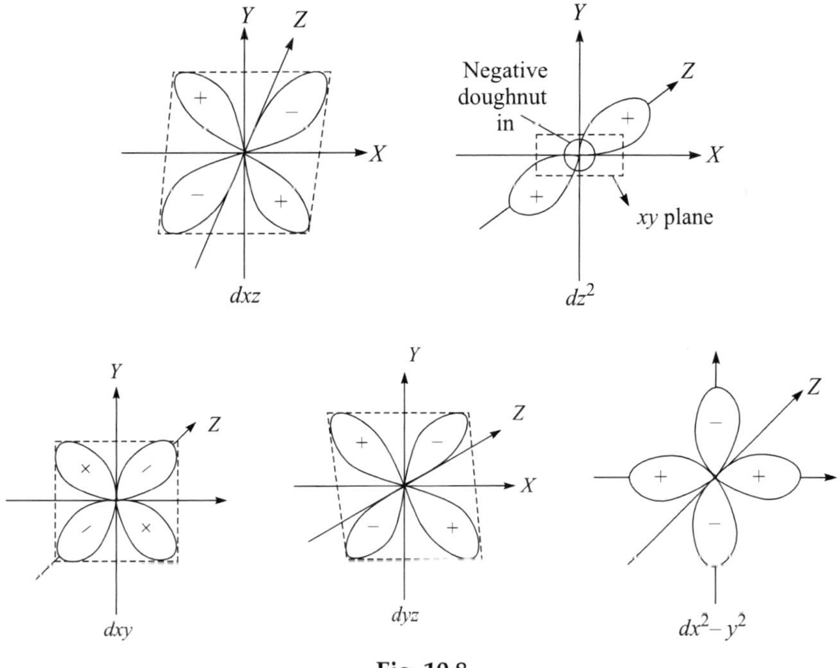

Fig. 10.8

All the *d*-orbitals are equivalent in energy in the absence of magnetic field and are said to be fivefold degenerate. The set of three orbitals, namely, *dxy*, *dyz*, *dzx* have their lobes lying symmetrically in between the co-ordinated axes indicated by the subscripts to *d*. For example, the lobs of *dxy* orbitals are pointing or lying in between the *x* and *y* axes. This set is referred to as t_{2g} set. The set of two orbitals, i.e., $d_{x^2-y^2}$ and d_{z^2} orbitals form, eg set having their lobes along the axes. For example, the lobes of $d_{x^2-y^2}$ orbital lie along the *x* and *y* axes while those of dz^2 orbital lie along *z*-axis.

10.1.4 Radial Probability Distribution

The angular probability graph indicates the most probable directions of the electrons but they do not give any indication of the probable distance of the electron from the nucleus. The probable distances of an electron are given by radial probability distribution diagram as given in Fig. 10.9.

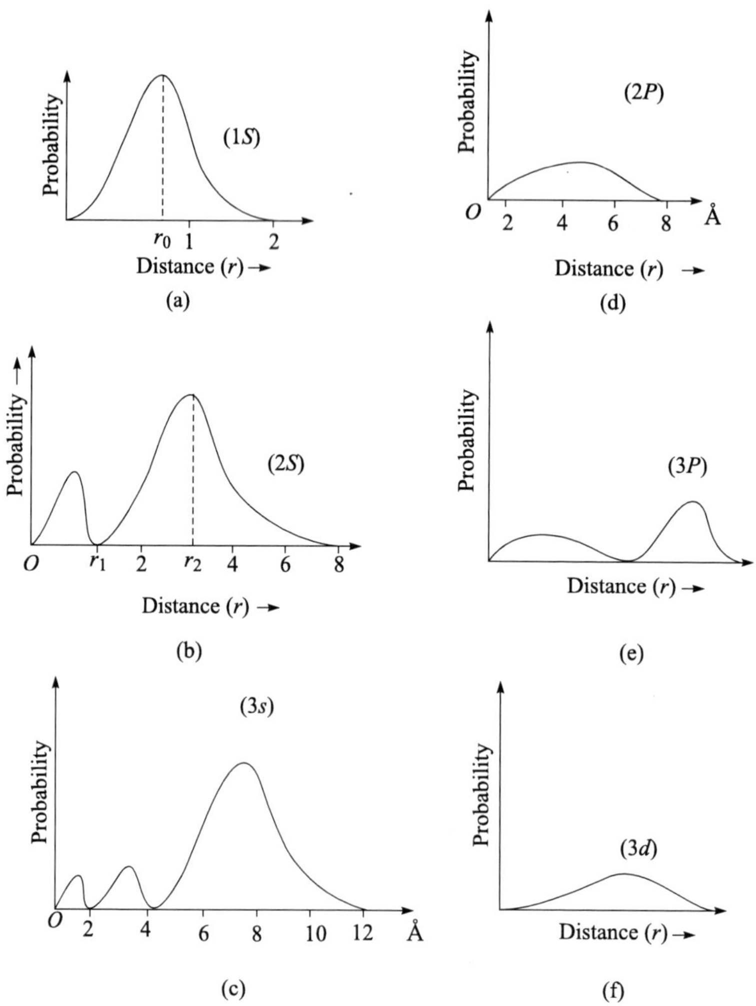

Fig. 10.9 *Radial probability distribution of 1s, 2s, 3s, 2p, 3p, 3d orbitals*

In these diagrams, the probability of finding an electron at a distance (r) from the nucleus is plotted against r (distance from the nucleus) is $1s, 2s, 3s, 2p\ 3p, 3d$ orbitals of hydrogen atom.

10.1.5 Hydrogen Atom Wave Function

There are a number of different ways of representing hydrogen atom wave functions graphically. We will illustrate some of these for the $1s$ group state. In atomic units is a decreasing exponential function of a single variable r and is simply plotted in Fig. 10.10.

$$\Psi_{1s}(r) = \frac{1}{\sqrt{\pi}} e^{-r}$$

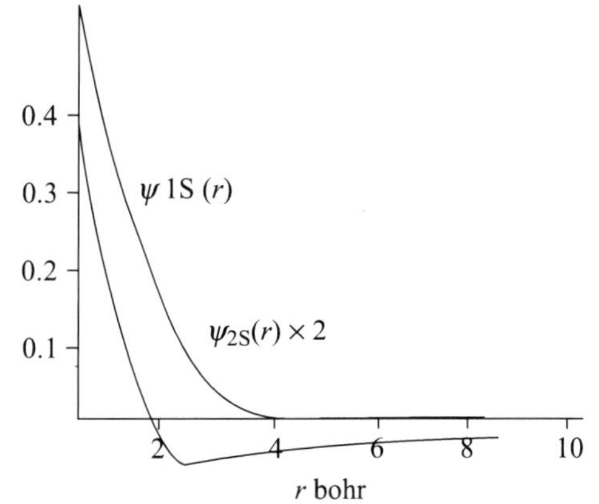

Fig. 10.10 *Wave functions for 1s and 2s orbitals for atomic hydrogen. The 2s-function (scaled by a factor of 2) has a node at r = 2 bohr*

According to Bohr's interpretation of wave function the probability per unit volume of finding the electron at the point (r, θ, ϕ) is equal to the square of the normalized wave function.

$$P_{1s}(r) = [\Psi_{1s}(r)]^2 = \frac{1}{\pi} e^{-2r}$$

This is represented in Fig. 10.11 by a scatter plot describing a possible sequence of observations of electron position. Although results of individual measurement are not predictable – a statistical pattern does emerge after a sufficiently large number of measurements.

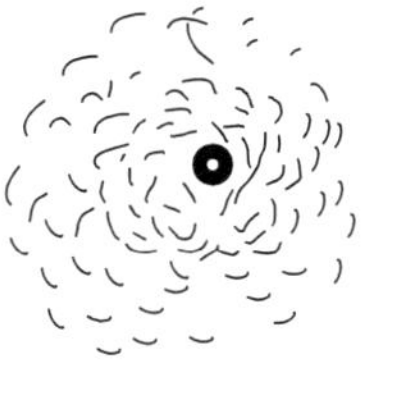

Fig. 10.11 *Scatter plot of electron position measurements in hydrogen is orbital*

The probability density is normalized such that

$$\int_0^{\infty} P_{1s}(r) 4\pi r^2 dr = 1$$

In some ways $P(r)$ does not provide the best description of the electron distribution, since the region around $r = 0$, where the wave function has its largest values is a relatively small fraction of the volume accessible to the electron. Larger radii (r) represent larger physical regions since in spherical polar coordinates, a value of r is associated with a shell of volume $4\pi r^2 dr$.

A more significant measure is therefore the radial distribution function,

$$D_{1s}(r) = 4\pi r^2 [\Psi_{1s}(r)]^2$$

which represents the probability density within the entire shell of radius r, normalized such that

$$\int_0^{\infty} D_{1s}(r) dr = 1$$

The functions and $P_{1s}(r)$ and $D_{1s}(r)$ are both shown in Fig. 10.12. Remarkably 1*s* RDF has its maximum at $r = a_0$ equal to the radius of the first Bohr orbital.

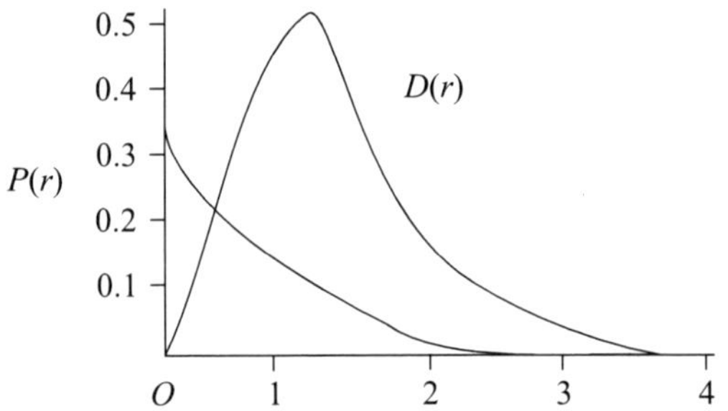

Fig. 10.12 *Density ρ(r) and radial distribution function D(r) for hydrogen is orbital*

Atomic Orbitals: According to quantum mechanics, an electron bound to an atom cannot possess any arbitrary energy or occupy any position in space. These characteristics can be determined by solving the time-independent Schrödinger's equation.

$$H\varphi = E\varphi \tag{1}$$

where $H \rightarrow$ Hamiltonian operator of the atom

We obtain a set of functions φ which are termed atomic orbitals. Their mathematical equations are shown in Table 10.1 for the 1*s* to the 3*d* orbitals inclusive. With each electron is associated an atomic orbital whose equation allows the position (or more precisely the probability (dP) of finding the electron within a given volume (dV) and the energy of the electron to be calculated.

$$dP = \varphi\varphi^{*}dV \tag{2}$$

$$E = \int \varphi H \varphi^{*} dV = \langle \varphi | H | \varphi \rangle \tag{3}$$

In the above equations φ^* is the complex conjugate of φ. In the case which we will cover, it is always possible to choose atomic orbitals which are mathematically real, so we will do this systematically.

Table 10.1 Some real atomic orbitals : z is the atomic number and a is the Bohr's radius ($a = h^2/4\pi^2 me^2 = 0.53 \times 10^{-8}$ cm)

$$\Psi_{1S} = \frac{1}{\sqrt{\pi}}\left(\frac{z}{a}\right)e^{3/2-zr/a}$$

$$\Psi_{2S} = \frac{1}{4\sqrt{2\pi}}\left(\frac{z}{a}\right)^{3/2}\left(2-\frac{zr}{a}\right)e^{-zr/2a}$$

$$\Psi_{2Pz} = \frac{1}{4\sqrt{2\pi}}\left(\frac{z}{a}\right)^{5/2} re^{-zr/2a}\cos\theta$$

$$\Psi_{2Px} = \frac{1}{4\sqrt{2\pi}}\left(\frac{z}{a}\right)^{5/2} re^{-zr/2a}\sin\theta\cos\varphi$$

$$\Psi_{2Py} = \frac{1}{4\sqrt{2\pi}}\left(\frac{z}{a}\right)^{5/2} re^{-zr/2a}\sin\theta\sin\varphi$$

$$\Psi_{3S} = \frac{1}{81\sqrt{3\pi}}\left(\frac{z}{a}\right)^{3/2}\left(27-18\frac{zr}{a}+2\frac{z^2r^2}{a^2}\right)re^{-zr/3a}$$

$$\Psi_{3Pz} = \frac{\sqrt{2}}{81\sqrt{\pi}}\left(\frac{z}{a}\right)^{5/2}\left(6-\frac{zr}{a}\right)re^{-zr/3a}\cos\theta$$

$$\Psi_{3Px} = \frac{\sqrt{2}}{81\sqrt{\pi}}\left(\frac{z}{a}\right)^{5/2}\left(6-\frac{zr}{a}\right)re^{-zr/3a}\sin\theta\cos\varphi$$

$$\Psi_{3Py} = \frac{\sqrt{2}}{81\sqrt{\pi}}\left(\frac{z}{a}\right)^{5/2}\left(6-\frac{zr}{a}\right)re^{-zr/3a}\sin\theta\sin\varphi$$

$$\Psi_{3d_z^2} = \frac{1}{81\sqrt{6\pi}}\left(\frac{z}{a}\right)^{7/2}\left(\frac{z}{a}\right)r^2e^{-zr/3a}(3\cos^2\theta-1)$$

$$\Psi_{3d_{xz}} = \frac{\sqrt{2}}{81\sqrt{\pi}}\left(\frac{z}{a}\right)^{7/2} r^2e^{-zr/3a}\sin\theta\cos\theta\cos\varphi$$

$$\Psi_{3dyz} = \frac{\sqrt{2}}{81\sqrt{\pi}}\left(\frac{z}{a}\right)^{7/2} r^2 e^{-zr/3a} \sin\theta\cos\theta\sin\varphi$$

$$\Psi 3dx^2 - y^2 = \frac{\sqrt{2}}{81\sqrt{2\pi}}\left(\frac{z}{a}\right)^{7/2} r^2 e^{-zr/3a} \sin^2\theta\cos 2\varphi$$

$$\Psi 3dxy = \frac{\sqrt{2}}{81\sqrt{2\pi}}\left(\frac{z}{a}\right)^{7/2} r^2 e^{-zr/3a} \sin^2 \sin 2\varphi$$

By extension, atomic orbital has also come to mean a volume, limited by an equiprobability surface wherein we have a high probability of finding an electron. Figure 10.13 depicts the shapes of some atomic orbitals and a scale showing their relative energies. It justifies a few comments.

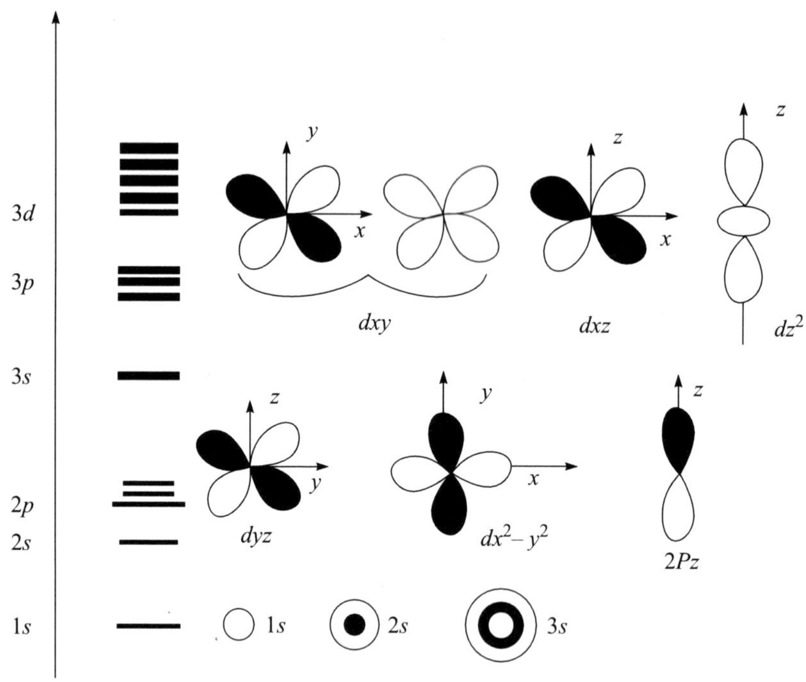

Fig. 10.13 *Shapes and approximate energies of some atomic orbitals*

1. The energy scale is approximate. First three shells are well separated from each other. However, the 4*s* and 3*d* orbitals have very similar energies. As a consequence, the 3*d*, 4*s* and 4*p* levels in the first row transition metals all function as valence orbitals. The *P*-orbitals are degenerate as are the five *d*-orbitals.
2. The orbitals of the same shell have more or less the same size. However, size increases with the principal quantum number. Thus, a 3*p*-orbital is more diffusive than a 2*p* orbital.

3. The sign shown inside each orbital lobe is the sign of the function (φ) within that region of space. Taken on its own this sign has no physical meaning, because the electron probability density is given by the square of ϕ. For this reason, we often distinguish between two different lobes by hatching or shading one of them, rather than using the symbols (+) or (–).
   Let us now compare a 1*s* and a 2*s* orbital. If we start at the nucleus and move away the 1*s* orbital always retains the same sign. The 2*s* orbital passes through a null point and change sign afterwards. The surface on which the 2*s* orbital becomes zero is termed a nodal surface. The number of nodal surfaces increases with increasing energy. Thus, the 1*s* orbital has none, the 2*s* orbital has one, and the 3*s* has two, etc.
4. Orbitals having the same azimuthal quantum number (*l*) have the same shape. All *S* orbitals have spherical symmetry and all *P*-orbitals have cylindrical symmetry. The dz^2 orbital is drawn differently from the other *d*-orbitals but being a linear combination of $d_{z^2-x^2}$ and $d_{z^2-y^2}$ orbitals, it is perfectly equivalent to them.
5. Outside the boundary, the function (φ) has very small values because its square, summed over all space from the boundary wall to infinity has a value of only 0.1. When we say that a molecular orbital is a linear combination of atomic orbitals, we imply that it is almost indestinguishable from φ_k in the neighbourhood of atom *k*. This is because we are then inside the boundary of φ_k and outside the boundary of $\varphi_1 (l \neq k)$, so that φ_k has finite values and contributions from φ_l are negligible. Therefore, a molecular orbital is broadly a series of atomic orbitals the size of each atomic orbital being proportional to its LCAO coefficient.
   Once the atomic orbitals are known, their occupancy is determined by
   1. *The Pauli exclusion principle:* Each orbital can only contain one electron of any given spin.
   2. *The Aufbau principle:* In the ground state (i.e., the lowest energy state), the lowest energy orbitals are occupied first.
   3. *Hund's rules:* When degenerate orbitals (orbitals having the same energy) are available, as many of them as possible will be filled using electrons of like spin.

10.1.6 Molecular Orbital

Valence bond theory fails to answer certain questions like why He_2 molecule does not exist and why O_2 is paramagnetic. Therefore, in 1932 H. Food and R.S. Mulliken came up with a theory known as molecular orbital theory to explain questions like above. According to molecular orbital theory individual atoms combine to form molecular orbitals, as the electrons of an atom are present in various atomic orbitals and are associated with several nuclei.

Electrons may be considered either of particle or of wave nature. Therefore, an electron in an atom may be described as occupying an atomic orbital or by a wave function, Ψ, which are solution to the Schrödinger's wave equation. Electrons in a molecule are said to occupy molecular orbitals. The wave function of a molecular orbital may be obtained by one of the two methods:

1. Linear combination of atomic orbitals
2. United atom method

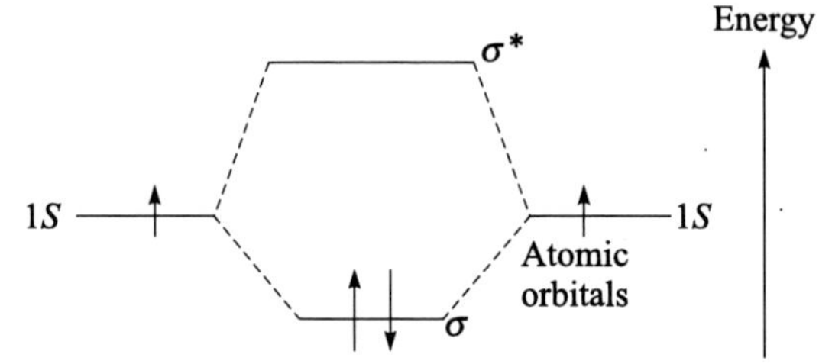

Fig. 10.14 *Molecular orbitals*

1. Linear Combination of Atomic Orbitals (LCAO): The formation of orbitals is because of linear combination (addition or subtraction) of atomic orbitals which combine to form molecule. Consider two atoms *A* and *B* which have atomic orbitals described by the wave functions, Ψ_A and Ψ_B.

If electron clouds of these two atoms overlap then the wave function for the molecule can be obtained by a linear combination of atomic orbitals, Ψ_A and Ψ_B, i.e., by subtraction or addition of wave function of atomic orbitals.

$$\Psi_{\text{MO}} = \Psi_A \pm \Psi_B$$

The above equation forms two molecular orbitals.

2. Bonding Molecular Orbitals: When addition of wave function takes place, molecular orbitals formed are called bonding molecular orbitals and are represented by

$$\Psi_{\text{MO}} = \Psi_A + \Psi_B$$

They have lower energy than atomic orbitals involved. It is similar to constructive interference occurring in phase because of which electron probability density increases resulting in the formation of bonding orbital. Molecular orbital formed by addition of overlapping of two *S* orbitals is represented by *S*.

10.1.7 Antibonding Molecular Orbitals

When molecular orbital is formed by subtraction of wave function, molecular orbitals formed called antibonding molecular orbitals and are represented by

$$\Psi_{\text{MO}} = \Psi_A - \Psi_B$$

They have higher energy than atomic orbitals. It is similar to destructive interference occurring out of phase resulting in formation of antibonding orbitals. It is represented by S^*. (*) is used to represent antibonding molecular orbital.

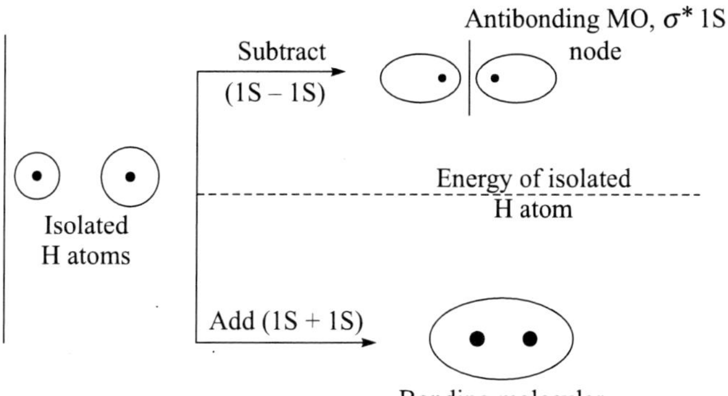

Fig. 10.15 *Formation of bonding and antibonding orbitals*

Therefore, combination of two atomic orbitals results in the formation of two molecular orbitals, bonding molecular orbital (BMO) whereas the other is antibonding molecular orbitals (ABMO).

BMO has lower energy and hence greater stability than ABMO. First BMOs are filled then ABMOs start filling because BMO has lower energy than that of ABMO.

Formation of molecular orbitals occurs by the combination of atomic orbitals of proportional symmetry and comparable energy. Therefore, a molecular orbital is polycentric and atomic orbitals is monocentric. Number of molecular orbitals formed is equal to the number of atomic orbitals.

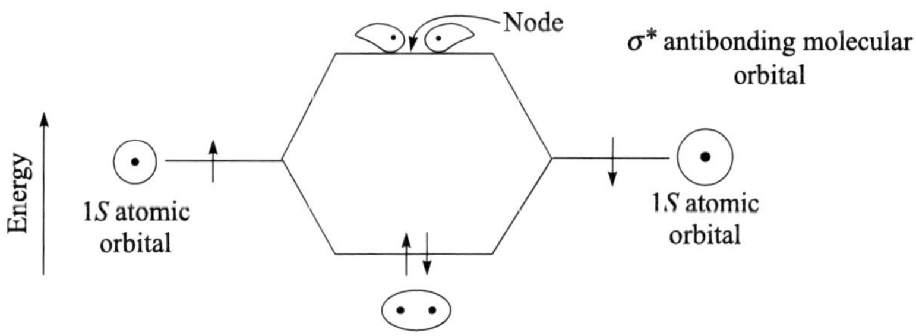

Fig. 10.16 *Energies of BMO and ABMO*

10.1.8 The MOs of a Homonuclear Diatomic Molecule

Calculations: Consider a homonuclear diatomic molecule A_2,whose two atoms A are identical. These interacting atomic orbitals which we will call ϕ_1 and ϕ_2 are chosen so as to be mathematically real. The following procedure is used to calculate the resulting molecular orbitals.

1. The two nuclei are held at a certain fixed distance from each other (i.e., we apply the Born-Oppenheimer approximation.

2. The time-independent Schrödinger's equation is written for the molecule, multiplied on the left hand side by Ψ, and integrated over all space.

$$H\Psi = E\Psi \tag{1}$$

$$\langle \Psi | H | \Psi \rangle = E\langle \Psi | \Psi \rangle \tag{2}$$

3. Each molecular orbital is expressed as a linear combination of atomic orbitals (LCAOs)

$$\Psi = c_1\varphi_1 + c_2\varphi_2 \tag{3}$$

In Eq. (3) we know φ_1 and φ_2, calculating an MO Ψ_1 therefore involves evaluating its associated energy, ε_i and the coefficients c_i1 and c_i2 of its LCAO expansion.

Incorporating Eq. (3) in Eq. (2) gives

$$\langle c_1\varphi_1 + c_2\varphi_2 [H] c_1\varphi_1 + c_2\varphi_2 \rangle = E\langle c_1\varphi_1 + c_2\varphi_2 | c_1\varphi_1 + c_2\varphi_2 \rangle \tag{4}$$

The linearity of integral allows the left-hand side of Eq. (4) to be expressed as

$$\langle c_1\varphi_1 + c_2\varphi_2 | H | c_1\varphi_1 + c_2\varphi_2 \rangle = \langle c_1\varphi_1 | H | c_1\varphi_1 \rangle + \langle c_1\varphi_1 | H | c_2\varphi_2 \rangle + \ldots$$

$$= c_1^2 \langle \varphi_1 [H] \varphi_1 \rangle + c_2^2 \langle \varphi_2 | H | \varphi_2 \rangle + \ldots$$

To express this more simply, let us set

$$\langle \varphi_i | H | \varphi_i \rangle = \alpha_i$$

$$\langle \varphi_i | H | \varphi_j = \beta_{ij}$$

$$\langle \varphi_i | \varphi_j \rangle = S_{ij}$$

where

$\alpha_i \rightarrow$ Coulomb integral

$\beta_{ij} \rightarrow$ Resonance integral

$S_{ij} \rightarrow$ Overlap integral

We are using normalized atomic orbital

So $S_{ij} = 1$

Furthermore, the two atoms are identical, so

$$\alpha_1 = \alpha_2 \; \beta_{12} = \beta_{21}$$

(In physical terms $\beta_{12} = \beta_{21}$ simply means that the force binding atom 1 to atom 2 is the same as the force binding 2 to 1).

Thus, Eq. (4) can be written as

$$(c_1^2 + c_2^2)\alpha + 2c_1c2\beta - E(c_1^2 + c_2^2 + 2c_1c_2s) = 0 \tag{5}$$

where α, β and S are parameters and c_1 and c_2 and E are unkown.

Let us now choose c_1 and c_2 so as to minimize E (variational method). To do this, we differentiate Eq. (5), and set the partial derivation a zero.

$$\frac{\partial E}{\partial c_1} = \frac{\partial E}{\partial c_2} = 0$$

Thus, obtaining the secular equations

$$\left.\begin{aligned}(\alpha - E)c_1 + (\beta - ES)c_2 = 0\\(\beta - ES)c_1 + (\alpha - E)c_2 = 0\end{aligned}\right\} \tag{6}$$

Thus, equations are homogenous in c_i. They have a nontrivial solution, if the secular determinant (i.e., the determinant of the secular equations) can be set to zero

$$\begin{bmatrix}\alpha - E & \beta - ES\\ \beta - ES & \alpha - E\end{bmatrix} = (\alpha - E)^2 - (\beta - ES)^2 = 0 \tag{7}$$

The solutions to Eq. (7) are

$$E_1 = \frac{\alpha + \beta}{1 + S} \quad \text{and} \quad E_2 = \frac{\alpha - \beta}{1 - S} \tag{8}$$

E_1 and E_2 are the only energies which an electron belonging to the diatomic molecule A_2 can have. Each anergy level (E_i) is associated with a molecular orbital (Ψ_i), whose coefficients may be obtained by setting $E = E_i$ in Eq. (6) and solving these equations taking into account the normalization condition.

$$\langle \Psi_i \mid \Psi_i \rangle = C_{i1}^2 + C_{i2}^2 + 2C_{i1}C_{i2}S = 1 \tag{9}$$

The solutions are

$$\Psi_1 = \frac{1}{\sqrt{2(1+S)}}(\varphi_1 + \varphi_2)$$

$$\Psi_2 = \frac{1}{\sqrt{2(1+S)}}(\varphi_1 - \varphi_2) \tag{10}$$

Figure 10.17 gives a pictorial representation of Eqs. (8) and 10.

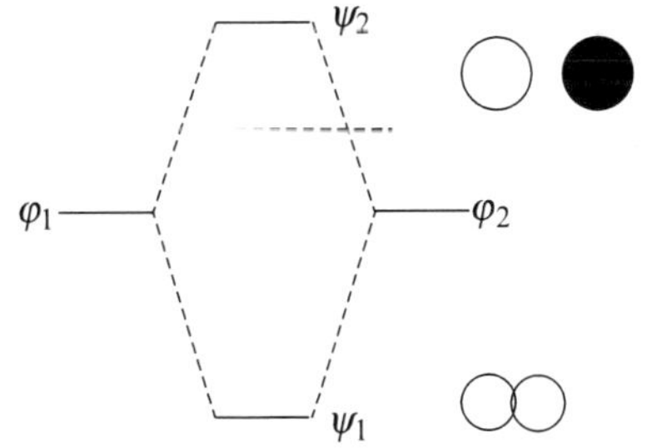

Fig. 10.17 *The molecular orbital of the homonuclear diatomic A_2. φ_1 and φ_2 are arbitrarily drawn as S orbitals. Note that the destabilization of Ψ_2 is greater than the stabilization of Ψ_1*

10.1.9 MOS of a Heteronuclear Diatomic Molecule

Calculation: A heteronuclear diatomic molecule is comprised of two different atoms, A and B. The two relevant atomic orbitals are then $\varphi_{A'}$ of energy φ_A and φ_B of energy α_B. The calculation is completely analogous to the case of the homonuclear diatomic given above.

For a heteronuclear diatomic molecule AB Eq. (7) where the secular determinant is set to zero becomes

$$(\alpha_A - \varepsilon)(\alpha_B - \varepsilon) - (\beta - ES)^2 = 0 \tag{11}$$

Equation (11) is a second order equation in ε which can be solved exactly. However the analogs of expressions, Eqs. (10) and (8) are rather unwieldy. For qualitative applications, they can be approximated as follows

$$E_1 \approx \alpha_A + \frac{(\beta - \alpha_{AS})^2}{\alpha_A - \alpha_B}$$

$$E_2 \approx \alpha_B + \frac{(\beta - \alpha_{BS})^2}{\alpha_B - \alpha_A} \tag{12}$$

$$\Psi_1 \approx N_1\left(\varphi_A + \frac{B - \alpha_{AS}}{\alpha_A - \alpha_B}\varphi_B\right)$$

$$\Psi_2 \approx N_2\left(\varphi_B + \frac{\beta - \alpha_{BS}}{\alpha_B - \alpha_A}\varphi_A\right) \tag{13}$$

where N_1 and N_2 are normalization coefficients, Eq. (12) assumes that ε_1 and ε_2 are not very different from α_A and α_B, respectively. Using this approximation it is possible to rewrite Eq. (11) in the form

$$\alpha_A - \varepsilon_1 = \frac{(\beta - \varepsilon_1 S)^2}{\alpha_\beta - \varepsilon_1} \approx \frac{(\beta - \alpha_A S)^2}{\alpha_B - \alpha_A} \tag{14}$$

which is equivalent to Eq. (12).

Equations (12) and (13) are shown pictorially in Fig 10.18.

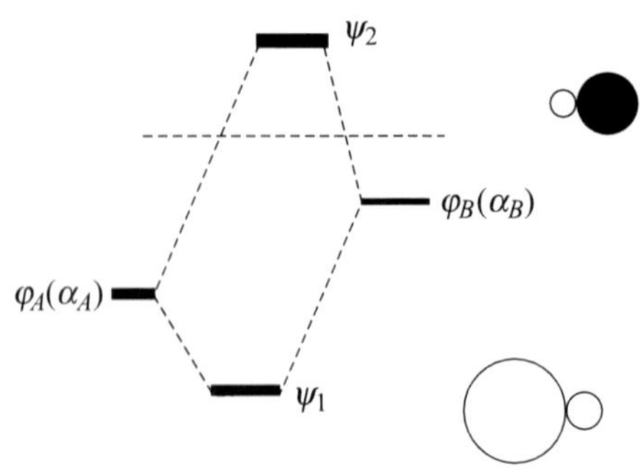

Fig. 10.18 *Molecular orbital of a heteronuclear diatomic molecule. φ_A and φ_B are arbitrarily shown as S-orbitals*

10.2 pi-MOLECULAR ORBITAL

pi bonds (π bonds) are covalent chemical bonds where two lobes of an orbital on one atom overlap with two lobes of an orbital on another atom and in which this overlap occurs laterally. The same plane is also a nodal plane for the molecular orbital of the π-bond.

Following are the conditions to formation of pi-molecular orbital.

1. The number of molecular orbitals (n) for a pi-system is equal to the number of contributing P orbitals. For the allyl system $n = 3$ we had three contributing P orbitals and thus three pi-molecular orbitals.
2. The lowest energy orbital always has zero nodes between the P-orbitals. That is in the lowest energy orbital. All phases of the contributing P-orbitals are aligned the same way.

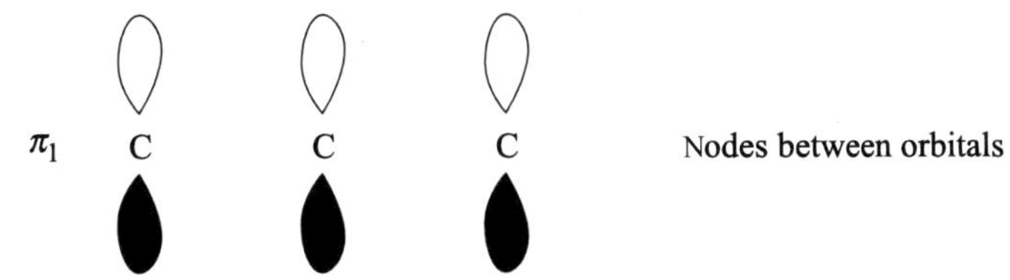

3. The number of nodes between P-orbitals increases by 1 for each successive energy level, such that the highest energy orbital has (n – 1) nodes.

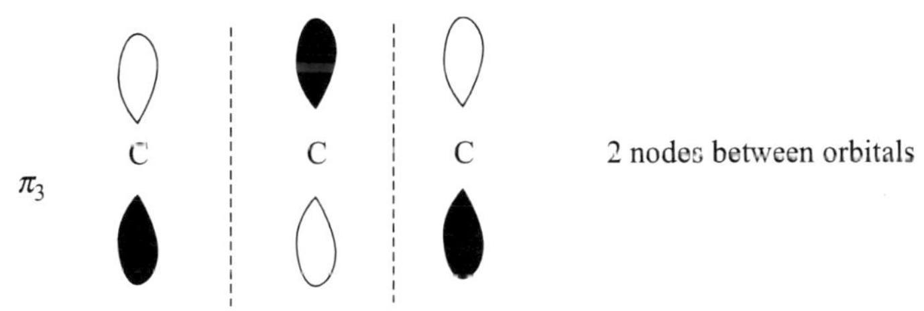

4. The lowest and highest energy orbitals are always the easiest pi-molecular orbitals to draw. It is helpful to draw them first.
5. The trick of drawing the orbital(s) of intermediate energy is to understand where to put the node(s). Nodes are positioned such that they are symmetrical relative to the centre. A system with 1 node has the node in the centre (right on the central atom in fact. A system with two nodes will have the nodes at equal distance relative to the centre).

Butadienyl πi-system (n = 4): As the name suggests, butadiene is composed of four carbons with two adjacent pi-bonds on conjugation and resonance these two pi-bonds are conjugated. All four P-orbitals are all aligned with each other and build up into a larger pi system. It is for this reason that we can describe the electron density in butadiene with resonance forms which show alternative distributions of pi-electron in the molecule.

Major resonance form

Minor resonance form

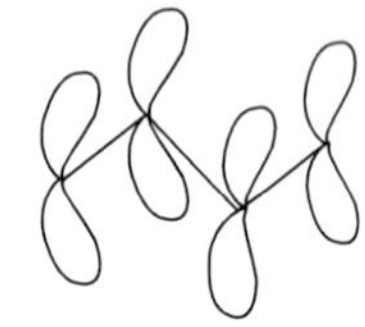

4 conjugated *P*-orbitals

Since butadiene consists of 4 individual *P*-orbitals the pi-system of butadiene will contain 4 pi-molecular orbitals.

10.2.1 The Lowest-energy Molecular Orbital (π_1)

The lowest energy molecular orbital will have *P*-orbitals with phases in complete alignment with each other.

Lowest energy pi-orbital in the butadienyl system has zero nodes.

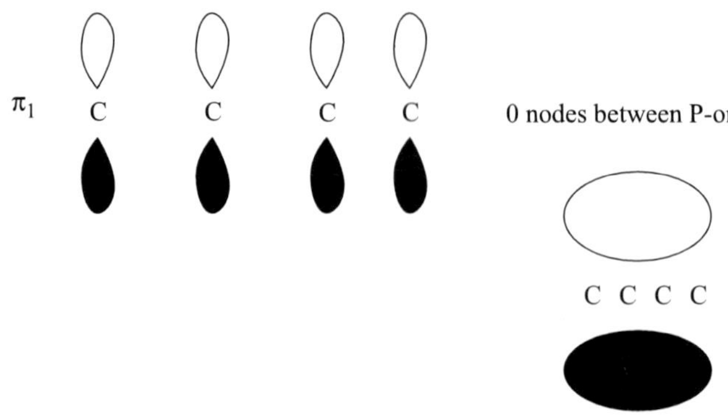

10.2.2 The Highest-energy Molecular Orbital (π_4)

This creates a pi-system with three nodes.

Highest-energy pi-orbital in the butadienyl system has three nodes.

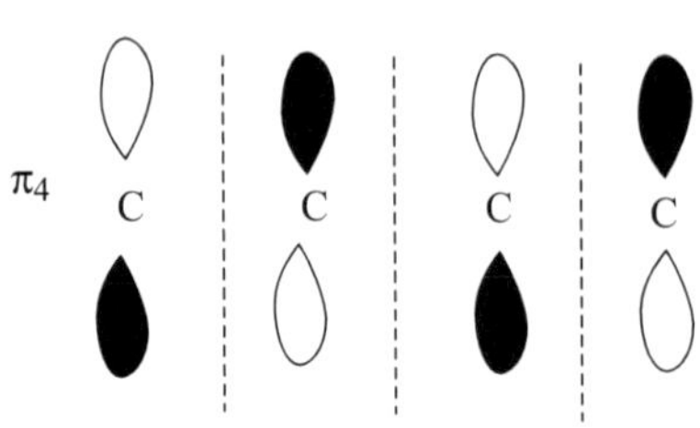

3 nodes between *P*-orbitals

The second lowest energy molecular orbital (π_2): The second-lowest energy molecular orbital in butadiene will have one node.

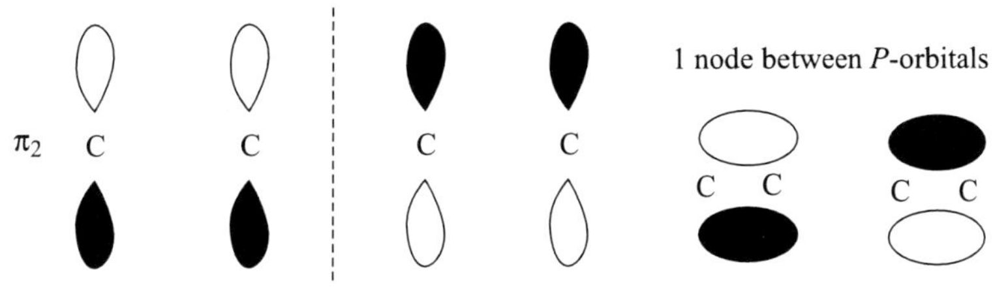

The third lowest energy molecular orbital (π_3): This has two nodes. The general principle is that they are placed symmetrically with respect to the centre.

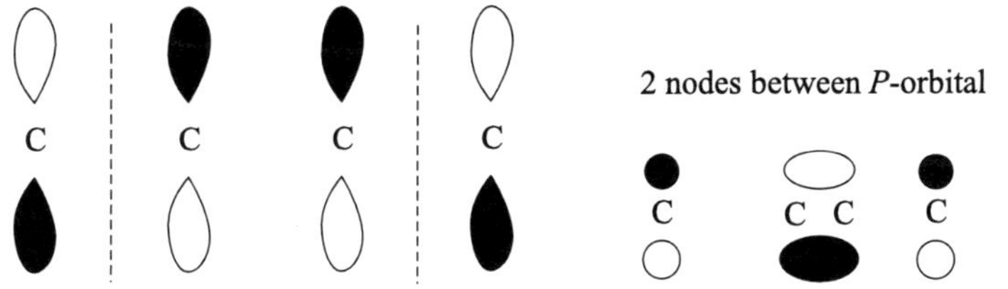

This gives us a 2-carbon pi-orbital in the centre flanked by two one-carbon orbitals on the sides.

10.2.3 Molecular Orbital Diagram for the Butadienyl System (n = 4)

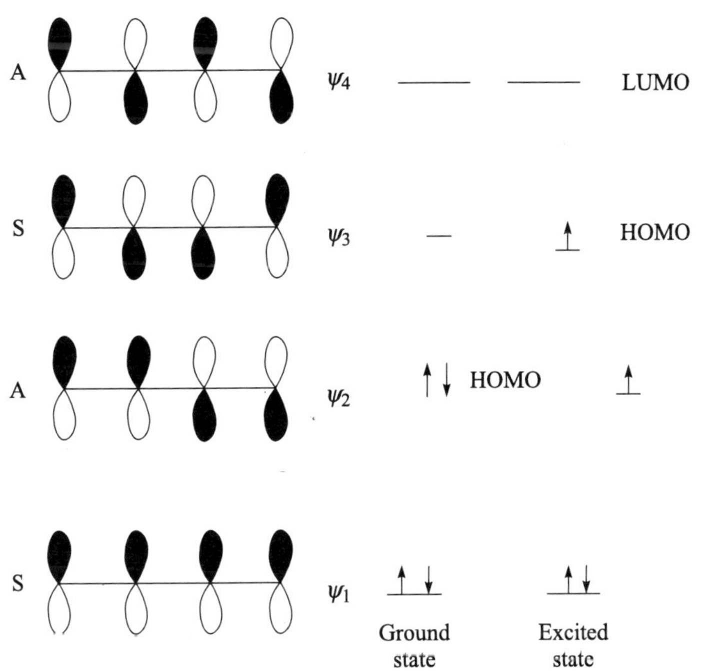

S – Symmetric
A – Antisymmetric

Butadiene has two double bonds with two electrons each, for a total of 4 pi-electrons. The highest occupied molecular orbitals is π_2. It is the electron donor nucleophile. Its electrons are going to come from that orbital. The LUMO is π_3. Here it is electron acceptor (electrophile) and the electrons will be donated to that orbital.

Benzene: Benzene is a planar molecule containing a ring of six carbon atoms each with a hydrogen atom attached. There are six *P*-orbitals in the pi-system of benzene.

The lowest-energy molecular orbitals of hexatriene and benzene have zero nodes.

This has zero nodes.

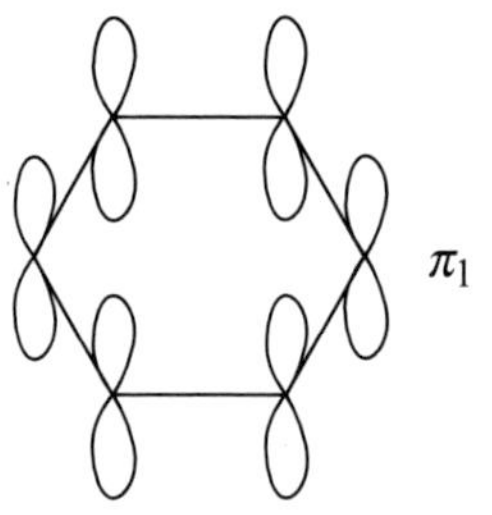

Fig. 10.19 *π_1 orbital*

1. All *P*-orbitals are aligned with phases pointing in the same direction.
2. There are no nodes between orbitals themselves. (There is a node in the plane of the molecule but that's inherent to all *P*-orbitals).
3. In this orbital, electrons are delocalized over the length of the molecule, resulting in the greatest lowering of energy.

The highest energy molecular orbitals have *P*-orbitals with completely alternating phases

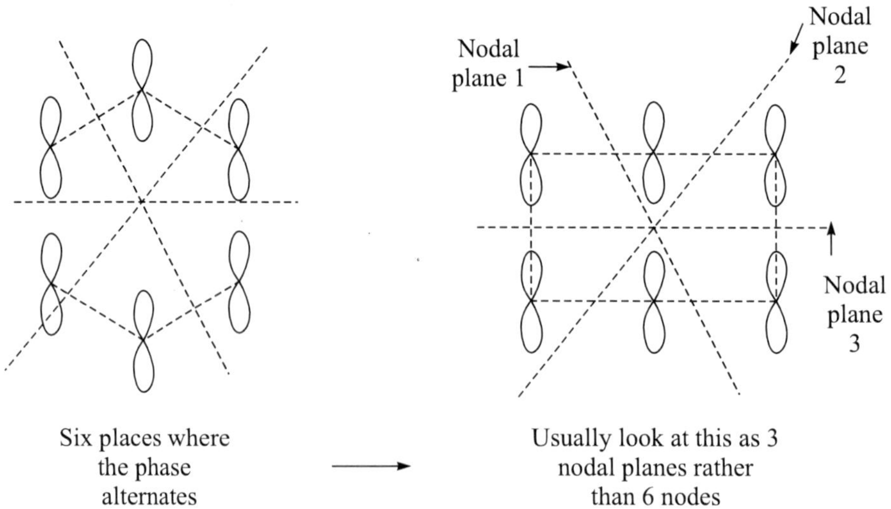

Fig. 10.20 *π_6 orbital*

This orbital has zero overlap between adjacent *P*-orbitals and therefore electrons in this orbital have the minimum possible delocalization. They are therefore of the highest energy.

Intermediate Energy Levels: It is impossible to draw a cyclic pi-system with one node, but we can draw a system with one nodal plane. Here, for instance, we have drawn a nodal plane that cuts through two of the single bonds.

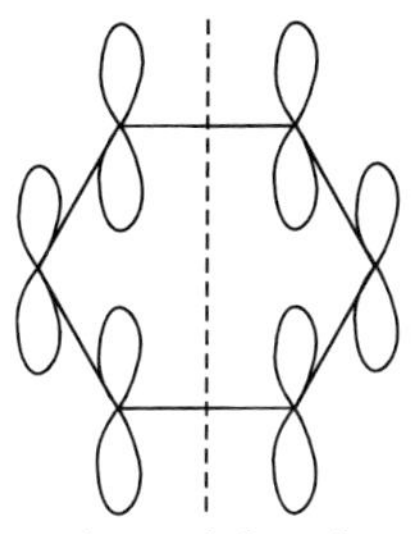

π_2 orbital

through bond

Benzene with one nodal plane.

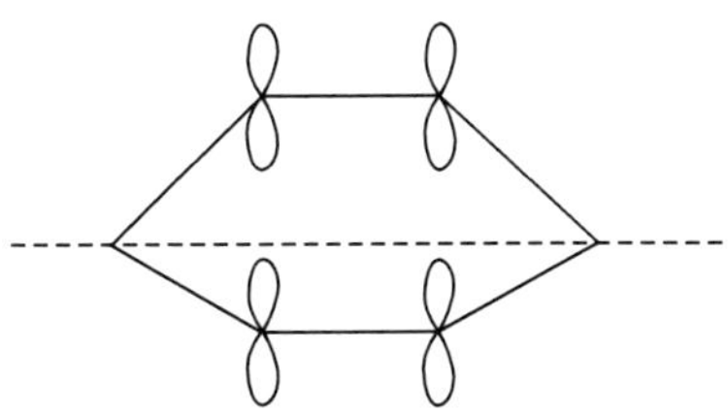

π_3 orbital

through atom

These two molecular orbitals (π_2 and π_3) have the same number of nodal planes and therefore have the same energy. So orbitals are degenerate.

π_4 and π_5 Orbitals: They have two nodal planes. These orbitals are of same energy.

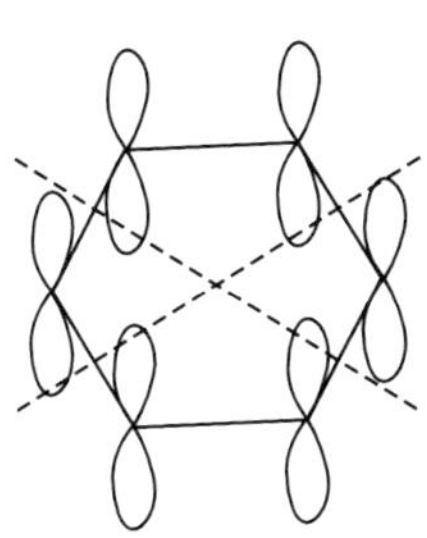

π_4 orbital

Through bond

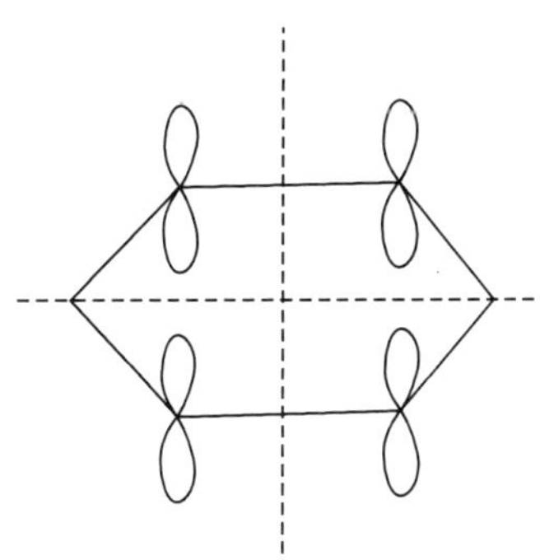

π_5 orbital

Through atom

These two orbitals are of equivalent energy.

The pi molecular orbital diagram of benzene

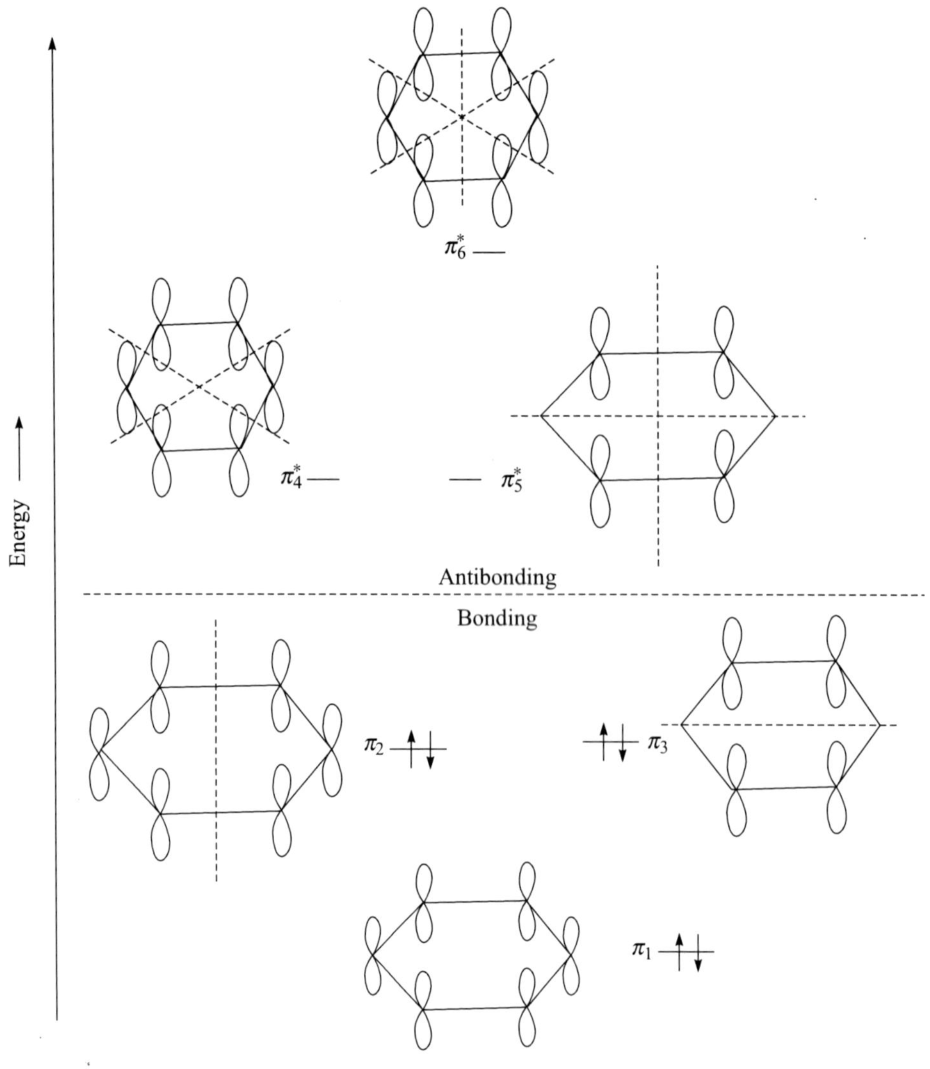

10.3 AROMATICITY

Aromaticity is a property of cyclic (ring shaped) planar (flat) structures with a ring of resonance bonds that gives increased stability compared to other geometric or connective arrangements with the same set of atoms.

There are four conditions a molecule must meet in order for it to be aromatic. If any of these conditions are violated no aromaticity is possible.

1. It must be cyclic (a ring of atoms)

For example:

Benzene
aromatic

Acyclic
1, 3, 5, hexatriene
(not aromatic)

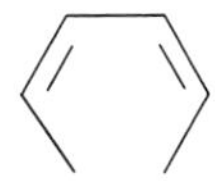

Cyclic
pyrrole
(Aromatic)

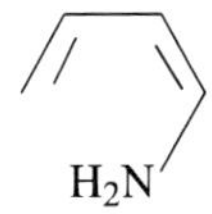

Acyclic
(not aromatic)

2. Every atom in the ring must be conjugated (*p*-orbitals at every atom in the ring). We can also say–

    every atom in the ring must have an available *p*-orbital, and

    every atom in the ring must be able to participate in resonance.

Remember that the available *p*-orbital condition applies not just to atoms that are part of a pi-bond, but also atoms bearing a lone pair, a radical or an empty *p*-orbital (e.g., carbocation).

For example:

Aromatic,
all ring atoms
conjugated

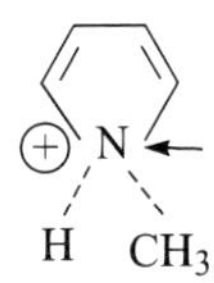

Not aromatic,
nitrogen atom
is not conjugated

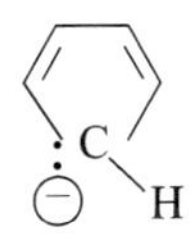

Aromatic,
all ring
atoms
conjugated

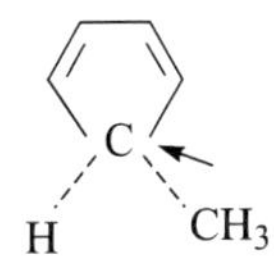

Not aromatic,
carbon is not
conjugated
(SP^3)

3. The molecule must have $[4n + 2]$ π-electrons

For example:

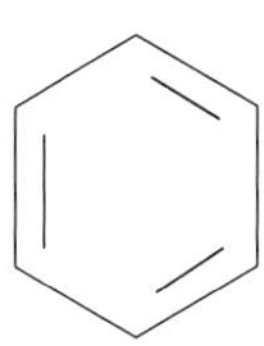

Benzene
→ Cyclic
→ Conjugated
→ 6π electrons
Aromatic

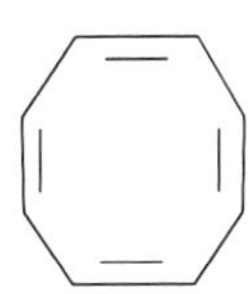

Cyclooctatetraene
→ Cyclic
→ Conjugated
→ 8π electrons
Not aromatic

4. The molecule must be planar (all atoms in the molecule lie in the same plane). Generally, if the first three conditions are met then it is usually safe to assume that the molecule is flat.

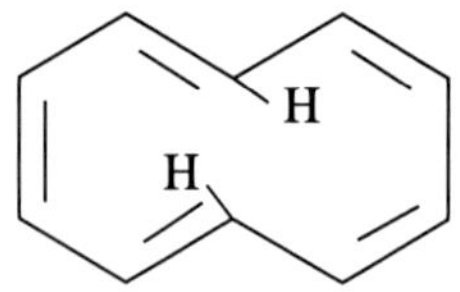

Not aromatic

Although it is cyclic, conjugated and has 10 $\pi-1e^0$. It is not flat due to repulsions between hydrogens that arise when it is in the flat conformation. Since it is not flat so it is not aromatic.

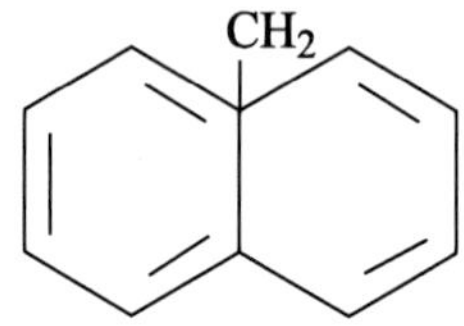

Aromatic

However, replacing the hydrogens with bonds to a bridging carbon allows all C – C π bonds to be in the same plane, so it is aromaitc.

Why $4n + 2\pi$ Electrons? The German chemist Erich Huckel was the first to recognize in 1931, that an aromatic compound must have an odd number of pairs of π-electrons. It is derived from the Huckel molecular orbital (HMO) theory and states that planar monocyclic completely conjugated hydrocarbons will be aromatic when the ring contains $4n + 2\pi$ electrons. According to Huckle's molecular orbital theory, a compound is particularly stable if all of its bonding molecular orbitals are filled with paired electrons. This is true of aromatic compounds, meaning they are quite stable. With aromatic compounds, 2 electrons fill the lowest energy molecular orbitals and 4 electrons fill each subsequent energy level (the number of subsequent energy levels is denoted by n), leaving all bonding orbitals filled and no anti-bonding orbitals occupied. This given a total of $4n + 2\pi$ electrons.

For example: Benzene has 6π electrons. Its first 2π electrons fill the lowest energy orbital and it has 4π electrons remaining. These 4 fill in the orbitals of the succeeding energy level. Notice how all of its bonding orbitals are filled but none of the antibonding orbitals have any electrons.

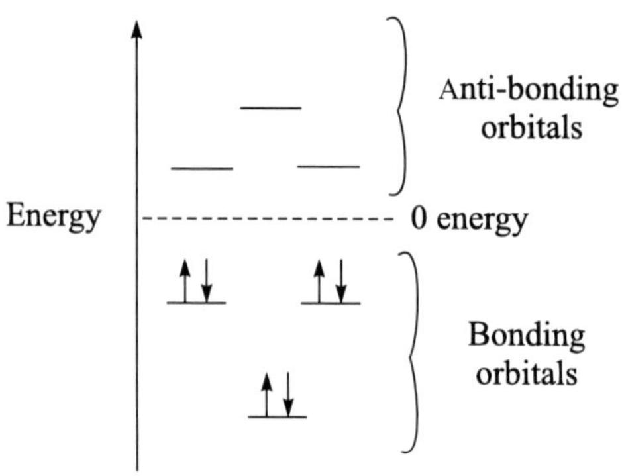

Molecular orbital diagram for benzene

$$4n + 2 = 6$$
$$4n = 4$$
$$n = 1$$

For benzene we find that that $n = 1$ which is a positive integer, so the rule is followed.

10.4 CRYSTAL FIELD THEORY (CFT)

The valence bond theory could not explain the stability of coordination compounds. It also failed to throw light on the differences between strong and weak ligands. Therefore, scientists proposed the crystal field theory. It is a model for the bonding interaction between transition metals and ligands. It describes the effect of attraction between the positive charge of the metal cation and negative charge on the non-bonding electrons of the ligand. When the ligands approach the central metal ion, the degeneracy of electronic orbital states usually *d* or *f* orbitals are broken due to the static electric field produced by a surrounding charge distribution. CFT successfully accounts for some magnetic properties colours and hydration energies of transition metal complexes, but it does not attempt to describe bonding. The electrons in the *d*-orbitals of the central metal ion and those in the ligand repel each other due to repulsion between like charges. Therefore, the *d*-electrons closer to the ligands will have a higher energy than those further away, which results in the *d* orbitals splitting in energy. This splitting is affected by:

- The nature of the metal's ion.
- The metal's oxidation state (a higher oxidation state leads to a larger splitting).
- The arrangement of the ligands around the metal ion.
- The nature of the ligands surrounding the metal's ion.

10.4.1 Crystal Field Splitting

It is the process of the splitting of degenerate level in the presence of ligand. The difference between the energy of t_{2g} and e_g level is denoted by Δ_0. There are some ligands producing strong fields and causing large crystal field splitting. On the other hand, others produce very weak fields. Thus, the crystal field splitting depends on the field produced by the ligand and the charge on the metal ion.

For example:

$$I^- < Br^- < Cl^- < SCM^- < F^- < OH^- < C_2O_4^{2-} < H_2O < NCS^- < \varepsilon DTA < NH_3 < en < CN^- < CO$$

Crystal Field Stabilization Energy: It is assumed that the sum of the energies of the five *d*-orbitals in the free state must be equal to the sum of the energies of the five *d*-orbitals in octahedral configuration.

Splitting occurs in such a way that there is no net change in energy for the fully occupied orbitals. If we take energies of free state as zero then

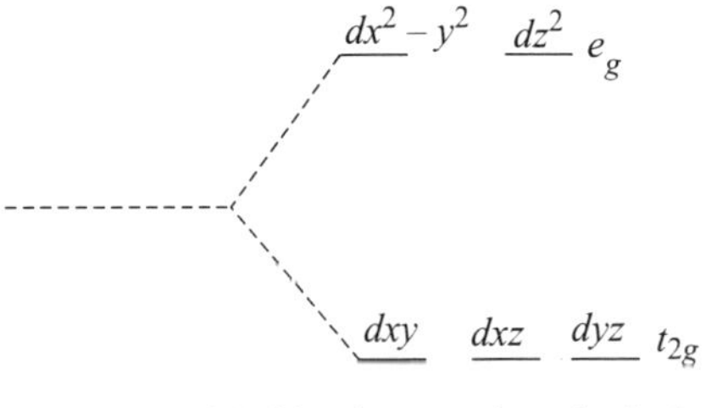

Fig. 10.21 *Crystal field splitting of octahedral complex*

$$Ed_{xy} + Ed_{xz} + Ed_{yz} + E_{dx^2-y^2} + Ed_{z^2} = 0 = 5E \quad (1)$$

$$\therefore \quad 3dxy + 2E_{dz^2} = 0$$

$$[Ed_{xy} = Ed_{xz} = Ed_{yz} \text{ and } Ed_{z^2} = Ed_{x^2-y^2}] \quad (2)$$

Also,

$$Ed_{z^2} - Ed_{xy} = \Delta_0 \text{ (octahedral)} \quad (3)$$

Now multiply Eq. (3) by two and substract from Eq. (2).

$$2Ed_{z^2} - 2Ed_{xy} = 2\Delta_0$$

Now,

$$3Ed_{xy} + 2E_{dz^2} - 2E_{dz^2} + 2E_{dxy} = -2\Delta_0$$

$$5E_{dxy} = -2\Delta_0$$

$$E_{dxy} = -\frac{2}{5}\Delta_0$$

$$= -0.4\Delta_0$$

$$E_{dxy} = -4D_q \; (\Delta_0 = 10D_q)$$

$$\therefore \quad E_{dxy} = E_{dxz} = E_{dyz} = -0.4\Delta_0 = -4D_q$$

Similarly, $E_{dz^2} - E_{dxy} = \Delta_0$ (From Eq. 3)

$$\therefore \quad E_{dz^2} - (-0.4\Delta_0) = \Delta_0$$

$$E_{dz^2} = \Delta_0 - 0.4\Delta_0$$

$$= 0.6\Delta_0$$

$$E_{dz^2} = 6D_q$$

$$E_{dz^2} = E_{dx^2-y^2} = 0.6\Delta_0 = 6D_q$$

This means that presence of an electron in t_{2g} level will stabilize the complex to the extent of $0.4\Delta_0$ or $4D_q$. This stabilization energy is called crystal field stabilization energy. The presence of an electron in e_g level will destabilise the complex to the extent of $0.6\Delta_0$ or $6D_q$ (Fig. 10.23).

Factors Affecting the CFSE: The pairing energies for first-row transition metals are relatively constant. Therefore, the difference between strong and weak field or low and high spin cases comes down to the magnitude of the crystal field splitting energy (Δ).

These following factors affect the crystal field stabilization energy (Δ).

1. **Geometry:** Δ_0 is larger than Δt. In almost all cases, Δt is smaller than P (pairing energy) so tetrahedral complexes are always weak field (high spin) square planar complexes may either be weak or strong fields.

2. **Oxidation state of metal cation:** A greater charge on cation results in a greater magnitude of Δ. Because greater charge pulls ligands more strongly towards the metal. Therefore, influences the splitting of the energy levels more.

3. **Size of the metal cation:** For second- and third-row transition metal ions, Δ_0 is larger than the first row. Less steric hindrance between the ligands, better overlap orbitals.

4. **Identity of the ligands:** A spectrochemical series has been developed and sorted by the ability to split metal *d*-orbitals.

   $I^- < Br^- < S^{2-} < Cl^- < SCM^- < NO_3^- < N^{3-} < F^- < OH^- < C_2O_4^{2-} < H_2O < NCS^- < CH_3CN <$ Pyridine $< NH_3 <$ ethylenediammine $< NO_2^- < PPh_3 < CN^- < CO$

   As the ligands decrease in size their ability to split the *d*-orbitals increases. The larger and bulkier ligands exhibit more steric hindrance and approach the metal less effectively.

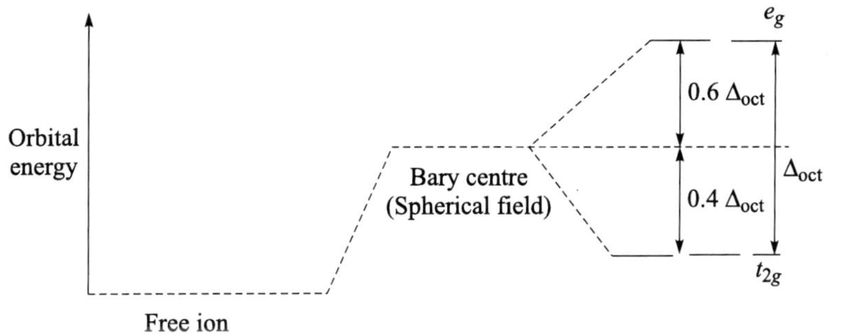

Fig. 10.22 *Octahedral crystal field stabilization energy*

10.4.2 Crystal Field Theory for Octahedral Coordinate Compounds

In case of an octahedral coordination compound having six ligands surrounding the metal atom, we observe repulsion between the electrons in *d*-orbitals and ligand electrons. The repulsion is experienced more in the case of $d_{x^2-y^2}$ and d_{z^2} orbitals as they point towards the axes along the direction of ligand. Hence, they have higher energy than average energy in spherical crystal field. On the other hand *dxy*, *dyz* and *dxz* orbitals experience lower repulsions as they are directed between the axes. Hence, these three orbitals have less energy than the average energy in spherical crystal field. Thus, the repulsion in octahedral coordination compound yields two energy levels.

t_{2g} – Set of three orbitals (*dxy*, *dyz* and *dxz*) with lower energy.

e_g – Set of two orbitals ($d_{x^2-y^2}$ and dz^2) with higher energy.

The splitting of degenerate level in the presence of ligand is known as crystal field splitting. The difference between energy of t_{2g} and e_g level is denoted by Δ_0 0 – stand for octahedral. Some ligands tend to produce strong field thereby causing large crystal field splitting. Thus, the crystal field splitting depends on the field produced by the ligand and the charge on the metal ion.

Filling of d-orbitals takes place in the following way.

The first three electrons are arranged in t_{2g} level as per the Hund's rule.

The fourth electron can either enter the t_{2g} level giving a configuration of t_{2g}^4, e_g^0 or can enter the e_g orbital giving a configuration of $t_{2g}^3, e_{g'}$. This depends on two parameters: magnitude of crystal field splitting (Δ_0), and pairing energy (ρ). The possibilities of two cases can better be explained as:

- $\Delta_0 > P$ electron enters the t_{2g} level giving a configuration of $t_{2g}^4 e_{g^0}$. Ligands producing this.
- $\Delta_0 < P$-electron enters the e_g level giving a configuration of $t_{2g}^3 e_{g'}$. Ligands producing this configuration are known as weak field ligands and form high spin complexes.

Table 10.2 Geometries and crystal field splitting diagrams

Name	Energy diagram
Octahedral	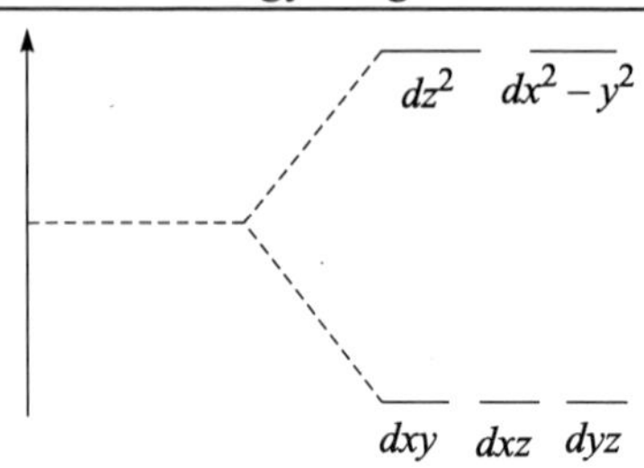
Pentagonal bipyramidal	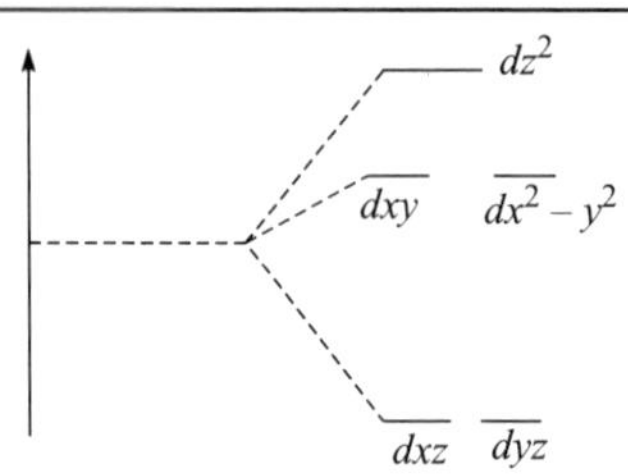
Square antiprismatic	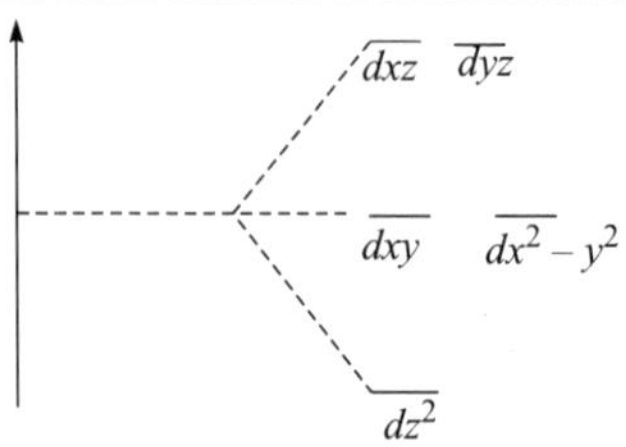
Square planar	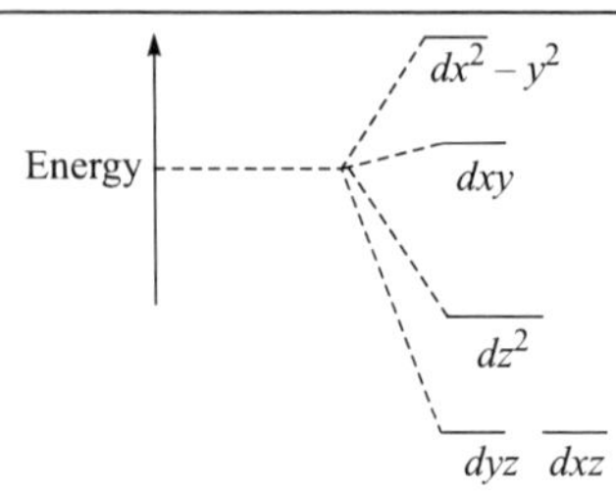

Contd...

Table 10.2 Contd.

Square pyramidal	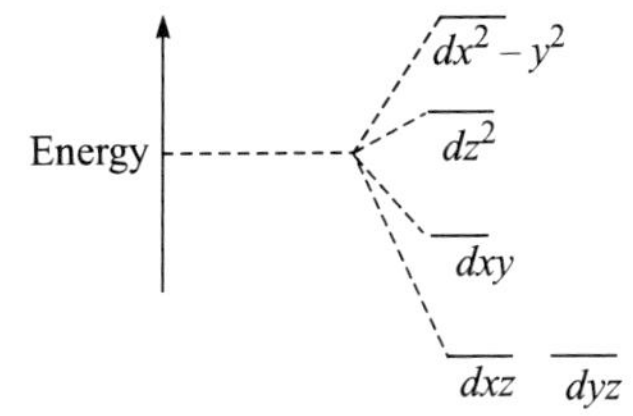
Tetrahedral	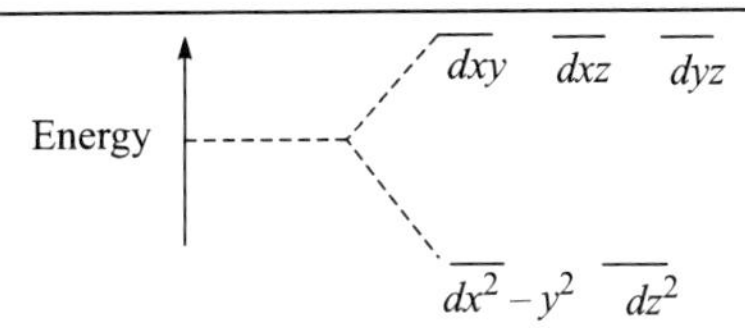

10.4.3 Electronic Structures of Metal Complexes

The *d*-orbital energy-level diagram in Table 10.3 predicts the electronic structures and some of the properties of transition metal complexes. We start with the Ti^{3+} ion which contains a single *d*-electron, and proceed across the first row of the transition metals by adding a single electron at a time. We place additional electrons in the lowest energy orbital available, while keeping their spins parallel as required by Hund's rule. As shown in Table 10.1 for $d^1 - d^3$ system such as $(Ti(H_2O)_6)^{3+}$, $[V(H_2O)_6)^{3+}$ and $[Cr(H_2O_6)^{3+}$ respectively, the electrons successively occupy the three degenerate t_{2g} orbital.

Table 10.3

d_1	Ti^{3+}	 High spin
d_2	$Ti^{2+}V^{3+}$	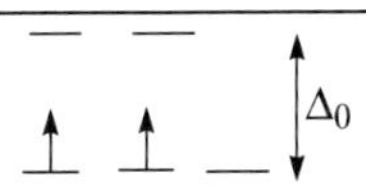
d_3	$V^{2+}Cr^{3+}$	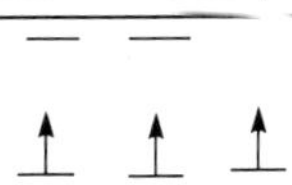
d_4	Cr^{2+}, Mn^{3+}	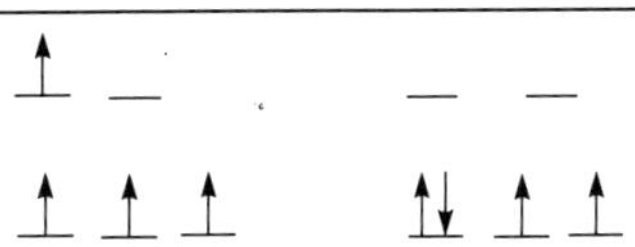 Low spin
d_5	Mn^{2+}, Fe^{3+}	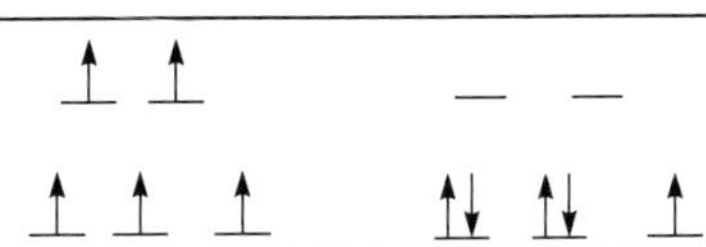
d_6	Fe^{2+}, CO^{3+}	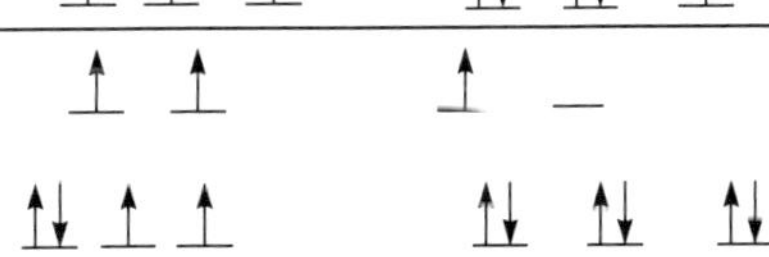

Contd...

d_7	CO^{2+}	↑ ↑ ↑ ↑↓ ↑↓ ↑ ↑↓ ↑↓ ↑↓
d_8	Ni^{2+}	↑ ↑ ↑↓ ↑↓ ↑↓
d_9	Cu^{2+}	↑↓ ↑ ↑↓ ↑↓ ↑↓
d_{10}	$Cu^{+}Zn^{2+}$	↑↓ ↑↓ ↑↓ ↑↓ ↑↓

Table 10.2 represents the possible electron configuration for octahedral d^n transition-metal complexes (n = 1 – 10). Two different configurations are possible for octahedral complexes of metal with d^4, d^5, d^6 and d^7 configurations. The magnitude of Δ_0 determines which configuration is observed.

With their spin parallel giving one, two and three unpaired electrons respectively. For example, chromium ion has a d^3 electron configuration or more succinctly Cr^{3+} is a d^3 ion.

When we reach the d^4 configuration, there are two possible choices for the fourth electron. It can occupy either one of the empty e_g orbitals or one of the singly occupied t_{2g} orbitals. Recall that placing an electron in an already occupied orbital results in electrostatic repulsion that increases the energy of the system. This increase in energy is called the spin-pairing energy (ρ). If Δ_0 is less than P, then the lowest energy arrangement has the fourth electron in one of the empty e_g orbitals. Because this arrangement results in four unpaired electrons, it is called a high spin configuration and a complex with this electron configuration such as the $[Cr(H_2O)_6]^{2+}$ ion is called a high-spin comlexes. Conversely, if Δ_0 is greater than P, then the lowest-energy arrangement has the fourth electron in one of the occupied t_{2g} orbitals. Because this arrangement results in only two unpaired electrons it is called a low-spin configuration and a complex with this electron configuration such as the $[Mn(CN)_6]^{3-}$ ion is called a low-spin complex. Similarly, metal ions with the d^5, d^6 or d^7 electron configurations can be either high spin or low spin, depending on the magnitude of Δ_0. In contrast only one arrangement of d electrons is possible for metal ions with $d^8 - d^{10}$ electron configurations.

For example: The $[Ni(H_2O)_6]^{2+}$ ion is d^8 with two unpaired electrons, the $[Cu(H_2O)_6]^{2+}$ ion is d^9 with one unpaired electrons and the $[Zn(H_2O)_6]^{2+}$ ion is d^{10} with no unpaired electrons.

REVIEW QUESTIONS

1. Using C.F.T. draw energy level diagram, write electronic configuration of the central metal atom/ion and determine the magnetic moment value in the following:
   (i) $[COF_6]^{3-}$, $[CO(H_2O)_6]^{2+}$, $[CO(CN)_6]^{3-}$
   (ii) $[FeF_6]^{3-}$, $[Fe(H_2O)_6]^{2+}$, $[Fe(CN)_6]^{4-}$
2. What is the relationship between observed colour of the complex and the wavelength of the light absorbed by the complex?
3. What different colours are observed in octahedral and tetrahedral complexes for the same metal and same ligands?
4. Deduce Schrödinger – time independent wave equation. Explain the terms involved in the expression and state its significance.
5. Explain Schrödinger's equation for quantum model of hydrogen atom.
6. Describe the physical significance of Schrödinger's wave function.
7. Apply Schrödinger wave equation for a particle in a box illustrating quantization of energy. Draw the radial plots for hydrogen atom.
8. Express Schrödinger's wave equation for 1, 3-butadiene and benzene using particle in a box model.
9. Discuss the application of particle in a box solution to conjugated butadiene and benzene system and write the wave equations.
10. Write a short note on applications of particle-box model to nanoparticle.
11. Discuss band theory to explain bonding in metals citing suitable examples.
12. Write a neat labelled MO diagram, explain the bonding in CO molecules. State the various features of Co molecule.
13. Draw a neat labelled MO diagram of dinitrogen and explain the bonding citing its electronic configuration.
14. What is bond order of a molecule? How is it calculated? Expalin bonding in HCl molecular with an MO diagram and state its characteristics.
15. What is a frost diagram? Draw frost diagram for benzene and cyclobutane and show their molecular orbital configurations.
16. Draw the molecular orbital diagram for oxygen molecule and explain its paramagnetic behaviour.
17. Write a neat labelled MO diagram, explain the bonding in NO molecule.
18. What is aromaticity? Explain aromaticity of benzene.
19. Discuss aromaticity of a compound. Explain the criteria for a compound to be considered aromatic.
20. Write a short note on Huckel's rule for aromaticity and frost diagrams.
21. Draw the radial probability distribution plots for hydrogen atom.
22. Distinguish between conductors and semiconductors.
23. What is aromaticity? Give an example of aromatic compound.
24. What are the criteria for a molecule to be aromatic?
25. Draw the molecular energy level diagram of
   (a) HCl (b) NO (c) O_2 (d) N_2 (e) H_2

Model Question Paper I

ENGINEERING CHEMISTRY

(*Subject Code:* 14 CHE 12/14 CHE 22)

Exam Duration: 3 hours *Marks: 100*

ANSWER THE FOLLOWING QUESTIONS

1. (a) What are reference electrodes? Explain the construction of a calomel electrode with its half-cell reactions and mention its advantages.

(b) Derive Nernst equation for single electrode potential. Explain the determination of single electrode potential using standard hydrogen electrode.

(c) Explain the following battery characteristics:

(i) Capacity

(ii) Cycle life.

(d) With a neat diagram, explain the construction, working and uses of Ni-MH battery. (6 + 5 + 4 + 5)

OR

(a) Explain a method for the determination of single electrode potential.

(b) What is an ion selective electrode? Explain the construction and working of glass pH electrode.

(c) Explain the construction, working and applications of Li-MnO_2 battery.

(d) What is a fuel cell? Give the construction, working and applications of CH_3OH—O_2 fuel cell. (4 + 5 + 5 + 6)

2. (a) Explain water line and pitting corrosion with the help of electrochemical theory.

(b) What is anodic metal coating? Explain the process of galvanizing.

(c) What is decomposition potential? How is it determined?

(d) Explain the process of electroplating of chromium. Give its applications. Indicate the reason for not employing chromium as anode. (5 + 4 + 4 + 7)

OR

(a) Explain the following types of corrosion:

1. Galvanic corrosion 2. Stress corrosion.

(b) What is anodizing? Explain anodizing of aluminium.

(c) Explain the role of the following factors on the nature of the electrodeposit:

(i) Current density

(ii) Throwing power.

(d) What is meant by electroless plating? Briefly explain the electroless plating of copper on PCB. (6 + 4 + 4 + 6)

3. (a) Define (i) Gross calorific value (ii) Net calorific value of a fuel. How are they related.

(b) What is knocking in IC engines? Explain the mechanism of knocking.

(c) What is biodiesel? How is it synthesized? Give its advantages over diesel.

(d) What is the union carbide process for synthesis of solar grade silicon. (5 + 5 + 5 + 5)

OR

(a) What is gross and net calorific value? Calculate the calorific value of a sample of coal from the following data:

Mass of coal = 0.95 g

Mass of water in copper calorimeter = 2000 g

Water equivalent of calorimeter = 700 g

Rise in temperature = 2.8°C

Specific heat of water = 4.187 kJ/kg/°C.

(b) Explain the mechanism of knocking in diesel engine. How can it be prevented?

(c) Mention the physical and chemical properties of silicon relevant to photovoltaic applications.

(d) Explain the design of solar modules, panels and arrays. (6 + 4 + 5 + 5)

4. (a) What is glass transition temperature? What are the three factors that influence the glass transition?

(b) Give the synthesis and an applications of:

1. Silicon rubber
2. Teflon

(c) Give the synthesis, properties and applications of Plexiglas.

(d) Give the synthesis, properties and applications of polycarbonate. (4 + 4 + 6 + 6)

OR

(a) Define number average, and weight average molecular mass of a polymer. Distinguish between M_n and M_w.

(b) Explain the mechanism of conduction in polyaniline.

(c) Give the synthesis, properties and applications of carbon and Kevlar fibres.

(d) Explain addition and condensation polymerization with examples. (6 + 4 + 6 + 4)

5. (a) Distinguish between scale and sludge.

(b) Why should the presence of CO_2 be avoided in boiler feed water? How is dissolved CO_2 removed from water by chemical means?

(c) Explain the precipitation method for preparation of nanomaterial with an example.

(d) Give the synthesis, properties and applications of nanocrystals. (5 + 4 + 5 + 6)

OR

(a) Describe the Winkler's method for the determination of dissolved oxygen. Give also the reactions involved.

(b) Explain the principle of desalination of water by ion selective electrodialysis technique.

(c) Give the structural features, properties and applications of buckminsterfullerene.

(d) What is a dendrimer? Give the structural properties responsible for biomedical applications of dendrimers. (5 + 4 + 6 + 5)

Model Question Paper II

ENGINEERING CHEMISTRY

(*Subject Code:* 14 CHE 12/14 CHE 22)

Exam Duration: 3 hours *Marks: 100*

ANSWER THE FOLLOWING QUESTIONS

1. (a) What are concentration cells? Calculate the emf of a galvanic cell consisting of two sliver electrodes immersed in silver nitrate solution of concentration 0.01 M and 10 M $AgNO_3$ at 30°C.

(b) Define standard electrode potential? Explain the origin of electrode potential.

(c) With a neat diagram, explain the construction, working and uses of Zn-Air battery.

(d) How do we classify batteries? What are the factors that affect the emf of a battery. (6 + 5 + 5 + 4)

OR

(a) What are amalgam and redox electrodes? Give one application of each.

(b) Explain the determination of pH using glass electrode.

(c) How a fuel cell is different from a battery? Mention the electrolytes used in different types of fuel cells.

(d) Mention the anode, cathode and electrolyte in the zinc-air battery. Give the anodic and cathodic reactions taking place during discharge. (4 + 5 + 6 + 5)

2. (a) Explain the electrochemical theory of corrosion with reference to iron.

(b) What is cathodic protection? Explain the sacrificial anode techniques.

(c) Give the advantages of electroless plating over electroplating.

(d) Explain the process of electroplating of chromium. Give its applications, also indicate the reasons for not employing chromium as anode. (5 + 4 + 4 + 7)

OR

(a) Explain the effect of the following factors on the rate of corrosion:

(i) Nature of corroding product

(ii) Polarization.

(b) What is phosphating? Explain the process.

(c) Explain the following factors influencing the rate of electrodeposit:

(i) Current density

(ii) Temperature.

(d) Explain the term overvoltage? What are the factors that affect overvoltage? (6 + 4 + 6 + 4)

3. (a) What is power alcohol? Give any three advantages over petrol.
   (b) What is cracking? Mention its significance. With a neat diagram, explain fluidized catalytic cracking.
   (c) Explain zone refining technique for purification of silicon. (6 + 4 + 6 + 4)

OR

   (a) Mention the physical and chemical properties of silicon relevant to photovoltaic applications.
   (b) Explain construction and working of a solar PV cell. Give its advantages and drawbacks. (6 + 4 + 5 + 5)

4. (a) Give the synthesis, properties and applications of Teflon.
   (b) Give the synthesis, properties and applications of polyurethane.
   (c) What is a polymer composite material? Give its properties and applications.
   (d) Give the synthesis, properties and applications of silicon rubber. (6 + 6 + 4 + 4)

OR

   (a) Explain the effect of structure of polymer on (i) crystallinity and (ii) chemical resistance of polymer.
   (b) Explain the free radical mechanism of polymerization of vinyl chloride.
   (c) What are the deficiencies of natural rubber? Explain the vulcanization process of rubber.
   (d) Describe the manufacture of:
      (i) Polystyrene
      (ii) Neoprene. (6 + 5 + 5 + 4)

5. (a) Explain the corrosion of water boilers due to dissolved oxygen CO_2 and $MgCl_2$.
   (b) Define COD. 20 cm^3 of wastewater requires 15.5 cm^3 of 0.02 M $K_2Cr_2O_7$ for its oxidation, calculate its COD.
   (c) Explain the sol-gel method for preparation of nanomaterial with an example.
   (d) Give the synthesis, properties and applications of carbon nanotube. (5 + 4 + 5 + 6)

OR

   (a) Explain the secondary and tertiary methods of sewage treatment.
   (b) Define BOD. Give a comparative account of BOD and COD.
   (c) Explain the chemical vapour condensation method for preparation of nanomaterials with an example.
   (d) What is a nanorod and a nanowire? Mention any two applications of each. (6 + 4 + 5 + 5)

Index